Mammals
OF THE
Neotropics

MAMMALS
OF THE
NEOTROPICS

The Northern Neotropics

VOLUME 1

Panama, Colombia, Venezuela,
Guyana, Suriname, French Guiana

With plates in color and black and white by Fiona Reid
and maps by Sigrid James Bonner

John F. Eisenberg

The University of Chicago Press
Chicago and London

John F. Eisenberg is the Katharine Ordway Professor of Ecosystem Conservation at the Florida Museum of Natural History, the University of Florida. He is the author of *The Mammalian Radiations*, also published by the University of Chicago Press, and the editor of several books on animal behavior and ecology.

The University of Chicago Press, Chicago 60637
The University of Chicago Press, Ltd., London

98 97 96 95 94 93 92 91 90 89   54321

Library of Congress Cataloging in Publication Data

Eisenberg, John Frederick.
    Mammals of the Neotropics : Panama, Colombia, Venezuela, Guyana, Suriname, French Guiana / John F. Eisenberg ; with plates by Fiona Reid and maps by Sigrid James Bonner.
       p.   cm.
    Includes index.
    Contents: v. 1. The northern Neotropics.
    ISBN 0-226-19539-2 (cloth)
    ISBN 0-226-19540-6 (paper)
    1. Mammals—South America.   2. Mammals—Panama.   I. Title.
QL725.A1E38   1989
599.098—dc19   88-27479
                                                                      CIP

To *Brenda,* who suffered long but offered great support and patience during the writing and proofreading of this manuscript

Contents

Acknowledgments

A work of this sort cannot be completed without the efforts of a great many people. So many have contributed so much that it is difficult to compile a full list. I take responsibility for the text and any errors therein, but many hands helped me finish the project.

I owe a great debt to Fiona Reid, who prepared the plates. They greatly enhance the final product. Ms. Reid herself acknowledges the invaluable assistance of Guy Tudor for the use of his extensive photo files and above all his artistic advice. She was also greatly aided by being permitted to use the American Museum of Natural History collection, and she thanks Guy Musser, Karl Koopman, and Dan Russell for their cooperation and assistance.

The text was typed and revised several times by Barbara Stanton, and I greatly appreciate her efforts. Sigrid James Bonner initially prepared the range maps and distributions for many of the species and helped considerably at the final stages. To Sigrid my profound thanks. Early completion of the range maps and most of the figures involved the care and diligence of Lyn Dorsey Rathbun. Chuck Gentry prepared the skull drawings, and Ms. Rathbun made the drawings of bat faces. Kent Redford was extremely helpful in obtaining bibliographical references and encouraged the project in all its phases. My colleague Mel Sunquist shared many thoughts on marsupial and carnivore biology.

Much of the initial work on the book was carried out while I was assistant director at the National Zoological Park. I thank all of my colleagues for their help. The text was written at the Florida Museum of Natural History, and I thank its former director, Wayne King, and the current director, Peter Bennett, for their encouragement.

Colleagues at the National Museum of Natural History were most kind in making collections available to me for inspection and in developing the range maps. I thank Richard Thorington, Henry Setzer, and C. O. Handley, Jr., for permission to use the collections. Colleagues in the United States Fish and Wildlife Service were extremely helpful in all phases of the publication's development. I thank Robert Brownell for sharing his expertise on cetaceans, Alfred Gardner for assisting with many problems concerning rodent distributions and taxonomy, and Don Wilson for his collaborative efforts. I should like to thank Guy Musser for his help at the American Museum of Natural History, Hugh Genoways and Duane Schlitter for aid at the Carnegie Museum, Bruce Patterson and Robert Timm for their efforts at the Field Museum of Natural History, and Oliver Pearson, James Patton, and William Lidicker at the Museum of Vertebrate Zoology, University of California, Berkeley. Special thanks are offered to Georgina Macc, who assisted me with the collection at the British Museum of Natural History. Charles Woods, curator of mammals at the Florida Museum of Natural History, generously allowed me to study the collection of small Caribbean cetaceans. Laurie Wilkins was most helpful in assisting with the museum records at Gainesville.

Aspects of the natural history of various taxa were clarified through discussions with colleagues at the Smithsonian Tropical Research Institute. I especially thank Yael Lubin for her information concerning the feeding habits of anteaters, G. G. Montgomery for general advice on edentates, and Nicholas Smythe for numerous conversations on caviomorph rodents.

The museum at Texas Tech University, Lubbock, Texas, was the backbone of our endeavor in Venezuela. Robert J. Baker and J. Knox Jones, Jr., were supportive in all aspects, and the voucher specimens from the 1975–79 research efforts are housed at that institution. The museum at Lubbock graciously allowed me to republish the keys to the genera of the phyllostomid bats included in this volume.

Colleagues at the National Zoological Park were helpful through the years by developing information on gestation, longevity, litter size, and feeding habits. Special thanks go to Devra G. Kleiman for her collaborative efforts on caviomorphs, carnivores, and chiropterans. Edwin Gould was helpful through the years with his knowledge of the Chiroptera. Katherine Ralls and Charles Handley, Jr., generously shared their measurements of the South American chiropterans with me. Christen Wemmer gladly offered much information on marsupials and carnivores; and Eugene Maliniak, Miles Roberts, and Larry Collins were helpful in ways too numerous to mention.

I should like especially to thank several of my foreign colleagues. Jorge Hernandez-Camacho generously shared information with me and checked my mammal list for Colombia. J. Ojasti and his students shared experiences in Venezuela and helped in numerous ways. Ambassador Edguardo Mondolfi helped me with a great many distributional problems in Venezuela, and I thank him profoundly. Tomas Blohm allowed many of us to work on his ranch in Venezuela; many of the studies carried out there greatly enhanced the natural history accounts for certain species. P. Charles-Dominique shared a great deal of information concerning his work in French Guiana.

Special thanks are due those individuals who reviewed distribution maps for me. I thank Karl Koopman for his aid on the Chiroptera, Alfred Gardner for his aid with the Rodentia, and the late Ralph Wetzel for his assistance with the Edentata, Carnivora, Artiodactyla, and Perissodactyla. Thanks to the American Museum of Natural History for permission to use some figures from Goodwin and Greenhall's monograph on the bats of Trinidad and Tobago. Philip Myers helped me enormously with a critique of the first draft, and Michael Mares offered excellent advice on the second draft. S. David Webb read draft two and gave useful advice on the palaeontology sections.

I wish also to extend my appreciation to my former students both predoctoral and postdoctoral who, through their efforts, taught me a great deal about South American mammals. Special thanks go to Nicholas Smythe, Nancy Muckenhirn, Christine Schonewald-Cox, Charles Brady, Peter August, Margaret Ann O'Connell, Ranka Sekulic, Theresa Pope, John G. Robinson, Carolyn Crockett, Rasanayagam Rudran, Jody Stallings, C. C. Smith, Mark Ludlow, Francisco Bisbal, Dennis Daneke, Jay Malcolm, Kathy Langtimm, José Fragosso, Rodrigo Medellin, and Kent Redford. Discussions with them and sharing their experiences with South American wildlife have made the whole project worthwhile.

All of us who are concerned with tropical biology and the conservation of tropical organisms are deeply indebted for many reasons to the three organizations and institutions whose names appear below. I am especially grateful that they have made a tangible expression of their concern through their support of the publication of the color plates in this book. My profound thanks and appreciation to

University of Florida, The Division of
    Sponsored Research
Friends of the National Zoo (FONZ)
World Wildlife Fund

Introduction

Background

I began fieldwork in the tropics of the Western Hemisphere in June 1960 when I crossed the Tropic of Cancer with my brother Thomas during a successful trapping expedition seeking *Liomys pictus* in Sinaloa and *Heteromys lepturus* in Veracruz, Mexico. The experience of the tropics profoundly influenced my subsequent research efforts. We were guided in our trapping endeavors by a newly published set of Hall and Kelson, *The Mammals of North America*, and I was impressed with the utility of this meticulously proofread two-volume work that clearly stated collection localities and habitat preferences. My present effort falls short of that 1959 classic, but it is a start.

Since that time in 1960, I have returned to the Neotropics to initiate several studies on mammalian behavior and ecology. In 1964 I spent several months at the Smithsonian Tropical Research Institute on Barro Colorado Island (BCI), Panama. Recall the efforts on BCI of C. R. Carpenter with howler monkeys and R. K. Enders with mammalian natural history published in the 1930s. My objective in 1964 was to document the free-ranging behavior of *Ateles geoffroyi*, but I was impressed with the opportunity for the study of mammalian behavior. Martin Moynihan and John Kaufmann had pioneered the study of Neotropical mammals in recently completed studies on *Aotus* and *Nasua*. I returned in 1965, after six weeks in Mexico, with Nicholas Smythe and Nancy Muckenhirn to initiate work on *Dasyprocta punctata* and *Saguinus geoffroyi*.

My own research efforts were productively diverted to Madagascar and Sri Lanka during the interval 1966–69, but I returned to Panama in 1970 with D. G. Kleiman to assist G. G. Montgomery and M. E. Sunquist in establishing their radiotelemetric studies on sloths and subsequently on anteaters. The results of their research efforts, achieved with the collaboration of Yael Lubin, have become classics in mammalian behavioral ecology.

When I first arrived in Panama in 1964 I met Richard Thorington, and we began a collaborative effort at censusing the mammal fauna of Barro Colorado Island. Through almost a decade we accumulated data that we subsequently published. My last trip to Panama in 1974 helped solidify some of our ideas, which resulted in a multiauthored handbook of wildlife techniques aimed at students of primatology but also applicable to diurnal tropical mammals.

In 1973 I began a series of trips to Colombia and Venezuela. My objective was to choose a field site for research on Neotropical vertebrates in a variety of habitats. This site was to be sufficiently distinct from Barro Colorado Island to allow for variation in habitat and degree of human disturbance, thus increasing our data base concerning mammalian responses to the heterogeneous habitats present in the Neotropics. Eventually I settled on northern Venezuela as a site for long-term studies. In part I was encouraged by C. O. Handley, Jr., who, with several colleagues, had recently completed a survey of the mammal fauna. Most important, I was encouraged by Edguardo and Ruth Mondolfi and Tomas and Cecilia Blohm, who aided me in numerous ways. In 1975 research commenced at two sites: Hato Masaguaral, in the state of Guárico, and Parque Nacional, Guatopo, in the state of Miranda. The initial effort involved K. Green, M. A. O'Connell, P. V. August, G. G. Montgomery, Y. Lubin, E. Morton, D. Marcellini, D. G. Kleiman, and R. Rudran. The research effort still continues.

After we began work in Venezuela, I became concerned because our group of researchers had access only to a very fragmented literature describing mammalian distributions, and confusion abounded concerning species descriptions. Thus was born the idea of developing a guide to the mammals of the northern Neotropics. I felt somewhat intimidated by the task but recognized the need.

Some 449 terrestrial and marine mammals have been identified in the northern Neotropics. I have trapped and observed in the field only 75 species, of which 31, or 41%, were bats and rodents. On the other hand, my seventeen years with the National Zoological Park gave me access to approximately 66 species, and other zoos offered about 30 more for comparison. In the zoological park setting at Washington, the emphasis was on primates and carnivores; thus 45, or 53%, of the species belonged to these two taxa. The net result was firsthand experience with about 116 species.

To augment my mammalogical experience, museums became increasingly important to me in the mid–1970s. Since bats constitute almost 45% of the terrestrial mammal fauna described in this volume, I tried to become acquainted with their rich variety. My efforts at the United States National Museum of Natural History were greatly aided by Don Wilson, who wrote a paper with me on relative brain size in bats. Thanks to the collection assembled by C. O. Handley, Jr., I had the opportunity to inspect some 236 species of mammals from Venezuela and Panama. In addition, I inspected some 217 species from five other museums. In total I have viewed in museums all but 13 of the species covered in this volume, and those were described from the literature.

I undertook writing a book on the mammals of northern South America with some concern; there are many people far more qualified for the task. One reason a book of this nature had not already been written is that we lack key information concerning limits of distribution and the validity of some of our current nomenclature. Certain areas in the region this book covers have not been well collected. Particularly vexing are the sigmodontine (cricetine) rodents and the bats. We are only beginning to appreciate the complexity of speciation events deriving from past isolations of tropical forest-adapted forms in refugia during Pleistocene climatic changes. Thus many may think the volume is somewhat premature.

On the other hand, there is a great need for a work that at least presents a state of the art for reference by students of mammalogy. Distribution and identification of mammals in Panama are covered in Hall (1981), and Husson (1978) produced an elegant volume dealing with the mammals of Suriname. Pittier and Tate (1932) prepared a preliminary list for Venezuela, which was updated by Aduze (1956) and Rohl (1959). Handley (1976) compiled the most recent and useful contribution on the distribution of mammals in Venezuela. Borrero H. (1967) presents a list of mammals and some species accounts for Co-

lombia; Diaz, Hernandez-Camacho, and Cadena-G. (1986) offer the most recent species list for this country. Work by Hershkovitz in northern Colombia has clarified a number of distributional problems. Yet this work is scattered and has appeared over a considerable span of time. Although Janzen (1983) presents a useful summary of Costa Rican natural history, no comparable volume exists for continental South America.

With the exception of Roth (1953), Mendez (1970), Tello (1979), Cabrera and Yepes (1960), Husson (1978), and Enders (1935), little has been written on the natural history of many South American mammal species. It is true that some species of mammals in northern South America are very poorly known, yet a considerable body of information has been assembled over the years, and I felt a preliminary synthesis was feasible.

The term Neotropics is well established in the literature and generally refers to the geographic region of subtropical Mexico, Central America, the Caribbean islands, and the continent of South America. When Alfred Russell Wallace formally proposed the term, he acknowledged the distinctiveness of the tropical fauna of South America. Of course, we all recognize that the temperate portion of the continent is not at all tropical. In trying to develop a title for a multivolume work dealing with the mammals of South America, I made use of Wallace's term; but I fully recognize that I have confined my investigations to the continent of South America and the republic of Panama. Though this is a somewhat arbitrary decision, the mammalian fauna of Central America and the Caribbean has already been dealt with by E. Raymond Hall in his *Mammals of North America*. Purists might quarrel with my decision and my choice of terminology.

Organization of the Book

For the most part the text is organized taxonomically, with a chapter devoted to each order. The first chapter, however, attempts to orient readers to the biogeographic history of South America and to contemporary habitats. The final two chapters deal with the problem of speciation and faunal affinities of mammals in the northern Neotropics and offer some general remarks for students on community ecology drawing from information presented in chapters 2–14. Within each chapter dealing with ordinal taxa the organization is similar. A diagnosis is presented defining members of each order according to anatomical characteristics. The diagnoses are a guide to key anatomical features in the contemporary Neotropical context but are not exhaustive. The ordinal distribution is then considered and, if warranted, remarks are made concerning taxonomy. Then follows a brief history of the evolution of the orders in northern South America.

Next each family within the order is considered in turn with a repeat of the diagnosis (field characters where possible), distribution, and remarks on natural history where warranted. Each genus is then taken up, with a more refined description, distribution for the genus, and remarks on natural history.

The species accounts are organized under each genus, with a physical description, notes on range and preferred habitat, and comments on natural history. This hierarchical organization attempts to avoid repetition; if one turns to a species account and finds some information lacking, one should look under the appropriate subheading at the generic level, the family level, and finally the ordinal level.

How to Use This Book

Latin binomials are the only practical way to refer to species, since five major European languages are spoken in the northern Neotropics. In addition, there are numerous Amerindian languages and dialects, all with different names for animals. For many bats and rodents there are no standard common names in any language. I have not attempted to coin new common names. I recognize that some common names are in wide usage within the northern Neotropics, especially at the family or genus level. Insofar as is possible, I have added these names in English, Spanish, and other languages commonly used in South America. Common names and scientific names are listed in the appropriate indexes.

The maps include dots and shading. The dots represent actual locations as determined from my inspection of museum collections or as published in the accounts marked with a bullet as "map references." The dots are certainly not exhaustive but do reflect broad geographic ranges. I have relied heavily on Husson (1978), Handley (1976), Hall (1981), Beebe (1919), and Anthony (1921) for Suriname, Venezuela, Panama, and Guyana, respectively. Location data for Colombia and French Guiana include a variety of sources marked with a bullet in each chapter's references. The map shading is in part an educated guess. For bat distributions I have followed Koopman (1982). The shading in Koopman's maps is a liberal interpretation of the possible range of a species. His published maps are very small, and more detailed shading would have been impossible. The llanos of Colombia and Venezuela exhibit a great deal of habitat heterogeneity. It has been my impression that many species of bat tend to be con-

centrated in the gallery forests associated with the larger river systems. The shading in the maps of this volume includes the llanos for many species in full recognition that their distribution is not continuous in these habitats. For several groups, especially the Cetacea, I have not attempted shading. For those maps where I have not determined an exact specimen locality but have included shading, I cite the authority in the legend.

Keys are included at the ends of most chapters, and I have tried to develop keys useful to the field mammalogist having a specimen in the hand and without access to museum collections. Some twenty-one plates and eighty-four figures are presented to aid in identification. Some taxa such as the Xenarthra (edentates), Carnivora, Primates, and "ungulates" need no keys given the plates and figures. The smaller marsupials, bats, and rodents are often difficult to identify. Not all keys lead to species identification; for field use, often only generic determinations are practical. For rodents and bats I have adapted some published keys. Some of the keys are restricted with respect to geographic locality, especially those for the sigmodontine (hesperomyine) rodents. The Panamanian rodent fauna is somewhat distinctive, and though several genera enter Colombia, this faunal assemblage presents special problems, and so these rodents are keyed to genus in a separate key.

No key is perfect. Most deal with adults; thus juvenile specimens present genuine problems in the field. Many museum workers genuinely feel that keys should be omitted from a volume like this, since I am writing an encyclopedic book for specialists as well as a practical volume for graduate students. A cautionary word: Any identification you may establish from this book should ultimately be checked with a museum curator having access to an appropriate series of the taxon in question.

Abbreviations are used in many tables and also when referring to collections. The following refer to museums: USNMNH, United States National Museum of Natural History; AMNH, American Museum of Natural History; CMNH, Carnegie Museum of Natural History; FMNH, Field Museum of Natural History; and MVZ, Museum of Vertebrate Zoology, Berkeley. The following abbreviations refer to linear measurements and weights: TL, total length; HB, head and body length; T, tail length; HF, hind foot length; E, ear length as measured from the notch to the tip; FA, forearm length, useful for bats; Wt, weight; and Wta, adult weight. Linear measurements for small mammals are given in millimeters (mm), and weights are in grams (g). Abbreviations in notations follow the following pattern: TL 120–31 (= TL 120 to 131).

The bat measurements in chapter 5 derive in the main from the Smithsonian Venezuela collections (Handley 1976). Shortened area names are employed for some of the tables in the species accounts: TFA, Territorio Federal Amazonas; Northern Venezuela, the north coast range of Venezuela and adjacent areas; Bolívar, the state of Bolívar; northwest Venezuela, the Maracaibo region of Venezuela and adjacent areas. These regions correspond to the biogeographic regions published by Eisenberg and Redford (1979).

Readers should also be prepared for alternative presentation of measurements. The scheme is not uniform throughout the volume because of a variety of circumstances. First, the standard small-mammal measurements TL, HB, T, HF, E are not useful across all orders. Among the Cetacea the standard measurements often do not apply. Second, for some species only a limited number of specimens are available, and those collected before 1950 may not include weights or even ear measurements on the museum tags. In cases where excellent series of measurements were obtained, they are included in a small table at the beginning of the species account. Where smaller series of specimens were available, the data are given in narrative style under the heading "Description."

Dental formulas correspond to a standard system: if a skull is viewed from one side, the upper and lower teeth are recorded in sequence according to an anterior-to-posterior notation. I = incisors, C = canines, P = premolars, M = molars. Thus the formula I 3/1, C 1/1, P 2/1, M 3/3 translates: viewing one side, there are three upper and one lower incisors, one upper and one lower canines, two upper and one lower premolars, and three upper and three lower molars. Remember that teeth tend to be bilaterally symmetrical.

In writing chapter 5, on the Chiroptera, I was fortunate to have access to measurements developed by C. O. Handley, Jr., and Katherine Ralls, who generously made these data available to me. I did not attempt to reproduce all their tables but rather employed a two-pronged approach. First, I included the entire set of Venezuelan bat measurements in one table, offering means for all standard measurements for the Venezuelan species (table 5.14). Second, I included smaller tables at the beginning of the species accounts for geographically restricted samples deriving from the larger set. These tables offer not only the mean values but also the standard deviations. In addition, some of the smaller tables under the species accounts in chapter 5 do not derive from the Handley-Ralls data set. Since we all

appreciate the problem of geographic variation in linear measurements within a species, this mode of presentation seemed practical for the Chiroptera, which represent almost 50% of the species included in this volume.

A mixed presentation of data under species accounts should not perturb readers. Our knowledge of species varies. The same style is employed in Smithers (1983).

In the descriptions of species I have usually referred to pelage colors using terms in common English usage. When working with a published description, I have used the author's terms, which often reflect the Ridgeway (1912) color terminology. I have tried to avoid this terminology in most species accounts, since the color standards are not widely available and museum skins can change shade because of the preservatives employed in preparation and through exposure to light.

Students will note that the original descriptions for species are referenced in Honacki, Kinman, and Koeppl (1982).

The species accounts include distributions for some of the nearer offshore islands—in particular, Trinidad, Tobago, and Isla Margarita. The tables in chapter 15 include species lists for these islands and the Dutch Antilles. Without exception, all insular forms are derived from the adjacent mainland.

I hope this book will prove useful, though I fully understand that the taxonomy for many groups is still in a state of confusion. I attempted to follow Honacki, Kinman, and Koeppl (1982) as a standard, and where I have departed I usually indicate in the text my reasons for doing so. Users of this book should realize that they have in their hands a progress report, since our knowledge of so many groups is incomplete. Nevertheless, I hope it can serve as a halfway house and promote more intensive research on everyone's part. I have tried to include most of the literature up to June 1987. When manuscripts were sent to me, I have included some references up to December 1988. I sincerely regret any omissions.

A Comment on Human Introductions

It is still a hotly debated subject—the entry of *Homo sapiens* into the Western Hemisphere. Archaeologists argue about how many human invasions occurred prior to 5,000 years before the present (B.P.) and about the dating of such events. It is enough to say that *H. sapiens* has been in the Western Hemisphere for some time (40,000 B.P. to 12,000 B.P.), and at least one group probably brought a domesticated, infrahuman mammal, *Canis familiaris*—the dog—after 10,000 B.P.

The European occupancy of North and South America commenced on a large scale after A.D. 1492 when Christopher Columbus returned to Spain with news of his landfall in the area now known as the Bahamas and Hispaniola. This of course was not the first European contact with the "New World," but it rapidly introduced European agronomy and animal husbandry into the Western Hemisphere.

In this book I discuss the introduction of Old World rodents into South America and its offshore islands. I do not discuss the distribution of common domestic livestock (e.g., the goat, *Capra hircus;* the cow, *Bos taurus;* the swine, *Sus scrofa;* the ass, *Equus asinus;* the horse, *Equus caballus;* and the sheep, *Ovis aries*), nor do I consider the distribution of introduced carnivores such as the domestic cat, *Felis catus;* the dog, *Canis familiaris;* or the mongoose, *Herpestes auropunctatus.* All these "European companioins" have been discussed in numerous publications and for the most part are easily recognized. If we are to preserve what is left of the original fauna, let us turn to what was here when our ancestors arrived. Apologies to our predecessors of some 12,000 to 40,000 years ago—you are recent arrivals also.

References

Aduze, P. 1956. Lista de los mamíferos señalados hasta el presente en Venezuela. *Mem. Soc. Cien. Nat. La Salle* 16 (43):5–18.

Anthony, H. E. 1921. Mammals collected by William Beebe at the British Guiana Tropical Research Station. *Zoologica* 3(13):265–90.

Beebe, W. 1919. The higher vertebrates of British Guiana with special reference to the fauna of the Bartica district, no. 7: List of Amphibia, Reptilia, and Mammalia. *Zoologica* 2:205–38.

Borrero H., J. I. 1967. *Mamíferos neotropicales.* Cali, Colombia: Universidad del Valle, Departamento de Biología.

Cabrera, A., and J. Yepes. 1960. *Mamíferos Sud Americanos*, 2d ed., 2 vols. Buenos Aires: Ediar.

Diaz-A. C., J. Hernandez-Camacho, and A. Cadena-G. 1986. Lista actualizada de los mamíferos de Colombia anotaciones sobre su distribución. *Caldasia* 15:471–501.

Eisenberg, J. F., and K. Redford. 1979. A biogeographic analysis of the mammalian fauna of Venezuela. In *Vertebrate ecology in the northern Neotropics*, ed. J. F. Eisenberg, 31–38. Washington, D.C.: Smithsonian Institution Press.

Enders, R. 1935. Mammalian life histories from Barro Colorado Island, Panama. *Bull. Mus. Comp. Zool. (Harvard)* 78(4):385–502.

Hall, R. 1981. *The mammals of North America.* 2d ed., 2 vols. New York: John Wiley.

Handley, C. O., Jr. 1976. Mammals of the Smithsonian Venezuelan project. *Brigham Young Univ. Sci. Bull., Biol. Ser.* 20(5):1–91.

Honacki, J. H., K. E. Kinman, and J. W. Koeppl, eds. 1982. *Mammal species of the world.* Lawrence, Kans.: Allen Press and Association of Systematics Collections.

Husson, A. M. 1978. *The mammals of Suriname.* Leiden: E. J. Brill.

Janzen, D., ed. 1983. *Costa Rican natural history.* Chicago: University of Chicago Press.

Koopman, K. 1982. Biogeography of the bats of South America. In *Mammalian biology in South America*, ed. M. A. Mares and H. H. Genoways, 273–302. Pymatuning Symposia in Ecology 6. Special Publication Series. Pittsburgh: Pymatuning Laboratory of Ecology, University of Pittsburgh.

Mendez, E. 1970. *Los principales mamíferos silvestres de Panama.* Panama City: Privately printed.

Pittier, H., and H. H. Tate. 1932. Sobre fauna Venezolana: Lista provisional de los mamíferos observada en el pais. *Bol. Soc. Venez. Cien. Nat.* 7:249–78.

Ridgeway, R. 1912. *Color standards and color nomenclature.* Washington, D.C.: Published by the author.

Rohl, E. 1959. *Fauna descriptiva de Venezuela.* Caracas: Nuevas Gráficas.

Roth, V. 1953. *Animal life in British Guiana.* Reprint, Georgetown: Daily Chronicle. Originally published 1941.

Smithers, R. H. N. 1983. *The mammals of the southern African subregion.* Pretoria: University of Pretoria.

Tello, J. 1979. *Mamíferos de Venezuela.* Caracas: Fundación La Salle de Ciencias Naturales.

1   An Introduction to Historical Biogeography and Contemporary Habitats of Northern South America

Historical Biogeography

The contemporary fauna of South America has held the interest of Western scientists and natural historians since the Spanish and Portuguese explorers first reached the continent in the fifteenth century. The area's distinctive fauna was recognized early on by Azara and Darwin, but it was not until paleontologists started to explore the fossil history of the mammalian fauna that they began to appreciate not only its unique aspects but the antiquity of the original assemblage. We have now come to appreciate that for millions of years South America was isolated and that only recently has a permanent land bridge connected North America and South America, allowing a more or less continuous faunal interchange (Simpson 1980; Webb 1976). During the Cretaceous era, South America, Africa, Antarctica, and Australia were in close proximity, if not joined together, composing the supercontinent of Gondwanaland. By the time of the Paleocene, South America was an island continent with a unique faunal assemblage that evolved in relative isolation from the more contiguous continental land masses (fig. 1.1).

There is still controversy over how often interchanges took place between North and South America before the Pliocene, but it is highly probable that some exchange occurred in the Eocene and the Miocene. One could envision, before the creation of the Panamanian land bridge, an archipelago between Central America and South America that permitted some exchange of mammals from the north to the south and vice versa. Clearly, before the Pliocene the exchange was limited (Reig 1981).

From the pioneering work of the Ameghino brothers (Ameghino 1906) and subsequently through the efforts of Scott (1937), Simpson (1948, 1967), Webb (1976, 1978), and Reig (1981), we have come to appreciate the structure of the ancient fauna of South America and how it was altered through sporadic in-

vasions from the north and finally catastrophically changed after the completion of the Pliocene land bridge.

Nonvolant mammals are especially useful in establishing interchange chronologies because, unlike bats, they do not readily disperse over water. Some of the earliest faunal assemblages reconstructed from the Paleocene and Eocene of southern South America attest to the unique character of the original fauna. The earliest assemblages indicate an advanced radiation of the Marsupialia and the Xenarthra. The marsupials were radiating to fill omnivore, insectivore, and carnivore trophic niches. The Xenarthra

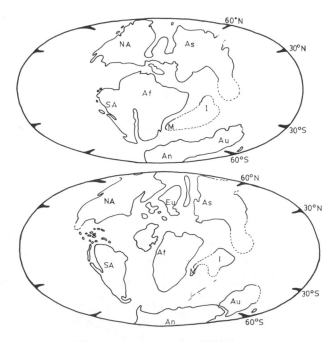

Figure 1.1. The relationship of continents during the Mesozoic era. *Above:* beginning of the separation in the early Cretaceous; *below:* separation at the time of the Paleocene/Cretaceous boundary (modified from Fenley 1979).

were beginning to differentiate into forms that subsequently would be insectivorous, omnivorous, and herbivorous. At the same time, the Litopterna and Notungulata were radiating to fill terrestrial large herbivore niches.

At the close of the Oligocene the beginnings of the caviomorph or hystricognath rodents can be discerned in the fossil record. By the Miocene the hystricognath rodents were the dominant small terrestrial herbivores, and some may have been specializing for arboreality. The primates first make their appearance in the late Oligocene and persist as an arboreal radiation specialized for frugivore, insectivore, and leaf-eating niches. Modern carnivores enter the scene by the time of the Miocene in the form of early procyonids (Linares 1981).

By the time of the Pliocene the Artiodactyla begin to enter South America, followed somewhat later by the Perissodactyla and the Proboscidea. The largest wave of carnivores also entered at this time. The latest entrants into South America were the true Insectivora and the Lagomorpha. Cricetine rodents probably entered South America in the Pliocene, though some workers think they may have come earlier. The point is still controversial (Reig 1981).

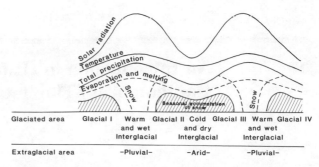

Figure 1.2. The interrelation of temperature, solar radiation, and precipitation during glacial phases (after Simpson 1965).

It is possible, given this three-step entry, to distinguish old endemic orders, the Marsupialia, Xenarthra, Notungulata, and Litopterna; the early invaders such as primates, hystricognath rodents, and procyonid carnivores; and finally the more recent invaders: Perissodactyla, including tapirs and horses; Artiodactyla, including deer, peccaries, and camels; and Carnivora such as felids, canids, and mustelids. This latest faunal movement has been referred to as "the great American interchange." It was not a one-way street with only northern forms entering South

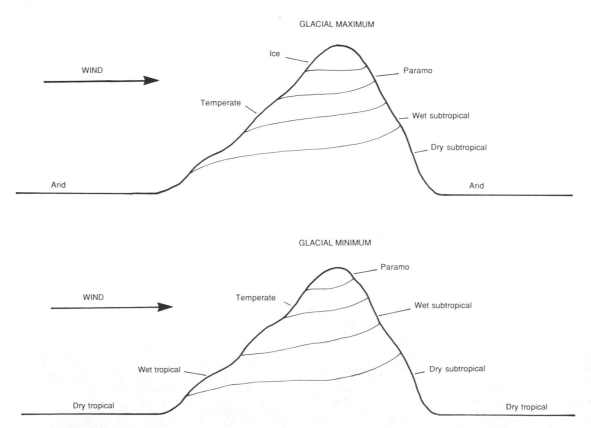

Figure 1.3. Representation of the Andes Mountains during glacial maximum and glacial minimum and the altitudinal distribution of vegetation zones (after Fenley 1979).

America; South American forms also penetrated the north, including ground sloths, armadillos, and glyptodonts. But the northern forms were more successful in their subsequent radiations in South America than were the South American forms moving in the opposite direction. An enormous number of extinctions took place throughout the interchange. The Litopterna and Notungulata disappeared from South America, along with glyptodonts and ground sloths. Some of the North American emigrants did not persist after becoming established in South America. Horses, proboscideans, and many of the larger camelids did not persist beyond the Pleistocene.

Pleistocene glacial perturbations had profound effects in North America as well. The giant ground sloths, glyptodonts, and hystricognath rodents, which had penetrated to Florida, all became extinct there at the conclusion of the Pleistocene. The establishment of northern elements in South America and the reestablishment of South American elements in North America is still in a dynamic state. Clearly, since the completion of the Pliocene land bridge at the Isthmus of Panama, Panama has become a crossroads. The intermingling of North American and Neotropical forms is still evident in the Isthmus, and no doubt the ranges of many northern forms are still expanding in South America (Webb and Marshall 1982).

Recently it has been hypothesized that speciation patterns in South America are in part a consequence of Pleistocene climatic changes. If chronology and paleoclimatic data were better known, one might be able to trace climatic changes even further into the past. But the significant feature bearing on present distributions is that the waxing and waning of glaciated areas in both the Northern Hemisphere and the Southern Hemisphere had a profound influence on world climate (fig. 1.2). Alternating wet and dry periods have influenced the distribution of multistratal tropical evergreen forest in South America and the extent of the savanna ecosystems (Webb 1978; Vuilleumier 1971). During arid cycles in South America the forest contracts, and it has been possible to identify montane forest refugia that could have separated contiguous populations of the same species and allowed speciation events to occur while they were isolated (Muller 1973) (Fig. 1.3). Since there were at least four great glaciations during the Pleistocene, it seems that expansion and contraction of forests has occurred at least four times. At present we are in a more mesic phase, and forests are probably again reaching a maximum. The distribution of bird species and other animals in South America conforms to hypothesized forest refugia (Haffer 1974; Prance 1982).

Physical Structure and Climate of the South American Continent

The continent of South America lies predominantly within the geographically defined tropics. It is dominated by a young (Miocene-Pliocene) mountain range in the west, the Andes, extending from the northernmost part of the continent to the southern tip (fig. 1.4). Older, worn-down mountains are identified as the Guyana shield and the Brazilian shield. The steep altitudinal gradient of the Andes provides a series of vegetation types eventually terminating above the tree line in the puna and paramos (fig. 1.3). The Andes exercise a direct influence on rainfall in the continent. At their southern end the mountains serve as an effective rain shadow, creating an arid-temperate to arid-tropical biotope to the east. A significant arid area that is not directly the result of a montane rain shadow is the coastal strip to the west of the Andes from southern Ecuador to northern Chile. Low rainfall derives from the failure of westerly winds in this latitudinal region, coupled with the cold Humboldt current. Two other minor areas are semiarid to arid: the first is the extreme east of Brazil between approximately 5° and 12° south latitude; the second is to the east of the Cordillera Orientale and surrounds the Gulf of Venezuela, including parts of northeastern Colombia and northwestern Venezuela.

Since the Andes are so far to the west, most of the great rivers of South America drain from west to east, including the Orinoco and Amazon systems in the north and the Río Negro and its tributaries in Argentina. A notable exception to this west-to-east drainage pattern involves the Río Paraná, which derives in part from the old Brazilian highlands and meanders in a southerly direction, capturing major drainages from the southern Andes of Bolivia and continuing south to exit as the Río de la Plata (fig. 1.4).

With a continent as varied in topography and geological history as South America, one may anticipate that the soils will also be highly variable. Some soils, especially in areas experiencing recent volcanic activity, are extremely fertile. Other soils over broad areas of the lowland tropics show a low natural fertility with a low cation-exchange capacity. Because so much of eastern tropical South America is very low and flat, there are numerous areas with impeded vertical drainage. During a large part of the year these low-lying areas may be inundated, and during the rest of the year they support limited tree growth. South American soils have been ably reviewed by Beek and Bramao (1969).

The climates of South America are variable. Given such a tremendous span of latitude, without consid-

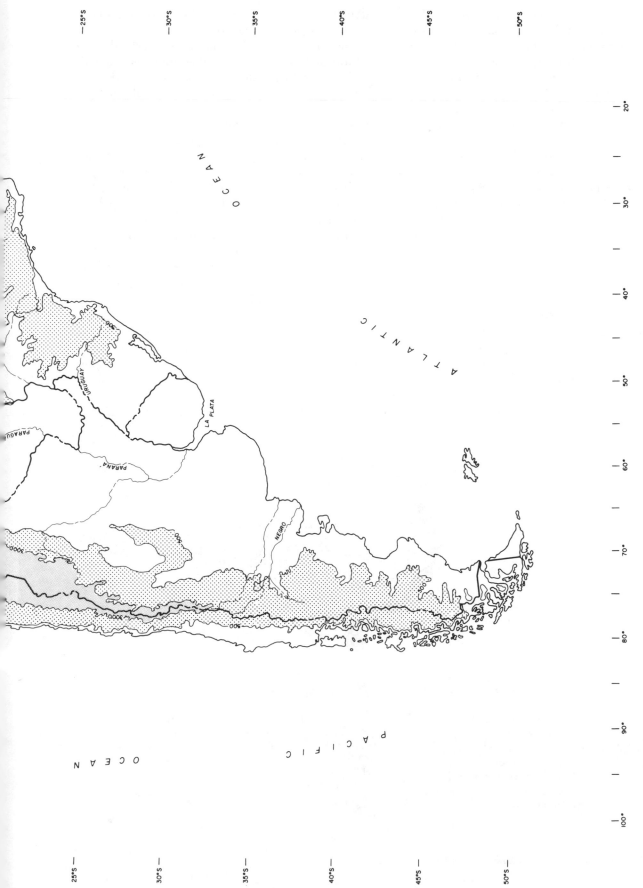

Figure 1.4. Map of South America indicating major elevational gradients and rivers.

eration of altitudinal effects, one-third of South America lies in the temperate zone. Over most of the northern part of the continent a tropical climate prevails. At high altitudes one rapidly passes into temperate and finally into tundra climates. As one proceeds south to the pole, temperate climates prevail below 23° south latitude. Even in areas of lowland tropics rainfall may be unevenly distributed throughout the year, creating a dry season that profoundly influences the form of vegetation. Dry deciduous forests and savanna vegetation systems characterize enormous tracts of South America, where the arid period exceeds three months. Over 20% of the South American continent shows strongly pulsed rainfall patterns, and this semiaridity results in a savanna vegetational climax and greatly lowers primary productivity.

Vegetation of the Neotropics

Figure 1.5 is a vegetation map of South America. One is immediately struck by the vast area dominated by dry deciduous tropical forest and savanna formations. Although we tend to think of the South American tropics in terms of multistratal tropical evergreen forest, in fact this forest type does not

dominate over vast areas of the continent. Let us turn to the northern Neotropics.

The area discussed in this volume extends from approximately 13° north to 0° latitude and from 85° to 50° west longitude (fig. 1.6). Its vegetation has been reviewed by Montenegro, Espinal, and Tosi (1963), Ewel and Madriz (1968), and Granville (1981). Within this region the major factors that define vegetation type are temperature, rainfall, and soil structure. Since the area is within the geographically defined tropics, at low elevations mean temperatures tend to be above 20° C, and annual fluctuation in temperature is minimal. Rainfall tends to be high, averaging above 1,500 mm per year, but the annual distribution can be markedly periodic, which in no small measure determines the nature of lowland vegetation. East of the Andes and north of the Río Guaviare in Colombia and northeast and north of the Orinoco in Venezuela, almost to the coast at elevations below 100 m, is an area characterized by open grasslands and clumps of trees in small tree islands or matas. This is referred to as the llanos. Above 100 m elevation, dry deciduous forest predominates. Within this area rainfall tends to be episodic, with a pronounced drought period extending from November to May. The pulse in annual rainfall re-

Tropical & subtropical rain forest
Tropical deciduous forest
South Brazilian forest & savanna
Palm forest
Subantarctic beech forest
Thorn forest
Savanna
Pampean regions
Pacific coastal deserts
Patagonian-Fuegian steppe
Montane zone
Desert scrub
Transitional vegetation

Figure 1.5. Vegetation map of South America (after Udvardy 1978).

sults in a preponderance of trees that shed their leaves annually. Only in the vicinity of large permanent streams or rivers does a gallery forest exist whose vegetational composition resembles that of typical lowland moist tropical evergreen forest. To the south of the Orinoco, dry deciduous forest predominates in a band approximately 25 to 40 km wide until one approaches the delta to the east (Fenley 1979).

Semidesert to desertlike conditions occur in very restricted localities in northern South America. Most prominent is the northern Caribbean coast of Colombia, including the state of La Guajira. In adjacent Venezuela this area extends around the mouth of Lake Maracaibo and to the Falcón peninsula. A small semidesert area includes parts of Isla Margarita and the peninsula just north of Cumaná. During the Pleistocene glacial maxima these desert areas are believed to have expanded, and most of the area characterized by dry deciduous tropical forest or the llanos proper was semidesert as recently as 30,000 years ago (Wijmstra and Van der Hammen 1966; Van der Hammen 1974).

Besides rainfall, elevation plays a dominant role in differentiating the other tropical life zones. At 10° north latitude temperature declines at a rate of approximately 2.7° C per 200 m elevation. The three cordilleras of the Andes that dominate the western portion of Colombia, and in particular the easternmost cordillera, are stratified into at least twelve zones depending on slope and exposure.

The form of the vegetation is also profoundly affected by vertical drainage. Where drainage is impeded by a perched water table on clay or by a nearly impermeable hardpan, rainwater can stand for a considerable time before evaporating or draining away. In the llanos, the alternation between standing water and extreme desiccation during prolonged droughts hinders tree growth, and thus many areas of the llanos are devoid of trees except the palm *Copernicia tectorum* (Beard 1953). During the dry season such areas, dominated by sedges, are extremely vulnerable to fires either started by lightning strikes or deliberately set by man. The palm is extremely resistant to fire, but burning further discourages the growth of shrubs. To what extent vast tracts of the llanos have been further altered by fire remains somewhat speculative, but the combination of fire and impeded vertical drainage maintains vast areas of the llanos below 100 m as open habitat (Sarmiento 1984).

The nature of vegetation cover in northern South America is also determined by soil fertility. Ultimately soil fertility depends on the parent rock underlying the soil. The parent rock upstream on the great river systems that drain this part of the world influences the fertility of alluvial deposits. In general, waters stemming from the Andes are richer in minerals than are rivers draining from the crystalline Guyana shield. The Guyana shield is very old (Eocene) and has eroded considerably. The granitic tepuis of southeastern Venezuela and Guyana stand as remnant inselbergs. The ancient Roraima sediments overlie the worn shield. Floristically, the plant assemblage coincident with the Roraima sediments is referred to as the Guyanan floral province, as opposed to the amazonian floral province. In reality, however, the picture is more complex, because sediments from the Andes are continually being deposited by the younger rivers that feed from the Andes east into the Orinoco or southeast into the Amazon (Sioli 1975; Herrera et al. 1978; Klinge et al. 1975; Fittkau et al. 1975).

The major rivers within the geographic limits of this volume that drain the Guyana shield remnants are the Río Negro, Río Caroní, Río Caura, Río Essequibo, Río Courantyne, Río Moroni, and Río Oyapock. The Negro discharges into the Amazon, the Caroní and Caura discharge into the Orinoco, and the Essequibo, Courantyne, Moroni, and Oyapock run north to the Atlantic (see fig. 1.6).

The major rivers draining the Andes in the northern Neotropics are the north-flowing Magdalena and Cauca, which discharge into the Caribbean. The Río Meta and Río Guaviare, as well as the Río Portuguesa-Apure, join the Orinoco on its course to the Atlantic. The Río Putumayo and the Río Caquetá join the Amazon and drain southeastern Colombia. In the same area the Río Içana and the Río Vaupés join the Río Negro. With the exception of these last two rivers, the Río Negro's main input derives from the Guyana shield, while the Orinoco's major input derives from the Andes. Both rivers, however, drain both the Andes and the Guyana shield. The Negro and Orinoco connect via the Río Casiquiare in southern Venezuela. As Alexander von Humboldt pointed out, the flow alternates directions depending on upstream discharge of the feeder rivers. Thus the fish faunas of the Orinoco and Amazonian systems have many species in common (see fig. 1.6).

The physical and chemical characteristics of rivers give indications of the ecosystems they drain. White-water rivers are not white but rather are the color of café au lait; they contain a great deal of sediment and in general drain the Andes or those portions of the Guyana region that are not geologically part of the Guyana shield. For example, on the fringes and within the Guyana shield area there are

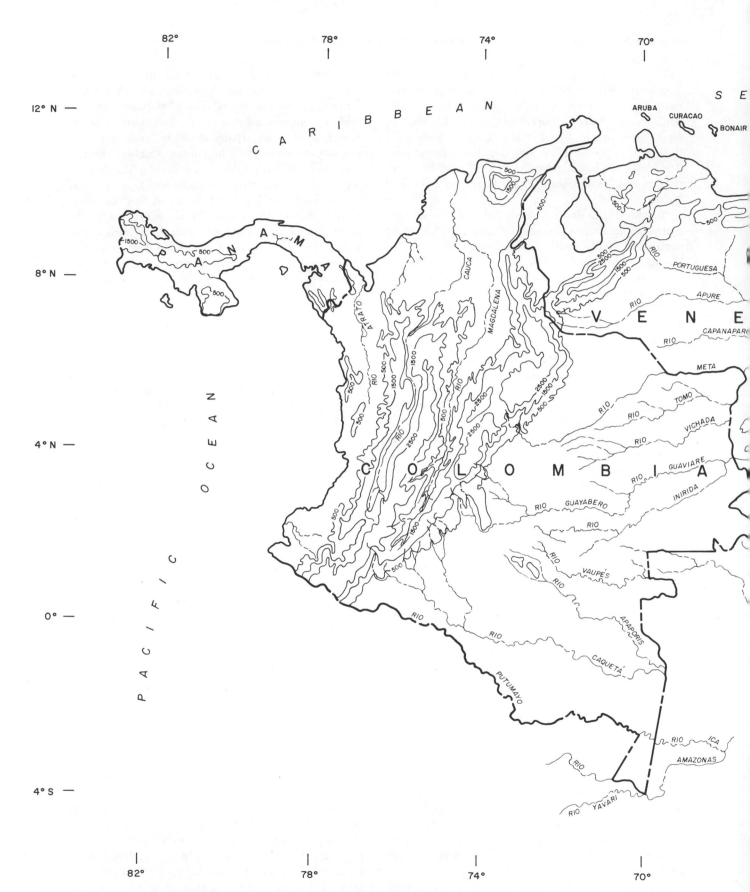

Figure 1.6. Major elevational features and river systems of the northern Neotropics.

low mountains of igneous origin and volcanoes of the Parima and Vijigua ranges that separate the headwaters of the Orinoco from those of the Río Uraricoera and some north bank tributaries of the Río Negro. The rivers draining these mountains are nutrient rich (Fittkau et al. 1975).

On the other hand, tributaries that originate from the Guyana shield usually either are clear (*krystal*) with low nutrient content or are black (*negro*) and contain a great amount of suspended alkaloids and tannins leached from leaves dropped by trees growing upstream. Black-water rivers often drain vegetation on white-sand soils. The low nutrient content of soils drained by black- and clear-water rivers results from the almost pure granite remnants of the Guyana shield. As a consequence, the waters have an extremely low pH (Janzen 1974).

The forest trees that grow on nutrient-poor soils have evolved a number of mechanisms to conserve nutrients. There is generally a dense root mat in the soil, and nutrient cycling from litter on the surface is made more efficient through mycorrhizal fungi that directly move nutrients to the roots. Furthermore, the trees themselves protect their leaves against insect attack by accumulating many secondary compounds. Hence when leaves are shed these secondary compounds leach out and contribute alkaloids to the runoff, making the downstream water black. The forests of the Guyana shield region often show a very low net productivity. If they are cut they recover very slowly. In conjunction with the low net primary productivity they generally have a low carrying capacity for terrestrial vertebrates and arthropods (Janzen 1974).

Characterizing the Guyana shield as an area of low productivity is somewhat misleading. Earlier I identified areas of recent volcanic activity near the Venezuelan-Brazilian border; streams that drain these regions tend to be highly productive whitewater systems (e.g., the Río Mavaca). Although it is possible to make generalized statements about certain areas, it is wise to remember that the Guyana shield area of southern Venezuela is in fact a mosaic of microhabitats (Eisenberg and Redford 1979; Eisenberg 1979). Vast differences in carrying capacity can occur between one valley system and another (see also Emmons 1984).

Biogeographic Regions

Although some mammals are ubiquitous in distribution, the mammalian fauna of northern South America shows several broad biogeographic patterns. First there are species that barely extend beyond Panama into northern Colombia; these species are more prominent in Central America and North America than in South America. They include such forms as the harvest mouse *Reithrodontomys*, white-footed mice of the genus *Peromyscus*, and the pocket gophers of the family Geomyidae. This is the first biogeographic area, the Panamanian.

The second biogeographic area is the region to the west of the eastern cordillera of the Andes, extending south to Peru. There are actually three cordilleras of the Andes that to some extent control the distribution of mammals in this region. The westernmost range defines a coastal strip continuous with Panama that has several species distributed southward to Peru without passing into eastern Colombia (table 1.1). This is the western region. The eastern cordillera is an effective barrier. The central cordillera is very weak as a barrier, and those species effectively crossing it seem to be broadly distributed to the eastern cordillera of the Andes.

The third faunal area is the Maracaibo basin. Ringed by mountains, the Maracaibo basin shows faunal similarities with northern Colombia and Panama rather than with the faunal assemblages to the east of the Andes, and it has few endemics.

The fourth biogeographic area is south of the Río Guaviare in Colombia and east of the Andes, extending into south-central Venezuela to the beginnings of the Guyana highlands. The faunal assemblage in this area is typical of the upper Amazon of Brazil and adjacent Peru. I have termed this the region of southern Colombia and adjacent Venezuela.

The fifth biogeographic area, north of the Río Guaviare and north of the Orinoco below the 100 m contour, is the llanos area of Colombia and Venezuela, a mosaiclike habitat dominated by savannas. The fauna within this area is in general derived from the adjacent Amazonian fauna or from relict faunal assemblages in the north coast range of Venezuela.

The sixth biogeographic area is the high Andes above the tropical lowland forest and into the paramos. It extends through the Andean chain at changing elevations as one proceeds toward the equator.

The seventh biogeographic region is the central Guyana highlands region, stretching south of the Orinoco in southern Venezuela, eastward to western Guyana, and along the mountain ranges separating the Guyanas from Brazil.

The eighth biogeographic region is the area east of the Río Essequibo, which shows distinct affinities to the fauna of northeastern Brazil. This is the eastern Guyanan region.

The ninth biogeographic region is extremely arid. It is situated around the Gulf of Venezuela and in

Table 1.1　Biogeographic Regions of Northern South America and Some Characteristic "Endemic" Nonvolant, Terrestrial Species

Panamanian Region[a]	Western Region[b]	Maracaibo Basin	Southern Colombia and Adjacent Venezuela
Marmosa mexicana	Caluromys derbianus	Sphiggurus vestitus	Cebuella pygmaea
Cryptotis endersi	Marmosa alstoni		Saguinus inustus
Saimiri oerstedii	Monodelphis adusta		Callimico goeldii
Ateles geoffroyi	Tamandua mexicana		Callicebus torquatus
Bassariscus sumichrasti	Choloepus hoffmanni		Cacajao melanocephalus
Procyon lotor	Cabassous centralis		?Atelocynus microtis
Sciurus variegatoides	Ateles fusciceps		Isothrix bistriatas
Syntheosciurus brochus	Cebus capucinus		Dactylomys dactylinus
Macrogeomys cavator	Alouatta palliata		
Nyctomys sumichrasti	Galictis allamandi		
Reithrodontomys creper	Tapirus bairdii		
Peromyscus mexicanus	Dasyprocta punctata		
Sphiggurus mexicanus	Proechimys semispinosus		
Diplomys labilis	Hoplomys gymnurus		
	Coendou rothschildi		

Llanos of Colombia and Venezuela	Andean Region	Central Guyanan Highlands	Eastern Guyanas
Dasypus sabanicola	Marmosa dryas	Marmosa tyleriana	Marmosa lepida (Pars?)
	Cryptotis thomasi	Sciurus flamifer	Euphractus sexcinctus
	Tremarctos ornatus	Proechimys amphicoricus	Saguinus midas
	Nasuella olivacea		Pithecia pithecia
	Pudu mephistophiles		Ateles paniscus
	Sciurus pucherani		Sciurillus pusillus
	Oryzomys minutus		Oryzomys delicatus
	Aepeomys lugens		Neacomys guianae
	Thomasomys aureus		Sphiggurus insidiosus
	Chilomys instans		Echimys chrysurus
	Echinoprocta rufescens		
	Agouti taczanowskii		

Northwestern Arid Region (Venezuela, Colombia)	North Coast Range (Venezuela)	Ubiquitous in Northern South America
Marmosa xerophila	Marmosa fuscata	Didelphis marsupialis
Calomys hummelincki	Marmosa marica	Myrmecophaga tridactyla
	Heteromys anomalus	Dasypus novemcinctus
	Proechimys guairae	Cerdocyon thous
		Procyon cancrivorus
		Potos flavus
		Eira barbara
		Lutra longicaudis
		Felis pardalis
		Felis yagouaroundi
		Felis wiedii
		Puma concolor
		Panthera onca
		Tayassu tajacu
		Odocoileus virginiana
		Mazama americana
		Oryzomys concolor
		Zygodontomys brevicauda
		Hydrochaeris hydrochaeris
		Agouti paca

[a] Species more or less confined to Panama.
[b] Species often occurring in the Panamanian region but widely distributed in western Colombia.

fragmented patches on the north coast of Venezuela to approximately 64° W. Isla Margarita shows affinities to this region in climate and vegetation. The vegetation and associated faunal assemblage may have been much more widespread during the Pleistocene glacial maximum that favored aridity. Since we are currently in a mesic phase, climatically speaking, we are really looking at relict xeric habitats. It follows that mammals adapted to xeric habitats would periodically be subjected to extremely small, fragmented refugia and be vulnerable to extinction. This may account for the lack of xeric-adapted forms in the northern Neotropics. The arid-adapted endemic mammals include just four species, *Myotis nesopolus, Rhogeessa minutilla, Marmosa xerophila*, and *Calomys hummelincki* (see also table 1.1).

The tenth biogeographic area is the complex north coast range of Venezuela, which acted as a refugium during Pleistocene drying cycles.

References

Ameghino, F. 1906. Les formations sédimentaires du Crétace supérieur et du Tertiaire de Patagonie avec une parallèle entre leurs faunes mammalogiques et celles de l'ancien continent. *Anal. Mus. Nac. Buenos Aires* 14:1–568.

Beard, J. S. 1953. The savanna vegetation of northern tropical America. *Ecol. Monogr.* 23:149–215.

Beek, K. G., and D. L. Bramao. 1969. Nature and geography of South American soils. In *Biogeography and ecology in South America*, vol. 1., ed. E. J. Fittkau, J. Illies, H. Klinger, G. H. Schwabe and H. Sioli, 82–112. The Hague: W. Junk.

Eisenberg, J. F. 1979. Habitat economy and society: Some correlations and hypotheses for the Neotropical primates. In *Primate ecology and human origins*, ed. I. S. Bernstein and E. O. Smith, 215–62. New York: Garland.

Eisenberg, J. F., and K. H. Redford. 1979. A biogeographic analysis of the mammalian fauna of Venezuela. In *Vertebrate ecology in the northern Neotropics*, ed. J. F. Eisenberg, 31–38. Washington, D.C.: Smithsonian Institution Press.

Emmons, L. 1984. Geographic variation in densities and diversities of nonflying mammals in Amazonia. *Biotropica* 16(3):210–22.

Ewel, J. J., and A. Madriz. 1968. *Zonas de vida de Venezuela*. Caracas: Technico del Instituto Interamericano de Ciencias Agrícolas.

Fenley, J. 1979. *The equatorial rainforest: A geological history*. London: Butterworths.

Fittkau, E. J., U. Irmler, W. J. Junk, F. Reiss, and G. W. Schmidt. 1975. Productivity, biomass, and population dynamics in Amazonian water bodies. In *Tropical ecological systems*, ed. F. B. Golley and E. Medina, 289–312. New York: Springer-Verlag.

Granville, J. T. 1981. Rainforest and xeric flora refuges in French Guiana. In *Biological diversification in the tropics*, ed. G. T. Prance, 159–81. New York: Columbia University Press.

Haffer, J. 1974. Avian speciation in tropical South America. *Publ. Nuttal Ornithol. Club* 14:1–390.

Herrera, R., C. F. Jordan, H. Klinge, and E. Medina. 1978. Amazon ecosystems: Their structure and functioning with particular emphasis on nutrients. *Interciencia* 3:223–32.

Janzen, D. 1974. Tropical black water rivers, animals and mast fruiting by the Dipterocarpaceae. *Biotropica* 6(2):69–103.

Klinge, H., W. A. Rodrigues, E. Brunig, and E. J. Fittkau. 1975. Biomass and structure in a central Amazon rain forest. In *Tropical ecological systems*, ed. F. B. Golley and E. Medina, 115–22. New York: Springer-Verlag.

Linares, O. J. 1981. Tres nuevos carnívoros prociónidos fosiles del Mioceno de Norte y Sudamérica. *Ameghiniana* 18:113–21.

Montenegro, E., L. S. Espinal, and J. Tosi. 1963. *Mapa ecológico de Colombia (1:1,000,000) y formaciones vegetales de Colombia*. Bogotá: Instituto Geográfico Augustin Codozzi.

Müller, P. 1973. *The dispersal centers of terrestrial vertebrates in the Neotropical realm*. The Hague: W. Junk.

Prance, G. T. 1982. A review of the Pleistocene climatic changes in the Neotropics. *Ann. Missouri Bot. Garden* 69:594–624.

Reig, O. 1981. *Teoría del origen y desarrollo de la fauna de mamíferos de América del Sur.* Monografie Naturae. Mar del Plata, Argentina: Museo Municipal de Ciencias Naturales Lorenzo Scaglia.

Sarmiento, G. 1984. *The ecology of Neotropical savannas.* Cambridge: Harvard University Press.

Scott, W. B. 1937. *A history of the land mammals of the Western Hemisphere.* New York: Doubleday.

Simpson, G. G. 1948. The beginning of the age of mammals in South America, part 1. *Bull. Amer. Mus. Nat. Hist.* 91:1–232

———. 1965. *The geography of evolution.* Philadelphia: Chilton.

———. 1967. The beginning of the age of mammals in South America, part 2, *Bull. Amer. Mus. Nat. Hist.* 137:1–259.

———. 1980. *Splendid isolation.* New Haven: Yale University Press.

Sioli, H. 1975. Tropical rivers as expressions of their terrestrial environments. In *Tropical ecological system*, ed. F. B. Golley and E. Medina, 275–88. New York: Springer-Verlag.

Udvardy, M. D. F. 1978. *World biogeographical provinces.* Sausalito, Calif.: CoEvolution Quarterly. (Based on M. D. F. Udvardy: International Union for Conservation of Nature and Natural Resources, Occasional Paper no. 18, 1975.)

Van der Hammen, T. 1974. The Pleistocene changes of vegetation and climate in tropical South America. *J. Biogeogr.* 1:3–26.

Vuilleumier, B. S. 1971. Pleistocene changes in the fauna and flora of South America. *Science* 173: 771–80.

Webb, S. D. 1976. Mammalian faunal dynamics of the great American interchange. *Paleobiology* 2:216–34.

———. 1978. A history of savanna vertebrates in the New World, part 2: South America and the great interchange. *Ann. Rev. Ecol. System.* 9:393–426.

Webb, S. D., and L. G. Marshall. 1982. Historical biogeography of recent South American land mammals. In *Mammalian biology in South America*, ed. M. A. Mares and H. H. Genoways, 39–52. Pymatuning Symposia in Ecology 6. Special Publication Series. Pittsburgh: Pymatuning Laboratory of Ecology, University of Pittsburgh.

Wijmstra, T. A., and T. Van der Hammen. 1966. Palynological data on the history of tropical savannas in northern South America. *Leidse Geolog. Mededelingen* 38:71–90.

Diagnosis and Comments on Reproduction

The dentition is heterodont, with easily distinguished incisors, canines, premolars, and molars, and the number of teeth often exceeds the basic eutherian number of forty-four. The bony palate is fenestrated, a diagnostic character for most species when compared with the eutherians. The auditory bullae are formed principally by the alisphenoid. Epipubic bones are present in all species and are di-agnostic for the order. The brain structure differs from that of eutherian mammals in that the corpus callosum is lacking.

Marsupial young have a brief intrauterine development and are born in an extremely undeveloped state. At birth the forelimbs are well developed, and the newborn is able to crawl to the female's teat area, where it attaches to a nipple and remains some four to seven weeks. In many marsupials the teat-

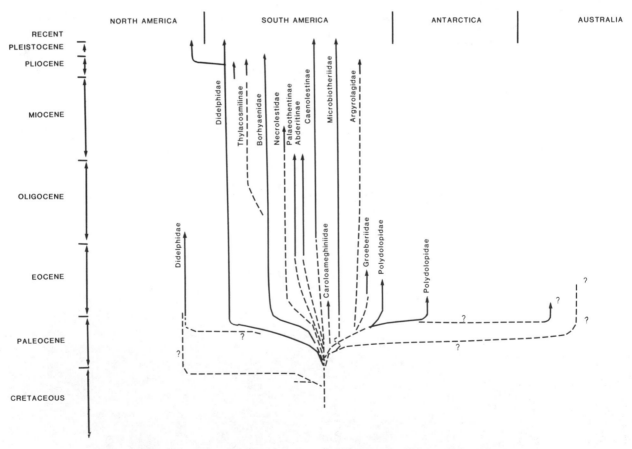

Figure 2.1. Evolutionary history of the Marsupialia. The Miocene radiation in Australia is not illustrated.

attachment phase terminates when the female deposits the young in a nest, where she nurses them for another three to seven weeks until they are fully weaned. In many species the teat area is enclosed in a pouch. The biology of New World forms is reviewed in the volume edited by Hunsaker (1977b).

Distribution

At present, marsupials in the Eastern Hemisphere are confined to the Australian region, and they have undergone an extensive adaptive radiation in Australia and New Guinea. Some species have colonized Celebes and the Moluccas. In the Western Hemisphere, the major adaptive radiation of marsupials occurred in South America. Most species are now found in South America, although many have extended their ranges into Central America. One species, *Didelphis virginiana*, has colonized temperate North America.

History and Classification

The oldest known marsupials are found in the mid-Cretaceous of North America. Records from South America are late Cretaceous, but the radiation, as exemplified by the fossil record in South America, suggests that earlier finds of marsupials may await discovery (Marshall, DeMuizon, and Sige 1983). Throughout the Tertiary in South America marsupials radiated into insectivore, frugivore, and carnivore niches (see Marshall 1982 for an extended account). Fossil marsupials first appear in Australia during the Oligocene, but clearly they must have arrived on the island continent before it became so distant from Antarctica. The recent discovery of fossil marsupials in the peninsula of Antarctica suggests that before the glaciation, when Antarctica, Australia, and South America were in closer proximity, Antarctica may have served as the bridge for the earliest marsupials' transit from South America to Australia (Woodburne and Zinmeister 1982) (see fig. 2.1).

The order Marsupialia is classically divided into thirteen families, eight of which are extant. Ride (1964) proposed that the Marsupialia be divided into four orders as indicated in table 2.1. In the New World there are three extant families: the Didelphidae, the Caenolestidae, and the Microbiotheriidae. In the area covered by this volume, northern South America, only the Didelphidae and Caenolestidae occur. The classification by Kirsch (1977) is followed here; a key to the genera is offered in table 2.2.

FAMILY DIDELPHIDAE
Opossums, Comadrejas

Diagnosis

The pouch may be either absent, as in the genera *Marmosa*, *Monodelphis*, and *Metachirus*, or well developed, as in *Didelphis*, *Philander*, *Chironectes*, and during lactation in *Caluromys*. There are five digits on each foot, and the hind foot has an opposable thumb. Members of this family have a long rostrum, large naked ears, and a tail that is often highly prehensile and usually almost hairless for at least the distal two-thirds. *Glironia* is exceptional in having its tail furred almost to the tip. This family exhibits a wide range of sizes. Some species may be as small as 80 mm in head and body length, while the larger species may reach 1,020 mm in total length. Pelage color is highly variable, ranging from gray to deep, rich brown. Some species exhibit banding of the dorsal pelage. The dental formula tends to be con-

Table 2.1   Two Alternative Classifications for the Marsupialia

Kirsch 1977	Ride 1964
Superorder Marsupialia	Superorder Marsupialia
Order Polyprotodontia	Order Marsupicarnivora
Suborder Didelphimorphia	
Superfamily Didelphoidea	Superfamily Didelphoidea
Family Didelphidae[a]	Family Didelphidae[a]
	(includes *Dromiciops*)
Family Microbiotheriidae[a]	
Family Thylacinidae	
Suborder Dasyuromorphia	
Superfamily Dasyuroidea	Superfamily Dasyuroidea
Family Dasyuridae	Family Dasyuridae
Family Myrmecobiidae	(includes *Myrmecobius*)
	and Family Thylacinidae
Suborder Peramelemorphia	Order Peramelina
Superfamily Perameloidea	
Family Peramelidae	Family Peramelidae
Family Thylacomyidae	
Suborder Notoryctemorphia	
Superfamily Notoryctoidea	
Family Notoryctidae	
Order Paucituberculata	Order Paucituberculata
Superfamily Caenolestoidea	
Family Caenolestidae[a]	Family Caenolestidae[a]
Order Diprotodonia	Order Diprotodonia
Superfamily Phalangeroidea	
Family Phalangeridae	Family Phalangeridae
Family Petauridae	
Family Burramyidae	
Family Macropodidae	Family Macropodidae
Superfamily Vombatoidea	
Family Vombatidae	Family Vombatidae
Family Phascolarctidae	
Superfamily Tarispedoidea	
Family Tarsipedidae	
	Marsupialia *incertae sedis*
	Family Notoryctidae

[a]Families present in South America.

Table 2.2   Key to the Families and Genera of Marsupials Inhabiting the Northern Neotropics

1   First lower incisors enlarged and project forward .... Caenolestidae:
                                                                                                                          Caenolestes obscurus
1' Lower incisors all approximately the same size .... Didelphidae
2   Hind feet webbed; coat color black and gray, exhibiting alternate bands horizontally arranged .... *Chironectes*
2' No webbing between toes of hind foot
3   Tail noticeably shorter than head and body; size small .... *Monodelphis*
3' Tail nearly as long as or longer than head and body 4
4   Black markings on shoulder contrast strongly with basic gray dorsal pelage; black shoulder markings join in
    center of back to display a single or double dorsal stripe .... *Caluromysiops*
4' Markings not as above 5
5   Tail furred for almost the entire length but scales visible on distal third; tail very thick at the base .... *Lutreolina*
5' Tail not as above 6
6   Face and or cheeks white; tail naked for its entire length .... *Didelphis*
6' Face and/or cheeks not distinctly white 7
7   Two white or buff spots on face above the eye
       a. Basic body color brown; spots buff; tail extremely long .... *Metachirus*
       b. Basic body color gray; spots white; tail about equal to head and body length .... *Philander*
7' No distinctive white spots on the face 8
8   Eyes circled by dark fur, creating a facial mask .... *Marmosa*
8' No facial mask; dark median stripe from nose to between the eyes; fur dense and woolly .... *Caluromys*

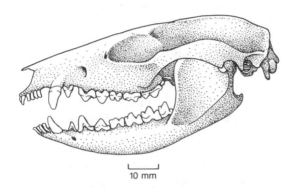

10 mm

Figure 2.2. Skull of *Didelphis marsupialis*.

servative, with fifty teeth typical: I 5/4, C 1/1, P 3/3, M 4/4. The postorbital bar is not developed (fig. 2.2). Karotypic data have been summarized by Reig et al. (1977) and Seluja et al. (1984).

Comment

Certain characters of *Caluromys*, *Caluromysiops*, and *Glironia* lend credence to the tendency to place them in their own family or subfamily (Caluromyidae) (Creighton 1984; Reig, Kirsch, and Marshall 1987).

Distribution

Species of this family occur widely over the northern Neotropics and occupy almost every habitat type except extremely high elevations and the desert areas surrounding the Gulf of Venezuela.

Natural History

Most members of this family are nocturnal. All occupy omnivore, insectivore, frugivore/insectivore, or carnivore feeding niches. Many members are strongly arboreal and seldom come to the ground,

especially those long-tailed forms that are specialized for multistratal tropical rain forests.

Marsupials are characterized by a unique mode of reproduction. All forms studied in the Western Hemisphere have only a yolk-sac placenta, and the shell membrane remains intact during the intrauterine growth phase. The young are born after a brief period of gestation, generally thirteen to fourteen days for the South American species. Upon passing out of the reproductive tract, the young climb unassisted to the nipple area, where each grasps a teat in its mouth and remains attached to it for a varying time during early growth and development. The mother transports the entire litter attached to her teats for approximately the first five to six weeks of development. The young then detach and begin a nest phase where the mother does not continuously transport them but returns to the nest to nurse them. The teat area may be enclosed in a pouch, or no pouch may be present (e.g., *Marmosa*, *Metachirus*, and *Monodelphis*).

Eisenberg and Wilson (1981) measured the cranial capacities of several species of didelphid marsupials. When comparing their cranial capacities with those of eutherian mammals they found that several didelphid species (e.g., *Monodelphis* and *Didelphis*) had extremely small brains, but many others had brain sizes near the norm for small eutherian mammals. The wide variety of relative brain sizes within the family Didelphidae strongly suggests that we should reexamine generalizations concerning behavioral capacities based on *Didelphis*.

By comparing the relative proportions of limb segment length, trunk length, and tail length, we can classify didelphid marsupials into groups reflecting different locomotor adaptations. Terrestrial adaptations include a shortened tail (e.g., *Monodelphis*), and rapid terrestrial locomotion involves increasing the relative mass of the hind leg (e.g., *Metachirus*). Arboreal adaptation is reflected by a long tail with a proportionately high mass and relatively large, powerful forelimbs (Grand 1983).

McNab (1978) studied the energetics of Neotropical marsupials and concluded that, in contrast to most Australian species, their metabolic rates fall in line with rates measured for many insectivorous and frugivorous eutherian mammals. It is true that no didelphid marsupial has a high metabolic rate, but it is untrue to say that they have lower metabolic rates than comparable-size eutherians specialized for similar diets.

The community ecology of didelphids has been receiving more attention recently. Fleming (1972) studied three sympatric species in Panama. P. Charles-Dominique (1983) completed an extensive comparative study of didelphid ecology in French Guiana. O'Connell (1979) investigated a didelphid community in northern Venezuela and analyzed reproductive performance, habitat preference, and substrate utilization. The behavior of didelphids has been reviewed by Papini (1986). Communication mechanisms were reviewed by Eisenberg and Golani (1977). The natural history of didelphid marsupials has been admirably reviewed in the article by Streilein (1982) and in books by Hunsacker (1977b) and Collins (1973).

Genus *Caluromys* J. A. Allen, 1900
Woolly Opossum, Zorrito de Palo

Description
The dental formula is I 5/4, C 1/1, P 3/3, M 4/4. The first upper premolar is very small, situated directly behind the canine, and there is a distinct gap between the first premolar and the much larger second premolar. The fur of the body is very thick, hence the common name woolly opossum. The dorsal pelage varies from brown to gray brown, and a brown stripe runs from the muzzle to between the ears. The tail is furred for more than one-third of its length; it is extremely long and fully prehensile. There are three recognized species (see plate 1).

Distribution
The genus is distributed from southern Veracruz in Mexico to Paraguay. The strong arboreal adaptations of this genus limit its distribution to areas of moist forest.

Natural History
The woolly opossums are distinctive when compared with other members of the family Didelphidae. Their litter size tends to be somewhat reduced,

and they have greater potential longevity in captivity. *Caluromys* is typified by a comparatively large encephalization quotient (Eisenberg and Wilson 1981). A pattern of more extended maternal care is suggested by the studies of Charles-Dominique (1983). Trapping records and radiotelemetry indicate that these strongly arboreal animals come to the ground infrequently. They tend to be mixed feeders, eating the pulp of various fruits and supplementing this diet with nectar, invertebrates, and small vertebrates. Vocalizations of *Caluromys* are similar to those of other didelphids. They hiss with an open-mouth threat during defensive behavior. Click sounds are made during male-female encounters and courtship and may grade into chirps (Eisenberg, Collins, and Wemmer 1975).

Caluromys derbianus and *Caluromys lanatus* appear not to overlap geographically, but *Caluromys philander* can co-occur with *C. lanatus*. The latter species is much larger than *C. philander*, and there is a strong suggestion that when congeners co-occur there is ecological segregation in feeding habits, indirectly indicated by the size differences (Eisenberg and Wilson 1981; Charles-Dominique et al. 1981).

Caluromys derbianus (Waterhouse, 1841)

Description
External measurements exhibit a wide range. Total length ranges from 760 to 587 mm, of which the tail may be 490–395 mm. The ear averages 40 mm and the hind foot 35 mm. Mean weight for six adults from Panama was 295 g. The basic dorsal pelage is a lighter brown than that of *C. lanatus*.
Chromosome number: $2n = 14$; FN $= 24$.

Range and Habitat
This species occurs from western Ecuador to Veracruz, Mexico, and apparently does not pass over the eastern cordillera of the Andes. To the east it is replaced by *Caluromys lanatus* (map 2.1). This species appears to be confined to lower elevations of moist evergreen tropical rain forests.

Natural History
This nocturnal, strongly arboreal species rarely descends to the ground. It forages in the upper strata of a mature rain forest, sheltering in tree cavities and supplementing its diet of ripe fruit with almost any invertebrate or small vertebrate it can capture. It is not known to exhibit torpor. The average litter size is three (Bucher and Hoffmann 1980), and the interval between estrous periods is twenty-eight days (Bucher and Fritz 1977). In Panama the breeding season commences with the onset of the dry season, generally in late January or early February (Enders 1966). Biggers (1962) recorded a similar

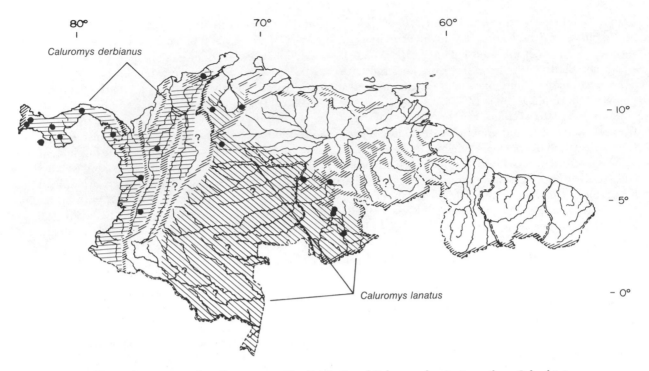

Caluromys derbianus

Caluromys lanatus

Map 2.1. Distribution of two species of woolly opossum. The distribution of *Caluromys lanatus* in southern Colombia is very poorly known.

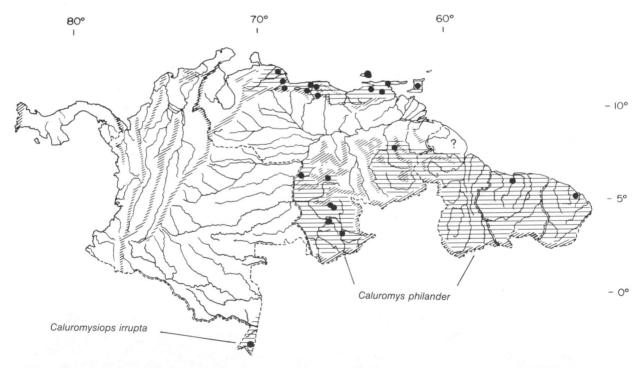

80°   70°   60°

Caluromys philander

Caluromysiops irrupta

Map 2.2. Distribution of *Caluromys philander* and *Caluromysiops irrupta*. *C. philander* is confined to the eastern part of northern South America. The black-shouldered opossum has been reported from southern Colombia (Simonetta 1979).

seasonality in Nicaragua. Sexual maturity is attained at seven to nine months (Fleming 1973; Collins 1973).

Caluromys lanatus (Illiger, 1815)

Description
Caluromys lanatus is nearly the same size as *C. derbianus*. Head and body length may reach 700 mm and tail length 437 mm; the hind foot averages 47 mm. Mean weight for two specimens from Venezuela was 356 g (see plate 1).
Chromosome number: $2n = 14$; FN = 24.

Range and Habitat
This species is distributed from eastern Colombia to Paraguay and apparently replaces *C. derbianus* to the east of the Andes. According to Handley (1976), *C. lanatus* is strongly arboreal and most often taken near streams or in other moist areas. In Venezuela no specimens were taken at an elevation exceeding 700 m, and all specimens were associated with multistratal evergreen tropical forests (map 2.1).

Natural History
This nocturnal arboreal specialist appears to have feeding habits similar to those of *C. derbianus*. Little is known about its reproductive biology, but Bucher and Fritz (1977) report that in captivity estrus occurs at an average interval of twenty-eight days. These authors remark that activity patterns and substrate use are similar to those of *C. derbianus*.

Caluromys philander (Linnaeus, 1758)

Description
In Suriname, head and body length ranges from 258 to 245 mm, while the tail ranges from 362 to 317 mm. One specimen recorded in Suriname attained a weight of 350 g. *C. philander* is usually smaller in Venezuela than in Suriname, with a mean weight of 170 g ($N = 3$).
Chromosome number: $2n = 14$; FN = 24.

Range and Habitat
The species ranges from northern Venezuela to Brazil. In Venezuela, Handley (1976) found *C. philander* to be highly arboreal; it was rarely taken on the ground and was strongly associated with moist habitats. Although most specimens were taken in multistratal evergreen tropical forests, the species is adaptable, and some were caught in orchards. *C. philander* can range up to 1,600 m in elevation (map 2.2).

Natural History
C. philander is the best-studied species of the genus (Atramentowicz 1982; O'Connell 1979; Charles-

Dominique et al. 1981; Charles-Dominique 1983). It is nocturnal and strongly arboreal, concentrating its activity in the upper canopy of a multistratal forest. It often shelters in tree cavities, where it constructs a nest of dead leaves. Although its diet includes ripe fruit, gum, nectar, small vertebrates, and invertebrates, it feeds primarily on fruit. It is not known to exhibit torpor.

In French Guiana litter size is about four; but six young were recorded by O'Connell (1979) in Venezuela. Gestation is assumed to be about fourteen days. The young remain in the mother's pouch for approximately eighty days, then enter a nest phase for another thirty days. During the nest phase the mother returns from her nocturnal forays to nurse the young. Dispersal from the natal nest occurs at approximately 130 days of age. In French Guiana a female does not breed until she is approximately one year old. Maternal care is prolonged in *Caluromys philander*. A female can produce three litters a year, but if there is a seasonal scarcity of food she will probably not rear more than one litter each year (Atramentowicz 1982).

Charles-Dominique (1983) describes a nightly home-range area of 0.3 to 1.0 ha. Home-range stability is a function of food availability. Interactions among adults tend to be agonistic except when a male is courting a female. There seems to be no strong territorial defense, and home ranges of adjacent adult females may show some overlap. Males' ranges broadly overlap each other and the ranges of several females. No permanent social grouping persists except the mother-young unit. Vocalizations during encounters have been analyzed by Charles-Dominique (1983). Both young and adults produce clicks, and clicking by suckling young apparently helps maintain maternal activity. Agonistic vocalizations include hisses and grunts. A distress scream is produced upon extreme disturbance. *Caluromys* will actively attempt to bite predators or human handlers.

Genus *Caluromysiops* Sanborn, 1951
Caluromysiops irrupta Sanborn, 1951
Black-shouldered Opossum

Description
Head and body length exceeds 200 mm, and the tail is longer than the head and body: TL 570; T 310; HF 67; E 37 (Peru). This species resembles the woolly opossum in that the tail is haired for over two-thirds of its length, but its color pattern is quite different—light gray on the dorsum and face with a distinctive black shoulder. The black markings on the shoulder join at the midline to form either a single dorsal stripe or a double stripe slightly off center of the midline (see plate 1).

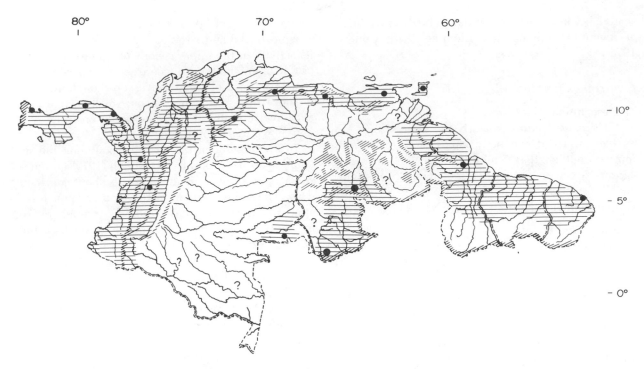

Map 2.3. Distribution of *Chironectes minimus*. The water oppossum has an extremely fragmented range but occurs widely over South America in areas of suitable habitat (Marshall 1978a).

Range and Habitat

The species is known from multistratal tropical evergreen forests in the Amazonian portion of Peru (map 2.2) and was reported from the extreme southeast of Colombia by Simonetta (1979). Simonetta offers a photograph, but there is no museum specimen to confirm this record.

Natural History

Little is known concerning the behavior and ecology of this rare opossum. Although its feeding habits are unknown from the wild, specimens in captivity show rather individualistic dietary preferences, taking a wide range of fruits as well as small rodents. Judging from its body proportions and behavior in captivity, this species is probably highly arboreal. Captives have lived more than seven years (Collins 1973).

Genus *Chironectes* Illiger, 1811
Chironectes minimus (Zimmerman, 1780)
Water Opossum, Comadreja de Agua

Description

Head and body length ranges from 270 to 325 mm, the tail from 355 to 395 mm, and weight from 540 to 790 g. The hind feet are webbed, and the tail is naked for most of its length. The pouch opens to the rear. This species is easily recognized by its peculiar color pattern, with alternating transverse gray and black markings on the dorsum (see plate 2). The face is generally black with pale gray above the eye, extending to the ear, and the underparts are white. The pelage is very dense.
Chromosome number: $2n = 22$; FN = 20.

Range and Habitat

The single species of *Chironectes* ranges from southern Mexico to northeastern Argentina and is invariably found in association with streams and lakes. It can occur at reasonably high elevations; in Venezuela has been taken at up to 1,860 m (map 2.3).

Natural History

Water opossums are nocturnal and adapted for foraging in water. They do not climb well. When Enders (1935) provided captives with branches they climbed awkwardly, and the stout tail was not very prehensile. Like other didelphids, they threaten by opening their mouths and hissing.

The animals swim well and are known to feed on fish and crustaceans. The pouch is extremely well developed, and when a female carrying pouch young swims, a strong sphincter closes the pouch so tightly that no water enters. Up to five young have been recorded, but two or three appears to be the typical litter size. When not foraging the animals use a bank burrow, in which they construct a nest of leaves (Marshall 1978a; Mondolfi and Medina Padilla 1957; Zetek 1930).

Table 2.3   Measurements for Two Species of *Didelphis*

	Didelphis marsupialis					*Didelphis albiventris*				
	Mean	Minimum	Maximum	N	Location	Mean	Minimum	Maximum	N	Location
TL	831.5	716	930	14	Venezuela	709.0	670	830	15	Venezuela
T	414.8	365	465	14	Venezuela	361.5	300	410	15	Venezuela
HF	61.0	45	70	14	Venezuela	54.0	47	68	15	Venezuela
E	49.9			14	Venezuela	52.0	46	60	15	Venezuela
Wt	1,496.0	665	2,090	4	Venezuela	832.5	600	1,000	4	Venezuela

Source: Mondolfi and Perez-Hernandez (1984).

Genus *Didelphis* Linnaeus, 1758
Large American Opossum, Rabipelado

Description and Taxonomy

This is the large opossum commonly encountered over most of North and South America. The dental formula is I 5/4, C 1/1, P 2/3, M 4/4. Head and body length ranges from 325 to 500 mm and the tail from 255 to 535 mm. Animals may weigh up to 5.5 kg. The pouch is well developed in the female. The pelage consists of two hair types, a dense underfur and long guard hairs that are generally white tipped, giving a shaggy appearance. The almost naked tail is darkly pigmented from the base for approximately one-third of its length, with the distal portion generally white to pink. The forelimbs tend to be black, and the face is white to yellow. Depending on the species, the basic dorsal color may be gray with varying amounts of black; melanistic forms are known from several parts of its range (see plate 2).

There are three recognized species: *Didelphis albiventris* (= *azarae*), *D. marsupialis*, and *D. virginiana*.

Distribution

The genus is distributed from the northeastern United States to Patagonia in Argentina. The species *D. virginiana* has been introduced to the western United States, where it occurs from southern British Columbia to California, and it ranges from Massachusetts, in the United States, to Costa Rica (Gardner 1973). Its behavior has been studied in detail by McManus (1970).

Didelphis marsupialis Linnaeus, 1758
Common Opossum, Rabipelado

Description

Total length ranges up to 1,000 mm, of which 482 mm is the head and body and 535 mm is the tail (see table 2.3). Females are generally smaller than males. *Didelphis marsupialis* tends to be darker than the northern *D. virginiana*, with black on the face although the cheeks are gray. The black ears distinguish it from *D. albiventris*.

Chromosome number: $2n = 22$; FN = 20.

Range and Habitat

D. marsupialis occurs from eastern Mexico to northern Argentina. It tolerates a variety of habitat types but is not found at extremely high elevations or in arid areas. It has been taken at up to 2,232 m elevation. In northern South America it is replaced in montane areas by *D. albiventris* (map 2.4).

Natural History

Didelphis marsupialis is nocturnal in its habits, and though it climbs well, it frequently forages on the ground, feeding on a wide range of fruits, invertebrates, and small vertebrates. It is an important reservoir for the parasite *Trypanosoma cruzi* in Venezuela (Telford et al. 1981). Animals usually are seen singly, but males actively court females during the breeding season, when two or more may be encountered. The female builds a leaf nest in a tree cavity or burrow, and after a fourteen-day gestation period she bears up to eight young. The teat number averages thirteen, but litter size varies with latitude; the smallest litters are typical near the equator. At the southern limits of its range in South America, litter size is appreciably larger. Mean litter size in Panama and western Colombia is 6.0, in northern Venezuela 6.7, and in southern Brazil 7.4 to 8.5. Females in captivity from southern Brazil average even larger litters, 10.7, suggesting that litter size may be determined in part by the food available to a breeding female (Motta, Araujo Carreira, and Franco 1983). Breeding may be seasonal in habitats having distinct seasonality in rainfall. In Panama and Venezuela breeding commences at the onset of the dry season (Fleming 1973; O'Connel 1979; Cordero 1983).

Given adequate shelter and a sustained food supply, the home range of a lactating female may be rather stable, but the animals are opportunistic feeders and readily shift home ranges to adapt to fluctuating resources. Core home range areas averaged 0.20 ha during the dry season in Guárico, Venezuela (O'Connell 1979, August 1981). Telford, Gonzalez,

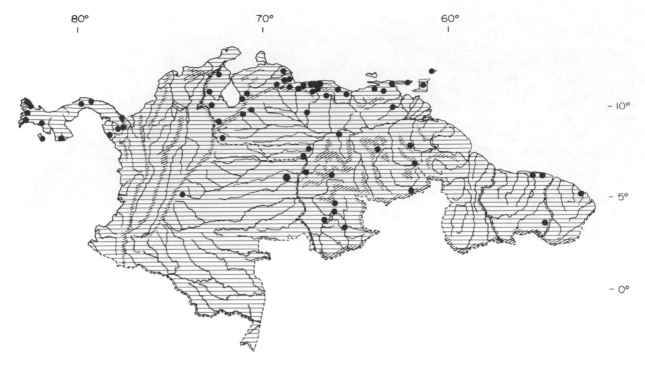

Map 2.4. Distribution of the South American common opossum. In suitable habitat at lower elevations *Didelphis marsupialis* is distributed throughout northern South America.

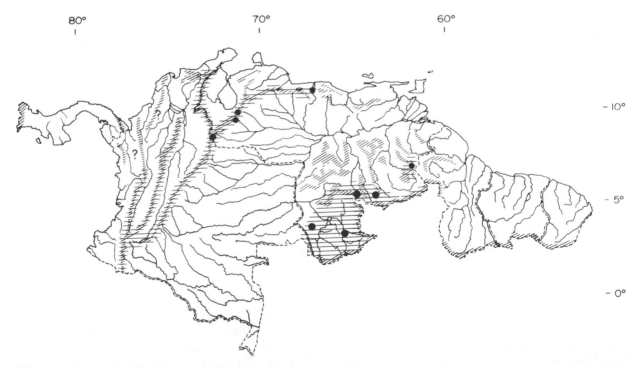

Map 2.5. Distribution of Azara's opossum. *Didelphis albiventris* replaces *D. marsupialis* at higher elevations (see Mondolfi and Perez-Hernandez 1984).

and Tonn (1979) found minimum home ranges of 0.8 to 0.9 ha for young animals and 2.2 to 2.5 ha for adults in Portuguesa, Venezuela. Recent radio-tracking data from Sunquist, Austad, and Sunquist (1987) in Venezuela indicate that extended home ranges are larger than the estimates derived from trap, mark, and release studies. Mean home ranges were 123 ha for males and 16 ha for females.

Didelphis albiventris Lund, 1840
Azara's Opossum, Comadreja Overa

Description
This species is approximately the same size as *D. marsupialis*. Total length averages 763 mm, of which the tail makes up 374 mm. Weight ranges from 625 to 1,100 g (see table 2.3). Varying degrees of black appear in the dorsal pelage, and the white venter contrasts strongly (Barlow 1965). White ears in the adult distinguish this species from *D. marsupialis*. Mondolfi and Perez-Hernandez (1984) recognize the population in southern Venezuela as a distinct subspecies.
Chromosome number: $2n = 22$; FN $= 20$.

Range and Habitat
D. albiventris occurs in western South America, from Colombia and Venezuela to Peru, and eastward through Brazil to northern Argentina. In northern South America it occurs at high elevations, and in Venezuela it was taken at 2,380 to 3,275 m. In the north it replaces *D. marsupialis* above 2,000 m (map 2.5).

Natural History
D. albiventris has not been the subject of detailed natural history studies. One assumes that in its dietary habits and activity patterns it resembles *D. marsupialis*. Teat number averages thirteen; litter size has been recorded at 4.2 in Colombia (Tyndale-Biscoe and Mackenzie 1976), but in Brazil and Argentina the litter size may reach twelve (Barlow 1965). Cerqueira (1984), working in northeast Brazil, found that the onset of reproduction was controlled by rainfall. Litter size varied from three to ten, and older females produced smaller litters. Home ranges averaged 0.57 ha in Argentina (Cajal 1981).

Genus *Lutreolina* Thomas, 1910
Lutreolina crassicaudata (Desmarest, 1804)
Thick-tailed Opossum, Coligrueso,
Comadreja Colorado

Description
In this medium-sized opossum, total length averages about 600 mm. Head and body length can reach 400 mm, with a tail length of 310 mm. weight may reach 540 g. The fur is dense and short, with a dorsal color from light to dark brown. The underparts are often a paler brown, sometimes tinted with red. In basic body form it suggests a weasel, since the legs are short and the neck long. The tail is very thick at the base, furred for almost its entire length, and not very prehensile (see plate 2).
Chromosome number: $2n = 22$; FN $= 20$.

Range and Habitat
The thick-tailed opossum is distributed from the Guyanas to Patagonia. It seems much more abundant in the southern parts of its range and occurs sporadically in Colombia, Venezuela, and the Guyanas (map 2.6) (Lemke et al. 1982). It prefers moist habitats and is often found near streams or lagoons.

Natural History
Thick-tailed opossums have not been the subject of a detailed field study. From captive studies we know that they can climb, but typically they forage terrestrially, preying on small mammals, reptiles, and insects. The female has up to nine teats, and a litter size of seven has been recorded. The animals den terrestrially and build grass nests (Marshall 1978b). Core home ranges average about 0.38 ha in Argentina, but extended home ranges have yet to be determined (Cajal 1981).

Genus *Marmosa* Gray, 1821
Mouse Opossum, Marmosa

Description
The dental formula is I 5/4, C 1/1, P 3/3, M 4/4. This genus is highly variable in size and includes some forty-seven species, twenty-one of which occur in the area under consideration (R. Pine, in Collins 1973). This latter number has been reduced to nineteen as a result of taxonomic revisions (G. K. Creighton, pers. comm.). Head and body length varies from 60 mm to over 200 mm, depending on the species, and the tail ranges from 100 to 281 mm and generally exceeds the head and body in length. The ear is usually slightly shorter than the hind foot. The tail is fully prehensile, and if its base is haired it is always for less than one-third of the length of the tail. The dorsal pelage can vary from red brown to gray, and the venter is usually paler, varying from tan to cream. There are no white spots on the face; the eye has dark brown hairs around it so that when viewed from the front the effect is of a contrasting mask.

Distribution
The genus is widely distributed from southern Veracruz, Mexico, to northern Argentina. Different species often occupy discrete altitudinal zones, and some are adapted to the Andean foothills.

Natural History, Identification, and Systematics

The mouse opossums exhibit an interesting adaptive radiation, with many species apparently having narrow habitat requirements while others tolerate a wide range of habitat types. Arboreal ability varies from species to species; some are strongly arboreal and seldom come to the ground, while others show a more scansorial tendency (O'Connell 1979; Handley 1976). The female does not develop a pouch. The teats are arranged in varying symmetrical patterns in the posterior ventral area, and the number of nipples is variable both within and between species. For the genus, the nipple number ranges from nine to nineteen (Osgood 1921; Tate 1933). The young are born in an extremely undeveloped state after a thirteen- to fourteen-day gestation and remain attached to the teats for almost the first thirty days of life (Eisenberg and Maliniak 1967). Litter size is very high in some species.

Marmosa robinsoni females exhibit ovarian senescence after fourteen to sixteen months of age (Barnes and Barthold 1969; pers. obs.). Given sexual maturity at six months, this senescence restricts breeding to a period of eight to ten months following maturity. It is doubtful if this species can breed more than once under more stringent climatic regimes where a long annual drought occurs (Eisenberg 1988). All species studied to date seem to be

Table 2.4   Field Key to the Species of *Marmosa*

A. Species occuring west of the eastern cordillera of the Andes
   1 Head and body length greater than 160 mm 2
   1′ Head and body length less than 160 mm 4
   2 Tail slightly longer than head and body length; only in western Panama . . . . *M. mexicana*
   2′ Tail greatly exceeds head and body length 3
   3 Tail furred from base to about one-third its length . . . . *M. cinerea, M. alstoni,* or *M. regina* [a]
   3′ Tail furred from base for about 10 mm; color pale brown, venter white; low elevations . . . . *M. robinsoni*
   4 Head and body length less than 160 mm but greater than 110 mm 5
   4′ Head and body length less than 110 mm 6
   5 Found at elevations above 2,000 m . . . . *M. impavida*
   5′ Found at elevations below 2,000 m; basal third of tail furred, distal half of tail white . . . . *M. phaea*
   6 Occurs in Panama and northwestern Colombia . . . . *M. invicta*

B. Species occurring north of the Orinoco and east of the eastern cordillera of the Andes (inclusive of)
   1 Head and body length greater than 160 mm 2
   1′ Head and body length less than 160 mm 3
   2 Tail furred from its base to about one-third of its length; distal portion of tail often somewhat depigmented . . . . *M. cinerea*
   2′ Tail furred from base for about 10 mm; tail uniformly colored . . . . *M. robinsoni*
   3 Head and body length less than 160 mm but greater than 110 mm 4
   3′ Head and body length less than 110 mm 7
   4 Restricted to semiarid areas around the Gulf of Venezuela . . . . *M. xerophila*
   4′ Not as above 5
   5 Confined to moist lowland habitats below 1,300 m . . . . *M. murina*
   5′ Generally ranges above 1,300 m 6
   6 Dorsum brown to wood brown . . . . *M. impavida*
   6′ Dorsum slate gray . . . . *M. fuscata*
   7 Tail approximately equal to head and body length . . . . *M. fuscata*
   7′ Tail greatly exceeds head and body length 8
   8 Occurs at elevations above 2,000 m . . . . *M. dryas*
   8′ Occurs at lower elevations . . . . *M. parvidens* or *M. cracens* [a]

C. Species occurring south of the Orinoco and east to the Atlantic
   1 Head and body length greater than 160 mm 2
   1′ Head and body length less than 160 mm 3
   2 Tail furred from its base to about one-third of its length; distal portion of tail often somewhat depigmented . . . . *M. cinerea*
   2′ Tail furred from base for about 10 mm; tail uniformly colored . . . . *M. robinsoni*
   3 Head and body length less than 160 mm but greater than 110 mm 4
   3′ Head and body length less than 110 mm 5
   4 Dorsum slate gray; occurs at elevations above 1,200 m . . . . *M. fuscata*
   4′ Dorsum light brown; occurs at elevations below 1,200 m . . . . *M. murina*
   5 Tail about equal to head and body length . . . . *M. lepida*
   5′ Tail greatly exceeds head and body length 6
   6 Occurs at elevations above 2,000 m . . . . *M. tyleriana*
   6′ Occurs at lower elevations . . . . *M. parvidens* and *M. emiliae* [a]

[a] Separable on distributions—see maps.

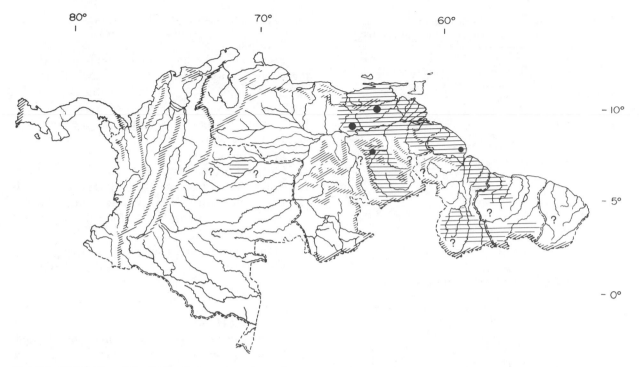

Map 2.6. Distribution of the thick-tailed opossum, *Lutreolina crassicaudata*. The thick-tailed opossum occurs in a fragmented range in Colombia, the Guyanas, and eastern Venezuela (Lemke et al. 1984; Marshall 1978b).

nocturnal. Most appear to eat fruit when it is available and feed heavily on invertebrates (Hunsaker 1977a).

Social life is rudimentary in *Marmosa*, and aside from a courting pair or a female and her dependent young, adults forage and nest alone. Communication between same-sex adults is agonistic. An open-mouth threat and hissing constitute a common defensive posture. Young animals chirp when detached from the teats, and the mother will return and push them under her body so they can reach the teat area. Males courting females approach and produce clicking sounds. Males mark by rubbing the chin, chest, or cheeks on the substrate. Both sexes urine mark and drag their cloaca on the substrate, leaving chemical traces that presumably coordinate reproductive behavior (Eisenberg and Golani 1977).

Several species that typically occupy montane habitats in northern South America include *Marmosa dryas*, *M. fuscata*, and *M. impavida*. Some species are adapted to rather arid habitats; the most extreme arid adaptation appears to be shown by *Marmosa xerophila*. The species of *Marmosa* fall generally into three size classes. Extremely small forms, generally with a head and body length less than 110 mm, include *Marmosa lepida*, *M. cracens*, *M. invicta*, *M. dryas*, *M. parvidens*, and *M. emiliae*. An intermediate size group ranges in head and body length from approximately 125 to 155 mm. This group includes *Marmosa impavida*, *M. xerophila*, *M. fuscata*, *M. murina*, *M. robinsoni*, and *M. phaea*. The larger species range from 180 to over 200 mm in head and body length and include *Marmosa alstoni*, *M. mexicana*, and *M. cinerea*.

Some caution is needed in using body size as a guide to identification, since individuals generally continue to grow for most of their lives. Thus a freely moving young specimen of *Marmosa cinerea* may show a head and body length of 140 mm but at full size can exceed 200 mm. In basing identifications on average measurements, care must be taken that one has an adult specimen in hand (see table 2.4).

Reig, Kirsch, and Marshall (1987) have proposed that the genus *Marmosa* may be divided into at least three valid genera.

Creighton (1984, pers. comm.) has proposed that *marmosa* may be divided into at least four genera. His classification is included in table 2.5. *Thylamys* Gray, 1843, is a southern genus and need not concern us in the area covered by this volume. Suffice it to say that species of this genus tend to have a woolly pelage and rounded premaxilla and store fat in their tails.

Creighton offers the following remarks concerning the morphological features separating the proposed new genera according to his outline. The genus *Micoures* Lesson, 1842, is distinctive in that most of the species are large, with head and body

Table 2.5   Provisional Classification of the Genus
Marmosa according to Creighton (1984)

Micoures
    M. alstoni = cinerea
    M. cinerea
    M. regina
    M. phaea
Marmosa
  The "*Murina* group"
    M. lepida
    M. mexicana
    M. murina
    M. robinsoni
    M. tyleriana
    M. xerophila
  The "*Microtarsus* group"
    M. dryas
    M. emiliae
    M. marica
Marmosops (= *noctivaga* group of *Marmosa*)
    M. cracens
    M. fuscata
    M. handleyi
    M. impavida
    M. invicta
    M. parvidens
Thylamys (= *elegans* group)
    M. elegans and relatives

Note: Only northern species are fully listed.

length exceeding 125 mm. The tail vastly exceeds the head and body length, averaging at least 1.3 times as long. The scales on the tail are rhomboid and coarse, with fourteen to sixteen rows per centimeter. The tail is never bicolored but is often whitish or mottled distally. The fur of the body extends at least 5 cm on the proximal portion of the tail. The tympanic bullae of the skull are small, and postorbital processes are prominent.

The genus *Marmosops* Matschie, 1916, also has rhomboidal scales on the tail, but they are much finer, with twenty two to twenty eight per centimeter. The bullae are small, but the postorbital processes are lacking. The lower canines are premolariform in shape, which is distinctive.

The genus *Marmosa* Gray, 1821, is subdivided into the *Murina* group and the *Microtarsus* group; the latter could be elevated to the genus *Grymaeomys* Burmeister, 1854. In the *Microtarsus* group the tail scales are square and very fine grained, as many as forty per centimeter, and are annular rather than spiral. The tail tends to be weakly bicolor, and the bullae are large. The *Murina* group of the genus *Marmosa* has tail scales that are rhomboidal and coarse, at about fifteen to twenty per centimeter, and spiral as in *Micoures*. Although fur may extend on the proximal portion of the tail, it never exceeds 2.5 cm. The premaxillaries exhibit an acute outline. Postorbital processes are present and prominent.

Marmosa alstoni (J. A. Allen, 1900)

Description
This is one of the larger mouse opossums, reaching 450 mm in total length: HB 185–200; T 230–50; HF 27–30; E 25–26. The tail greatly exceeds the head and body length, and approximately 30% of the base of the tail is haired (see plate 2). The dorsum is grayish to gray brown, contrasting sharply with the cream-colored venter.

M. alstoni resembles *M. cinerea*, which occurs to the east of the Andean cordillera. In Panama, *M. alstoni* co-occurs with *M. mexicana*, from which it may readily be distinguished by the color of the dorsal pelage. *M. robinsoni* co-occurs with *M. alstoni* in northern Colombia and Panama. In general *M. robinsoni* is more scansorial in its habits and replaces *M. alstoni* in dryer habitats.
Chromosome number: $2n = 14$; FN = 24.

Range and Habitat
Marmosa alstoni occurs from Belize south through the Isthmus to western Colombia. It occurs at lower elevations and is strongly associated with tropical forested habitats (map 2.7).

Natural History
See *M. cinerea*.

Marmosa cinerea (Temminck, 1824)

Measurements

	Mean	Minimum	Maximum	N
TL	360.29	284	423	7
T	213.86	175	252	7
HF	23.57	20	29	7
E	27.80	25	30	5
Wt	75.75	50	96	4

Location: Guyana, Paraguay, Suriname (AMNH)

Description
This is one of the largest species of the genus, having a maximum total length of 406 mm, with a tail of 220 mm in specimens from Suriname. The basal part of the tail is furred as in *M. alstoni*. The dorsum varies from dark brown to gray brown, and the underparts vary from buff to yellow. Over much of its range, a dark brown eye ring characterizes this species. The tail may show varying degrees of depigmentation on its distal portion.
Chromosome number: $2n = 14$; FN = 24.

Range and Habitat
This species occurs from eastern Colombia through Venezuela and the Guyanas, and south to northern Argentina. It is restricted to moist habitats, being associated with multistratal tropical evergreen forests at elevations below 1,000 m (map 2.7).

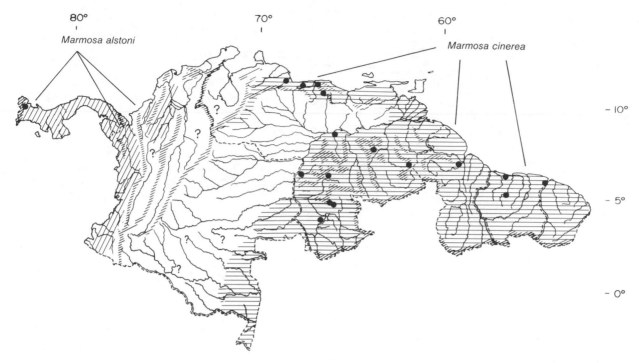

Map 2.7. Distribution of two mouse opossums.

Natural History

M. cinerea forages both arboreally and on the ground. In Venezuela it was trapped 47% of the time on the ground and 53% of the time in trees and bushes (Handley 1976). There are up to eleven teats; a litter of nine was recorded in Suriname. In northern Venezuela seasonality of breeding is tied to rainfall, with no reproduction during the winter dry season (O'Connell 1979).

Beach (1939) studied this species in captivity, observing a female with five young attached to the teats for several weeks. When detached from the teat the young utters a repetitive chirping cry, inducing the female to approach, grasp it with her forepaws, and push it under her venter, whereupon the young reattaches to the nipple.

The female builds a nest by transporting leaves in her mouth or with her prehensile tail.

Marmosa cracens Handley and Gordon, 1979

Description

This is a medium-sized species of *Marmosa*, with a total length averaging about 237 mm: HB 105; T 132; HF 16; E 25. Weights were recorded at 24 to 27 g (USNMNH). The dorsum is gray brown, contrasting sharply with the cream-colored venter (Handley and Gordon 1979).

Range and Habitat

The species is restricted almost exclusively to the state of Falcón in Venezuela. It occurs at low eleva-

tions, no specimens being taken above 170 m. Most were caught at the base of trees in moist foothill forests (Handley and Gordon 1979) (map 2.8).

Marmosa dryas Thomas, 1898

Description

In this very small species of *Marmosa*, head and body length ranges from 90 to 100 mm, tail length from 130 to 150 mm. Weight averages about 18 g. The dorsum is dark reddish brown, with the venter pale cream to pale brown.

Range and Habitat

This species occurs in the mountains of western Venezuela and eastern Colombia. It is associated with moist forest habitat at elevations over 2,000 m (map 2.9).

Natural History

This small *Marmosa* forages both arboreally and terrestrially. It is found in cloud forest over most of its range but may be more abundant in second growth (Aagaard 1982). In the mountains of northern Venezuela it is replaced by the equally small *M. marica*.

Marmosa emiliae Thomas, 1909

Description

This extremely small species of *Marmosa* has an average head and body length of 60 mm and a tail length of 112 mm (Husson 1978). The dorsum is chestnut brown, contrasting with the cream-colored venter.

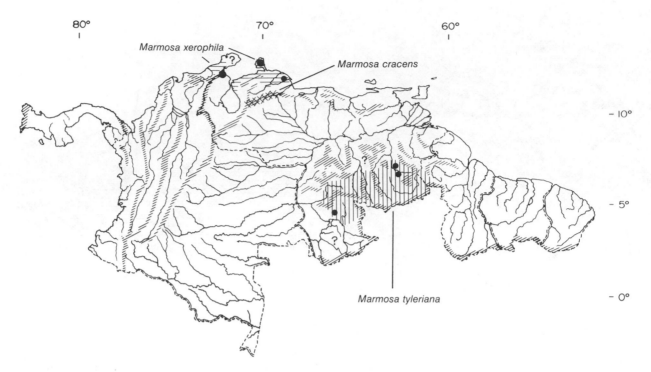

Map 2.8. Distribution of three mouse opossums in northern South America (see Handley and Gordon 1979).

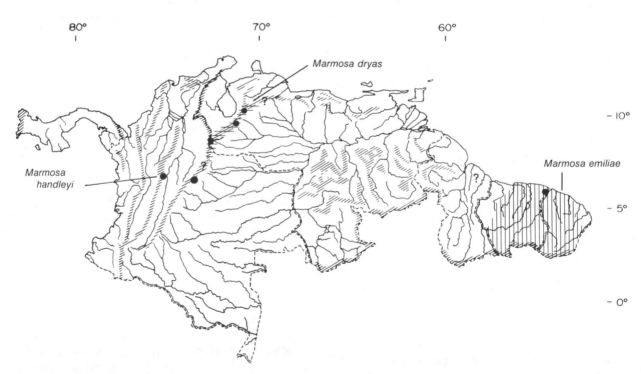

Map 2.9. Distribution of three mouse opossums in northern South America (Pine 1981; Husson 1978).

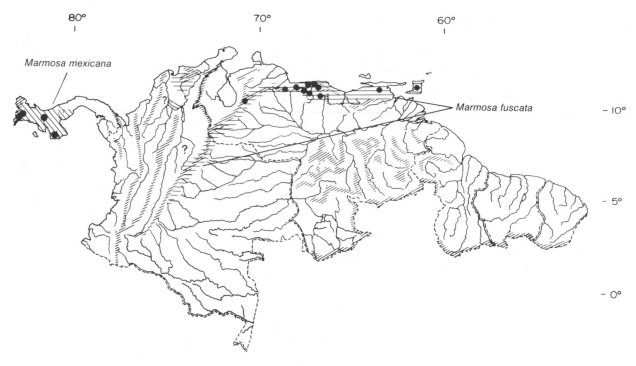

Map 2.10. Distribution of the dusky mouse opossum, *Marmosa fuscata*, and the Mexican mouse opossum, *Marmosa mexicana*.

Range and Habitat

The species is distributed in northeastern Brazil and extends into Suriname. Further research may indicate that this form is a subspecies of *M. micro-tarsus* (map 2.9).

Marmosa fuscata Thomas, 1896
Dusky Mouse Opossum

Description

This medium-sized species of *Marmosa* has a total length of 280–305 mm: HB 128–33; T 152–72; HF 17–19; E 24–26 (N = 4; Venezuela, USNMNH). The dorsum is slate gray, contrasting sharply with the cream-colored venter. Mean weight for four specimens was 40 g.
Chromosome number: $2n = 14$; FN = 24.

Range and Habitat

Marmosa fuscata ranges in the mountains of northern Venezuela, south into central Colombia. It also occurs on the island of Trinidad. All specimens have been taken at elevations over 1,000 m. The species is strongly associated with moist, forested habitats (map 2.10).

Natural History

This species was studied by O'Connell (1979) in northern Venezuela. It is nocturnal and forages both terrestrially and in low shrubs, feeding on fruits and invertebrates. Five young have been recorded. It is assumed that gestation and development for *M. fuscata* are similar to those described for *M. robinsoni*. In northern Venezuela, reproduction does not occur from December through March but commences in April with the onset of rains. At lower elevations in northern Venezuela *M. fuscata* is replaced by *M. murina*.

Marmosa handleyi Pine, 1981

Description

This newly described species is based on two specimens collected in Antioquia, Colombia, described by Pine (1981) (map 2.9). Total length is 233 to 271 mm: T 129–49; HF 17–20; E 20. This small to medium-sized opossum is dark brown dorsally and laterally, with cheeks creamy buff, chin white, venter cream to dirty white, and an eye mask present but faint (Pine 1981).

Marmosa impavida (Tschudi, 1844)

Description

This medium-sized species has a total length of about 300 mm (280–320): HB 125–50; T 155–87; HF 19–22; E 21–23. Weights range from 36 to 45 g (MVZ). The dorsum is dark brown, blending to gray brown on the sides; the venter is buffy to gray, and usually the midventer is cream.

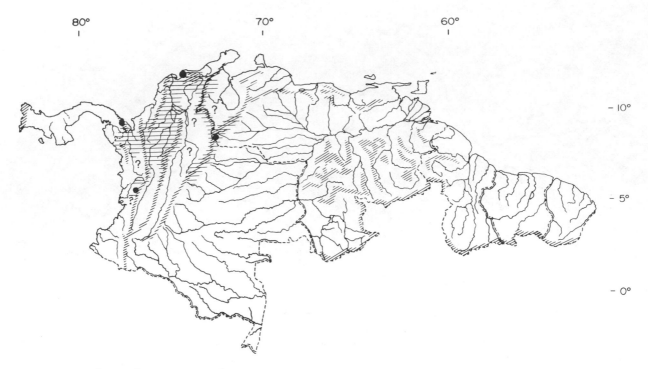

Map 2.11. Distribution of *Marmosa impavida*.

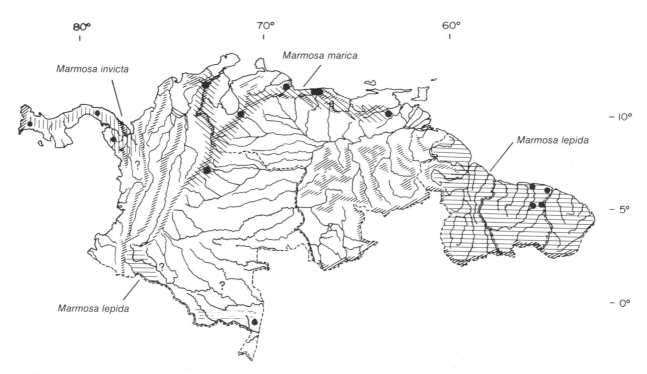

Map 2.12. Distribution of three species of mouse opossum (in part from Pine 1981; Husson 1978).

Range and Habitat

Marmosa impavida occurs from the montane portions of Panama to the mountains of western Venezuela and then is distributed south to Ecuador, Peru, and western Brazil. This species prefers moist evergreen forests. It is strongly arboreal and occurs at elevations over 2,300 m (map 2.11).

Marmosa invicta Goldman, 1912

Description

This small *Marmosa* averages about 240 mm in total length (230–59 mm): HB 106–21; T 124–38; HF 17; E 21–22. The dorsum is slate gray to dark brown; the venter is gray, but the chin may be white.

Range and Habitat

This small species of *Marmosa* occurs in Panama and probably extends into extreme northwestern Colombia. It may be closely related to *M. parvidens* (map 2.12).

Marmosa lepida (Thomas, 1888)

Description

This small mouse opossum has a total length of 208 to 262 mm and an average head and body length of 108 mm, with the tail 150 mm and the hind-foot 16 to 19 mm. The dorsum is a dark reddish brown, and the venter is cream or dirty white

Range and Habitat

This species is widely distributed from eastern Peru, Ecuador, and Bolivia, extending east to Suriname. It may penetrate extreme southern Colombia from Ecuador, but its presence there has not been confirmed (map 2.12).

Marmosa marica Thomas, 1898

Measurements

	Mean	Minimum	Maximum	N
TL	233.86	200	265	7
T	134.71	116	151	7
HF	16	15	18	7
E	19.57	15	22	7

Location: Colombia, Venezuela (USNMNH)

Description

This small *Marmosa* has body proportions resembling those of *M. lepida*. Total length is 200 to 236 mm: T 116–36; HF 15–17; E 15–21 (N = 6; Venezuela, USNMNH). The dorsum is pale red brown, contrasting with a deep cream venter; the bases of the ventral hairs are gray. The dorsal fur is woolly.

Range and Habitat

Present distribution is confined to the northern montane region of Venezuela. It is taken over a range of elevations from 18 to 2,000 m (map 2.12).

Natural History

This small mouse opossum is strongly arboreal; 83% of all captures in northern Venezuela were in trees or low shrubs. It is strongly associated with moist habitats and often exploits the interface between evergreen forest and savanna (Handley 1976).

Marmosa mexicana Merriam, 1897
Mexican Mouse Opossum

Measurements

	Mean	Minimum	Maximum	N
TL	283.25	279	287	4
T	164	152	170	4
HF	20.5	20	21	4
E	21.5	21.5	22	3

Location: Veracruz, Mexico (FMNH)

Description

This is one of the larger species of *Marmosa*, attaining a total length of 386 mm. Head and body length for large specimens may exceed 190 mm, and the tail is approximately equal to the head and body length. The dorsum is red brown, contrasting with the yellow to buffy venter.
Chromosome number: $2n = 14$.

Range and Habitat

This species is distributed from Tamaulipas, Mexico, to extreme western Panama (map 2.10). It is found in moist tropical forests.

Natural History

Gewalt (1968) studied this species in captivity, where a female specimen raised a litter of eleven young. Retrieval behavior is identical to that described for *M. cinerea*. Defensive behavior includes opening the mouth wide and hissing. The animal will grasp insect prey with its forepaws, which have remarkable manipulative powers.

Marmosa murina (Linnaeus, 1758)
Marmosa Ratón

Measurements

	Mean	Minimum	Maximum	N
TL	275.29	221	358	14
T	161.71	133	212	14
HF	18.75	14	25	14
E	20.67	20	23	9
Wt	26	13	44	5

Location: Mount Duida, Venezuela (AMNH); Suriname (CMNH)

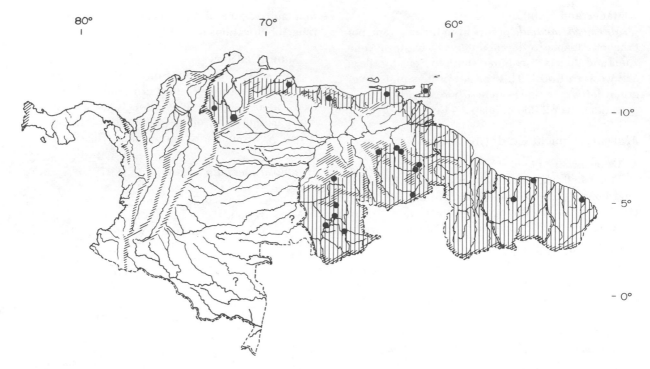

Map 2.13. Distribution of *Marmosa murina* in northern South America.

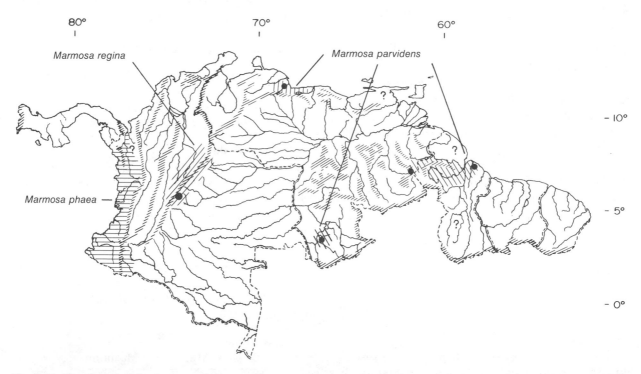

Map 2.14. Distribution of three species of mouse opossum in northern South America (*Marmosa parvidens* reviewed by Pine 1981).

Description

Total length averages 332 mm (250–358), with head and body length 110 to 146 mm; the tail is longer than the head and body (140–212 mm) (Mount Duida, Venezuela). *Marmosa murina* has a pale buff dorsal pelage contrasting with cream underparts. The black facial mask is prominent, and the tail may be faintly bicolor.

Chromosome number: $2n = 14$; $FN = 24$.

Range and Habitat

This species is distributed from northeastern Brazil west to Ecuador and Peru and north to adjacent parts of Venezuela and the Guyanas. In Venezuela it extends through the north coast range to the Maracaibo basin. It is found at elevations below 1,300 m and is strongly associated with moist habitats and tropical evergreen forest (map 2.13).

Natural History

This mouse opossum is arboreal, nocturnal, and insectivorous, but it is versatile in its habitat exploitation and is frequently trapped on the ground, sometimes near human dwellings. The female is tolerant of the male only during estrus; copulation may last several hours, and gestation takes thirteen days. The litter size averages 5.8, and the teat number is eleven. The species' behavior in captivity was analyzed by Eisentraut (1970). Maternal behavior is similar to that described for *M. cinerea*; the female constructs a leaf nest by transporting nesting material with her prehensile tail. Young are weaned at about 12 g body weight.

Marmosa parvidens Tate, 1931

Measurements

	Mean	Minimum	Maximum	N
TL	235.67	225	248	6
T	136.67	129	145	6
HF	16.33	15	18	6
E	18.83	15	20	6

Location: Venezuela, Bolivia, Colombia (AMNH)

Description

This species rarely exceeds 250 mm in total length (227–93), and the tail greatly exceeds the head and body length (134–60 mm); HF 15–18; E 15–20. The dorsum is pale brown to chocolate brown, the venter is gray to white. Handley notes that the coat has a shaggy appearance that distinguishes it from *M. cracens* (Handley and Gordon 1979).

Range and Habitat

The range of this species appears to be highly fragmented. It occurs in the Guyanas, Venezuela, and south through Amazonian Brazil to Peru. In Venezuela it occurred at elevations below 1,000 m, but Pine (1981) has records up to 2,000 m elevation. It is strongly associated with moist habitats (map 2.14).

Natural History

This species appears to forage both in the trees and on the ground, and it prefers moist tropical forest. Embryo counts vary from six to seven; teat number is seven (Pine 1981). Further research may prove that it is very closely related to *M. emiliae*.

Marmosa phaea Thomas, 1899

Description

This medium-sized mouse opossum has a head and body length averaging 137 mm (130–45): T 180–200 mm; HF 23–26; E 23. The basal third of the tail is haired, and the distal half is depigmented. The dorsum is brown and the venter orange to cream.

Range and Habitat

This species occurs from Ecuador north to western Colombia. It is strongly associated with elevations over 1,000 m. Tate (1933) considered this species allied to the *M. cinerea* group (map 2.14).

Marmosa regina Thomas, 1898

Description

This large species of *Marmosa* approximates *M. alstoni* in body proportions: HB 230; T 220; HF 30; E 25. On the other hand, it differs in that its dorsal color is cinnamon rather than grayish. Tate (1933) places this species in the *M. cinerea* grouping. It may well prove to be a subspecies of *M. alstoni*.

Range and Habitat

M. regina has been described from the region of Bogotá, Colombia (map 2.14).

Marmosa robinsoni Bangs, 1898

Description

External measurements show a rather wide range: HB 110–57; T 135–83; HF 18–23; E 20–24. Males tend to be considerably larger than females. A large male may show a total length of over 350 mm, with the tail exceeding 200 mm. The dorsum is a light brown, and the venter tends to be white. Dark coloration around the eyes gives the effect of a little black mask (see plate 2).

Chromosome number: $2n = 14$; $FN = 24$.

Comment

Includes *Marmosa mitis* (Honaki, Kinman, and Koeppl 1982).

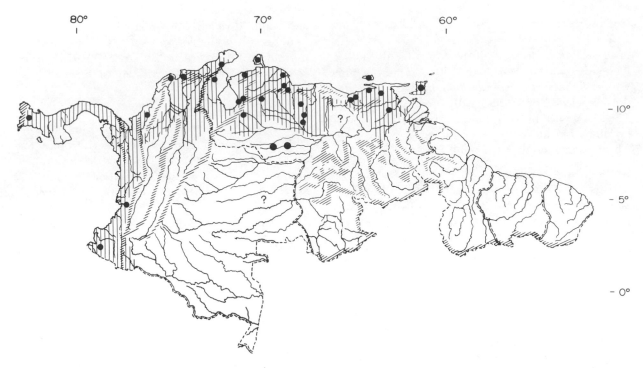

Map 2.15. Distribution of *Marmosa robinsoni* in northern South America.

Range and Habitat

The species' range extends from Panama across Colombia and Venezuela; it is not recorded from the Guyanas. This opossum has a broad habitat tolerance and has been taken at from sea level to 1,260 m. Most specimens in Venezuela were taken at below 500 m elevation. It is widely tolerant of seasonally dry, deciduous habitats and ranges down into the llanos of Colombia and Venezuela (map 2.15).

Natural History

This species appears to forage for fruit and insects both on the ground and in vines and trees. In the llanos of Venezuela it frequently shelters under the fronds of the palm *Copernicia tectorem*. Home-range size is influenced by flooding in the llanos. August (1981) estimated a mean home range of 0.36 ha in the dry season and 0.17 ha in the wet season when flooding occurs.

Its reproductive biology has been studied by Fleming (1973) in Panama and by O'Connell (1979) in Venezuela. The onset of reproduction is influenced by the rainy season. Captive studies of reproduction include Eisenberg and Maliniak (1967) and Barnes and Barthold (1969). A male courts a female by approaching and emitting clicks. While copulating the male maintains a grip with his prehensile tail wrapped around a branch. Copulation lasts several hours. Gestation time is thirteen to fourteen days.

The female builds a leaf nest in a suitable cavity, and while rearing young she may show extreme site fidelity. From birth to five weeks of age, she carries the young continuously. They remain attached to the teats until approximately twenty-one days of age. They may then begin to detach and crawl on the female's venter, but they do not enter the nest phase until about thirty-eight days of age. Their eyes open at thirty-nine to forty days, and they then remain in the nest while the female forages alone. The young

Table 2.6   Developmental Data for Captive-Bred
Marmosa robinsoni Young

Age (days)	Developmental Characteristics
Newborn	Total length 8 to 12 mm; blind, naked, and ear pinnae not visible.
20 (approx.)	Young first detach themselves from teats and crawl about on female's venter. Dorsal pigmentation first noted.
22	First dorsal hairs evident; dentition visible.
34	Young begin to crawl on female's back. Full dorsal pelage complete.
38	Venter is covered with white hair.
39–40	Eyes open (Barnes and Barthold 1969 found this to occur at 35 days).
40	Young first leave nest alone. Auditory meatus open.
50–58	First solid food taken.
65	Young are fully weaned.
69	Young nest solitarily.

Source: Eisenberg and Maliniak (1967).

begin to leave the nest and venture forth at from forty to fifty days of age. The young are fully weaned at sixty-five days and shortly thereafter disperse and establish solitary nest sites (table 2.6). The female has fifteen teats. In Panama, litter size ranged from six to thirteen (Fleming 1973). In the llanos of Venezuela, litter sizes ranged from thirteen to fifteen, with an average size of fourteen (O'Connell 1983).

Marmosa tyleriana Tate, 1931

Measurements

	Mean	Minimum	Maximum	N
TL	282	257	301	9
T	166.78	145	177	9
HF	19.11	17	21	9
E	18.25	15	20	8

Location: Venezuela (AMNH)

Description
Although reminiscent of a larger *M. dryas*, this species appears to be allied to *M. murina*. Total length averages 268 to 301 mm: T 145–77; HF 18–20; E 17–20 (Auyan Tepui, Venezuela) The dorsum is dark brown, and the venter is grayish brown, with the distal half of the tail cream colored.

Range and Habitat
This opossum is found in southern Venezuela in montane regions. It was first taken at Mount Duida at elevations above 2,000 m (map 2.8).

Natural History
The species has only four teats, indicating one of the smallest litter sizes for any species of *Marmosa* (Tate 1933).

Marmosa xerophila Handley and Gordon, 1979

Description
This intermediate-sized *Marmosa* has a total length averaging 294 mm, of which the head and body make up 131 mm: Tl 260–314; T 144–81; HF 17–20; E 24–28 (N = 7; USNMNH). The dorsum is wood brown, contrasting sharply with the white venter (Handley and Gordon 1979).

Range and Habitat
The species is known only from a restricted locality in the drier areas around the mouth of Lake Maracaibo in Colombia and Venezuela at elevations below 90 m (map 2.8).

Natural History
This species is strongly adapted to dry habitats. It was taken 81% of the time in trees in dry, deciduous forest (Handley and Gordon 1979).

Concluding Comments
The systematics of *Marmosa* is only partially understood. Clearly the genus exhibits a stunning adaptive radiation, for most tropical habitats are occupied by several species of this genus. When species co-occur they are usually graded in size, suggesting a fine subdivision of feeding niches.

Genus *Metachirus* Burmeister, 1854
Metachirus nudicaudatus (E. Geoffroy, 1803)
Brown Four-eyed Opossum, Cuica de Cola Rata

Description
This form is easily distinguished from the gray four-eyed opossum *Philander* even though both have a buff patch over each eye. The dental formula is I 5/4, C 1/1, P 3/3, M 4/4. The dorsum is brown, fading to light gray to white on the venter (see plate 1). The tail is proportionately much longer than is the tail of *Philander*, and fur extends only about 20 mm on its base. Head and body can exceed 260 mm, and the tail may be over 330 mm. There is no pouch, but rather a teat area where the young attach, comparable to the condition in *Marmosa*. Chromosome number: $2n = 14$; FN = 24.

Range and Habitat
This species is distributed from southern Nicaragua to southern Brazil. Although found in multistratal tropical evergreen forests, it also appears to effectively exploit second-growth edge habitats and low-stature forests. The long tail suggests that it is arboreally adapted, yet it is frequently caught on the ground (map. 2.16).

Natural History
Brown four-eyed opossums are completely nocturnal. They forage terrestrially and arboreally, and their powerful hindquarters are specialized for running rapidly (Grand 1983). They have an omnivorous diet, including fruits, small vertebrates, and invertebrates. Litter sizes range from one to nine, with a mean of five. The average teat number for the female is nine (Osgood 1921).

Husson (1978) reports that this species is preyed upon by the owl *Asio clamator*.

Genus *Monodelphis* Burnett, 1830
Short-tailed Opossum

Description
The dental formula is I 5/4, C 1/1, P 3/3, M 4/4. Head and body length ranges from 110 to 140 mm, and tail length from 45 to 65 mm. The tail is approximately half the head and body length, and this characteristic immediately distinguishes the short-tailed opossums from any other opossums within their

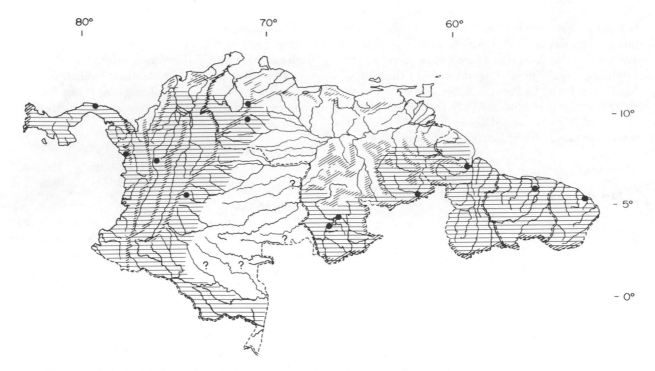

Map 2.16. Distribution of the brown four-eyed opossum, *Metachirus nudicaudatus*, in northern South America.

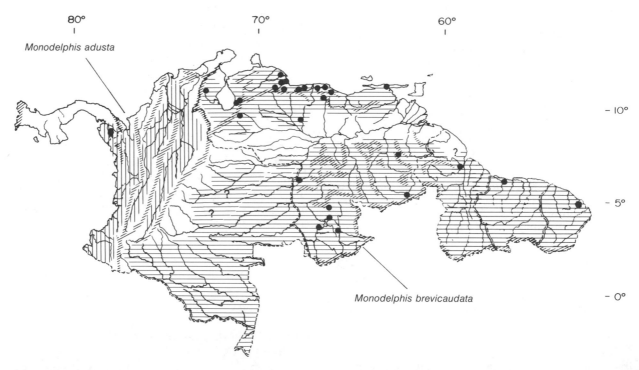

Map 2.17. Distribution of two species of short-tailed opossum in northern South America (Reig et al. 1977 assert that populations near the Orinoco are specifically distinct).

Figure 2.3. *Monodelphis brevicaudata.*

range (fig. 2.3). Color varies widely depending on the species, from gray to chestnut brown. Some species in Brazil have black dorsal stripes.

Distribution

The genus *Monodelphis* is widely spread from southwestern Panama to Argentina.

Comment

Some species of this genus may be both nocturnal and diurnal.

Monodelphis adusta (Thomas, 1897)

Description

This small short-tailed opossum has a head and body length of 108 mm: T 60; HF 17; E 14; Wt 35 g. The tail is only slightly prehensile. The upperparts are dark brown, darkest along the dorsal midline. In some specimens a dorsal midline stripe is discernible, but this is not invariant. The throat is gray and the venter a darker gray. The midline of the venter is marked by a buff-white stripe from the pectoral region to the teat area.

Range and Habitat

This species is found from southwestern Panama and west of the western cordillera of the Andes in Colombia to Ecuador (map 2.17).

Natural History

The habits of this species are very poorly known. It is assumed to be less arboreal than most opossums and apparently feeds on invertebrates, fruits, and small vertebrates.

Monodelphis brevicaudata (Erxleben, 1777)

Measurements

	Mean	Minimum	Maximum	N
TL	219.63	170	242	8
T	80.25	60	90	8
HF	22	18	24	8
E	19.63	16	21	8
Wt	54.13	24	78	8

Location: Suriname (CMNH)

Description

This small short-tailed mouse opossum averages 84 g in weight (67–95 g): TL 170–242; T 60–90; HF 18–24; E 16–21 (Suriname, CMNH); TL 173–210; T 61–83; HF 15–20; E 10–13 (Auyan Tepui, Venezuela, AMNH). It is gray brown above, grading to reddish on the sides. The venter varies from pale brown with a violet cast to cream, but its color fades in a dead specimen. The proximal portion of the tail is furred for about 20 mm.

Chromosome number: $2n = 18$; FN = 30.

Range and Habitat

This species occurs in eastern Colombia, Venezuela, the Guyanas, and south to Brazil. It tolerates a variety of habitat types, occurring at up to 1,200 m in elevation. Although it frequents multistratal tropical evergreen forests, it also may be found in edge habitats around clearings. It is less abundant in dry deciduous forests (map 2.17).

Natural History

This species is predominantly terrestrial and crepuscular in its habits. Up to seven young are born in a single litter. In northern Venezuela the breeding season extends from May through August. The behavior patterns and breeding biology are probably similar to those of *M. domestica*, a species well studied in captivity (Trupin and Fadem 1982; Fadem and Cole 1985; Fadem et al. 1982; Streilein 1982).

Concluding Comments

Monodelphis touan is probably a junior synonym of *M. brevicaudata*. Recently Reig et al. (1977) separated *M. orinoci* from *M. brevicaudata*, suggesting that species of *Monodelphis* near the Orinoco are distinct from the northern forms. *Monodelphis americana* is reputed to extend from southeastern Brazil to French Guiana (Honacki, Kinman, and Koeppl 1982), but I cannot confirm this. Husson (1978) recognizes only *M. brevicaudata* in Suriname. Pine and Handley (1984) recognize *M. americana* as a valid species and concur with Reig et al. (1977) in separating *M. orinoci* from *M. brevicaudata*. The systematics of the genus *Monodelphis* are still in flux, but field studies of this genus offer great promise (Pine, Dalby, and Matson 1985).

Pine (1976) recognizes four color patterns within the genus *Monodelphis*. Species exhibiting similar patterns need not be closely related. Within the area covered in this volume, three distinct color patterns may be found. Two dark dorsal stripes on either side of the midline characterizes *M. americana*, probably not within the range covered by this volume. A dark brown dorsum grading to reddish on the flanks with a lighter venter of variable color is typical for *M. brevicaudata* and *M. orinoci*. A uniformly brown

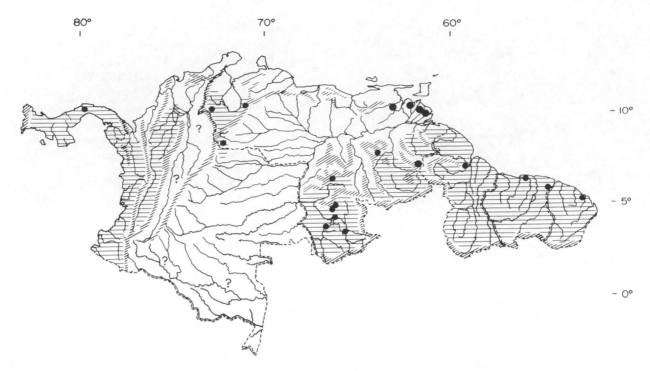

Map 2.18. Distribution of the gray four-eyed opossum, *Philander opossum*, in northern South America.

dorsum with a faint dorsal stripe in the midline of the dorsum is characteristic of *M. adusta*.

Genus *Philander* Tiedemann, 1808
Philander opossum (Linnaeus, 1758)
Gray Four-eyed Opossum, Chucha Mantequera, Zorro de Cuatrojos

Measurements

	Mean	Minimum	Maximum	N
TL	563.43	509	618	7
T	291	255	316	
HF	43.86	40	49	
E	39.2	36	45	5
Wt	421.2	280	570	

Location: Venezuela (CMNH)

Description
The genus *Philander* is used in accordance with Honacki, Kinman, and Koeppl (1982). *Metachirops* is considered a junior synonym. The dental formula is I 5/4, C 1/1, P 3/3, M 4/4. Head and body length can range from 300 to 335 mm, with the tail 278 to 262 mm. The mean weight for eight specimens was 330 g, but animals over 400 g are not uncommon (USNM). The most diagnostic feature in the intact animal is a white spot over each eye. *Metachirus nudicaudatus* has a similar spotted pattern, but the spots are buff. The tail of *Philander* is bicolor, the base being dark and the distal portion white, and its

base is densely furred. Coat color is grizzled or blackish gray, in contrast to the brown dorsal coat of *Metachirus* (see plate 1).
Chromosome number: $2n = 22$; FN = 20.

Range and Habitat
The genus *Philander* occurs from Tamaulipas, Mexico, to northeastern Argentina. Over this range only two species are recognized, *Philander opossum* and *P. mcilhennyi*. The latter species is known only from its type locality in Loreto, Peru (map 2.18). In Venezuela, *Philander* exhibits a predominant preference for moist habitats. It is generally associated with multistratal evergreen tropical forests, but it can invade croplands. Most specimens were taken at below 400 m elevation.

Natural History
Philander opossum forages a great deal on the ground or in the understory of a multistratal forest. It is strictly nocturnal in its habits. Where it co-occurs with *Didelphis*, *Didelphis* always exceeds *Philander* in body size. Although *Philander* feeds on the pulp of ripe fruits and on nectar, it also is an active predator. Gewalt (1969) describes its predatory behavior in captivity. In Panama, *Philander* has been implicated as preying actively on frogs.
In French Guiana *Philander* has a mean litter size of 4.2 young; in Panama Fleming (1973) recorded an average litter size of 4.6. The species in Panama shows a reproductive quiescence at the end of the

rainy season (December to January). Compared with *Caluromys*, *Philander* has a higher reproductive rate. Sexual maturity is attained earlier, often at less than seven months, and the rearing phase of a *Philander* female is somewhat shortened. The phase of teat attachment by the young is as short as sixty days, and the nest phase lasts for eight to fifteen days before dispersal (Charles-Dominique 1983).

This nocturnal animal has a home-range pattern characteristic of didelphids in that there is broad overlap between the ranges of neighboring adults. There is no clear-cut defense of a territory. Home-range stability depends on the availability of adequate resources. Apart from mating, contact among adults is minimal (Charles-Dominique 1983). *Philander* actively defends itself when attacked by a potential predator. Husson (1978) notes the boa *Corallus enydris* as a predator.

Preliminary data on the vocalizations of *Philander* are presented in Charles-Dominique (1983). Clicks, chirps, and hisses are employed in communication and are comparable to the sounds noted for *Caluromys* and *Marmosa*. *Philander* has a relatively large brain, approximating the condition of *Caluromys* (Eisenberg and Wilson 1981).

FAMILY CAENOLESTIDAE
Rat Opossums

Diagnosis
This family contains three genera and seven species. The dental formula is I 4/3, C 1/1, P 3/3, M 4/4. Rat opossums are easily distinguished from members of the family Didelphidae in that the first lower incisors are enlarged and project forward. The remaining incisors are very reduced and show a single cusp (fig. 2.4). Members of the Caenolestidae are ratlike in appearance, with head and body length ranging from 90 to 135 mm while the tail ranges from 65 to 135 mm. The rostrum is long and narrow,

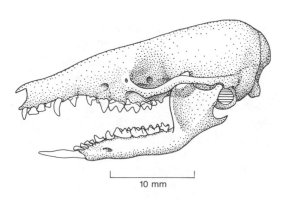

| | | 10 mm | |

Figure 2.4. Skull of *Caenolestes obscurus*.

and the eyes are small. The tail is covered with short hairs.

Distribution
The species of this family are confined to the high montane paramos of the Andes or the moist, temperate lowlands of Chile and Argentina. They range from western Colombia to southern Chile.

There are five recognized species within the genus *Caenolestes*, found in the montane areas of Colombia, Venezuela, and Ecuador to central Peru. They are confined to alpine meadows in the Andes. A single species occurs within the area covered by this volume, *C. obscurus*.

Natural History
Little material is available concerning the behavior of these animals in the field. The family has an ancient history, being recognizable in the Eocene of South America, and underwent an adaptive radiation in the Oligocene with numerous species that subsequently became extinct (Marshall 1980). The extant forms are relicts in the high montane meadows of the Andes and in the temperate lowlands of Chile (Osgood 1924). In these habitats they use runways where they feed upon small invertebrates.

Genus *Caenolestes* Thomas, 1895
Caenolestes obscurus Thomas, 1895
Ratón Comadreja

Measurements

	Mean	Minimum	Maximum	N
TL	235.75	232	242	4
T	118.25	114	120	4
HF	23	22	24	4
E	14	14	14	4

Location: Colombia (FMNH)

Description
Head and body length ranges from 93 to 135 mm and tail length from 93 to 127 mm. The dorsal pelage is dense and soft. Although the tail is haired, the hairs are sparse (see plate 2). There is no pouch. The rostrum is elongated, and the ears and eyes are greatly reduced in size. The dorsum is dark brown tending toward lighter brown on the venter.

Range and Habitat
C. obscurus is distributed from the Andes of western Venezuela to the Andes of northern Colombia. Specimens are generally taken at above 2,300 m (map 2.19).

Natural History
This species is adapted to cool, wet habitats. The animals use runways at interfaces between grassland

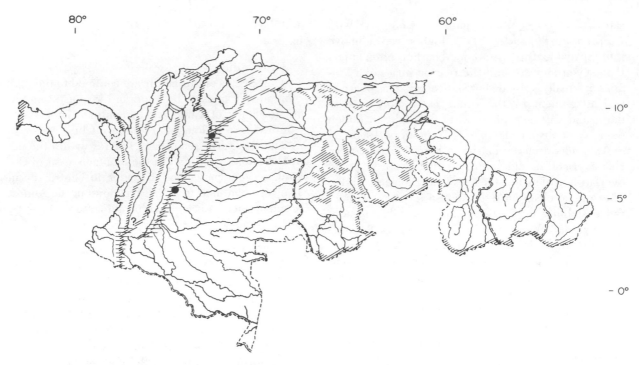

80° 70° 60°

−10°

−5°

−0°

Map 2.19. Distribution of the rat opossum, *Caenolestes obscurus*, in northern South America.

and scrub forest. When disturbed they hiss and open their mouths wide. In captivity they feed on insects, earthworms, and baby rats. When moving rapidly they bound, with forelimbs and hind limbs striking the ground alternately. The tail is not prehensile but serves as a prop when they sit upright. The female has four teats not enclosed in a pouch (Kirsch and Waller 1979).

Barkley and Whitaker (1984) studied *C. fuliginosus* in Peru, where lepidopteran larvae, centipedes, and arachnids made up 75.5% of the diet by volume. They conclude that *Caenolestes* is an opportunistic feeder that prefers invertebrate larvae but will eat vertebrate prey, fruit, and other vegetable matter.

References

• References used in preparing distribution maps

Aagaard, E. M. J. 1982. Ecological distribution of mammals in the cloud forests and paramos of the Andes, Mérida, Venezuela. Ph.D. diss., Colorado State University.

Atramentowicz, M. 1982. Influence du milieu sur l'activité locomotrice et la reproduction de *Caluromys philander* (L.). *Rev. Ecol. (Terre et Vie)* 36:373–95.

August, P. 1981. Population ecology and community structure of small mammals in northern Venezuela. Ph.D. diss., Boston University.

————. 1984. *Population ecology of small mammals in the llanos of Venezuela: Contributions in mammalogy in honor of Robert L. Packard*. Special Publications of the Museum 22. Lubbock: Texas Tech University Press.

Barkley, L. J., and J. O. Whittaker, Jr. 1984. Confirmation of *Caenolestes* in Peru with information on diet. *J. Mammal.* 65:328–30.

Barlow, J. C. 1965. Land mammals from Uruguay: Ecology and zoogeography. Ph.D. diss., University of Kansas.

Barnes, R. D., and S. W. Barthold. 1969. Reproduction and breeding behaviour in an experimental colony of *Marmosa mitis* Bangs (Didelphidae). *J. Reprod. Fert.*, suppl. 6:477–92.

Beach, F. A. 1939. Maternal behavior in the pouchless marsupial *Marmosa cinerea*. *J. Mammal.* 20:315–22.

• Beebe, W. 1919. The higher vertebrates of British Guiana with special reference to the fauna of the Bartica district, no. 7. List of Amphibia, Reptilia, and Mammalia. *Zoologica* 2:205–38.

Biggers, J. D. 1967. Notes on the reproduction of the wooly opossum *Caluromys derbianus* in Nicaragua. *J. Mammal.* 48:678–80.

Bucher, J. E., and I. Fritz. 1977. Behavior and maintenance of the wooly opossum (*Caluromys*) in captivity. *Lab Anim. Sci.* 27:1007–12.

Bucher, J. E., and R. S. Hoffmann. 1980. *Caluromys derbianus*. *Mammal. Species* 140:1–4.

• Cabrera, A. 1958. *Catálogo de los mamíferos de América del Sur.* Vol. 1. Buenos Aires: Museo Argentino de Ciencias Naturales "Bernardino Rivadavia," Zoología.

Cajal, J. L. 1981. Estudios preliminares sobre el area de acción en marsupiales (Mammalia-Marsupialia). *Physis* (Buenos Aires), sec. C, 40(98):27–37.

Cerqueira, R. 1984. Reproduction de *Didelphis albiventris* dans le nord-est Brésil. *Mammalia* 48: 95–104.

Charles-Dominique, P. 1983. Ecology and social adaptations in didelphid marsupials: Comparison with eutherians of similar ecology. In *Advances in the study of mammalian behavior,* ed. J. F. Eisenberg and D. G. Kleiman, 395–422. Special Publication 7. Shippensburg, Pa.: American Society of Mammalogists.

• Charles-Dominique, P., M. Atramentowicz, M. Charles-Dominique, H. Gerard, A. Hladik, C. M. Hladik, and M. F. Prevost. 1981. Les mammifères frugivores arboricoles nocturnes d'une forêt guyanaise: Interrelations plantes-animaux. *Rev. Ecol. (Terre et Vie)* 35:341–435.

Collins, L. R. 1973. *Monotremes and marsupials: A reference for zoological institutions.* Washington, D.C.: Smithsonian Institution Press.

Cordero, R, G. A. 1983. *Características demográficas de una población de rabipelado,* Didelphis marsupialis Linnaeus, 1758 *(Marsupialia, Didelphidae) en un bosque húmedo al norte de Venezuela.* Trabajo de Ascenso. Caracas: Universidad Central de Venezuela.

Creighton, G. K. 1984. Systematic studies on opossums (Didelphidae) and rodents (Cricetidae). Ph.D. diss., University of Michigan, Ann Arbor.

Eisenberg, J. F. 1988. Reproduction in polyprotodont marsupials and similar-sized eutherians with a speculation concerning the evolution of litter size in mammals. In *Evolution of life histories of mammals,* ed. M. S. Boyce, 291–311. New Haven: Yale University Press.

Eisenberg, J. F., L. R. Collins, and C. Wemmer. 1975. Communication in the Tasmanian devil (*Sarcophilus harrisii*) and a survey of auditory communication in the Marsupialia. *Z. Tierpsychol.* 37:379–99.

Eisenberg, J. F., and I. Golani. 1977. Communication in Metatheria. In *How animals communicate,* ed. T. Sebeok, 575–99. Bloomington: Indiana University Press.

Eisenberg, J. F., and E. Maliniak. 1967. Breeding the murine opossum *Marmosa* in captivity. *Int. Zoo Yearb.* 7:78–79.

Eisenberg, J. F., and D. E. Wilson. 1981. Relative brain size and demographic strategies in didelphid marsupials. *Amer. Nat.* 118:1–15.

Eisentraut, M. 1970. Beitrag zur Fortpflanzungsbiologie der Zwergbeutelratte *Marmosa murina* (Didelphidae, Marsupialia). *Z. Säugetierk.* 35: 159–73.

Enders, R. K. 1935. Mammalian life histories from Barro Colorado Island, Panama. *Bull. Mus. Comp. Zool. (Harvard)* 78(4):385–502.

———. 1966. Attachment, nursing, and survival of young in some didelphids. In *Comparative biology of reproduction in mammals,* ed. J. W. Rowlands, 195–203. New York: Academic Press.

Fadem, B. H., and Cole, E. A. 1985. Scent-marking in the grey short-tailed opossum (*Monodelphis domestica*). *Anim. Behav.* 33:730–38.

Fadem, B. H., G. L. Trupin, J. L. Vandeberg, and V. Hayssen. 1982. Care and breeding of the gray, short-tailed opossum. *Lab. Anim. Sci.* 32:405–9.

Fleming, T. H. 1972. Aspects of the population dynamics of three species of opossums in the Panama Canal Zone. *J. Mammal.* 53:619–23.

———. 1973. The reproductive cycles of three species of opossums and other mammals in the Panama Canal Zone. *J. Mammal.* 54:439–55.

Gardner, A. L. 1973. The systematics of the genus *Didelphis* (Marsupialia: Didelphidae) in North and Middle America. Special Publications of the Museum 4. Lubbock: Texas Tech University Press.

Gewalt, W. 1968. Kleine Beobachtungen an selteneren Beuteltieren im Berliner Zoo. V. Zwergbeutelratte (*Marmosa mexicana* Merriam 1897). *Zool. Gart.,* n.s., 35:288–303.

———. 1969. Kleine Beobachtungen an selteneren Beuteltieren im Berliner Zoo. VI. Vieraugenbeutelratte (*Metachirops opossum* [L. 1758]). *Zool. Gart.,* n.s., 37:248–53.

• Goodwin, G. G., and A. M. Greenhall. 1961. A review of the bats of Trinidad and Tobago. *Bull. Amer. Mus. Nat. Hist.* 122(3):187–302.

Grand, T. 1983. Body weight: Its relationship to tissue composition, segmental distribution of mass, and motor function. 3. The Didelphidae of French Guyana. *Aust. J. Zool.* 31:299–312.

• Hall, E. R. 1981. *The mammals of North America.* 2d ed., 2 vols. New York: John Wiley.

• Handley, C. O. Jr. 1966. Checklist of the mammals of Panama. In *Ectoparasites of Panama,* ed. R. L. Wetzel and V. J. Tipton, 753–95. Chicago: Field Museum of Natural History.

• ———. 1976. Mammals of the Smithsonian Venezuelan project. *Brigham Young Univ. Sci. Bull., Biol. Ser.* 20(5);1–90.

• Handley, C. O., Jr., and L. K. Gordon, 1979. New

species of mammals from northern South America: Mouse opossums, the genus *Marmosa*. In *Vertebrate ecology in the northern Neotropics*, ed. J. F. Eisenberg, 65–71. Washington, D.C.: Smithsonian Institution Press.

Honacki, J. H., K. E. Kinman, and J. W. Koeppl, eds. 1982. *Mammal species of the world*. Lawrence, Kans.: Allen Press and Association of Systematics Collections.

Hunsaker, D. 1977a. The ecology of New World marsupials. In *The biology of marsupials*, ed. D. Hunsaker, 95–156. New York: Academic Press.

———., ed. 1977b. *The biology of marsupials*. New York: Academic Press.

• Husson, A. M. 1978. *The mammals of Suriname*. Leiden: E. J. Brill.

Kirsch, J. A. W. 1977. The classification of marsupials with special reference to karyotypes and serum proteins. In *The biology of marsupials*, ed. D. Hunsaker, 1–50. New York: Academic Press.

• Kirsch, J. A. W., and P. F. Waller. 1979. Notes on the trapping and behavior of the Caenolestidae. *J. Mammal.* 60:390–95.

Lemke, T. O., A. Cadena, R. H. Pine, and J. Hernandez-Camacho. 1982. Notes on opossums, bats, and rodents new to the fauna of Colombia. *Mammalia* 46(2):225–34.

McManus, J. J. 1970. The behavior of captive opossum *Didelphis marsupialis virginiana*. *Amer. Midl. Nat.* 84:144–69.

McNab, B. K. 1978. The comparative energetic of Neotropical marsupials. *J. Comp. Physiol.* 125:115–28.

• Marshall, L. G. 1978a. *Chironectes minimus*. *Mammal. Species* 109:1–6.

• ———. 1978b. *Lutreolina crassicaudata*. *Mammal. Species* 91:1–4.

———. 1980. *The systematics of the South American marsupial family Caenolestidae*. Fieldiana, Geology, n.s., 5. Chicago: Field Museum of Natural History.

———. 1982. Evolution of South American marsupialia. In *Mammalian biology in South America*, ed. M. A. Mares and H. H. Genoways, 251–72. Pymatuning Symposia in Ecology 6. Special Publication Series. Pittsburgh: Pymatuning Laboratory of Ecology, University of Pittsburgh.

Marshall, L., C. DeMuizon, and B. Sige. 1983. Late Cretaceous mammals (Marsupialia) from Bolivia. *Geobios* 16:739–45.

• Mondolfi, E., and G. Medina Padilla. 1957. Contribución al conocimiento del "Perrito de Agua" (*Chironectes minimus* Zimmermann). *Mem. Soc. Cient. Nat. La Salle* 17:140–55.

Mondolfi, E., and R. Perez-Hernandez. 1984. Una nueva subespecie de Zarigüeya del grupo *Didelphis albiventris*. *Acta Cient. Venez.* 35:407–13.

Motta, Maria de Fatima, J. C. DeZonne de Araujo Carreira, and A. M. R. Franco. 1983. A note on the reproduction of *Didelphis marsupialis* in captivity. *Mem. Inst. Oswaldo Cruz Rio* 78:507–9.

O'Connell, M. A. 1979. Ecology of didelphid marsupials from northern Venezuela. In *Vertebrate ecology in the northern Neotropics*, ed. J. F. Eisenberg, 73–87. Washington, D.C.: Smithsonian Institution Press.

———. 1983. *Marmosa robinsoni*. *Mammal. Species* 203:1–6.

Osgood, W. H. 1921. A monographic study of the American marsupial *Caenolestes*. *Field Mus. Nat. Hist., Zool. Ser.* 14:1–162.

———. 1924. Review of living caenolestids with description of a new genus from Chile. *Field Mus. Nat. Hist., Zool. Ser.* 14(2):165–73.

Papini, M. R. 1986. Psicología comparada de los marsupiales. *Rev. Latinoamericana Psicología* 18(2):215–46.

Pine, R. 1976. *Monodelphis umbristriata* is a distinct species of opossum. *J. Mammal.* 57:785–87.

———. 1981. Review of the mouse opossum *Marmosa parvidens* and *M. invicta* with a description of a new species. *Mammalia* 45:56–70.

Pine, R. H., Dalby, P. L., and J. O. Matson. 1985. Ecology, postnatal development, morphometrics, and taxonomic status of the short-tailed opossum, *Monodelphis dimidiata*, an apparently semelparous annual marsupial. *Ann. Carnegie Mus.* 54(6):195–231.

Pine, R. H., and C. O. Handley, Jr. 1984. A review of the Amazonian short-tailed opossum *Monodelphis emiliae*. *Mammalia* 48:239–45.

Reig, O. A., A. L. Gardner, N. O. Bianchi, and J. L. Patton. 1977. The chromosomes of the Didelphidae (Marsupialia) and their evolutionary significance. *Biol. J. Linnean Soc. London* 9:191–216.

Reig, O. A., J. A. W. Kirsch, and L. G. Marshall. 1987. Systematic relationships of the living and Neocenozoic American "opossum-like" marsupials. In *Possums and opossums: Studies in evolution*, ed. M. Archer, 1–89. Chipping Norton, N.S.W., Australia: Surrey Beatty.

Ride, W. D. L. 1964. A review of Australian fossil marsupials. *J. Proc. Roy. Soc. West Aust.* 47:97–131.

Seluja, G. A., M. V. DiTomaso, M. Brun-Zorilla, and H. Cordoso. 1984. Low karyotypic variation in two didelphids. *J. Mammal.* 65:702–7.

• Simonetta, A. M. 1979. First record of *Caluromysiops* from Colombia. *Mammalia* 43:247–48.

• Streilein, K. E. 1982. Behavior, ecology, and distribution of South American marsupials. In *Mammalian biology in South America*, ed. M. A. Mares and H. H. Genoways, 231–50. Pymatuning Symposia in Ecology 6. Special Publication Series. Pittsburgh: Pymatuning Laboratory of Ecology, University of Pittsburgh.

Sunquist, M. E., S. N. Austad, and F. Sunquist. 1987. Movement patterns and home range patterns in the common opossum, *Didelphis marsupialis. J. Mammal.* 68(1):173–76.

• Tate, G. H. H. 1933. A systematic revision of the marsupial genus *Marmosa. Bull. Amer. Mus. Nat. Hist.* 66(1):1–250.

• ———. 1939. The mammals of the Guiana region. *Bull. Amer. Mus. Nat. Hist.* 76(5):151–229.

Telford, S. R., Jr., J. Gonzalez, and R. J. Tonn. 1979. Densidad, area de distribución y movimiento de poblaciones de *Didelphis marsupialis* en los llanos altos de Venezuela. *Bol. Dirección Malariologia Saneamiento Ambiental* 19 (3–4):119–28.

Telford, S. R., Jr., R. J. Tonn, J. J. Gonzalez, and P. Betancourt. 1981. Dinámica de las infecciones tripanosómicas entre la comunidad de los bosques tropicales secos en los llanos altos de Venezuela. *Bol. Dirección Malariologia Saneamiento ambiental* 21 (3–4):196–209.

Trupin, G. L., and B. H. Fadem. 1982. Sexual behaviour of the gray short-tailed opossum (*Monodelphis domestica*). *J. Mammal.* 63:409–14.

Tyndale-Biscoe, H., and R. B. Mackenzie. 1976. Reproduction in *Didelphis marsupialis* and *D. albiventris* in Colombia. *J. Mammal.* 57:249–65.

Woodburne, M. O., and W. J. Zinmeister. 1982. Fossil land mammals from Antarctica. *Science* 218:284–86.

Zetek, J. 1930. The water opossum *Chironectes panamensis. J. Mammal.* 11:470–71.

3   Order Xenarthra (Edentata)

Diagnosis

The name Xenarthra is used here because I concur with Emery (1970) that the Palaenodonta are not closely related or ancestral to the Xenarthra. In spite of the older ordinal name Edentata, not all members lack teeth; however, the tooth number is often reduced and tooth structure is simplified. Among the existing families the incisors are absent as well as the canines, but the two-toed sloth has a presumptive premolar that is caniniform. The premolars and mo-

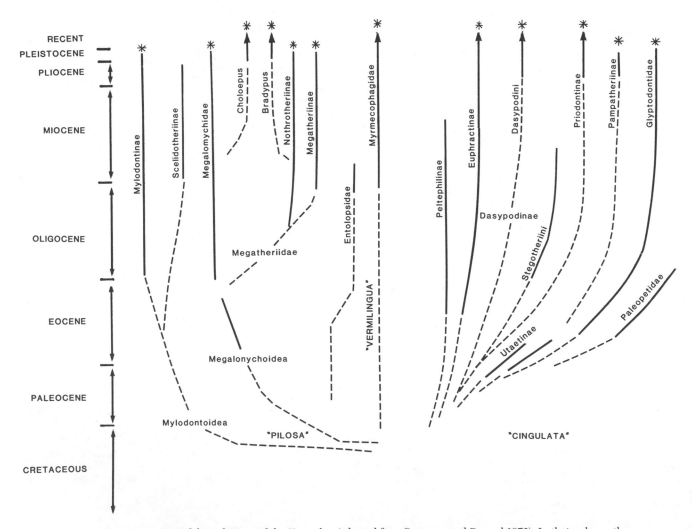

Figure 3.1. Phylogenetic tree of the radiations of the Xenarthra (adapted from Patterson and Pascual 1972). In their scheme the Chlamyphorini are subsumed in the Euphractinae and the Tolypeutini within the Priodontinae. Asterisk means survived to Pleistocene.

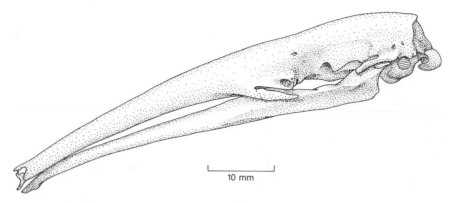

10 mm

Figure 3.2. Skull of *Myrmecophaga*.

lars, when present, have no enamel. The tympanic bone is ring shaped in many forms. All forms have trunks reinforced by broadened ribs and by accessory zygapophyses on the thoracic and lumbar vertebrae, hence the ordinal name Xenarthra. Many of the living forms show great specialization for vastly different ways of life, making a simple diagnosis for the order difficult.

Distribution

The edentates are entirely confined to the New World. One species, the nine-banded armadillo, *Dasypus novemcinctus*, has extended its range in Recent times to the southeastern United States and the southern portions of the Great Plains. All other species, if they range across the Panamanian land bridge, generally do not extend farther north than southern Mexico. To the south one species of armadillo ranges to Patagonia.

History and Classification

Edentates appear to have originated in South America and to have undergone their adaptive radiation there (Reig 1981). Although the suborder Palaenodonta is first noted in the Paleocene of North America, this is now believed to be either an early offshoot from South American stock or an entirely separate taxon. The more recent edentates first appear in the Paleocene of South America. These specimens are believed to belong to the armadillo family, the Dasypodidae. Six families of the Xenarthra are now extinct, including the giant ground sloths of the families Megalonychidae and the Megatheriidae (Patterson and Pascual 1972) (see fig. 3.1). The glyptodonts (Glyptodontidae) were somewhat similar to armadillos in having bony scutes forming a turtlelike carapace, but they were an herbivo-

rous, independent offshoot that first appeared in the Eocene and persisted to the Pleistocene (Gillette and Ray 1981). The glyptodonts, as well as some of the giant ground sloths, crossed from South America to North America in the late Miocene or early Pliocene at the completion of the Panamanian land bridge. They subsequently became extinct in North America at the close of the Pleistocene (Webb and Marshall 1982).

The biology of the Xenarthra is admirably reviewed in the volume edited by Montgomery (1985b). Captive maintenance has been summarized by Meritt (1976). The extant xenarthrans are included in four families, the Myrmecophagidae, the Bradypodidae, the Choloepidae, and the Dasypodidae. All four families are represented in the area covered by this volume.

FAMILY MYRMECOPHAGIDAE
Anteaters

Diagnosis
Teeth are totally lacking, and the rostrum is extremely elongated (fig. 3.2). The long tongue, extended in feeding on ants and termites, is coated with a copious sticky saliva produced by greatly enlarged submaxillary glands. The forefoot has an enlarged third digit with a strong claw, and the other digits are reduced or absent. The hind feet have four or five digits each. The body is covered with hair; the tail is long and also haired to varying degrees.

Distribution
Species of anteaters occur from southern Mexico to northern Argentina. Ranges are reviewed by Wetzel (1985a).

Natural History
Members of this family are highly adapted for feeding on ants and termites. The strong claws are

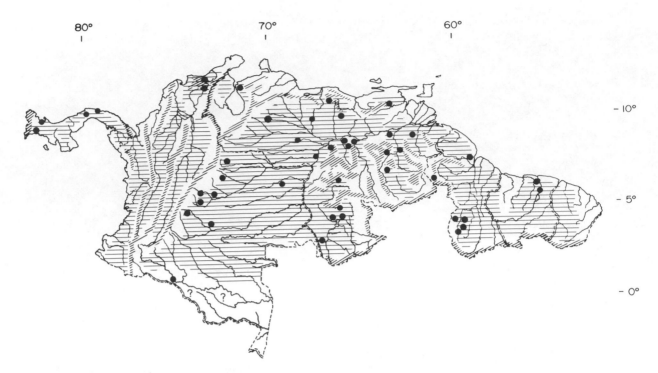

Map 3.1. Distribution of *Myrmecophaga tridactyla*.

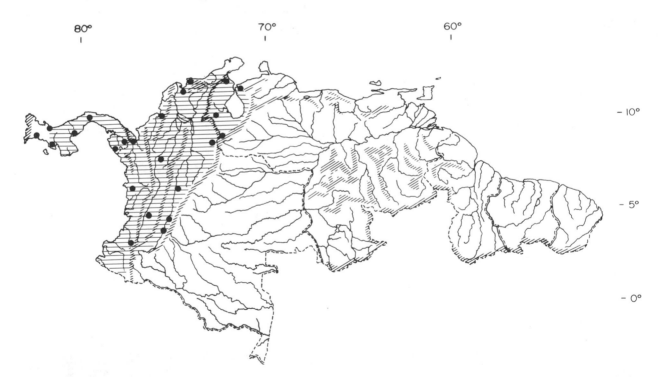

Map 3.2. Distribution of *Tamandua mexicana*.

used to open the nests, and the long, extensible tongue aids in feeding. Females produce only a single young at a time, and parental care is highly developed. The young is frequently carried on the mother's back when she changes resting areas or shortly before weaning, when it accompanies her to feeding sites. Relative to their body size anteaters have a low metabolic rate, apparently an adaptation for feeding on ants and termites, a ubiquitous resource of low nutritional quality (McNab 1982; Redford 1987).

Genus *Myrmecophaga* Linnaeus, 1758
Myrmecophaga tridactyla Linnaeus, 1758
Giant Anteater, Oso Hormiguero

Description

In this largest of the New World anteaters, the head and body length ranges from 1,000 to 1,300 mm, tail length from 650 to 900 mm, and weight from 22 to 39 kg. The giant anteater is strikingly marked (see plate 3). Hairs on the tail are long, giving it a fanlike appearance. A black stripe extends from the throat across the shoulder to the midline, bordered dorsally and ventrally by a white stripe. On the shoulder the black stripe is demarcated by gray to dark brown. The forelegs are generally light tan to the wrists, where there is a dark band before the manus. Flanks, back, and tail are brown to gray brown. Its large size, fanlike tail, and distinctive coloration immediately distinguish the giant anteater from the other anteaters.

Range and Habitat

Giant anteaters range from southern Guatemala to northern Argentina (map 3.1). They tolerate a range of habitats at lower elevations but prefer areas that support high densities of ant and termite mounds, usually areas with reasonably well drained soil. Where ant and termite mounds are abundant in savanna regions of southeastern Brazil, central Venezuela, and Colombia, giant anteaters may reach high densities.

Natural History

The giant anteater's activity rhythm seems to vary over its range, depending on the temperature reached at midday. In savanna regions near the equator, with high daily temperatures, it may be active from dusk until just before dawn, with a major feeding period at night. The giant anteater is a very selective feeder, visiting many termite mounds or ant nests in the course of feeding. It makes a small breach with the large third claw, and feeding is usually brief. Cessation of feeding seems to be determined by the time it takes for the soldier caste of ants or termites to arrive at the breach and defend the nest, generally by noxious chemical secretions (Redford 1983, 1985a).

The giant anteater can climb, but it almost invariably forages on the ground. A single young is born after a gestation period of approximately 190 days. Giant anteaters do not dig burrows but bivouac in sheltered places, generally dense clumps of shrubs. The dependent young is carried on the mother's back when she transports it to a new bivouac site; older young may be carried on feeding excursions. Home-range size varies with the carrying capacity of the habitat. In areas with high densities of termites, home ranges may be less than 1 km^2; in areas with lower carrying capacity, they may exceed 25 km^2. Spacing behavior among like-sexed adults is most pronounced in areas of high carrying capacity, where the individuals have small home ranges. With low carrying capacity and large home ranges, there is considerable overlap among individual ranges (Montgomery and Lubin 1977; Redford 1985a; Shaw, Carter, and Machado-Neto 1985, Shaw, Machado-Neto, and Carter 1987; Montgomery 1979, 1985a).

Genus *Tamandua* Gray, 1825
Lesser Anteater, Oso Colmenero

Description

The head and body length varies from 470 to 770 mm, and the tail length from 400 to 675 mm. The body is covered with a dense coat of hair, and the prehensile tail is haired dorsally for part of its length but naked for the distal two-thirds. The middle digit of the forefoot bears an extremely large claw. The color pattern is variable; one variant is pale yellow on the face, head, and forearms, with a black vest extending to the midbody and pale yellow hindquarters (see plate 3). At another extreme, the basic body color may be a uniform honey brown. Melanistic specimens are known for *T. tetradactyla*.

Distribution

The genus ranges from southern Veracruz, Mexico, to northern Argentina.

Natural History

Species of *Tamandua* forage both arboreally and terrestrially, and their arboreal abilities permit a versatile foraging strategy. They opportunistically den in tree cavities or in the burrows of armadillos. The amount of arboreal activity covaries positively with soil moisture and standing water. They are specialized for feeding on ants and termites.

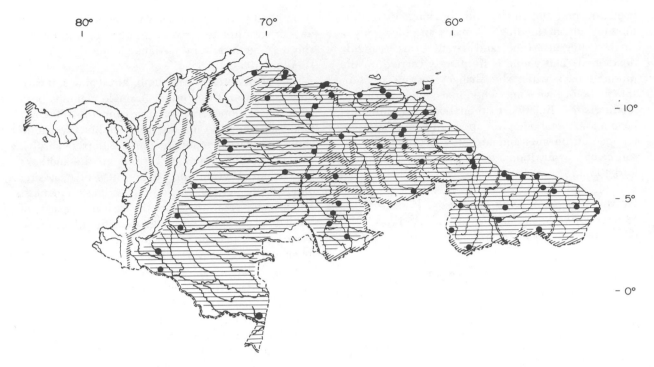

Map 3.3. Distribution of *Tamandua tetradactyla*.

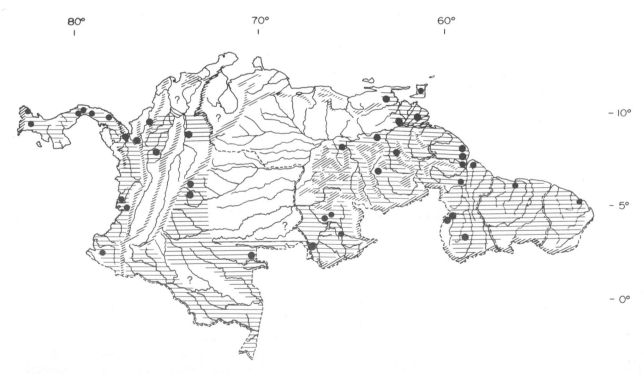

Map 3.4. Distribution of *Cyclopes didactylus*.

Tamandua mexicana (Saussure, 1860)

Description

The head and body length is approximately 563 mm; tail length, 544 mm; weight, 3.2 to 5.4 kg. Over most of its range *T. mexicana* exhibits the black vest markings described in the genus account (see plate 3).

Range and Habitat

The species is distributed from southern Veracruz, Mexico, and south through Colombia, west of the eastern Andean cordillera to Peru (map 3.2). It is present in the Maracaibo basin in Venezuela but is replaced to the east in Venezuela and Colombia by *T. tetradactyla*. It occurs at low elevations in a variety of habitats from gallery forests to multistratal tropical evergreen forests.

Natural History

Depending on temperature and the incidence of direct sunlight, *Tamandua mexicana* can be active during both the day and the night; generally it rests at midday. In Panama, *Tamandua mexicana* may have a home range of approximately 70 ha. There is considerable home-range overlap, but adult females tend to be spaced. The animals forage both terrestrially and arboreally, and there may be seasonal shifts in activity related to the relative abundance of terrestrial and arboreal ants or termites. In forested areas they frequently shelter in hollow trees. The single young is placed in a nest, usually in a secure place such as a hollow tree, to which the mother returns for nursing after her foraging bouts. When the young is older it frequently accompanies the mother, riding on her back to foraging sites. The strongly prehensile tail greatly assists in arboreal locomotion (Montgomery and Lubin 1977; Lubin 1983).

Tamandua tetradactyla Linnaeus, 1758

Description

Tamandua tetradactyla is slightly larger than *T. mexicana;* head and body length averages 593 mm and tail length 511 mm. Weight ranges from 3.4 to 7 kg. Over much of its range the coat shows no black vest but is a uniform honey color (see plate 3). However, the black markings do appear in some subspecies over parts of its range (Wetzel 1982).

Range and Habitat

This species replaces *T. mexicana* to the east of the Andes (map 3.3). Its range extends across Venezuela and the Guyanas, south to Uruguay and northern Argentina. Broadly tolerant of a range of habitats, it can occur in gallery forests adjacent to savannas as well as in multistratal tropical evergreen forests.

Natural History

T. tetradactyla often inhabits gallery forests adjacent to savannas. Under these conditions, if the soil is well drained, it may opportunistically den in armadillo burrows. In parts of the Venezuelan llanos, where the carrying capacity is low, the home range may reach 375 ha. Under such conditions home-range overlap is extensive and little spacing behavior is noted, probably because large areas are indefensible. Though it is an arboreal specialist, the degree of arboreality expressed depends on the density of ant and termite nests on the ground or in the trees. A single young is born annually; maternal care patterns are similar to those for *T. mexicana* (Montgomery and Lubin 1977).

Genus *Cyclopes* Gray, 1821
Cyclopes didactylus (Linnaeus, 1758)
Pygmy or Silky Anteater

Description

This is the smallest of the anteaters; total length ranges from 360 to 450 mm and the tail is 180 to 262 mm. Weight is rarely more than 400 g. The color is golden brown, with some of the dorsal hairs silver tipped. The pelage is very dense.

Range and Habitat

Cyclopes occurs from extreme southern Mexico south to Brazil. It is generally confined to moist multistratal tropical evergreen forests and is absent from savanna areas (map 3.4).

Natural History

This arboreal specialist has an extremely prehensile tail. It is almost totally nocturnal and rarely descends to the ground, occurring where the forest canopy is continuous (Sunquist and Montgomery 1972). A single young is born after an unknown gestation period. Females are strictly spaced with little home-range overlap, but the home range of a single male may overlap those of several females (Montgomery 1983). Best and Harada (1985) note that these animals are opportunistic feeders on ants and seem to track the relative abundance of their prey, taking ubiquitous species more often than rare species of ants.

FAMILY BRADYPODIDAE
Three-toed Sloths

Diagnosis

The exact identification of the tooth types remains in some doubt; thus the dental formula is usually listed as 5/4–5. The teeth are cyclindrical and grow throughout life (fig. 3.3). Head and body length

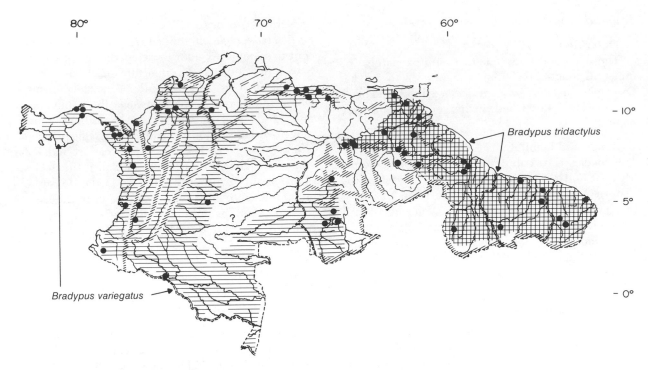

Map 3.5. Distribution of two species of *Bradypus*.

ranges from 400 to 800 mm and weight from 2.25 to 5.50 kg; the stout tail is approximately 68 mm long. The neck contains eight or nine cervical vertebrae. The forelimbs are slightly longer than the hind limbs, and each forefoot has three long claws. The hair is long and rather stiff; individual hairs may support algae and thereby have a blue-green color. In addition to algal symbionts, moths have been recorded in the fur of members of this family (Waage and Best 1985).

Distribution

Three-toed sloths extend from eastern Honduras across South America to northern Argentina.

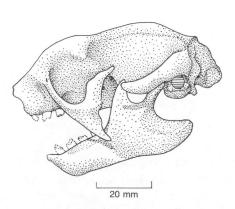

Figure 3.3. Skull of *Bradypus variegatus*.

Natural History

Three-toed sloths are active both diurnally and nocturnally. They are highly specialized as selective browsers on the leaves of trees; the stomach is compartmentalized, and fermentative reduction occurs both in the stomach and in the intestines. Three-toed sloths are characterized by a low metabolic rate and a low core body temperature, apparently an adaptation for feeding on leaves that are abundant but have a low nutrient content (McNab 1978). They locomote slowly while suspended beneath a tree branch by all four limbs. A single young is produced after a gestation period of approximately six months.

Genus *Bradypus* Linnaeus, 1758
Three-toed Sloth, Perezoso de Tres Dedos
Bradypus variegatus Schinz, 1825
Brown-throated Sloth

Description

External measurements are recorded under the diagnosis for the family. Weight averages 4.3 kg. The coat consists of long, coarse hairs overlying a short, dense underfur. The face is white with a brown stripe on each side enclosing the eye, the throat is brown, and the body varies from pale brown to yellowish. On the middorsum of the male the hair is extremely short, generally black surrounded by a yellow band (see plate 3).

Range and Habitat

Bradypus variegatus is distributed from southern Honduras through Panama and western Colombia, south through the Amazon to northern Argentina (map 3.5). It is replaced north of the Amazon and east of the Orinoco by *B. tridactylus*. The three-toed sloth prefers multistratal tropical evergreen forests at low elevations, especially those with a continuous canopy.

Natural History

The three-toed sloth is active mainly during the day but can be active at night (Sunquist and Montgomery 1973). It feeds exclusively on leaves from a wide variety of trees and lianas. Home-range use is highly predictable for a given individual. There is considerable overlap among neighbors, but except for a courting couple, only a single adult or a female with young occupies a given tree at any time (Montgomery and Sunquist 1978).

In Panama, home ranges average about 1.5 ha. A single young is born after a six-month gestation period and is carried by the female until approximately four months of age. Apparently the young animal adopts its mother's dietary preferences. She will dislodge the young when it reaches approximately 15% of her body weight and move to another part of her home range. The young will continue to use a portion of the mother's range until it is approximately a year old, then it usually disperses. Mother and young maintain contact in part by vocalizations. The distinctive whistle of a distressed sloth has been described by Montgomery and Sunquist (1974). Like the two-toed sloth, the animal descends to the ground to defecate at the base of its feeding tree. The natural history of this species has been described in detail by Montgomery and Sunquist (1975, 1978).

Bradypus tridactylus Linnaeus, 1758
Pale-throated Sloth

Description

This species is similar in size and proportions to *B. variegatus*. Head and body length averages 500 mm and tail length 31 to 75 mm. The throat is white to buff; the dorsal color varies but frequently exhibits a marked speckled pattern, immediately distinguishing it from the basic brown to yellow brown of *B. variegatus* (see plate 3).

Range and Habitat

This species is found to the east of the Andes and south of the Orinoco River, although it crosses at the delta. It occurs in south-central Venezuela, eastward through the Guyanas, and south to the Amazon and the Río Negro in Brazil (map 3.5) (Wetzel 1982).

Natural History

The natural history patterns of this sloth have been described by Beebe (1925). Its habits are very similar to those described for *B. variegatus*.

FAMILY CHOLOEPIDAE
Two-toed Sloths

Diagnosis

This family was formerly subsumed under the Bradypodidae, but behavioral and anatomical differences are so profound that it deserves a family rank of its own. Webb (1985) places the two-toed sloths within the family Megalonychidae, a group that has a rich fossil record. Nomenclature for the tooth types remains in doubt; the dental formula is given as 5/4–5. The anterior teeth in both the upper and lower jaw are distinctly caniniform (see fig. 3.4), and a wide gap separates the upper caniniform tooth from the remaining teeth. The short neck has six cervical vertebrae, the forefoot has only two claws, and the tail is absent or vestigial. The color varies from dark brown to pale yellow (see plate 3), and algal symbionts in the fur have been reported for this family.

Distribution

The species of two-toed sloths occur from northern Nicaragua to the Amazon region of Peru and Brazil.

Natural History

Highly specialized for arboreal life, the two-toed sloth exhibits adaptations similar to those of *Bradypus* in locomoting while suspended from a branch (Goffart 1971). In conformity with its leaf-eating habits, it has a relatively low metabolic rate.

Genus *Choloepus* Illiger, 1811
Perezoso de Dos Dedos
Choloepus hoffmanni Peters, 1859

Description

Head and body length ranges from 540 to 700 mm, and weight averages 5.7 kg. This species is usually pale yellow with brownish limbs.

Range and Habitat

From northern Nicaragua, its range extends west of the Andes to Ecuador (map 3.6).

Natural History

Two-toed sloths are nocturnal. Although specialized for feeding on leaves, they eat a considerable amount of fruit. They have a larger home range than the three-toed sloth, often 2–3 ha (Montgomery and Sunquist 1975, 1978). In nature adult females seem to vastly outnumber adult males, suggesting that

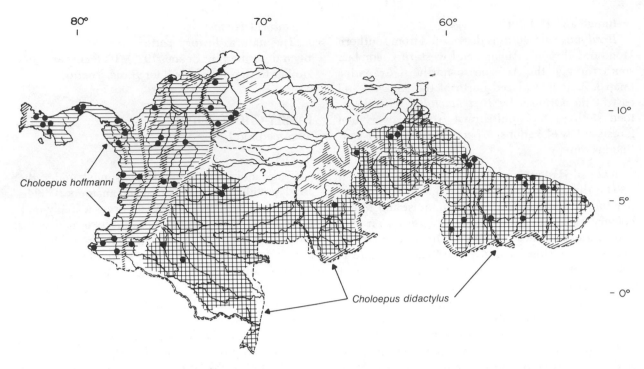

Map 3.6. Distribution of two species of *Choloepus*.

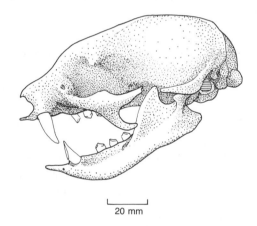

```
20 mm
```

Figure 3.4. Skull of *Choloepus hoffmanni*.

there is some spacing among males and perhaps competition for access to females.

Choloepus didactylus (Linnaeus, 1758)

Description
This species of two-toed sloth is somewhat larger than *C. hoffmanni*. Head and body length ranges from 600 to 860 mm, the vestigial tail from 14 to 15 mm, and weight from 4.0 to 8.4 kg. The fur is generally tan to buffy brown (plate 3).

Range and Habitat
The species occurs east of the Andes in southern Colombia, Venezuela, and along both banks of the

Amazon in Brazil to the Amazon basin of Colombia, Ecuador, and Peru (map 3.6).

Natural History
The natural history of *Choloepus didactylus* has not been studied in the wild but is assumed to be similar to that of *C. hoffmanni*. A single young is born after a gestation period of approximately eleven months. The young is dependent on the female for four to five months and is carried until it reaches approximately 15% of the mother's body weight (Eisenberg and Maliniak 1985).

FAMILY DASYPODIDAE
Armadillos, Quirquinchos

Diagnosis
The dental formula is highly variable and will be given in the generic accounts (Wetzel 1985b). The rostrum is long (fig. 3.5). Members of this family are characterized by numerous bony dermal scutes in regular arrangements, forming movable bands in the midsection and on the tail as well as immovable shields on the forequarters and hindquarters. The bony scutes are covered by horny epidermal scales. Sparse hairs appear between the bands and on the underside of the animal, which is not armored.

Distribution
The armadillos are distributed from Oklahoma in the United States south to the Strait of Magellan in Chile and were recently introduced into Florida.

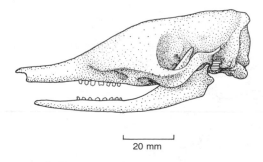

20 mm

Figure 3.5. Skull of *Dasypus novemcinctus*.

The armadillos originated in South America and radiated extensively. The rich fossil record shows that they entered North and Central America during the late Miocene and Pliocene as the Panamanian land bridge neared completion (Webb and Marshall 1982; Webb 1985).

Natural History

The dermal scutes apparently are an antipredator defense. Armadillos' dietary habits vary greatly depending on the species. Some, such as *Cabassous*, feed almost exclusively on ants and termites; others (*Euphractus*) eat fruits, invertebrates, and small vertebrates. All have extremely well developed claws and are capable burrowers. Members of the genus *Dasypus* frequently forage on the surface, whereas others, such as *Cabassous*, are adapted for a fossorial existence.

The feeding habits of the armadillos have been reviewed by Redford (1985b). Species adapted for a fossorial life and specialized for feeding on ants and termites (e.g., *Priodontes* and *Cabassous*) have very low metabolic rates (McNab 1982). The distribution of most armadillos is strongly governed by the mean low temperature of an area, since armadillos have a high thermal conductance and readily lose body heat at low temperatures. Though burrowing protects an individual from the extremes of ambient temperatures, only four species of armadillos have significantly occupied the warmer portions of the temperate zone (McNab 1982). Further information on this widely varying group is in the species accounts.

Genus *Dasypus* Linnaeus, 1758
Long-nosed Armadillo, Cachicamo

Description

The rostrum is long (see fig. 3.5), and the dental formula is 7–9/7–9. The dark brown carapace comprises scapular and pelvic shields, with six to eleven movable bands separating the two shields. The ears are long and have no scales or scutes. The long tail, generally exceeding 55% of the head and the body length, tapers to a slender tip. The proximal two-

thirds of the tail is covered with rings, each formed by two or more rows of scales and scutes. The forefoot bears four long claws, the longest on the second and third digits. The hind foot bears five claws, the longest on the third digit.

Distribution

The genus *Dasypus* is distributed from the south-central United States to the Río Negro in Argentina.

Dasypus novemcinctus Linnaeus, 1758
Nine-banded Armadillo

Description

Head and body length ranges from 395 to 573 mm and the tail from 290 to 450 mm. Weight ranges from 3.2 to 4.1 kg. The number of movable bands varies from eight to ten, but the modal number is nine in northern South America (see plate 4).

Range and Habitat

The nine-banded armadillo has the largest range of any armadillo species, occurring from the central and southern United States to approximately 32° south latitude (map 3.7). It occupies a wide range of habitat types at low elevations, from dry deciduous forests to multistratal tropical evergreen forests, and even ranges into the semiarid llanos of Venezuela and Colombia and the caatinga of Brazil.

Natural History

Depending on season and temperature, the armadillo may be active during both day and night, but it avoids extremes of temperature and is generally inactive at midday. The long claws permit the animal to dig efficiently, and it constructs its own dwelling burrows. It transports nesting material by raking leaves under the body and then, clasping them with its forepaws, hopping bipedally to the burrow (Eisenberg 1961). It forages by scratching in the forest litter and probing with its long, sensitive nose. Nine-banded armadillos eat a wide variety of foods. Insect larvae and ants can predominate in their diets (Redford 1986), but small vertebrates and some plant food are also included (Taber 1954; Redford 1985b).

Dasypus novemcinctus has a delayed implantaton in the northern part of its range, so the time from insemination to birth can exceed 240 days, but the true gestation is approximately 70 days. Four young are typically produced at birth and are identical quadruplets, indicating that one egg divides into four separate blastocysts. Armadillos tend to forage alone, but there is considerable home-range overlap. Though several individuals may use the same burrow, they may be members of a family group. Home-range size in *Dasypus* varies considerably depending on the carrying capacity of the habitat and

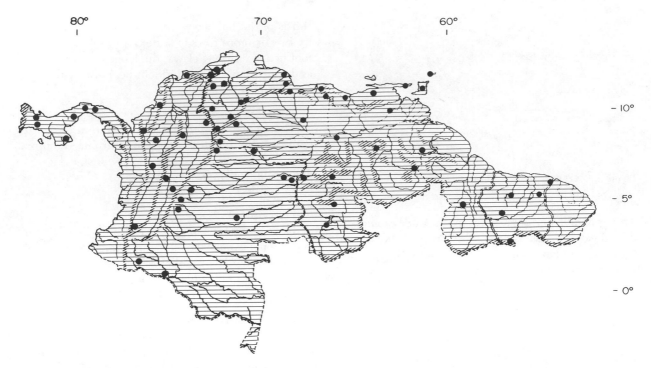

Map 3.7. Distribution of *Dasypus novemcinctus*.

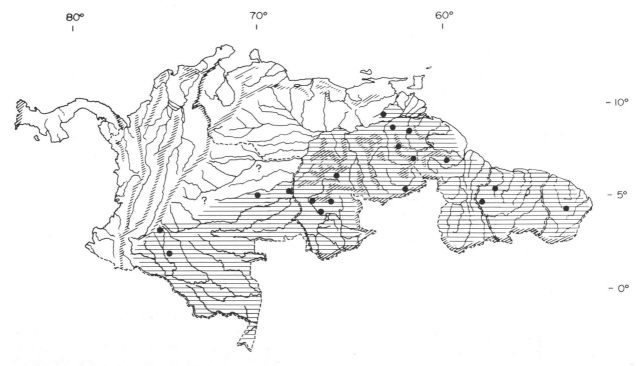

Map 3.8. Distribution of *Dasypus kappleri*.

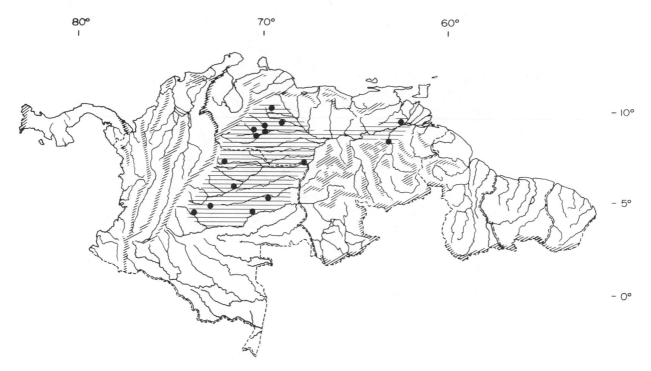

Map 3.9. Distribution of *Dasypus sabanicola*.

may be as small as 3.4 ha. At lower carrying capacities it can exceed 15 ha (Wetzel and Mondolfi 1979).

Dasypus kappleri Krauss, 1862
Greater Long-nosed Armadillo

Description
There are seven to nine teeth in both the upper and lower jaws. Head and body length ranges from 510 to 575 mm, tail length from 325 to 483 mm, and weight from 8.5 to 10.5 kg. There are seven to eight movable bands in the midsection between the shields covering the forequarters and the hindquarters, and large projecting scales or spurs are arranged in transverse rows on the proximal posterior surface of the hind leg. These scales on the hind legs are themselves diagnostic of the species, but its great size easily distinguishes it from *D. novemcinctus* (see plate 4).

Range and Habitat
Dasypus kappleri is confined to the Amazon and Orinoco basins. It occurs through the Guyanas but not north of the main channel of the Orinoco (map 3.8). This large armadillo appears to prefer forested areas and does not extend into the northern savannas.

Natural History
Little is known concerning the natural history of *Dasypus kappleri*. It has a reduced litter size, typically bearing two young. Its activity patterns, di-

etary preferences, and foraging strategies are presumed to be similar to those of *D. novemcinctus*. Barreto, Barreto, and D'Alessandro (1985) reported the stomach contents of one specimen where 14% was vertebrate and 86% invertebrate material. The species constructs burrows in well-drained soil and does not appear to be gregarious (Wetzel and Mondolfi 1979).

Dasypus sabanicola Mondolfi, 1968
Northern Lesser Long-nosed Armadillo, Cabasu

Description
Head and body length ranges from 253 to 314 mm and tail length from 175 to 205 mm. The weight averages 1.5 kg, with a range of 1 to 2 kg. There are seven to nine movable bands in the midsection, with eight as a modal number. This species is very similar in appearance to *D. novemcinctus*, but it is smaller and typically has fewer movable bands (Mondolfi 1968), and its ears tend to be shorter. Whereas *D. novemcinctus* has ears typically 50% of the head length, in *D. sabanicola* the value is approximately 40% (Wetzel and Mondolfi 1979) (see plate 4).

Range and Habitat
This small long-nosed armadillo typically inhabits the llanos of central Colombia and Venezuela. It is absent from multistratal tropical evergreen forests but abundant in the gallery forest and shrub savannas of the upper Orinoco drainage system (map 3.9).

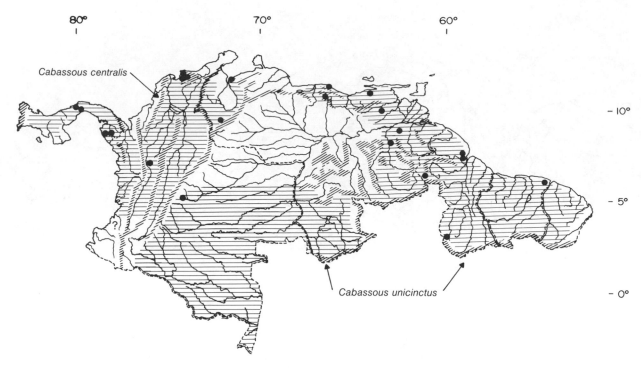

Map 3.10. Distribution of two species of *Cabassous*.

It can co-occur with *D. novemcinctus* but is easily distinguished by its smaller size and shorter ears.

Natural History

In activity patterns, diet, and burrowing ability, this small savanna-dwelling form resembles *D. novemcinctus*. Exactly how niche space is partitioned between the two species where they co-occur is imperfectly understood.

The ecology and behavior of *D. sabanicola* have been reported in detail by Ferguson-Laguna (1984), whose study in the state of Apure, Venezuela, has elucidated a number of important aspects of this species' biology. Animals usually shelter in burrows of their own construction and build nests of leaves as does *D. novemcinctus*. Burrows may have one or two openings. Home ranges vary from 1.7 to 11.6 ha. The diet consists primarily of invertebrates; ants, termites, and coleopterans predomiante. In Apure the animals appear to breed from October to March. Though details of reproduction are not well understood, typically four monozygotic young are born, as in *D. novemcinctus*, and are tended by the female in a burrow system.

Baretto, Baretto, and D'Alessandro (1985) report one specimen where 89% of the stomach contents was termites. Small armadillo species co-occur with *D. novemcinctus* south of the Amazon in savanna areas; *D. sabanicola* has a small ecological counterpart in the southern savannas in the form of *D. hybridus*, which occurs in Uruguay, southern Para-

guay, and northern Argentina. In the drier savanna areas of Brazil, the comparable small long-nosed armadillo is *D. septemcinctus*. The standard reference for the long-nosed armadillos is Wetzel and Mondolfi (1979).

Genus *Cabassous* McMurtrie, 1831
Naked-tailed Armadillo, Cabasu

Description

The four species within this genus differ little in external morphology except for size (Wetzel 1980). The number of teeth is highly variable, ranging between 7/8 and 10/9. There are no teeth in the premaxillary bone. Head and body length ranges from 300 to 490 mm and the tail from 90 to 200 mm. The snout is very short and broad, the ears are moderately large and funnel shaped, and the eyes are extremely small. The forefeet have five claws, the middle one extremely large and sickle shaped. The dorsal plates are arranged in transverse rows for the entire length of the body. The slender tail is distinctive, with either no armor or small, widely spaced thin plates; hence the common name naked-tailed armadillo. The tail alone serves to distinguish this genus from all other armadillos (see plate 4).

Distribution

The genus *Cabassous* is distributed from southern Guatemala south to northern Argentina, typically in multistratal tropical evergreen forests. It prefers rather moist habitats with well-drained, loose soil (Wetzel 1980).

Cabassous centralis (Miller, 1899)

Description
Head and body length is approximately 340 mm; tail length, 130 to 180 mm; weight, 2.0 to 3.5 kg. This is the smaller of the two species occurring within the range covered by this volume (see plate 4).

Range and Habitat
Cabassous centralis occurs from southern Mexico, through Panama, to Ecuador west of the eastern cordillera of the Andes. It is replaced to the east by *C. unicinctus* (map 3.10).

Natural History
Little is known concerning the behavior of *Cabassous* in the field. Apparently it is nocturnal and specialized for feeding on terrestrial ants and termites. Its long tongue can be extruded to a great length when feeding. It is highly specialized for burrowing, and a great deal of its activity may take place underground. The few observations that have been made were of animals digging into rotten logs, apparently seeking termites.

If the animal is disturbed while foraging on the surface, it usually tries to burrow into the ground to escape. When disturbed it produces a low buzz or growl, which is startlingly loud, since many armadillos make very soft sounds (Wetzel 1982). A single young is produced after an unknown gestation period.

Cabassous unicinctus (Linnaeus, 1758)
Cabasu de Orejas Largas

Description
This is one of the larger species of the genus, as specimens from Venezuela and Suriname show: head and body length may range from 347 to 445 mm, and tail length from 165 to 200 mm. In external appearance it is similar to the preceding species, but their ranges do not overlap and identity can easily be determined by the area.

Range and Habitat
C. unicinctus ranges to the east of the Andes in Colombia, Venezuela, and the Guyanas and south to the Brazilian shield. Its preferred habitat appears to be similar to that described for *C. centralis* (map 3.10).

Natural History
The sparse information available indicates that this species' diet and foraging strategies are similar to those of *C. centralis* (Wetzel 1982).

Genus *Priodontes* F. Cuvier, 1825
Priodontes maximus Kerr, 1792
Giant Armadillo, Tatu-guasu

Description
The tooth number is highly variable, and teeth are lost as the animal matures. At any time up to twenty teeth may be present in one side of the jaw. In this largest extant species of armadillo, the head and body length ranges from 750 to 1,000 mm, and the tail is approximately 500 mm. Adults may weigh up to 30 kg. Its size alone distinguishes this species from all others (see plate 4). The ears and eyes are rather small. The claws on the forefeet are extremely powerful; the claw of the third fore toe is very large and may measure up to 203 mm along the curve. The back is covered with transverse rows of small plates; eleven to thirteen of the bands are movable. The tail is completely covered with a series of close-set plates, but they are not arranged in bands.

Range and Habitat
The giant armadillo tolerates a wide range of habitat types, from multistratal tropical evergreen forests to savannas. It prefers well-drained soils. Its range extends east of the Andes in Colombia, Venezuela, and the Guyanas and south to northern Argentina (map 3.11).

Natural History
The giant armadillo is specialized for feeding on colonial ants and termites. Barreto, Barreto, and D'Alessandro (1985) report that in three individuals from Colombia ants made up 0% to 78% of the stomach contents while termites composed 22% to 85% The animal digs a very large burrow that may be used as a resting site for several days (Carter 1983), and an individual often has several burrows within its home range. When digging into a colonial termite mound, the armadillo makes a characteristic large excavation. In its foraging it is much more destructive of ant or termite mounds than is the giant anteater. A single young is born after an undetermined gestation period. The giant armadillo has been exterminated over many parts of its former range, since it is highly prized for food (Redford 1985b; Wetzel and Mondolfi 1979).

Genus *Euphractus* Wagler, 1830
Euphractus sexcinctus (Linnaeus, 1758)
Six-banded Armadillo, Tatu-puyu

Description
Head and body length averages about 446 mm, and the tail averages 231 mm. Weight averages 5.4 kg. The ears are extremely short, and the muzzle is somewhat short and broad at the tip. There are five

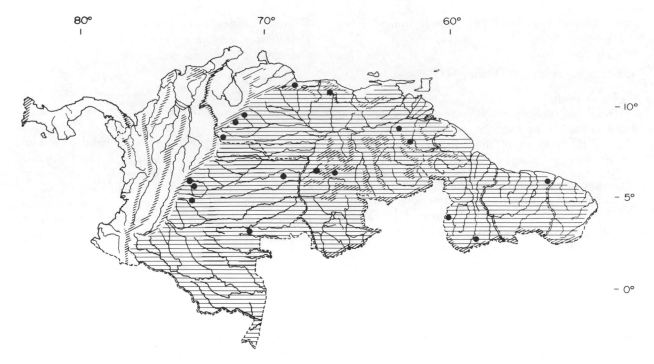

Map 3.11. Distribution of *Priodontes maximus*.

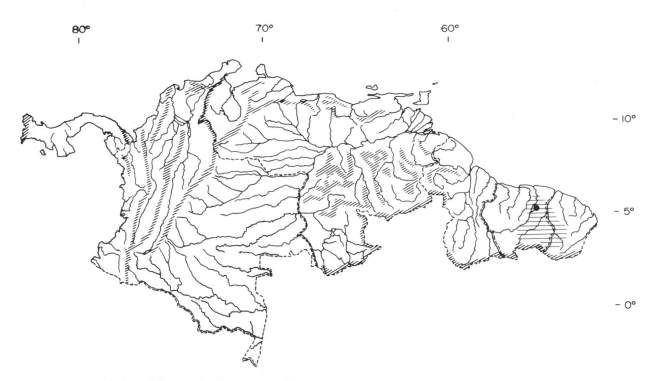

Map 3.12. Distribution of *Euphractus sexcinctus*.

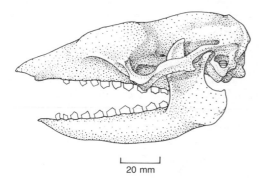

Figure 3.6. Skull of *Euphractus sexcinctus*.

toes on the forefeet, and the second toe bears the longest claw (see plate 4 and fig. 3.6). The scutes on the pectoral and pelvic portions of the body are in transverse rows, but those on the middle of the body are distinctly arranged into six movable bands. The scutes on the tail are arranged in transverse rows for almost its entire length. Sparse white hairs generally show between the movable bands.

Range and Habitat

The six-banded armadillo occurs chiefly south of the Amazon. It is absent from Colombia and Venezuela, and within the range covered by this volume it makes its first appearance in the Guyanas (map 3.12). From then on south it is abundant through Brazil to northern Argentina. It is frequently found in savannas, though it can exploit tropical rain forest habitats as well.

Natural History

The six-banded armadillo seems to have a typical omnivore feeding pattern, eating vertebrates and plants as well as invertebrates. Use of ants and termites appears to be minimal. One to three young are born after a sixty- to sixty-five-day gestation period. The female is intolerant of other adults during the rearing phase. The jaws can be opened very wide in this species, and young may be carried from one nest to another in the mother's mouth. The animals tend to be nocturnal or crepuscular; like other members of the armadillo family, they construct burrows for shelter and for rearing young. A pregnant female generally builds a nest by carrying leaves and other material in her mouth (Gucwinska 1971; Redford and Wetzel 1985).

References

• References used in preparing distribution maps

Barreto, M., P. Barreto, and A. D'Alessandro, 1985. Colombian armadillos: Stomach contents and infection with *Trypanosoma cruzi*. *J. Mammal.* 66:188–93.

• Beebe, W. 1919. The higher vertebrates of British Guiana with special reference to the fauna of the Bartica district, no. 7. List of Amphibia, Reptilia, and Mammalia. *Zoologica* 2:205–38.

———. 1925. The three-toed sloth, *Bradypus cuculliger* Wagler. *Zoologica* 7:1–67.

Best, R. C., and A. Y. Harada. 1985. Food habits of the silky anteater (*Cyclopes didactylus*) in the central Amazon. *J. Mammal.* 66(4):780–81.

Carter, T. S. 1983. The burrows of giant armadillos. *Säugetierk. Mitl.* 31:47–53.

• Defler, T. R. 1983. A remote park in Colombia. *Oryx* 17:15–17.

Eisenberg, J. F. 1961. The nest-building behavior of armadillos. *Proc. Zool. Soc. London* 137:322–24.

Eisenberg, J. F., and E. Maliniak. 1985. Maintenance and reproduction of the two-toed sloth *Choloepus didactylus* in captivity. In *The evolution and ecology of armadillos, sloths, and vermilinguas*, ed. G. G. Montgomery, 327–32. Washington, D.C.: Smithsonian Institution Press.

Emery, R. J. 1970. A North American Oligocene pangolin and other additions to the Pholidota. *Bull. Amer. Mus. Nat. Hist.* 142(6):457–510.

Ferguson-Laguna, A. 1984. El cachicamo sebanero: Su biología y ecología. *Acta Cient. Venez.* Caracas: Fondo Editorial.

Gillette, D. D., and C. E. Ray. 1981. *Glyptodonts of North America*. Smithsonian Contributions to Paleobiology 40. Washington, D.C.: Smithsonian Institution Press.

Goffart, M. 1971. *Function and form in the sloth*. Oxford: Pergamon Press.

• Goodwin, G. G., and A. M. Greenhall. 1961. A review of the bats of Trinidad and Tobago. *Bull. Amer. Mus. Nat. Hist.* 122(3):191–301.

Gucwinska, H. 1971. Development of six-banded armadillos, *Euphractus sexcinctus* at Wroclaw Zoo. *Int. Zoo. Yearb.* 11:88–89.

• Hall, E. R. 1981. *The mammals of North America*. 2d ed., 2 vols. New York: John Wiley.

Handley, C. O., Jr. 1966. Checklist of the mammals of Panama. In *Ectoparasites of Panama*, ed. R. L. Wetzel and V. J. Tipton, 753–95. Chicago: Field Museum of Natural History.

———. 1976. Mammals of the Smithsonian Venezuelan project. *Brigham Young Univ. Sci. Bull., Biol. Ser.* 20(5):1–90.

• Husson, A. M. 1978. *The mammals of Suriname*. Leiden: E. J. Brill.

Lubin, Y. D. 1983. *Tamandua mexicana*. In *Costa Rican natural history*, ed. D. H. Janzen, 494–96. Chicago: University of Chicago Press.

McNab, B. 1978. Energetics of arboreal folivores:

Physiological problems and ecological consequences of feeding on a ubiquitous food supply. In *The ecology of arboreal folivores*, ed. G. G. Montgomery, 153–62. Washington, D.C.: Smithsonian Institution Press.

———. 1982. The physiological ecology of South American mammals. In *Mammalian biology in South America*, ed. M. Mares and H. Genoways, 187–207. Pymatuning Symposia in Ecology 6. Special Publication Series. Pittsburgh: Pymatuning Laboratory of Ecology, University of Pittsburgh.

Meritt, D. A. 1976. The nutrition of edentates. *Int. Zoo Yearb.* 16:38–46.

Mondolfi, E. 1968. Descripción de un nuevo armadillo del género *Dasypus* de Venezuela. *Mem. Soc. Cienc. Nat. La Salle* 27:149–67.

Montgomery, G. G. 1979. El grupo alimenticio del oso hormiguero. *Conciencia* 6:3–6.

———. 1983. *Cyclopes didactylus*. In *Costa Rican natural history*, ed. D. H. Janzen, 461–63. Chicago: University of Chicago Press.

———. 1985a. Movements, foraging and food habits of the four extant species of Neotropical vermilinguas (Mammalia; Myrmecophagidae). In *The evolution and ecology of armadillos, sloths, and vermilinguas*, ed. G. G. Montgomery, 365–77. Washington, D.C.: Smithsonian Institution Press.

———., ed. 1985b. *The evolution and ecology of armadillos, sloths, and vermilinguas*. Washington, D.C.: Smithsonian Institution Press.

Montgomery, G. G., and Y. D. Lubin. 1977. Prey influences on movements of Neotropical anteaters. In *Proceedings of the 1975 Predator Symposium*, ed. R. L. Philips and C. Jonkel, 103–31. Missoula: Montana Forest and Conservation Experiment Station, University of Montana.

Montgomery, G. G., and M. E. Sunquist. 1974. Contact-distress calls of young sloths. *J. Mammal.* 55:211–13.

———. 1975. Impact of sloths on Neotropical forest energy flow and nutrient cycling. In *Tropical ecological systems*, ed. F. B. Golley and E. Medina, 69–98. New York: Springer–Verlag.

———. 1978. Habitat selection and use by two-toed and three-toed sloths. In *The ecology of arboreal folivores*, ed. G. G. Montgomery, 329–59. Washington, D.C.: Smithsonian Institution Press.

Patterson, B., and R. R. Pascual. 1972. The fossil mammal fauna of South America. In *Evolution, mammals, and southern continents*, ed. A. Keast, F. C. Erk, and B. Glass, 247–309. Albany: State University of New York Press.

Redford, K. H. 1983. Mammalian myrmecophagy: Feeding, foraging, and food preference. Ph.D. diss., Harvard University.

———. 1985a. Feeding and food preferences in captive and wild giant anteaters (*Myrmecophaga tridactyla*). *J. Zool.* (London) 205:559–72.

———. 1985b. Food habits of armadillos (Xenarthra: Dasypodidae). In *The evolution and ecology of armadillos, sloths, and vermilinguas*, ed. G. G. Montgomery, 429–37. Washington, D.C.: Smithsonian Institution Press.

———. 1986. Dietary specialization and variation in two mammalian myrmecophages. *Rev. Chilena Hist. Nat.* 59:201–8.

———. 1987. Ants and termites as food: Patterns of mammalian myrmecophagy. In *Current mammalogy*, vol. 1, ed. H. Genoways, 349–400. New York: Plenum Press.

Redford, K. H., and R. M. Wetzel. 1985. *Euphractus sexcinctus*. *Mammal. Species* 252:1–4.

Reig, O. 1981. *Teoría del origen y desarrollo de la fauna de mamíferos de América del Sur*. Monografie Naturae. Mar del Plata, Argentina: Museo Municipal de Ciencias Naturales Lorenzo Scaglia.

Shaw, J. H., T. S. Carter, and J. C. Machado-Neto. 1985. Ecology of the giant anteater *Myrmecophaga tridactyla* in Serra da Canastra, Minas Gerais, Brazil: A pilot study. In *Evolution and ecology of armadillos, sloths, and vermilinguas*, ed. G. G. Montgomery, 379–84. Washington, D.C.: Smithsonian Institution Press.

Shaw, J. H., J. Machado-Neto, and T. S. Carter. 1987. Behavior of free-living anteaters (*Myrmecophaga tridactyla*). *Biotropica* 19:255–59.

Sunquist, M. E., and G. G. Montgomery. 1972. Activity pattern of a translocated silky anteater (*Cyclopes didactylus*). *J. Mammal.* 54:782.

———. 1973. Activity patterns and rate of movement of two-toed and three-toed sloths. *J. Mammal.* 54:946–54.

Taber, F. W. 1954. Contribution on the life history and ecology of the nine-banded armadillo. *J. Mammal.* 26:211–26.

• Tate, G. H. H. 1939. The mammals of the Guiana region. *Bull. Amer. Mus. Nat. Hist.* 76(5):151–229.

Waage, J. K., and R. C. Best. 1985. Arthropod associates of sloths. In *The evolution and ecology of armadillos, sloths, and vermilinguas*, ed. G. G. Montgomery, 297–311. Washington, D.C.: Smithsonian Institution Press.

Webb, S. D. 1985. The interrelationships of tree sloths and ground sloths. In *The evolution and ecology of armadillos, sloths, and vermilinguas*, ed. G. G. Montgomery, 105–12. Washington, D.C.: Smithsonian Institution Press.

Webb, S. D., and L. G. Marshall. 1982. Historical biogeography of recent South American land mammals. In *Mammalian biology in South America*, ed. M. Mares and H. Genoways, 39–52.

Pymatuning Symposia in Ecology 6. Special Publication Series. Pittsburgh: Pymatuning Laboratory of Ecology, University of Pittsburgh.

• Wetzel, R. M. 1975. The species of *Tamandua* Gray (Edentata, Myrmecophagidae). *Proc. Biol. Soc. Washington* 88:95–112.

• ———. 1980. Revision of the naked-tailed armadillos, genus *Cabassous. Ann. Carnegie Mus.* 49:323–57.

• ———. 1982. Systematics, distribution, ecology and conservation of South American edentates. In *Mammalian biology in South America*, ed. M. Mares and H. Genoways, 345–75. Pymatuning Symposia in Ecology 6. Special Publication Series. Pittsburgh: Pymatuning Laboratory of Ecology, University of Pittsburgh.

——— 1985a. The identification and distribution of recent Xenarthra (= Edentata). In *The evolution and ecology of armadillos, sloths, and vermilinguas*, ed. G. G. Montgomery, 5–21. Washington, D.C.: Smithsonian Institution Press.

———. 1985b. Taxonomy and distribution of armadillos, Dasypodidae. In *The evolution and ecology of armadillos, sloths, and vermilinguas*, ed. G. G. Montgomery, 23–46. Washington, D.C.: Smithsonian Institution Press.

• Wetzel, R. M., and D. Kock. 1973. The identity of *Bradypus variegatus* (Mammalia, Edentata). *Proc. Biol. Soc. Washington* 86:25–34.

• Wetzel, R. M., and E. Mondolfi. 1979. The subgenera and species of long-nosed armadillos, Genus *Dasypus* L. In *Vertebrate ecology in the northern Neotropics*, ed. J. F. Eisenberg, 43–63. Washington, D.C.: Smithsonian Institution Press.

Wetzel, R. M., K. H. Redford, and J. F. Eisenberg. n.d. Dasypodidae. In *Mammals of South America*, vol. 1, ed. A. L. Gardner. Chicago: University of Chicago Press. Forthcoming.

Diagnosis

This diagnosis treats only the hedgehogs (Erina-ceidae), moles (Talpidae), shrews (Soricidae), and solenodons (Solenodontidae). For the purposes of this volume, the order Insectivora has been defined in a restricted sense, excluding the tenrecs (Tenreci-dae), golden moles (Chrysochloridae), tree shrews (Tupaiidae), and elephant shrews (Macroscelidea), which are considered to be ordinal ranks in their own right (see McKenna 1975; Eisenberg 1981). The proposed classification followed here is outlined under McKenna and Eisenberg in table 4.1.

Modern insectivores are characterized by a rather small body size. Members of the order as so defined have long pointed snouts and relatively small brains dominated by large olfactory bulbs. In the Western Hemisphere only the moles, solenodons, and shrews are represented. The solenodons are antillean in their distribution. The moles are strongly adapted for a fossorial life, with forefeet and claws enlarged for digging. Although shrews are also specialized to some extent for digging in loose soil, the develop-ment of musculature and shortening of the humerus so characteristic of the moles are not morphological specializations of shrews. Shrews have a small to minute body size, and many species have moder-ately long tails. The pelage is soft and velvety. The eyes are very reduced, as is the external ear. Fur-ther diagnostic features will be discussed at the fam-ily level.

Distribution

At present the Insectivora, as so defined for this volume, are distributed in North America, South America, Eurasia, and Africa. Members of the order did not cross into South America until the comple-tion of the Panamanian land bridge in the Pliocene.

History and Classification

The Recent families of insectivores originated on the northern continents. The hedgehogs, family Erina-ceidae, first appeared in the Eocene and remain conservative in many of their morphological fea-tures. The shrews (Soricidae) and moles (Talpidae) appeared in the early Oligocene.

FAMILY SORICIDAE

Shrews, Musaranas

Diagnosis

The first upper incisors are long and hooked, with a cusp projecting ventrally at the base, and the first lower incisors are directed forward, forming a pin-cerlike grip (fig. 4.1). The skull is narrow, and there is no zygomatic arch. Instead of a tympanic bulla there is a bony ring. The rostrum is long and pointed. The eyes are extremely small, and though pinnae are present, they are reduced. Normally there are five toes on the forepaws and hind feet.

Table 4.1   Alternative Classifications for the Insectivora

Simpson 1945	McKenna 1975; Eisenberg 1981
Order Insectivora	Order Tenrecomorpha
Family Tenrecidae	Family Tenrecidae
	Subfamily Tenrecinae
Family Potomogalidae	Subfamily Potomogalinae
Family Chrysochloridae	Family Chrysochloridae
	Order Erinaceomorpha
Family Erinaceidae	Family Erinaceidae
	Order Soricomorpha
Family Soricidae	Family Soricidae
Family Solenodontidae	Family Solenodontidae
Family Talpidae	Family Talpidae
	Order Macroscelidea
Family Macroscelididae	Family Macroscelididae

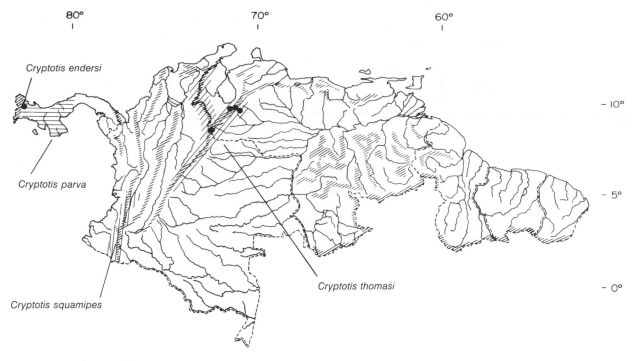

Map 4.1. Distribution of four species of *Cryptotis* (Choate 1970; Hall 1981).

Distribution

Members of the family Soricidae are found in Eurasia, Africa, North America, and northern South America.

Natural History

Shrews are specialized for feeding on small invertebrates in the litter of the forest floor. They forage beneath the leaf litter, and their activity extends into the day as well as the night. Shrews are represented in the northern Neotropics by a single genus, *Cryptotis*.

Genus *Cryptotis* Pomel, 1848
Least Shrew, Musarana de Cola Corta

Description

The dental formula is I 3/1, C 1/1, P 2/1, M 3/3. The total length is less than 130 mm, and the tail does not exceed 50% of the head and body length. The external ear is greatly reduced. There are twelve species within the genus, and six occur in the area covered by this volume.

Distribution

The genus is distributed from the northeastern portion of North America south through Panama, and in the Andes to Peru (maps 4.1 and 4.2).

Comment on Systematics

All the species of least shrews are extremely similar in external appearance. Appropriate identification often requires skull characters, but when the location is known it is possible to assign a specific identification with some certainty (Choate 1970).

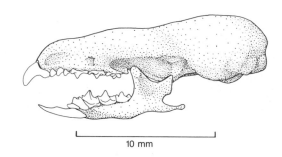

Figure 4.1. Skull of *Cryptotis parva*.

Table 4.2   Key to the Species of *Cryptotis* Occurring in Northern South America

1 Venter (pale buff to buff) markedly paler than dorsum (gray brown) . . . . *C. thomasi*
1′ Venter only slightly if at all paler than dorsum 2
2 Pelage pale gray both above and below . . . . *C. montivaga* [a]
2′ Pelage dark brown or black above and below 3
3 Size large (palatal length 9.2 to 10 mm) . . . . *C. squamipes*
3′ Size small (palatal length 8.3 to 8.6 mm) . . . . *C. avia*

Source: Adapted from Choate and Fleharty (1974).
Note: For the identification of the Panamanian species, see text.
[a] To date recorded from Ecuador.

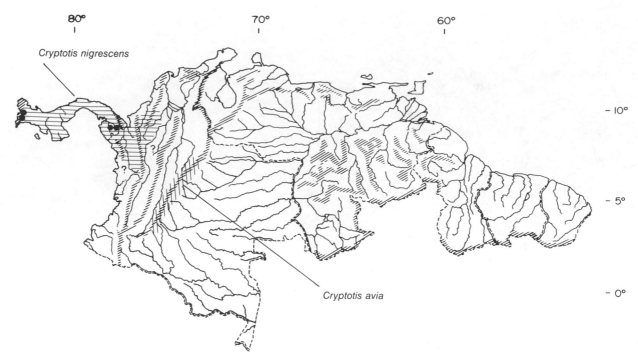

CHAPTER FOUR

Map 4.2. Distribution of two species of *Cryptotis* (Handley 1976; Cabrera 1958; Hall 1981).

Keys to the species are provided by Choate and Fleharty (1974) (see also table 4.2).

Cryptotis avia G. M. Allen, 1923

Description
This small species has a palatal length of 8.3 to 8.6 mm; total length is less than 100 mm. The venter is the same color as the dorsum—dark brown to black.

Range and Habitat
The species is confined to the higher elevations of the southern portion of the eastern cordillera of the Andes (map 4.2).

Cryptotis endersi Setzer, 1950

Description
This is one of the larger species of short-tailed shrews; total length exeeds 109 mm: HB 73 mm; T 36 mm.

Range and Habitat
This species is known only from the area of Bocas del Torro in northeastern Panama (map 4.1).

Cryptotis nigrescens (J. A. Allen, 1895)

Description
Total length ranges from 82 to 108 mm, the tail from 21 to 26 mm, and the hind foot from 11 to 12 mm (*N* = 4; USNMNH). There is a distinct differ-

ence in pelage color between summer and winter. This dark-colored shrew is blackish on the dorsum in winter (Mexico) and slightly paler beneath; in summer it is more grayish. It is distinguishable from *C. parva* by its slightly larger body size, darker color, and darker venter.

Range and Habitat
C. nigrescens occurs from central Mexico through Panama to northwestern Colombia. The extent of its distribution in the westernmost cordillera of the Andes is unknown (map 4.2).

Natural History
Throughout its range this species occupies a variety of habitats from xeric oak forest to evergreen tropical forests. Its elevational range extends from 70 m (Yucatán) to 3,400 m (Panama) (Choate 1970).

Cryptotis parva (Say, 1823)

Description
This is one of the smallest species, with total length from 70 to 103 mm and tail less than 45% of the head and body length. The dorsum is brown to black; the venter is paler than the dorsum and in some species is white (fig. 4.2). In Central America the white venter distinguishes *C. parva* from *C. nigrescens*. The Panamanian subspecies can co-occur with *C. endersi* (map 4.1), which can be distinguished from *C. parva* by its larger size, total length

greater than 109 mm, and tail more than 45% of the head and body length.

Range and Habitat
This species occurs from New England in the United States to western Panama (map 4.1).

Natural History
Only one species of the genus, *Crypotis parva*, has been studied in detail. This species tolerates a range of habitat types from clearings to mountain forests. This small mammal is so inconspicuous that most of our information concerning its biology comes from captive studies. It is one of the most socially tolerant of the shrews in that the male apparently remains with the female through the rearing of the litter. These shrews feed on insect larvae, earthworms, spiders, and the internal organs of larger insects; apparently the hard exoskeleton is discarded. Females attain sexual maturity at ninety days of age; reproduction may be year-round in the southern part of the range. Gould (1969) described development of the young and vocalizations. Mean litter size is about five (one to eight), and gestation takes twenty-one to twenty-three days (Davis and Joeris 1945; Conaway 1958; Broadbrooks 1952; Whitaker 1974). The Central American subspecies of *C. parva* may have smaller litters and breed throughout the year (Choate 1970).

Cryptotis squamipes (J. A. Allen, 1912)

Description
This large species of *Cryptotis* has a palatal length of 9.2 to 10 mm and a total length greater than 106 mm. The dorsum and venter do not contrast in color, varying from dark brown to black.

Range and Habitat
This shrew apparently replaces *C. nigrescens* in the southern portion of the western cordillera of the Colombian Andes (map 4.1).

Cryptotis thomasi (Merriam, 1897)

Description
Total length is 118 to 129 mm; T 35–39; HF 14–17; E 6–8; Wt 11.7–15.9 g (Venezuela, USNMNH). In Venezuela Aagard (1982) found this species to vary in total length from 117 to 131 mm; weight ranged from 9 to 15 g. The dorsum is dark brown, almost black, and the venter is lighter than the dorsum.

Range and Habitat
The species is found in the Andes of Venezuela and presumably in the higher montane areas of the Andes rimming the western side of Lake Maracaibo (map 4.1).

Natural History
Aagard (1982) found this species moderately abundant at between 3,500 and 4,100 m in the Venezuelan Andes. It was the most common catch in the paramos, and within this habitat breeding takes place in the wet season (April to December).

References
• References used in preparing distribution maps

Aagaard, E. M. J. 1982. Ecological distribution of mammals in the cloud forests and paramos of the Andes, Mérida, Venezuela. Ph.D. diss., Colorado State University.

Figure 4.2. *Cryptotis parva.*

Broadbrooks, H. E. 1952. Nest and behavior of a short-tailed shrew, *Cryptotis parva*. *J. Mammal.* 33:241–43.

• Cabrera, A. 1958. *Catálogo de los mamíferos de América del Sur.* Vol. 1. Buenos Aires: Museo Argentino de Ciencias Naturales "Bernardino Rivadavia," Zoología.

• Choate, J. R. 1970. Systematics and zoogeography of Middle America shrews of the genus *Cryptotis*. *Univ. Kansas Publ. Mus. Nat. Hist.* 19:195–317.

Choate, J. R., and E. D. Fleharty. 1974. *Cryptotis goodwini. Mammal. Species* 44:1–3.

Conaway, C. H. 1958. Maintenance, reproduction, and growth of the least-shrew in captivity. *J. Mammal.* 39:507–12.

Davis, W. B., and L. Joeris. 1945. Notes on the life history of the little short-tailed shrew. *J. Mammal.* 26:136–38.

Eisenberg, J. F. 1981. *The mammalian radiations: An analysis of trends in evolution, adaptation, and behavior.* Chicago: University of Chicago Press.

Gould, E. 1969. Communication in three genera of shrews (Soricidae): *Suncus, Blarina,* and *Cryptotis. Comm. Behav. Biol.,* part A, 3:11–31.

• Hall, E. R. 1981. *The mammals of North America.* 2d ed., 2 vols. New York: John Wiley.

• Handley, C. O., Jr. 1976. Mammals of the Smithsonian Venezuelan project. *Brigham Young Univ. Sci. Bull., Biol. Ser.* 20(5):1–90.

• Honacki, J. H., K. E. Kinman, and J. W. Koeppl, eds. 1982. *Mammal species of the world.* Lawrence, Kans.: Allen Press and Association of Systematics Collections.

McKenna, M. C. 1975. Toward a phylogenetic classification of the mammalia. In *Phylogeny of the primates,* ed. W. P. Luckett and F. S. Szalay, 21–46. New York: Plenum Press.

Simpson, G. G. 1945. The principles of classification and a classification of mammals. *Bull. Amer. Mus. Nat. Hist.* 85:1–350

Whitaker, J. O., Jr. 1974. *Cryptotis parva. Mammal. Species* 43:1–8.

5   Order Chiroptera (Bats, Murciélagos)

Diagnosis

This order includes the only true flying mammals. Some other mammals glide, but the bats really fly. A wing membrane extends from each side of the body and hind leg to the forearm, where it is supported by the fingers, which are elongated to create a large surface area for the membrane's support. There may be an additional membrane between the hind legs, sometimes enclosing the tail. The clavicle is well developed. The hind leg has become rotated to support the wing membrane, so that the knee is directed laterally and backward. The sternum is usually keeled for attachment of the massive pectoral muscles that are used in the power stroke during flying (see fig. 5.1).

In his classification of bats Miller (1907) used the morphology of the humerus as an indicator of specialization for flight. Family diagnoses often refer to the relative development of the greater and lesser tuberosities (trochiter and trochin) on the proximal end of the humerus.

Distribution

At present bats are found on all continents except Antarctica. The families Pteropodidae (= Pteropidae), Rhinopomatidae, Nycteridae, Megadermatidae, Rhinolophidae, Hipposideridae, Craesonycteridae, Mystacinidae, and Myzopodidae are confined to the Old World (Honacki, Kinman, and Koeppl 1982 include the Hipposideridae within the Rhinolophidae), while the families Thyropteridae, Natalidae, Furipteridae, Mormoopidae, Phyllostomidae (= Phyllostomatidae), and Noctilionidae are confined to the New World. The families Emballonuridae, Vespertilionidae, and Molossidae are worldwide in their distribution. Bats show their greatest species richness and numbers in the subtropical and tropical regions of the world (see figs. 5.2 and 15.1).

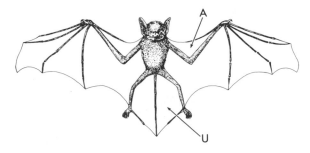

Figure 5.1. External anatomy of a bat: A = antebrachial membrane (propatagium); U = uropatagium (interfemoral membrane). Redrawn from Husson (1978).

History and Classification

Bats are poorly represented in the early fossil record. Their small size and delicate bones apparently reduce the probability of preservation. The earliest bat known is from the Eocene of North America (Jepsen 1970). This fossil clearly indicates that the bat was completely volant; thus the early ancestors of bats showing preflight adaptations are as yet unknown or unrecognized in the fossil record. The earliest records of the phyllostomid bats are from the Miocene of Colombia.

The bats are typically divided into two suborders, the Megachiroptera and the Microchiroptera. The megachiropterans are entirely Old World and are represented by a single family, the Pteropodidae. These are the Old World fruit bats, specialized for feeding on fruits, pollen, and nectar. Their ecological equivalent in the New World tropics is the family Phyllostomidae. The latter family, together with sixteen other families, is included in the Microchiroptera (see fig. 5.2).

The megachiropterans are chiefly distinguished from the microchiropterans by the following characters: the second finger of a megachiropteran is capable of some independent movement and generally has a claw; and the postorbital process of the skull is well developed, helping to support the very large eyes that are typical of megachiropterans.

In the microchiropterans the second finger lacks a claw and is not capable of independent movement. Usually a median fold of skin termed the tragus projects into the pinna, or external ear. All microchiropterans show specializations of the auditory nerve and larynx that correlate with their highly evolved ability to echolocate, which serves them in orientation and prey capture. Only one genus of the Megachiroptera, *Rousettus*, employs echolocation in its orientation. Bats of this genus produce the echolocating pulse by tongue clicks, not from the larynx as do the Microchiroptera. The standard reference works for the Chiroptera are Wimsatt (1970a,b,

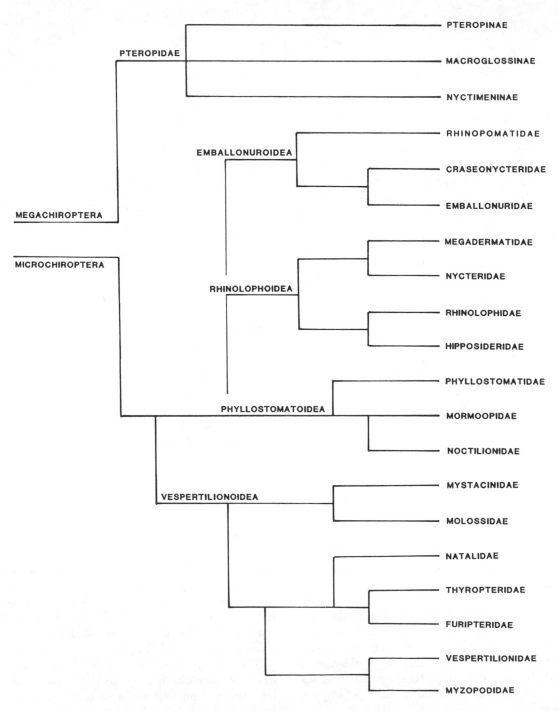

Figure 5.2. Cladogram indicating the relationship of the families of the Chiroptera (Eisenberg 1981).

Table 5.1   Key to the Families of New World Bats

1   Nose leaf or conspicuous folds or plates of skin on chin present 2
1′   Nose leaf and folds or plates of skin absent 3
2   Folds of skin on chin; no nose leaf present . . . . Mormoopidae
2′   Nose leaf present; incisors 2/2, 2/1, or 2/0 . . . . Phyllostomidae [a]
3   Tail perforates uropatagium and is enclosed in a sheath for half its length; upper lip is not noticeably cleft; claws of hind feet are not noticeably long; most species have glandular sacs in the wing or tail (see text) . . . . Emballonuridae
3′   Tail and other characters not as above 4
4   Conspicuous suckers on forefeet and hind feet . . . . Thyropteridae
4′   Sucker disks absent from hind feet 5
5   Tail usually is longer than head and body but is completely enclosed in an interfemoral membrane; ears funnel shaped; tragus triangular . . . . Natalidae
5′   Tail and other characters not as above 6
6   Thumb (first finger) very reduced and enclosed in a membrane or absent . . . . Furipteridae
6′   First finger present 7
7   Upper lip divided; tail shorter than interfemoral membrane but perforates the uropatagium; claws of hind feet very long . . . . Noctilionidae
7′   Not as above 8
8   Tail extends beyond the interfemoral membrane; incisors 1/2–3; tooth number 28–32 . . . . Molossidae
8′   Not as above; incisors 1–2/2–3, upper incisors widely spaced at base; tooth number 28–38 . . . . Vespertilionidae

[a]The nose leaf is vastly modified in *Desmodus, Centurio,* and *Sphaeronycteris* (see text). The Desmodontinae have a postcanine dental formula less than 4/4, and the upper incisors are bladelike (see text).

1977) and Kunz (1982). Keys to the South American families are given in tables 5.1 and 5.2. Additional measurements for the species of phyllostomids included in this volume are given in table 5.14.

Natural History of Bats

Bat Form and Function

Comparing bat morphologies can often give us some indication of habits and adaptation (Findley and Wilson 1982), since wing form is correlated with differences in flight patterns. The long, flat narrow wings of the Molossidae correlate with rapid, long-range flight. *Myotis* has short, broad wings that reflect its brief periods of slow, maneuverable flight (Vaughan 1959). Jaw morphology reflects dietary adaptations: nectar and pollen feeders have a relatively long rostrum, and bats habitually feeding on large fruits usually have a very short rostrum and wide gape. Freeman (1979) suggested that the tooth and jaw structure of molossid bats reflect specializations for either "hard" prey such as beetles (e.g., *Molossus*) or soft prey such as moths (e.g., *Nyctinomops*). In the latter group the jaws are relatively thin and the teeth are smaller.

Many species of bats exhibit size dimorphism; frequently the female is larger than the male. Ralls (1976) advanced the idea that the larger size of females in most bat species could reflect selection through competition among females for scarce resources, but selection probably favors larger females because they can produce larger young that presumably have a higher probability of surviving. Myers

(1978) concluded that large size in females may relate to problems of flight during pregnancy, since a larger size and longer forearm permit a larger wing surface to provide extra aerodynamic lift.

Bat Physiology

The basal metabolic rates of all mammals tend to decrease with increasing body size. Nevertheless, there are two broad subdivisions of bat species when basal metabolic rate is regressed against body weight. In one group, which includes the Vespertilionidae and the Molossidae, basal metabolic rate tends to be somewhat low. The second grouping of bats, with higher basal metabolic rates, includes most of the phyllostomids and pteropodids. In spite of their lower resting metabolic rate, the vespertilionids actively colonize and occupy habitats with lower mean annual temperatures than the phyllostomids require. The northern species of vespertilionids also hibernate during winter. The lower metabolic rate of the small vespertilionids may be viewed as an energy-conserving mechanism and an overall adjustment on the part of most to lower their body temperatures when roosting during the day. The larger phyllostomid bats tend to maintain a more constant body temperature over a range of environmental temperatures (McNab 1969). On the other hand, small phyllostomids are unable to maintain a constant body temperature at extremely low temperatures (McNab 1982).

Tropical environments do exhibit seasonality in abundance of both fruit and insects. Thus frugivorous and insectivorous bats will tend to adjust their reproductive patterns to times of the year when their food supply is abundant (Fleming, Hooper and Wilson 1972; Wilson 1979; Bradbury and Vehrencamp 1977b). Bats adapted to the temperate zone hibernate during months of low food supply, an option not taken by the tropical bats.

Activity Patterns and Nocturnal Movements

Although bats are commonly known to be active at night, nonchiropterologists seldom realize that the time of active flight may vary significantly between species (Brown 1968). *Carollia perspicillata* shows an abrupt onset of activity in the early evening, declines to low levels of activity about midnight, and has a second burst of activity just before dawn. *Sturnira lilium*, on the other hand, has a low activity level early in the evening and a higher level later on, before dawn. A similar pattern is shown by *Artibeus lituratus* (Erkert 1982). Different activity rhythms among bats in a community may in fact have resulted from an effort to reduce interspecific scramble competition for food (e.g., *Arbiteus jamaicensis* and *A. lituratus;* Bonaccorso 1979). On the other hand, differences in activity rhythms might relate to predator avoidance. In small insectivorous species, however, the bimodal activity rhythm suggests two bouts of feeding, with the first meal digested during

Table 5.2   Comparison of Morphological Features of New World Bat Families and Subfamilies

	Nose Leaf Present	Chin or Cheek Flaps Present	Stripes on Dorsum	Stripes on Face	Tail Vestigial or Absent	Uropatagium Reduced	Uropatagium Virtually Absent	Sacs in Antebrachial Membrane	Glandular Sac at Base of Tail
Emballonuridae	−	−	+/−	−	−	−	−	+/−	+/−
Emballonurinae	−	−	+/−	−	−	−	−	+/−	−
Diclidurinae	−	−	−	−	−	−	−	−	+
Noctilionidae	−	−	−	−	−	−	−	−	−
Mormoopidae	−	+	−	−	−	−	−	−	−
Phyllostomidae	+/−	−	+/−	+/−	+/−	+/−	+/−	−	−
Phyllostominae	+	−	−	−	−	+/−	−	−	−
Glossophaginae	+	−	−	−	+	+	+	−	−
Carolliinae	+	−	−	−	−	+	+	−	−
Sturnirinae	+	−	−	−	+	+	+	−	−
Stenoderminae	+/−	+/−	+/−	+/−	+/−	+/−	+/−	−	−
Desmodontinae	+[a]	−	−	−	+	+	+	−	−
Natalidae	−	−	−	−	−	−	−	−	−
Furipteridae	−	−	−	−	−	−	−	−	−
Thyropteridae	−	−	−	−	−	−	−	−	−
Vespertilionidae	−	−	−	−	−	−	−	−	−
Molossidae	−	−	−	−	−	−	−	−	−

	Tufts of Fur on Forearms	Tail Perforates Interfemoral Membrane	Dorsal Surface of Uropatagium Densely Haired	Dorsal Surface of Uropatagium Sparsely Haired	Back Covered with Naked Skin	Thumb Reduced or Absent	Tail Projects beyond Uropatagium	Tail > HB	Suckers on Forefeet and Hind Feet
Emballonuridae	+/−	+/−	−	+/−	−	+/−	−	−	−
Emballonurinae	+/−	+/−	−	+/−	−	−	−	−	−
Diclidurinae	−	+	−	−	−	+/−	−	−	−
Noctilionidae	−	+	−	−	−	−	−	−	−
Mormoopidae	−	+/−	−	+/−	+/−	−	−	−	−
Phyllostomidae	−	+/−	−	+/−	−	−	−	−	−
Phyllostominae	−	+/−	−	+/−	−	−	−	−	−
Glossophaginae	−	+/−	−	+/−	−	−	−	−	−
Carolliinae	−	+	−	+/−	−	−	−	−	−
Sturnirinae	−	−	−	+/−	−	−	−	−	−
Stenoderminae	−	−	−	+/−	−	−	−	−	−
Desmodontinae	−	−	−	+/−	−	−	−	−	−
Natalidae	−	−	−	−	−	−	−	+	−
Furipteridae	−	−	−	−	−	+	−	−	−
Thyropteridae	−	−	−	−	−	−	−	−	+
Vespertilionidae	−	−	+/−	+/−	−	−	−	−	−
Molossidae	−	−	−	−	−	−	+	−	−

[a] Highly modified.

a resting interval and the second meal digested during the day. This would imply an attempt to maintain an energy balance throughout a twenty-four-hour cycle.

During their nocturnal forays microchiropteran bats employ echolocation. The insectivorous and carnivorous bats have the most highly refined echolocation system both in call form and in the neurological adaptations needed to perceive the echoes from their intended prey. The literature on echolocation in bats is vast (see Novick and Dale 1971; Gould 1977; Griffin 1958; Griffin, Webster, and Michael 1960; Fenton 1982). Foraging in frugivorous species seems to place less stringent requirements on the precision of the echolocation system.

Frugivorous bats typically expend energy in searching for appropriate fruit trees, but once they find them they establish a night roost some distance from the feeding tree, which they exploit intermittently throughout the night. The night roost is at a distance from a fruiting tree to reduce competitive interactions with other bats and probably also to lessen exposure to predators that may be attracted to the tree. The demonstration by Morrison (1978a) that *Artibeus jamaicensis* is less active under a full moon strongly suggests that predation is an ever-present risk during bat foraging journeys. Fleming (1982) reviews the literature on the foraging strategies of frugivorous bats.

Roosting Sites

Bats rest in both diurnal and nocturnal roosts, opportunistically using caves, cracks, and tree cavities. More than one species may use the same cavity. Bats that roost in sparse foliage or on branches or tree trunks frequently are solitary, but those that roost in dense foliage sometimes are found in larger groups, especially if they roost high in the canopy. The habit of roosting on smooth surfaces like the insides of large leaves has often enhanced selection for foot disks such as those shown by *Thyroptera discifera*.

Many phyllostomids are "tent-making" bats. By biting leaves in particular patterns, they cause them to fold over and thereby provide a shelter. Tent-making bats include *Ectophylla alba, Artibeus cinereus,* and *Uroderma bilobatum.*

Night roosts are used between foraging bouts and are usually some distance from the feeding site. Such night roosts are carefully chosen to minimize predation risk, so thick-crowned trees are often preferred.

Bat Diets

Bats have specialized for different foraging strategies and different prey. Broadly speaking, they can be classified according to the scheme developed by Wilson (1973b). Foliage gleaners prey on insects that are feeding, resting, or moving on vegetation. Aerial insectivores catch flying insects. Frugivores feed extensively on fruit but also take insects at certain seasons (Ayala and D'Alessandro 1973). Nectarivorous bats have long tongues and usually long rostra. Although they feed on pollen and nectar, some of these species are important plant pollinators, and co-evolved system of bats and plants has been discerned (Howell 1974; Sazima 1976). Pollinating bats also feed on fruits and insects. Carnivorous bats eat vertebrates such as frogs, lizards, rodents, birds, and other bats. Sanguivorous bats feed on blood and are found only in the Neotropics (see Desmodontinae below). Data concerning the feeding habits of the Phyllostomidae are reviewed by Gardner (1977a).

Reproduction and Life Span

Mating systems are highly variable within the Chiroptera. Species such as *Saccopteryx bilineata, Artibeus jamaicensis,* and *Phyllostomus hastatus* tend to form harems, with males actively defending temporary groups of females against incursions by other males. Although vespertilionid bats living in temperate and subtropical latitudes mate in autumn and delay fertilization, this is not true of vespertilionids in the tropics. On the other hand, delayed embryonic development does occur in some species, and the delay length is under the control of the female. Flemming (1971) showed that in *Artibeus jamaicensis* mating occurs at the end of the dry season in Panama and births occur four months later, coinciding with a peak in the abundance of fruits. The female has a postpartum estrus, but the blastocysts implant and then enter a diapause for three months before developing normally for four months, so that the young are born at the end of the dry season just before a peak in the availability of large fruits. The delay in development allows births to occur when the young have optimum conditions for foraging.

In seasonally arid habitats the timing of reproduction is influenced by rainfall. With the onset of rains both insects and fruit become abundant. In less seasonal habitats bats may reproduce in any month, but subtle environmental factors may favor reproduction during a particular time, thus causing a marked peak. Other effects can also influence the timing of reproduction (Arata and Vaughan 1970; Wilson 1973a, 1979).

Insectivorous bats such as vespertilionids and molossids usually have a briefer gestation and a shorter lactation period than the phyllostomid frugivores (Kleiman and Davis 1979). Most species of bats produce a single young, but there are exceptions: *Eptesicus fuscus* and *Lasiurus cinereus* commonly produce twins, and *Lasiurus borealis* may have triplets. Most vespertilionids attain sexual maturity in less than a year; some, such as *Eptesicus fuscus*, mature as early as four months. The larger phyllostomids probably do not reach sexual maturity until they are nearly a year old. Vespertilionid bats, in spite of their small size, can attain a considerable age. *Eptesicus fuscus* has lived at least nineteen years, the vampire bat eighteen years, and *Artibeus jamaicensis* seven years (Tuttle and Stevenson 1982).

Relative Brain Size

Bat brains vary considerably in size. The relatively large brains of the New World phyllostomid bats are convergent with the patterns shown by the Pteropodidae. Foraging strategies that involve finding rich food resources isolated in small pockets are correlated with a large brain weight relative to body mass. The highly specialized aerial insectivores have the lowest brain/body weight ratios (Eisenberg and Wilson 1978).

Concluding Remarks

The natural history of New World bats has been described in numerous publications. Useful summaries are contained in Goodwin and Greenhall (1961), Husson (1962, 1978), and Baker, Knox Jones, and Carter (1976, 1977, 1979). Kunz (1982) contains much useful information on the ecology of the Chiroptera. Koopman (1982) has summarized distribution patterns for the South American species; Linares (1986) has done so for Venezuela; and Baker, et al. (1982) have summarized the data on karyotypes.

The range maps have been crosshatched in conformity with Koopman (1982). In some cases I have indicated recent changes deriving from new collections. The measurements attributed to the USNMNH for the most part are taken with permission from unpublished data assembled by C. O. Handley, Jr., and Katherine Ralls.

FAMILY EMBALLONURIDAE
Sac-winged Bats, Sheath-tailed Bats

Diagnosis

A distinguishing but not unique feature of this family of bats is that the tail is enclosed in a sheath of the interfemoral membrane, and the tip perforates the membrane's upper surface to lie free upon it; hence the name sheath-tailed bats. The Noctilionidae also exhibit this feature. The dental formula is highly variable in the family, from I 2/3, C 1/1, P 2/2, M 3/3 in *Emballonura* to I 1/2, C 1/1, P 2/2, M 3/3 in *Taphozous*. The premaxillary bones do not meet anteriorly. Many species have an opening on the antebrachial membrane over a gland field, and scent

is produced during some of their wing-waving displays. Other species (the diclidurines) have a gland field surrounding the tail in the uropatagium. During resting, the third finger exhibits a reflexed proximal phalanx (see fig. 5.3). Although the trochiter is well developed, it is smaller than the trochin and does not articulate with the scapula (Sanborn 1937). A key to the species is given in table 5.3.

Distribution

This family is widely distributed in both the Old World and New World tropics and includes some twelve genera and fifty species. In the Neotropics it extends from Sonora, Mexico, south through the Isthmus to southern Brazil.

Natural History

These bats are entirely nocturnal, and most are specialized for feeding on insects. Strategies of in-

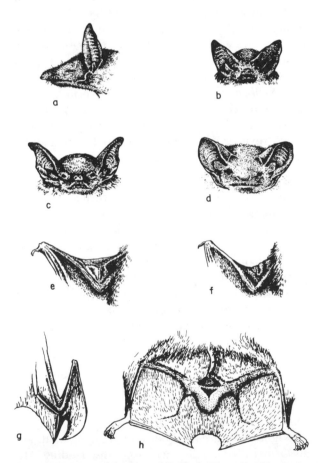

Figure 5.3. External features of the Emballonuridae: (*a*) Head of *Rhynchonycteris naso*. (*b*) Face of *Peropteryx macrotis*. (*c*) Face of *Saccopteryx bilineata*. (*d*) Face of *Diclidurus albus*. (*e*) Position of glandular sac in the antebrachial membrane of *Saccopteryx bilineata*. (*f*) Glandular sac near the leading edge of the antebrachial membrane of *Peropteryx macrotis*. (*g*) Dorsal view of the third finger of *Saccopteryx* demonstrating reflexed proximal phalanx. (*h*) Dorsal view of the interfemoral membrane of *Diclidurus albus*; note the glandular sac at the tip of the tail. (Redrawn with permission from Goodwin and Greenhall 1961.)

Table 5.3   Key to the Common Adult Emballonuridae of Northern South America

1   Glandular sacs not present in either the interfemoral membrane or the antebrachial membrane 2
1' Sacs present in either the interfemoral membrane or the antebrachial membrane 4
2   Tufts of whitish hairs on the forearms; muzzle pointed; two faint stripes on the dorsum . . . . *Rhychonycteris naso*
2' Not as above 3
3   Dorsal pelage tawny . . . . *Centronycteris maximiliani*
3' Dorsal pelage smoky gray to black . . . . *Cyttarops alecto*
4   Glandular sac present in the interfemoral membrane (Diclidurinae) 5
4' Glandular sac present in the antebrachial membrane of males 8
5   Head and body length exceeds 75 mm; forearm exceeds 65 mm; color pale gray . . . . *Diclidurus ingens*
5' Head and body length less than 75 mm 6
6   Dorsal pelage light brown; head and body length less than 60 mm . . . . *Diclidurus isabellus*
6' Not as above; head and body length greater than 60 mm 7
7   Dorsal pelage white; wings yellow; head and body length averages greater than 69 mm . . . . *Diclidurus albus*
7' Dorsal pelage pale; head and body length greater than 65, less than 69 mm . . . . *Diclidurus scutatus*
8   Three upper incisors present; sac in the antebrachial membrane long, extending from the bend of the elbow to the anterior edge of
    the antebrachial membrane . . . . *Cormura brevirostris*
8' Fewer than three upper incisors 9
9   Distal portion of wings white; ears connected at the base by a membrane . . . . *Peropteryx leucoptera*
9' Not as above 10
10 Two longitudinal white stripes present on the dorsum (*Saccopteryx*) 11
10' No dorsal stripes (*Peropteryx*) 13
11 Head and body length exceeds 45 mm . . . . *Saccopteryx bilineata*
11' Head and body length less than 45 mm 12
12 Head and body length less than 45 mm; hind foot less than 8 mm . . . . *Saccopteryx canescens*
12' Head and body less than 45 mm; hind foot exceeds 8 mm . . . . *Saccopteryx leptura*
13 Head and body length exceeds 55 mm . . . . *Peropteryx kappleri*
13' Head and body length less than 55 mm . . . . *Peropteryx macrotis*

sect capture are variable, but many species seem to be adapted for aerial pursuit (see Bradbury and Emmons 1974; Bradbury and Vehrencamp 1976a,b, 1977a,b). The usual number of young is one.

SUBFAMILY EMBALLONURINAE
Genus *Rhynchonycteris* Peters, 1867
Rhynchonycteris naso (Wied-Neuwied, 1820)
Sharp-nosed Bat, Murciélago de Trompa

Measurements

	Sex	Mean	S.D.	N
TL	M	55.83	2.79	24
	F	58.25	2.01	24
HB	M	39.92	4.19	24
	F	40.75	2.61	24
T	M	15.92	2.50	24
	F	17.50	1.77	24
HF	M	7.29	0.55	24
	F	7.42	0.78	24
E	M	13.75	1.07	24
	F	14.08	1.10	24
FA	M	36.83	0.93	32
	F	38.56	1.05	31
Wta	M	3.83	0.46	19
	F	3.78	0.40	11

Location: TFA, Venezuela (USNMNH)

Description
The dental formula is I 1/3, C 1/1, P 2/2, M 3/3. Head and body length ranges from 37 to 43 mm. In Venezuela, total length averaged 56.5 mm for males and 59.2 mm for females; tails made up 15.4 mm and 16.8 mm of the total length. Weight ranged from 3.8 to 3.9 g (see table 5.14). The muzzle is elongated and pointed (fig. 5.3). There are no conspicuous wing sacs, so characteristic of other bat genera in this subfamily. The dorsum is grizzled yellow, and the venter is paler. The dorsal surfaces of the forearms bear tufts of whitish hairs; this character is diagnostic. Chromosome number: $2n = 22$; FN = 36.

Range and Habitat
This species ranges from southern Veracruz, Mexico, to southeastern Brazil and is widely distributed at low elevations (map 5.1). It is almost always associated with moist areas near multistratal evergreen forests, generally at elevations below 500 m.

Natural History
These bats tend to roost in small, single-species colonies of about ten to twenty-four, on tree trunks, in tree cavities, or in rock caves. When roosting they are often aligned in vertical rows with individuals about 100 mm apart. Several males occur in a roosting group, and there appears to be no harem formation or defense. These bats are aerial insectivores (Husson 1978; Goodwin and Greenhall 1961), and

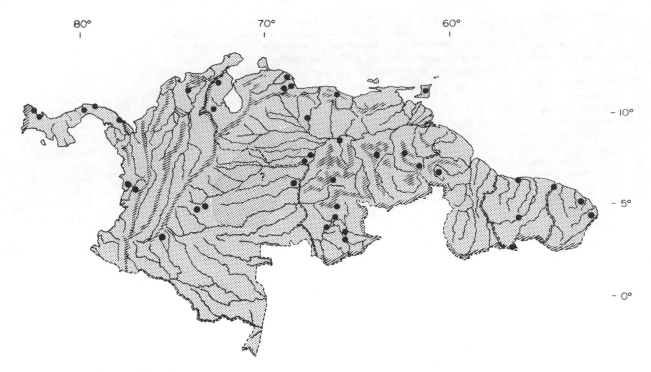

Map 5.1. Distribution of *Rhynchonycteris naso*.

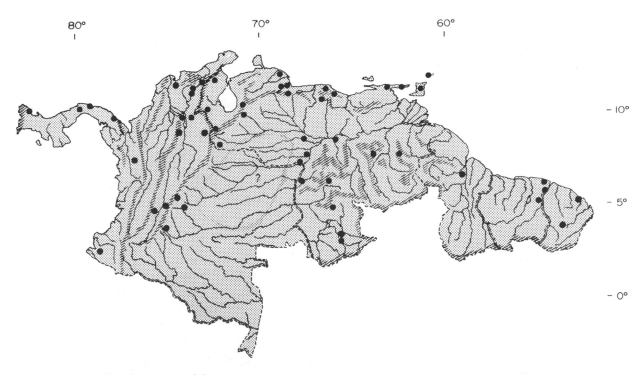

Map 5.2. Distribution of *Saccopteryx bilineata*.

they tend to feed over water, flying only a short distance above the surface.

Genus *Saccopteryx* Illiger, 1811
White-lined Bat, Murciélago Rayado

Description
The dental formula is I 1/3, C 1/1, P 2/2, M 3/3. Members of this genus possess a glandular sac in the antebrachial membrane, quite close to the forearm, which is more prominent in the male (see fig. 5.3). A pair of white dorsal stripes courses the length of the body on either side of the median axis. The skull is highly domed, and the postorbital process is well developed. The premaxillary bones do not meet anteriorly (see fig. 5.4).

Distribution
The genus is broadly distributed from southern Mexico to southeastern Brazil and west to southeastern Peru.

Natural History
The sac-winged bats tend to have specific roosts that may include tree trunks or cave walls. They are aerial insectivores. Mating and foraging systems have been extensively studied by Bradbury and Vehrencamp (1976a,b, 1977a,b).

Saccopteryx bilineata (Temminck, 1838)
Two-lined Bat, Murciélago de Listas

Measurements

	Sex	Mean	S.D.	N
TL	M	73.88	3.00	24
	F	76.00	3.68	42
HB	M	52.29	2.96	24
	F	54.36	3.74	42
T	M	21.58	2.21	24
	F	21.58	2.67	43
HF	M	12.25	0.85	24
	F	12.11	0.79	43
E	M	17.96	1.85	24
	F	17.51	1.71	43
FA	M	47.13	1.03	27
	F	49.16	1.37	42
Wta	M	8.91	0.54	22
	F	9.32	0.99	18

Location: TFA, Venezuela (USNMNH)

Description
Head and body length averages 53.4 mm for males and 55.3 mm for females; tail 19.6 mm for males, 19.6 mm for females; ear 16.6 mm for males, 16.2 mm for females (see table 5.14). Weights range from 8.5 to 9.3 g. The dorsal fur is black, with distinctive white lines running longitudinally on

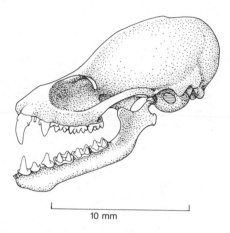

Figure 5.4. Skull of *Saccopteryx bilineata*.

each side of the midline; the venter is paler, almost gray (see plate 5).
Chromosome number: $2n = 26$; FN = 36.

Range and Habitat
This species ranges from Colima, Mexico, south to southeastern Brazil (map 5.2). The bats usually forage near streams and in moist areas; they prefer multistratal evergreen forests but may forage in clearings and rarely range above 500 m in elevation.

Natural History
Roosting colonies average about a dozen individuals. These bats may roost with other species of bats in hollow trees or caves. Males defend harems; they have well-developed scent glands in their antebrachial membranes and emit scent while flapping their wings in ritualized combat (Bradbury and Emmons 1974).

Saccopteryx canescens Thomas, 1901

Measurements

	Sex	Mean	S.D.	N
TL	M	54.67	4.16	3
	F	62.50	1.29	4
HB	M	39.33	4.73	3
	F	44.75	1.89	4
T	M	15.33	0.58	3
	F	17.75	2.06	4
HF	M	8.00	0.00	3
	F	7.00	0.00	4
E	M	13.33	2.52	3
	F	13.75	0.96	4
FA	M	37.55		2
	F	39.40	0.63	4
Wta	M	3.55		2
	F	3.90		1

Location: Northern Venezuela (USNMNH)

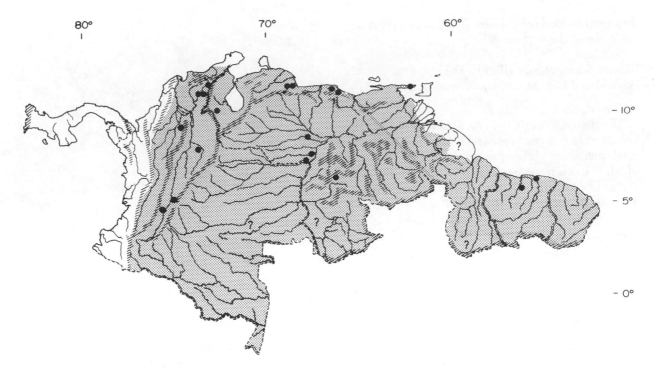

Map 5.3. Distribution of *Saccopteryx canescens*.

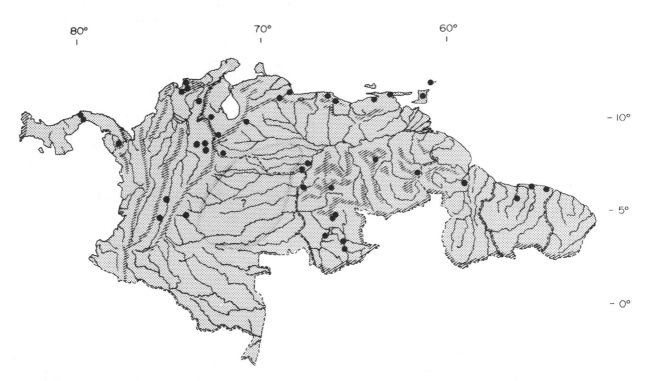

Map 5.4. Distribution of *Saccopteryx leptura*.

Description

Head and body length averages 38.6 mm for males, 44 mm for females; weights average 3.2 g for males and 3.5 g for females (see table 5.14). This species is much smaller than *S. bilineata* but neraly the same size as *S. leptura*. The hind foot of *S. canescens* is considerably shorter than that of *S. leptura*, being 7.2 mm for both sexes, whereas the tail is slightly longer in *S. canescens* than in *S. leptura*. Tail length averages 16.1 mm for males and 17.6 mm for females; forearms are 35.9 mm for males and 38.6 mm for females. The fur extends onto the wing membranes both above and below for about a third of the distance from the body to the elbow. The dorsal surface of the interfemoral membrane is furred to the distal part of the tail. The basic color of the dorsal pelage varies from gray brown to brown and has a grizzled appearance. The two white longitudinal stripes may be indistinct in some specimens, and the striped pattern is not as bold as in *S. bilineata*. The venter is only slightly lighter than the dorsum. Chromosome number $2n = 24$; FN $= 38$.

Range and Habitat

This species is distributed from eastern Colombia across northern South America to northern Brazil and Peru (map 5.3). It frequents multistratal evergreen forests as well as open fields and occurs at low elevations.

Natural History

This small bat has not been the subject of any detailed investigations. Often found near streams and moist areas and in clearings, it co-occurs over much of its range with *S. leptura*, a better-studied species, but the exact way they partition resources is poorly understood. In Venezuela *S. canescens* commonly occurs below 150 m elevation; *S. leptura* can be taken at.up to 500 m (Handley 1976).

Saccopteryx leptura (Schreber, 1774)
Murciélago de Listas y de Cola Corta

Measurements

	Sex	Mean	S.D.	*N*
TL	M	57.40	3.21	5
	F	59.91	4.85	11
HB	M	42.20	3.83	5
	F	43.82	5.56	11
T	M	15.20	3.19	5
	F	16.09	2.26	11
HF	M	8.20	0.84	5
	F	8.45	1.04	11
E	M	13.60	1.67	5
	F	14.82	1.72	11
FA	M	37.57	0.99	6
	F	39.43	1.77	12
Wta	M	4.72	0.67	6
	F	5.74	1.51	5

Location: TFA, Venezuela (USNMNH)

Description

Head and body length averages 43.8 mm for males and 45 mm for females. Weight averages 4.4 g for males and 5.7 g for females (see table 5.14). In general, *S. leptura* is heavier and slightly larger than *S. canescens*, but both species are smaller than *S. bilineata*. The scent sac opening on the dorsal side of the antebrachial membrane is quite large in the male. The upperparts of *S. leptura* are brown. The two longitudinal stripes are usually distinct, but when the pelage is worn they may not be sharply set off.

Chromosome number: $2n = 28$; FN $= 38$.

Range and Habitat

This species extends from Chiapas, Mexico, south through the Isthmus to eastern Brazil and adjacent Peru (map 5.4). It occurs in moist habitats and is strongly associated with multistratal evergreen forests at elevations below 500 m.

Natural History

Roosting groups range from one to nine bats. There is some shifting in the composition of the social group; males tend to defend individual females during breeding, when they exist as monogamous pairs (Bradbury and Vehrencamp 1977a).

Genus *Cormura* Peters, 1867
Cormura brevirostris (Wagner, 1843)
Wagner's Sac-winged Bat

Measurements

	Sex	Mean	S.D.	*N*
TL	M	66.33	2.52	3
	F	69.80	1.79	5
HB	M	51.00	1.00	3
	F	54.20	2.28	5
T	M	15.33	2.08	3
	F	15.60	1.95	5
HF	M	9.00	1.00	3
	F	8.80	0.84	5
E	M	15.33	1.53	3
	F	16.20	1.10	5
FA	M	45.20	1.04	4
	F	46.48	1.60	6
Wta	M	7.93	1.21	3
	F	10.15		2

Location: TFA, Venezuela (USNMNH)

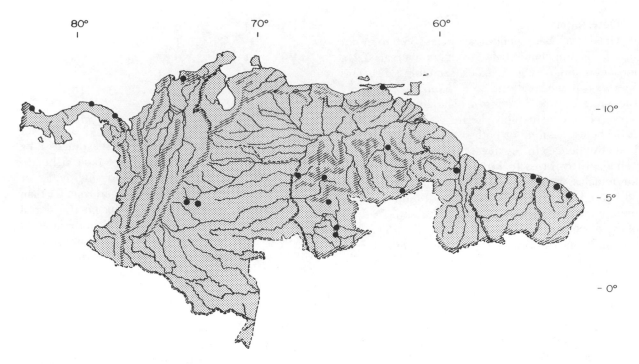

Map 5.5. Distribution of *Cormura brevirostris*.

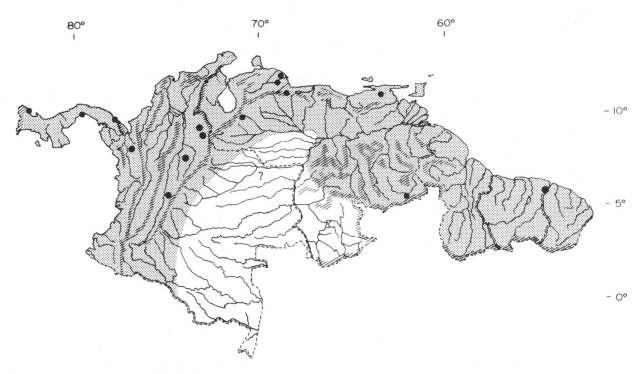

Map 5.6. Distribution of *Peropteryx kappleri*.

Description

The dental formula for the genus is I 3/3, C 1/1, P 2/2, M 3/3, and this alone distinguishes it from other emballonurid bats encountered within its range. Head and body length averages 54 mm for males and 54.2 mm for females; mean weights are 8.6 g for males and 10.2 g for females (see table 5.14). The sac in the antebrachial membrane opens via a rather long aperture from near the anterior edge of the membrane to the elbow. The wing membrane extends to the ankle, and this character alone separates the genus from *Peropteryx*. The upperparts are buffy brown to black with no longitudinal dorsal stripes; the venter does not contrast with the dorsum. Chromosome number: $2n = 22$; $FN = 40$.

Range and Habitat

The genus *Cormura* is widely distributed from Nicaragua south through the Isthmus to Amazonian Brazil (map 5.5). These rare bats are associated with streams and moist areas, preferably multistratal tropical evergreen forests. Most specimens in Venezuela were taken at below 500 m elevation (Handley 1976).

Natural History

This rare bat has not been the subject of a detailed field study. Specimens have been collected singly, roosting in hollow fallen trees or under the projecting ends of fallen trees (Handley 1976).

Genus *Peropteryx* Peters, 1867

Peters's Sac-winged Bat

Description

The dental formula is I 1/3, C 1/1, P 2/2, M 3/3. Head and body length for the genus ranges from 45 to 55 mm, and tail length is 12 to 18 mm. Adults weigh from 4 to 10 g. The opening to the gland in the wing is near the anterior edge of the antebrachial membrane, thus distinguishing the genus from *Saccopteryx* (fig. 5.3).

Distribution

The genus is distributed from southern Mexico to southeastern Brazil.

Natural History

These sac-winged bats may roost in shallow caves or rock crevices as well as in the hollows of dead trees. They are aerial insectivores.

Peropteryx kappleri Peters, 1867

Measurements

	Sex	Mean	S.D.	N
TL	M	71.93	2.20	14
	F	74.74	2.34	23
HB	M	58.21	2.67	14
	F	61.09	2.41	23
T	M	13.71	1.20	14
	F	13.65	1.07	23
HF	M	10.86	0.86	14
	F	11.39	0.84	23
E	M	18.07	0.92	14
	F	18.00	1.13	23
FA	M	48.30	1.12	16
	F	50.34	1.16	25
Wta	M	7.73	1.21	13
	F	8.50	1.16	21

Location: Northern Venezuela (USNMNH)

Description

This is the largest species of the genus. Head and body length averages 58.2 mm for males and 60.9 mm for females, with forearm length 48.3 mm for males and 50.3 mm for females (see table 5.14). Weight averages 7.7 g for males and 8.5 g for females. There are two color phases, ranging from light brown to dark brown, but in both cases the underparts are slightly lighter. There is no white line on the dorsum.

Range and Habitat

This species ranges from southern Veracruz across northern South America to eastern Brazil and Peru (map 5.6). It tolerates dry situations and though it prefers evergreen forest, does forage considerably in open fields. It may range to moderate elevations of 850 m in Venezuela (Handley 1976; Husson 1978).

Natural History

Roosting colonies are small, ranging from one to seven, and roosts are in caves and boulder crevices. The mating system is based on a monogamous pair, and the male defends his female against intruding males (Bradbury and Vehrencamp 1977a).

Peropteryx macrotis (Wagner, 1843)

Measurements

	Sex	Mean	S.D.	N
TL	M	61.41	1.94	22
	F	66.13	2.73	32
HB	M	48.00	2.71	22
	F	51.19	3.68	32
T	M	13.41	1.40	22
	F	14.94	1.81	32
HF	M	8.95	0.49	22
	F	9.13	0.55	32
E	M	15.36	0.73	22
	F	16.31	0.82	32
FA	M	42.96	0.90	26
	F	45.87	1.22	36
Wta	M	5.60	0.39	26
	F	6.81	0.76	30

Location: Bolívar, Venezuela (USNMNH)

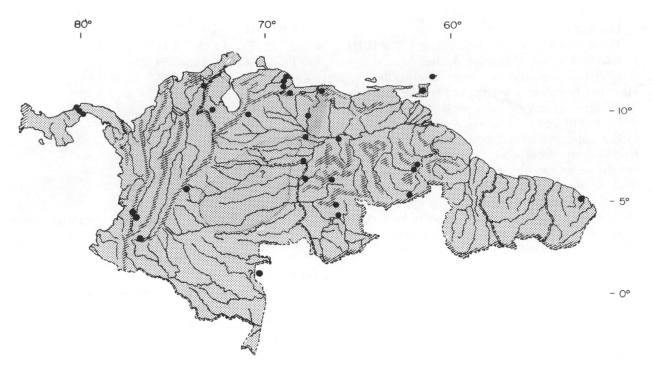

Map 5.7. Distribution of *Peropteryx macrotis*.

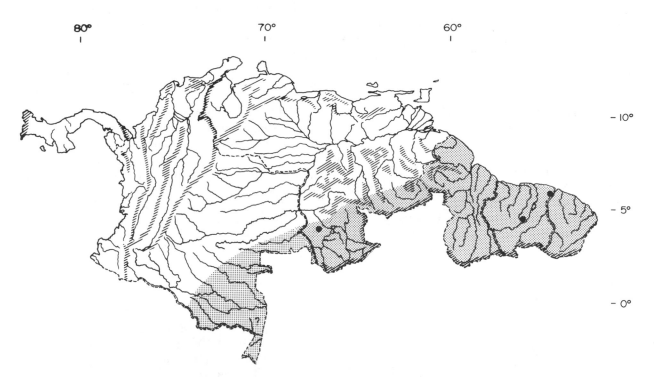

Map 5.8. Distribution of *Peropteryx leucoptera*.

Description

In this account *Peropteryx trinitatis* is considered a subspecies of *P. macrotis*. In general the Trinidadian subspecies is slightly smaller (see table 5.14). Males range from 56.5 to 60.5 mm in total length and females from 64.4 to 62.2 mm. Head and body length ranges from 46.4 to 47.2 mm in males and 49.37 to 51.09 mm in females (see table 5.14). The dorsal fur varies from buffy brown to reddish brown to very dark brown, and the venter is only slightly paler.

Chromosome number: $2n = 26$; FN = 48.

Range and Habitat

This species ranges from Oaxaca, Mexico, southward to Peru and eastward to Paraguay and southeastern Brazil (map 5.7). It prefers moist sites and multistratal evergreen forest interspersed with savannas and generally does not occur above 800 m elevation in Venezuela (Handley 1976).

Natural History

This bat roosts in limestone and coral caves, sometimes with other bat species. Roosting groups may be as large as six individuals. It is entirely insectivorous, apparently feeding on the wing (Goodwin and Greenhall 1961).

Peropteryx leucoptera Peters, 1867

Description

This species was formerly assigned to the genus *Peronymus*. It is approximately the size of *Peropteryx kappleri*, with forearm length ranging from 40 to 44 mm in males and 42 to 47 mm in females. Means for seven specimens were TL 59; HB 56; T 13; E 16; FA 43 (Suriname; CMNH and Husson 1962). The upperparts are reddish brown to dark brown with the underparts slightly paler, and the distal portions of the wings are white. In addition to the wing coloration, the best distinguishing feature for this species is that the ears tend to be joined at the bases by a very low membrane. In *Peropteryx kappleri* the ears are entirely separate (see fig. 5.3).

Chromosome number: $2n = 48$; FN = 62.

Range and Habitat

This species ranges from Peru and eastern Brazil north to the Guyanas (map 5.8). It is found at low elevations. Ochoa-G. (1984b) recorded two specimens from TFA, Venezuela.

Natural History

Very little is known concerning the natural history of this species; it is assumed to be an aerial insectivore like other members of this family. Roosts are in hollow trees or in hollow rotten logs on the ground.

Genus *Centronycteris* Gray, 1838
Centronycteris maximiliani (Fischer, 1829)
Shaggy-haired Bat

Description

The dental formula is I 1/3, C 1/1, P 2/2, M 3/3. Head and body length averages 52 mm (range 50–62mm), tail length 18 to 23 mm, and forearm length 45 mm. Measurements for one specimen from Suriname were TL 71; HB 47; T 24; HF 7; E 18; Wt 15 g (CMNH). The long, lax dorsal pelage is tawny, with no dorsal stripes, and the venter does not contrast with the dorsum. This small bat does not have a sac in the antebrachial membrane.

Chromosome number: $2n = 28$; FN = 48.

Range and Habitat

This species occurs from southern Veracruz, southwest of the central cordillera of the Andes to Ecuador and thence across Brazil, reappearing in Venezuela and the Guyanas (map 5.9). It prefers lower elevations.

Natural History

This bat tends to roost in hollow trees. It appears to be a slow flier and has a rather regular pattern of foraging in its home range, a feature shared with other emballonurids.

Genus *Balantiopteryx* Peters, 1867
Balantiopteryx infusca (Thomas, 1897)

Description

The dental formula is I 1/3, C 1/1, P 2/2, M 3/3. Total length averages 69 mm: HB 42; T 27; E 11, FA 41 (Albuja 1982). The wing sac is centrally placed in the antebrachial membrane. The dorsum is dark brown to dark gray with the venter paler, and dorsal stripes are absent.

Range and Habitat

The genus has been recorded from Central America, but on the southern continent it is known from the lowlands of western Ecuador and may range to southwestern Colombia.

Natural History

The species *B. infusca* is poorly known. *B. io* roosts in caves, and colonies in Costa Rica can consist of several hundred individuals.

SUBFAMILY DICLIDURINAE
Genus *Cyttarops* Thomas, 1913
Cyttarops alecto Thomas, 1913

Description

The dental formula is I 1/3, C 1/1, P 2/2, M 3/3. Head and body length ranges from 50 to 55 mm, the

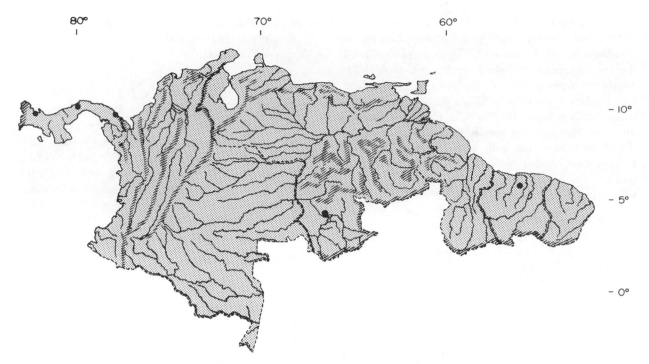

Map 5.9. Distribution of *Centronycteris maximiliani*.

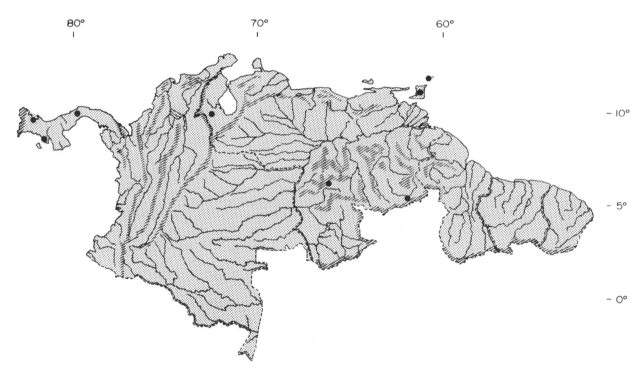

Map 5.10. Distribution of *Diclidurus albus*.

tail from 20 to 25 and forearm from 45.8 to 47.2 mm. Females are slightly larger than males. There are no antebrachial glands, and no gland is associated with the tail as in the genus *Diclidurus*. The ultimate phalanx of the thumb is free of the propatagium. The nostrils open through short, diverging tubes (Starrett 1972). The long, soft fur is almost black, but grayer on the lower back and shoulders, and the venter is gray to black.

Range and Habitat
This rare bat is known disjunctly from Costa Rica, Guyana, and near Belém, Brazil (map 5.12).

Natural History
This bat is found in lowland forests roosting under palm fronds in groups of one to four. It is insectivorous (Starrett 1972).
Chromosome number: $2n = 32$; FN = 60.

Genus *Diclidurus* Wied, 1820
Ghost Bat

Description
The dental formula is I 1/3, C 1/1, P 2/2, M 3/3. Head and body length ranges from 50 to 80 mm, tail length from 15 to 25 mm, and forearm length from 45 to 66 mm. Species of this genus are unique in having white or whitish gray pelage. The base of the hair tends to be slate color, and the wings are depigmented in the white species. These bats do not have a wing sac, but where the tail penetrates the interfemoral membrane there is a small pouch and associated glandular tissue (see fig. 5.3). The four species found in his region appear to be graded in size, with *Diclidurus ingens* the largest and *D. isabellus* the smallest.

Distribution
The genus is distributed from southern Mexico across northern South America to Brazil.

Natural History
The habits of these bats have been poorly recorded; Wilson (1973b) classified them as aerial insectivores. They are relatively rare in collections.

Diclidurus albus Wied, 1820
Ghost Bat, Murciélago Blanco

Measurements

	Sex	Mean	S.D.	N
TL	M	92.80	6.10	5
	F	96.50	3.73	6
HB	M	73.00	5.83	5
	F	75.67	3.72	5
T	M	19.80	0.45	5
	F	20.83	0.75	6
HF	M	12.60	0.55	5
	F	13.00	0.00	6
E	M	16.20	0.45	5
	F	17.00	0.63	6
FA	M	62.68	1.26	5
	F	65.37	1.14	6
Wta	M	15.48	2.14	5
	F	18.50		1

Location: Northwestern Venezuela (USNMNH)

Description
Head and body length averages 70.4 mm for males and 71.4 mm for females. Forearm length averages 62.5 mm for males and 64.4 mm for females. Only *D. ingens* is larger. The dorsal pelage is rather long, and the general appearance is of a white bat, though the bases of the hairs are gray. The flying membranes are yellowish white (see plate 5).

Range and Habitat
This rare bat is distributed broadly from Nayarit, Mexico, south through the Isthmus to eastern Brazil (map 5.10). No specimens have been taken in Venezuela above 850 m. It prefers moist areas and tolerates man-made clearings.

Natural History
In Trinidad this bat has been seen roosting between the leaves of tall coconut palms. It is not colonial and appears to roost individually. It is known to feed on insects (Goodwin and Greenhall 1961).

Diclidurus ingens Hernandez-Camacho, 1955

Description
This is the largest species of the genus. One female collected in Venezuela had a head and body length of 80.6 mm, with a forearm length of 71.4 mm (see table 5.14). Size alone should distinguish this rare species from other members of the genus.

Range and Habitat
The species ranges from southeastern Colombia and adjacent Venezuela to the Guyanas, extending south to northwestern Brazil (map 5.11). It is associated with multistratal evergreen forests and prefers moist areas. In Venezuela specimens have been taken at below 200 m elevation.

Diclidurus isabellus (Thomas, 1920)

Measurements

	Sex	Mean	S.D.	N
TL	M	80.38	2.80	21
	F	81.33	3.01	6
HB	M	59.67	2.56	21
	F	59.33	2.25	6

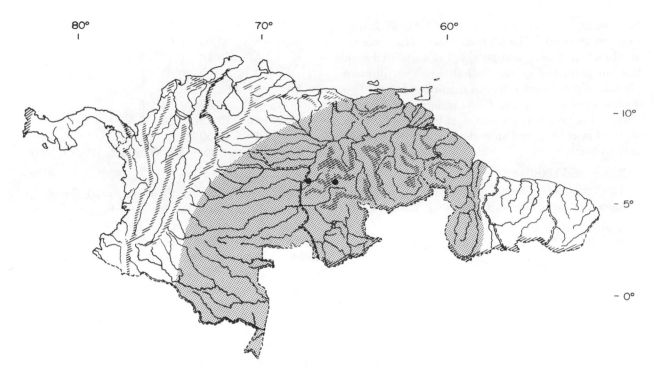

Map 5.11. Distribution of *Diclidurus ingens*.

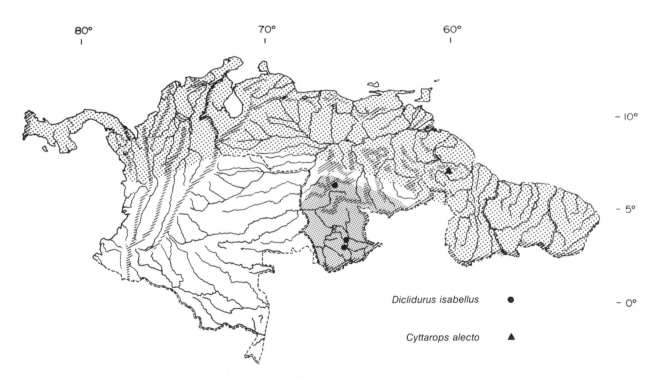

Diclidurus isabellus   ●

Cyttarops alecto   ▲

Map 5.12. Distribution of *Diclidurus isabellus* and *Cyttarops alecto* (based on Starret 1972; Koopman 1982).

T	M	20.71	2.10	21
	F	22.00	1.41	6
HF	M	12.19	0.93	21
	F	12.83	0.75	6
E	M	17.71	0.96	21
	F	17.67	1.37	6
FA	M	57.93	1.02	21
	F	59.10	1.73	5
Wta	M	13.36	0.95	21
	F	15.36	1.63	4

Location: TFA, Venezuela (USNMNH)

Description

This is the smallest species within the genus in the areas covered by this volume. Head and body length ranges from 59.6 to 59.3, and weights from 13.3 to 15.6 g. Males and females are nearly the same size. The dorsum is whitish but grades to pale brown posteriorly.

Range and Habitat

This species occurs from northwestern Brazil into southern Venezuela (map 5.12).

Natural History

This bat is strongly associated with wet habitats and multistratal evergreen forests. In Venezuela it was not collected above 155 m (Handley 1976).

Diclidurus scutatus Peters, 1869

Description

The species is larger than *D. isabellus* but smaller than *D. albus*. Head and body length averages 68 mm for males and 65 mm for females; weight averages 13.6 g for males and 12.9 g for females (see table 5.14). The male is clearly larger than the female. Where the tail emerges from the membrane there are two distinct glandular pouches, one immediately adjacent to the tip of the tail, the other separated by a distinct ridge. The second pouch extends almost to the end of the interfemoral membrane.

Range and Habitat

This species is distributed in the Amazonian portions of Peru, Brazil, adjacent Venezuela, and the Guyanas (map 5.13). It is strongly associated with moist habitats and broadly tolerant of human activity. In Venezuela it occurs at 99 to 850 m elevation (Handley 1976).

FAMILY NOCTILIONIDAE
Bulldog Bats, Murciélagos Pescadores

Diagnosis

This family includes only one genus and two species. The dental formula is I 2/1, C 1/1, P 1/2, M 3/3 (see fig. 5.5). There is no prominent nose leaf, and

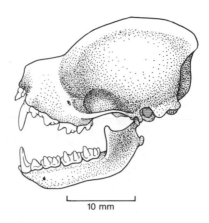

Figure 5.5. Skull of *Noctilio leporinus*.

the upper lip is medially divided (see fig. 5.6). The tail is considerably shorter than the interfemoral membrane, and its tip protrudes through the membrane on the dorsal side. There are three phalanges on the third finger, and the claws of the feet are greatly enlarged. This suite of characters serves to distinguish the family from other Neotropical bats (see fig. 5.6).

Distribution

This family is distributed from Sinaloa, Mexico, to northern Argentina.

Natural History

Noctilio leporinus is known as the fishing bat because it flies near the surface of water and seizes minnows in its claws. The other species, *N. albiventris*, is not specialized for fishing but feeds primarily on aquatic insects.

Noctilio albiventris Desmarest, 1818
Lesser Bulldog Bat

Measurements

	Sex	Mean	S.D.	N
TL	M	91.73	4.33	26
	F	88.07	2.84	46
HB	M	72.46	3.37	26
	F	68.61	3.22	46
T	M	19.27	3.13	26
	F	19.46	2.27	46
HF	M	18.65	1.44	26
	F	17.20	1.00	46
E	M	24.76	1.39	25
	F	24.46	1.66	46
FA	M	60.63	2.93	25
	F	58.95	6.03	50
Wta	M	31.70	5.17	21
	F	23.92	2.90	39

Location: TFA, Venezuela (USNMNH)

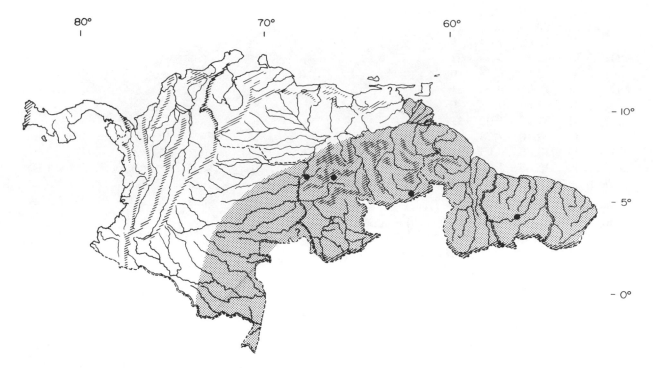

Map 5.13. Distribution of *Diclidurus scutatus*.

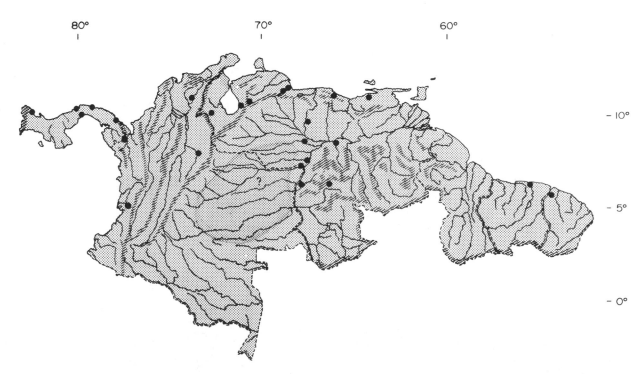

Map 5.14. Distribution of *Noctilio albiventris*.

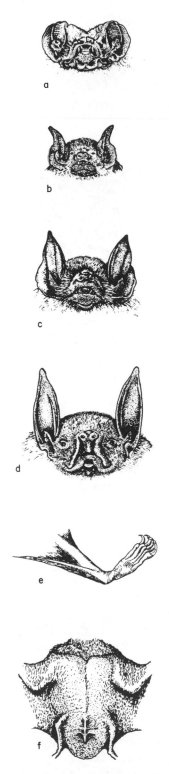

Figure 5.6. External features of the Noctilionidae and Mormoopidae: (*a*) Face of *Mormoops megalophylla;* note the complex chin leaf. (*b*) Face of *Pteronotus davyi.* (*c*) Face of *Pteronotus parnellii;* note the small chin leaf. (*d*) Face of *Noctilio leporinus;* note the cleft upper lip. (*e*) Foot of *Noctilio leporinus;* note the extremely long claws. (*f*) Back of *Pteronotus davyi;* note the attachment of the wing membranes along the dorsal midline. (Redrawn with permission from Goodwin and Greenhall 1961.)

Description

Noctilio albiventris, in its basic morphology, is similar to *N. leporinus* but much smaller. Males average 74.9 mm in head and body length, while females average 71.6 mm. Weight for males averages 28.7 g and that for females 23 g (see table 5.14). See figure 5.6 for anatomical details. The dorsum is brown, though some specimens may be light brown, or bright red on the head and shoulders; the venter is paler.

Chromosome number: $2n = 34$; FN = 58–62.

Range and Habitat

Noctilio albiventris ranges from Nicaragua south to northern Argentina (map 5.14). It prefers lower elevations; in Venezuela bats were not taken above 300 m. It is strongly associated with streamside habitats and tolerates a wide variety of vegetation types, using areas modified for croplands as well as evergreen forests.

Natural History

This bat often feeds over water and appears to primarily eat aquatic insects (Whitaker and Findley 1980). There is an early feeding phase at dusk, followed by a second activity peak after midnight. The bats roost in hollow trees. Timing of reproduction may be strongly influenced by insect abundance. Only a single young is born each year. The natural history has been reviewed in detail by Hood and Pitocchelli (1983).

Noctilio leporinus (Linnaeus, 1758)
Fishing Bat, Murciélago Pescadore

Measurements

	Sex	Mean	S.D.	N
TL	M	128.52	6.01	21
	F	125.50	7.28	16
HB	M	97.24	8.34	21
	F	96.38	7.89	16
T	M	31.29	4.36	21
	F	29.13	3.18	16
HF	M	33.48	1.60	21
	F	31.88	1.36	16
E	M	27.48	1.03	21
	F	27.50	1.51	16
FA	M	85.52	1.63	19
	F	85.39	1.98	16
Wta	M	66.40	3.34	9
	F	64.90	6.64	4

Location: Northern Venezuela (USNMNH)

Description

The dental formula is I 2/1, C 1/1, P 1/2, M 3/3 (see fig. 5.5). In this large bat, males' head and body

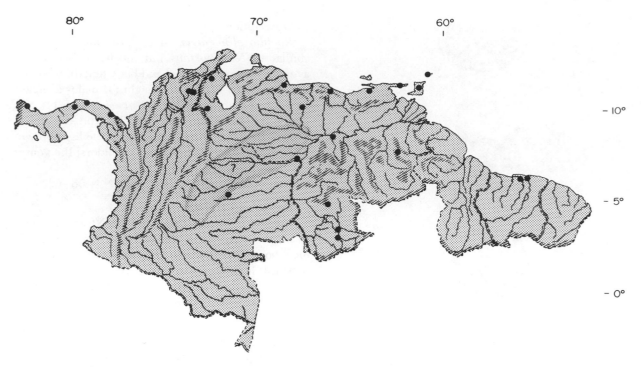

Map 5.15. Distribution of *Noctilio leporinus*.

length averages 96.9 mm, while that of females averages 91.3 mm. Weight for males averages 67 g, that for females 56 g (see table 5.14). As with *N. albiventris*, the male is larger, unlike members of many other Neotropical bat families. The dorsum is orange brown (plate 5).
Chromosome number: $2n = 34$; FN = 58–62.

Range and Habitat
The fishing bat extends from Sinaloa, Mexico, to northern Argentina (map 5.15). It is usually found below 200 m elevation and is strongly associated with streams or moist areas. In Venezuela it was netted in savannas, pastures, and marshes about half the time but was also associated with evergreen and deciduous tropical forest types (Handley 1976).

Natural History
This bat will roost in colonies of up to seventy-five, frequently in a hollow tree. Foraging tends to be individualistic, but occasionally two bats may be seen flying together over water. While fishing they fly very close to the surface and can detect small fish by echolocation. As fish swim close to the surface they distort the water, betraying their presence as the bat scans the pond for ripples. The bat then seizes the fish in the long claws of its hind feet. Only a single young is born at a time, usually one per year. Females apparently form nursery colonies, and breeding may be highly synchronous in strongly seasonal habitats (Hood and Knox Jones 1984; Goodwin and Greenhall 1961).

FAMILY MORMOOPIDAE

Diagnosis
This family was formerly included in the Phyllostomidae. It was defined at the family rank and reclassified by J. D. Smith (1972). Although almost all members of the family Phyllostomidae have a prominent nose leaf, the mormoopid bats do not possess such a structure. The lips are expanded and adorned with flaps and folds (fig. 5.6), and there are often bristlelike hairs on the sides of the lips. The nostrils are incorporated into the expanded upper lip and often have ridges and protuberances above or between them (see fig. 5.6). The trochiter is well developed but smaller than the trochin and does not articulate with the scapula. Two genera are recognized by Smith, *Pteronotus* and *Mormoops*. *Pteronotus* is divisible into three subgenera, *Phyllodia*, *Chilonycteris*, and *Pteronotus*. Identifications are outlined in table 5.4. External features are illustrated in figure 5.6. The skull of *Pteronotus* is portrayed in figure 5.7.

Distribution
This family is confined to the New World tropics, ranging from southern Mexico to northeastern Brazil. The Antilles have been extensively colonized by species of this family.

Natural History
These bats have been classified as aerial insectivores by Wilson (1973b). They primarily roost in

Table 5.4   Key to the Common Mormoopidae of Northern South America

1 Hairless skin of wings meets in the midline of the back (subgenus *Pteronotus*) 2
1' Skin of wings does not meet in the midline of back; back presents the usual furred appearance 3
2 Head and body length greater than 60 mm . . . . *Pteronotus suapurensis*
2' Head and body length less than 60 mm . . . . *Pteronotus davyi*
3 Ears joined by flap of skin extending over the crown of the head; chin leaf prominent . . . . *Mormoops megalophylla*
3' Ears not connected by skin flap 4
4 Head and body length greater than 59 mm . . . . *Pteronotus parnellii*
4' Head and body length less than 59 mm . . . . *Pteronotus personatus*

Note: See family account.

caves or tunnels. A single young is born after a gestation period of at least sixty days. The echolocation pulses of mormoopid bats are distinctive and separable from those of the phyllostomids. The high-frequency sounds mormoopids produce during echolocation resemble more closely the sounds emitted by rhinolophids than those of any of the phyllostomids. The individual pulses are long and loud, without major frequency modulation, and have three harmonics that contribute qualitatively different echoes, allowing the bat to discriminate two-dimensional cues while echolocating (Fuzenssery and Pollak 1984). Sound is emitted from the mouth, not through the nose as in phyllostomids.

Genus *Pteronotus* Gray, 1838
Subgenus *Phyllodia*
Pteronotus parnellii (Gray, 1843)
Mustached Bat

Measurements

	Sex	Mean	S.D.	N
TL	M	95.06	4.70	65
	F	92.36	4.40	45
HB	M	72.73	4.32	64
	F	69.78	4.91	45
T	M	22.48	2.40	64
	F	22.58	1.94	45
HF	M	14.62	0.96	65
	F	14.78	1.02	45
E	M	23.35	1.74	65
	F	23.18	1.54	45
FA	M	59.68	1.44	106
	F	59.83	1.66	79
Wta	M	19.32	1.86	59
	F	17.30	1.99	28

Location: Northern Venezuela (USNMNH)

Description

The dental formula is I 2/2, C 1/1, P 2/3, M 3/3 (see fig. 5.7). This is one of the largest species of the genus. Males' head and body length averages 71.7 mm while females' is 70.4 mm, and weight averages 20.4 g for males and 19.6 g for females (see table 5.14). The dorsum ranges from dark to light brown

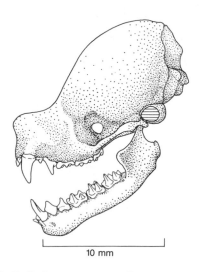

Figure 5.7. Skull of *Pteronotus parnellii*.

depending on age and molt. This description includes *Pteronotus* (= *Chilonycteris*) *rubiginosa*. Chromosome number: $2n = 38$; FN = 60.

Range and Habitat

This species ranges from southern Sonora, Mexico, south through the Isthmus and across the northern Neotropics to Brazil. It is found at up to 1,500 m altitude in Venezuela, but in the main it occurs below 500 m. It generally lives in moist areas but tolerates both multistratal evergreen forest and dry deciduous forest (Handley 1976) (map 5.16).

Natural History

These bats often roost in caves and may co-occur with other species of mormoopids and phyllostomids. In Venezuela they breed once a year. Mating takes place in January, when both sexes roost together, and the young are born in May when insects are most abundant. They feet mostly on Lepidoptera and Coleoptera. Insects are found by echolocation, but the echolocation pulse is unique in that the major portion of the call is not frequency modulated. Apparently the bats determine the distance and direction of their prey from the Doppler shift in pitch of the echo. This echolocation system is convergent with the Old World families Hipposideridae and Rhinolophidae (Herd 1983; Novick 1977).

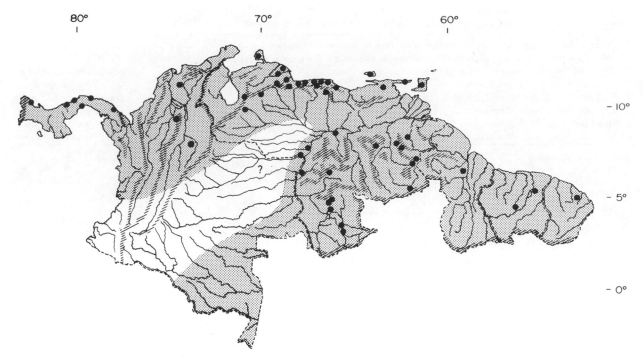

Map 5.16. Distribution of *Pteronotus parnellii*.

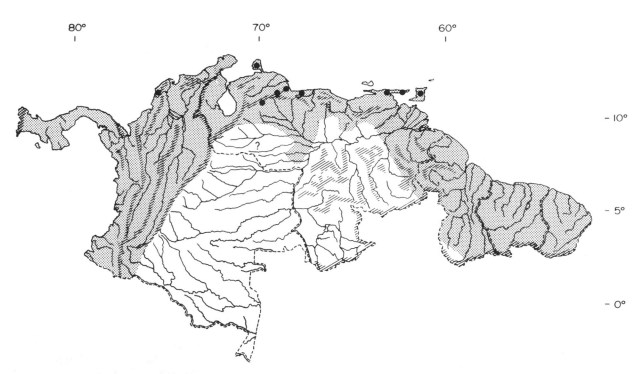

Map 5.17. Distribution of *Pteronotus davyi*.

Subgenus *Pteronotus*
Naked-backed Bat

Description
This subgenus is easily separated from the other mormoopids because the skin of the wings meets in the midline, giving a naked-backed appearance (fig. 5.6). There are two species, *Pteronotus davyi* and *Pteronotus gymnonotus* (= *P. suapurensis*).

Pteronotus davyi Gray, 1838

Measurements

	Sex	Mean	S.D.	N
TL	M	79.42	2.91	31
	F	80.50	2.87	34
HB	M	57.48	2.80	31
	F	58.12	2.27	34
T	M	21.94	0.93	31
	F	22.38	1.74	34
HF	M	12.06	0.68	31
	F	12.09	0.79	34
E	M	17.87	1.02	31
	F	18.06	0.92	34
FA	M	46.49	0.85	66
	F	46.98	1.13	65
Wta	M	9.22	0.74	31
	F	9.05	0.62	31

Location: Northern Venezuela (USNMNH)

Description
The dental formula is I 2/2, C 1/1, P 2/3, M 3/3. This is the smallest species of the naked-backed bats. Head and body length averages 56.4 mm for males and 57.9 mm for females; weight averages 9.3 g for males and 9.6 g for females (see table 5.14 and plate 5). The dorsal fur is dark brown, and the venter is somewhat paler.
Chromosome number: $2n = 38$; FN = 60.

Range and Habitat
This species is distributed from southern Sonora, Mexico, south to northwestern Peru and northeastern Brazil. In Venezuela it most frequently occurs below 500 m elevation. It shows a broad tolerance for habitat types; most specimens in Venezuela occur in dry areas. The bats frequently occur in dry deciduous thorn forests, but they also range into multistratal wet evergreen forest (Handley 1976) (map 5.17).

Natural History
This species prefers to roost in damp caves, often with several other species including *Pteronotus parnellii* and various phyllostomid bats.

Pteronotus gymnonotus Natterer, 1843
Greater Naked-backed Bat

Measurements

	Sex	Mean	S.D.	N
TL	M	87.9	2.52	18
	F	88.3	2.63	13
T	M	23.6	1.7	18
	F	23.7	1.2	13
HF	M	12.9	0.64	18
	F	13.0	0.55	14
E	M	18.9	1.35	18
	F	19.0	1.18	14
FA	M	51.13	1.4	25
	F	51.93	1.0	23
Wta	M	12.62	0.7	18
	F	13.48	3.2	12

Location: Northern Venezuela (USNMNH)

Description
This species is similar in color to *Pteronotus davyi* but considerably larger. Head and body length averages 64.3 mm for males and 64 mm for females, and weight averages 12.6 g for males and 13.9 g for females.
Chromosome number: $2n = 38$; FN = 60.

Range and Habitat
This species ranges from southern Veracruz, Mexico, south to Peru, and across northern South America to the Guyanas and southwestern Brazil (map 5.18).

Natural History
No specific observations have been recorded on this species other than that it generally occurs below 400 m in Venezuela and prefers dry deciduous forests. Note that Smith (1972) synonymized *P. suapurensis* with *Pteronotus gymnonotus*.

Subgenus *Chilonycteris*
Pteronotus personatus (Wagner, 1843)
Mustached Bat

Description
The species resembles *Pteronotus parnellii* in morphology, color, and dentition but is smaller. With its haired back, it cannot be confused with *P. davyi* and *P. gymnonotus*. Head and body length for males averages 53.5 mm, while females are larger, with an average length of 58 mm. Weight averages 8 g for males and 7 g for females (see table 5.14). Chromosome number: $2n = 38$; FN = 60.

Range and Habitat
This species ranges from northern Colombia south through Peru and east to Brazil and Suriname. Its

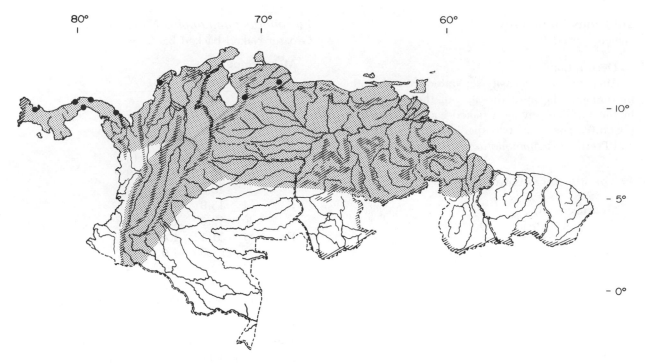

Map 5.18. Distribution of *Pteronotus gymnonotus*.

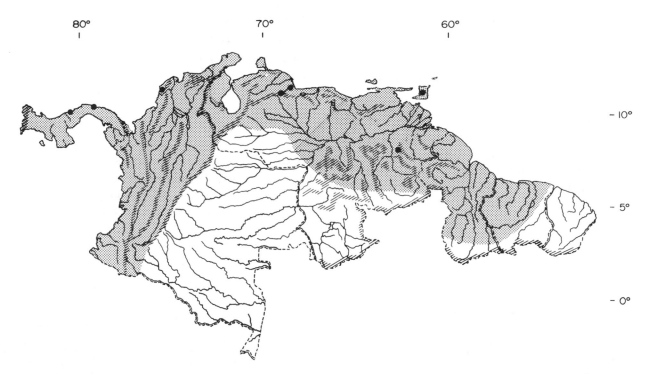

Map 5.19. Distribution of *Pteronotus personatus*.

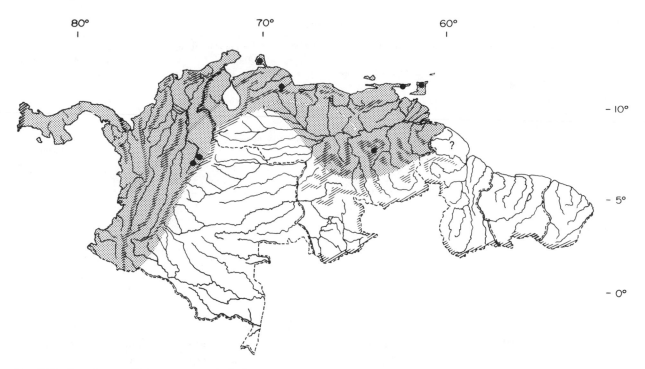

Map 5.20. Distribution of *Mormoops megalophylla*.

range continues northward from northern Colombia to southern Mexico, reaching southern Sonora. It generally occurs at elevations below 400 m and tolerates both multistratal evergreen forests and dry deciduous tropical forests (Handley 1976) (map 5.19).

Genus *Mormoops* Leach, 1821
Mormoops megalophylla (Peters, 1864)
Leaf-chin Bat

Measurements

	Sex	Mean	S.D.	N
TL	M	94.27	1.90	11
	F	94.73	3.07	11
HB	M	68.55	1.69	11
	F	69.18	2.36	11
T	M	25.73	1.68	11
	F	25.55	1.97	11
HF	M	12.73	0.65	11
	F	12.82	0.40	11
E	M	16.73	0.65	11
	F	16.91	0.70	11
FA	M	55.50	0.72	11
	F	54.53	1.96	11
Wta	M	16.77	1.57	11
	F	16.43	2.20	10

Location: Northwestern Venezuela (USNMNH)

Description
The dental formula is I 2/2, C 1/1, P 2/3, M 3/3. Males have an average head and body length of 68 mm, while females are slightly larger at 69.6 mm (see table 5.14). Weights range from 17.2 g for males and 16.5 g for females (see table 5.14). This reddish-brown bat has a highly ornamented face with a prominent chin leaf. A flap of skin connecting the ears is diagnostic and separates this genus from other mormoopid bats (fig. 5.6).
Chromosome number: $2n = 38$; FN = 62.

Range and Habitat
This species ranges from southern Texas through the Isthmus of Panama and across northern Colombia and northern Venezuela. It has colonized many of the Caribbean islands. It is frequently found in dry deciduous thorn forests but also forages in moist areas, primarily below 400 m elevation (map 5.20).

Natural History
Although it exploits dry deciduous forest, this species roosts in moist caverns. It may occur in colonies of up to twenty, often with other species of mormoopids and phyllostomids (Goodwin and Greenhall 1961).

FAMILY PHYLLOSTOMIDAE (= PHYLLOSTOMATIDAE)
American Leaf-nosed Bats

Taxonomy and Identification
This diverse family of bats is divided here into seven subfamilies. I have retained the Sturnirinae as distinct from the Stenoderminae. In table 5.15

Table 5.5   Field Key to the Subfamilies of the Phyllostomidae

1   Nose leaf rudimentary, face not wrinkled; calcar absent; wing attaches near the knee joint; upper incisors broader than canines
.... Desmodontinae
1' Nose leaf prominent or face exhibits prominent wrinkles; upper incisors smaller than canines 2
2   Muzzle long and narrow 3
2' Muzzle short and broad 5
3   Ear when pressed forward does not reach the nose tip .... Glossophaginae
3' Ear when pressed forward reaches the tip of the nose or beyond 4
4   Single papilla on chin or large central papilla on chin surrounded by smaller papillae .... Carolliinae
4' Y- or V-shaped shield on chin .... Phyllostominae
5   White facial stripes and often median dorsal stripe present .... Stenoderminae
5' No white facial stripes 6
6   Face naked, wrinkled, or naked with hornlike structure on nose 8
6' Face not naked, no wrinkles 7
7   Interfemoral membrane absent .... Sturnirinae
7' Interfemoral membrane present .... Ametrida
8   Face exhibits extreme wrinkling; fold of skin on throat .... Centurio
8' Face with folds; hornlike structure on nose; fold of skin on throat .... Sphaeronycteris

Note: See also table 5.15.

the Sturnirinae are subsumed in the Stenoderminae (Knox Jones and Carter 1976). True vampire bats are considered a subfamily, that is the Desmodontinae of the Phyllostomidae (Honacki, Kinman, and Koeppl 1982). The Phyllonycterinae (= Brachyphyllinae) are endemic to the Greater and Lesser Antilles and are not covered by this volume. The subfamilies are easily distinguished: If the muzzle is long and narrow, indicating a mixed-feeding or carnivorous strategy, one need only estimate the relative length of the ear to break specimens clearly into two groups. If the ear, when pressed forward, does not reach the end of the rostrum, the species belongs to the subfamily Glossophaginae. If the ear reaches the tip of the nose or beyond, then the chin should be inspected. If there is a single papilla on the chin, either standing alone or surrounded by smaller papillae, the species belongs to the subfamily Carolliinae; if there is a Y- or V-shaped shield on the chin, it belongs to the subfamily Phyllostominae.

If the bat in hand has a short, broad muzzle, one should immediately inspect the face; if it bears white stripes, the species is referable to the subfamily Stenoderminae. Only the genera *Pygoderma, Ectophylla, Ametrida, Sphaeronycteris,* and *Centurio* lack the white facial stripes within this subfamily, though they are faint in some species of *Artibeus.* Once one eliminates these five genera, then if there are no white facial stripes, no interfemoral membrane is present, and the foot is rather hairy, the species in question probably belongs to the subfamily Sturnirinae (see table 5.5). This "in-the-hand" key should be cross-checked with table 5.15, which includes a complete key to the subfamilies and genera.

Diagnosis

For this description the subfamily Desmodontinae (the true vampire bats) and some stenodermines are aberrant, since the nose leaf is very reduced, but their morphology is so distinctive that they cannot be mistaken for other species of bats (see Desmodontinae, below). The New World leaf-nosed bats are characterized by a skull without a postorbital process. The third finger has three complete bony phalanges (fig. 5.8). The most distinctive feature is the nose leaf that uniquely defines almost all members of this family. The tooth number is highly variable from twenty-six to thirty-four, reflecting in part the diverse feeding specializations exhibited by the family as a whole.

Distribution

This family is confined to the Western Hemisphere and is distributed from southern California and Arizona, as well as the Gulf coastal plain of Texas, south through the Isthmus to northern Argentina. It includes some 51 genera and over 140 species that are for the most part adapted to lower elevations and the tropical and subtropical areas of the New World.

Natural History

These bats show a variety of feeding specializations, though many feed on fruit (Carvalho 1961). According to Wilson (1973b) most genera, because they include a great deal of fruit in their diets, may be considered frugivores. On the other hand, some are mixed feeders, eating fruit as well as insects, and their insect searching seems to involve close inspection of leaves; thus they are termed foliage gleaners. Genera with this feeding habit include *Mi-*

cronycteris, *Tonatia*, *Mimon*, and *Phylloderma*. Some bats are specialized for feeding on nectar and pollen: most notable are *Glossophaga*, *Anoura*, and *Lionycteris*. Two genera belonging to the subfamily Desmodontinae are specialized for feeding on blood, including *Diphylla* and *Desmodus*. Some bats are active carnivores, preying upon lizards, birds, and small mammals; these include *Phyllostomus*, *Trachops*, *Chrotopterus*, and *Vampyrum*. *Artibeus*, *Carollia*, and *Sturnira* are specialized for frugivory, as are many stenodermine species (Gardner 1977a).

Because many phyllostomid bats are foliage gleaners, nectarivores, or frugivores, their food does not require precise localization by echolocation, in contrast to that of the aerial insectivores. The echolocating pulse of phyllostomids generally is of low amplitude, rather brief, and highly frequency modulated. Because of these acoustical characters, these bats are sometimes referred to as "whispering bats."

Most species of phyllostomids produce a single young. Some are highly seasonal in the timing of births, especially in areas subject to seasonal aridity. Birth often occurs at a time of maximum fruit or insect abundance (Wilson 1979).

An analysis of bat communities in Panama by Bonaccorso (1979) and Humphrey and Bonaccorso (1979) indicates that abundance varies considerably depending on the habitat. Some species are specialists in second-growth habitats, others are specialized for feeding along streams or creeks, and still others are specialists in mature forests. Regardless of the habitat preference, only three or four species make up the most abundant forms. Over 70% of all bats netted in a mature forest in Panama may belong to just 4 species: *Artibeus jamaicensis*, *A. lituratus*, *Glossophaga soricina*, and *Carollia perspicillata*.

Within any habitat type, species of bats belonging to the same trophic category seem to avoid direct competition by specializing for different vertical strata or different foods. It appears from the work of Gardner (1977a) and Bonaccorso (1979) that feeding categories are not as easily demarcated as previously thought. Wilson's trophic categorization into frugivores, foliage gleaners, and so on, has been refined in Bonaccorso's publication and further refined in the paper by Humphrey, Bonaccorso, and Zinn (1983). Many of the so-called frugivores are highly opportunistic feeders, eating insects at one time of year and switching to fruit at another. Opportunism seems more characteristic of most medium-sized phyllostomids (Humphrey, Bonaccorso, and Zinn 1983).

The biology of phyllostomid bats has been elegantly summarized in three volumes edited by R. J. Baker, J. Knox Jones, and D. Carter (1976, 1977, 1979). A key to the subfamilies of the Phyllostomidae is included in table 5.5). Chromosomal data included in the species accounts are from Baker et al. (1982). Owen (1988) is revising *Artibeus*.

SUBFAMILY PHYLLOSTOMINAE

Diagnosis
This diverse subfamily exhibits a wide range of sizes, including the largest Neotropical bat, *Vampyrum spectrum*. The rostrum is long, and when pressed forward the ears reach the tip of the nose or extend beyond. The interfemoral membrane shows little reduction in the insectivorous forms (*Micronycteris*), but in other genera it may be greatly reduced. External features are illustrated in figures 5.8 and 5.9. A key to the genera appears in table 5.6.

Distribution
The subfamily is widespread in suitable habitat from the western Gulf coast and southwestern United States to northern Argentina.

Genus *Micronycteris* Gray, 1866
Little Leaf-nosed Bat

Description
The dental formula is I 2/2, C 1/1, P 2/3, M 3/3 (see fig. 5.10). The tail extends only to approximately the middle of the interfemoral membrane. These leaf-nosed bats are small, with the forearm less than 41 mm long. The head and body length ranges from 42 to 52 mm, and tail length is about 12 mm. In some species the ears are connected by a band of skin, which can be an aid in species recognition (fig. 5.8). The dorsum is usually some shade of brown, with buff below. Keys to the species are included in table 5.16.

Distribution
The genus is distributed from central Mexico south across northern South America to Brazil.

Natural History
These small leaf-nosed bats are highly insectivorous. Although they occasionally take fruit, they are not highly specialized for a frugivorous diet but are foliage gleaners (Gardner 1977a).

Micronycteris brachyotis (Dobson, 1879)
Yellow-throated Bat

Description
This bat is of intermediate size: TL 57–75; T 7–14; HF 10–18; E 12–19; FA 39–43; Wt 9–15 g (Medellin, Wilson, and Navarro L. 1985). There is

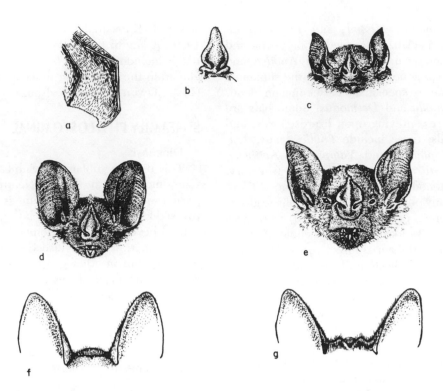

Figure 5.8. External features of the Phyllostominae: (*a*) Dorsal view of the third finger of *Micronycteris sylvestris* demonstrating three phalanges. (*b*) Nose leaf of *Micronycteris sylvestris*, a character typifying the Phyllostomidae. (*c*) Face of *Micronycteris brachyotis*. (*d*) Face of *Micronycteris hirsuta*. (*e*) Face of *Phyllostomus discolor*. (*f*) Condition of the connecting band between the ears of *Micronycteris megalotis*. (*g*) Condition of the connecting band between the ears of *Micronycteris minuta*. (Redrawn with permission from Goodwin and Greenhall 1961.)

Table 5.6   Key to the Genera of the Phyllostominae

1	Single lower incisor on each side, 2/1	2
1′	At least two lower incisors	4
2	Head and body length greater than 80 mm	*Chrotopterus*
2′	Head and body length less than 80 mm	3
3	Nose leaf rather long, greater than 14 mm	*Mimon*
3′	Nose leaf rather short, less than 14 mm	*Tonatia*
4	Head and body length greater than 130 mm; no tail; interfemoral membrane vastly reduced	*Vampyrum*
4′	Head and body length less than 130 mm	5
5	Nose leaf equals tragus in length (greater than 25 mm)	*Lonchorhina*
5′	Nose leaf does not equal tragus in length	6
6	Wartlike protrusions evident on both upper and lower lips; head and body length greater than 75 mm	*Trachops*
6′	Wartlike protrusions not evident on upper lip	7
7	Only two lower premolars present	*Phyllostomus*
7′	Three lower premolars present	8
8	Head and body length greater than 80 mm	*Phylloderma*
8′	Head and body length less than 80 mm	9
9	Tail extends to end of uropatagium	*Macrophyllum*
9′	Tail extends to middle of uropatagium	10
10	One upper incisor on each side	*Barticonycteris*
10′	Two upper incisors on each side	*Micronycteris*

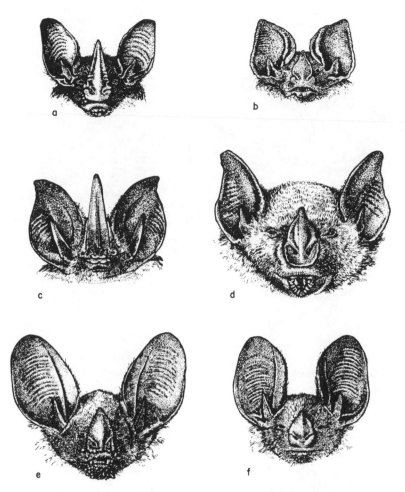

Figure 5.9. (*a*) Face of *Mimon crenulatum;* note the elongate nose leaf. (*b*) Face of *Micronycteris sylvestris.* (*c*) Face of *Lonchorhina aurita;* the nose leaf reaches its maximum size in this genus. (*d*) Face of *Phyllostomus hastatus.* (*e*) Face of *Trachops cirrhosus.* (*f*) Face of *Tonatia bidens.* (Redrawn with permission from Goodwin and Greenhall 1961.)

no band connecting the ears over the top of the head, a character it shares with *M. silvestris* and *M. nicefori.* The dorsum is light brown, with the venter slightly paler.

Chromosome number: $2n = 32$; FN = 60.

Range and Habitat

This species ranges from Oaxaca, Mexico, through the Isthmus, across northern South America and south to Amazonian Brazil. In northern Venezuela most specimens were caught below 150 m. It is strongly associated with moist evergreen habitats (Handley 1976) (map 5.21).

Natural History

This bat appears to roost in the hollow trunks of trees. A colony may contain up to ten individuals. One male may occur with nine females, suggesting a polygynous mating system (Medellin, Wilson, and Navarro L. 1985). Although it takes some fruits, insects make up the bulk of its diet. Bonaccorso (1979)

reports it is most commonly caught 3–12 m off the ground.

Micronycteris hirsuta (Peters, 1869)

Description

This is the largest species of *Micronycteris:* TL 78; HB 61–64; T 14–17; HF 11–13; E 25–27; FA 43–44; Wt 12–14 g. The ears are connected by a band over the top of the head that is low in profile and does not have a notch, which distinguishes *M. hirsuta* from *M. megalotis* and *M. minuta* (see fig. 5.8). The dorsum is gray brown, scarcely contrasting with the venter.

Chromosome number: $2n = 28–30$; FN = 32.

Range and Habitat

This species ranges from Honduras south through the Isthmus to Amazonian Peru and Brazil. It prefers lower elevations and concentrates its activity near streams or moist areas (map 5.22).

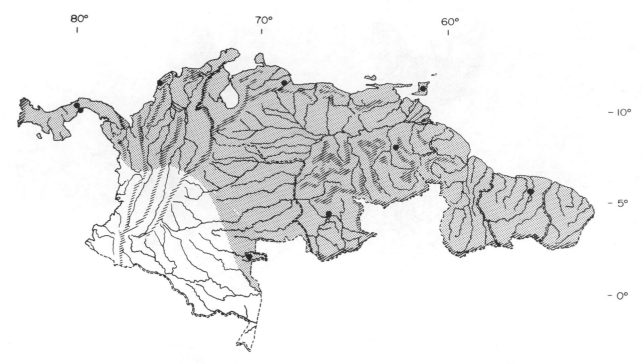

Map 5.21. Distribution of *Micronycteris brachyotis*.

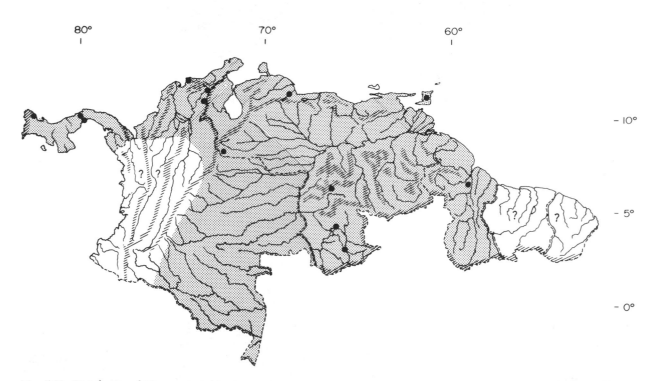

Map 5.22. Distribution of *Micronycteris hirsuta*.

Natural History

This bat will nest in hollow trees and under bridges, often with other species of bats. The timing of birth may correspond with abundance of the insects and fruits it feeds on.

Micronycteris megalotis (Gray, 1842)

Measurements

	Sex	Mean	S.D.	N
TL	M	57.14	3.78	22
	F	57.60	2.85	15
HB	M	43.77	4.10	22
	F	43.53	2.75	15
T	M	13.36	2.44	22
	F	14.07	2.58	15
HF	M	9.95	0.79	22
	F	10.20	0.77	15
E	M	22.27	1.12	22
	F	22.33	1.11	15
FA	M	33.02	1.01	23
	F	33.70	0.92	14
Wta	M	5.41	0.79	16
	F	5.61	0.87	12

Location: TFA, Venezuela (USNMNH)

Description

Head and body length averages 43.8 mm for males and 44.6 mm for females, and weight averages 5 g for males and 5.7 g for females. The ears of this pale brown bat are connected by a band that is high and notched in the center (fig. 5.8). *M. microtis* is considered a junior synonym of *M. megalotis*.
Chromosome number: $2n = 40$; FN = 68.

Range and Habitat

This species ranges from southern Tamaulipas in Mexico south to Peru, east across Colombia and Venezuela to French Guiana, and thence south to Brazil. This species does not occur above 800 m ele-

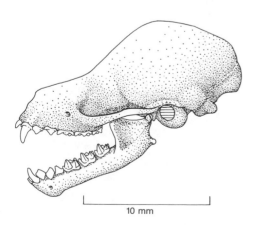

10 mm

Figure 5.10. Skull of *Micronycteris magalotis*.

vation in Venezuela. It is broadly tolerant of both multistratal evergreen forests and dry thorn forests (map 5.23).

Natural History

This bat forages near streams, and it roosts in hollow trees, logs, caverns, or even houses in groups of up to twelve. It is a mixed feeder, taking fruits when in season as well as insects.

Micronycteris minuta (Gervais, 1856)

Measurements

	Sex	Mean	S.D.	N
TL	M	60.00	2.45	13
	F	60.33	2.81	12
HB	M	48.23	2.55	13
	F	48.83	2.48	12
T	M	11.77	1.09	13
	F	11.50	1.31	12
HF	M	11.45	0.97	13
	F	11.58	1.24	12
E	M	22.23	1.17	13
	F	21.67	0.98	12
FA	M	34.81	0.80	14
	F	35.04	0.77	14
Wta	M	7.76	0.59	9
	F	8.63	0.50	7

Location: Northwestern Venezuela (USNMNH)

Description

Head and body length for males averages 47.4 mm; females are larger, averaging 49 mm. Average weight for males is 6.9 g and for females 7.2 g (see table 5.14). A band connects the ears over the head but is not raised as in *M. megalotis*, and a deep notch divides the band into equal halves (fig. 5.8). Though this bat is pale brown, the light basal color of the dorsal hairs may yield the effect of a buff dorsal midline stripe.
Chromosome number: $2n = 28$; FN = 50.

Range and Habitat

This species ranges from Nicaragua south to Amazonian Brazil and Peru. It is broadly distributed over northern South America. In Venezuela, most specimens occur below 500 m. It will forage in man-made clearings and dry deciduous forest as well as in multistratal evergreen forest (map 5.24).

Natural History

Fleming, Hooper, and Wilson (1972) found 76% insects and 24% fruit by volume in the stomachs of specimens taken in Costa Rica and Panama. The bats roost in hollow trees and caves, often with other species (Goodwin and Greenhall 1961).

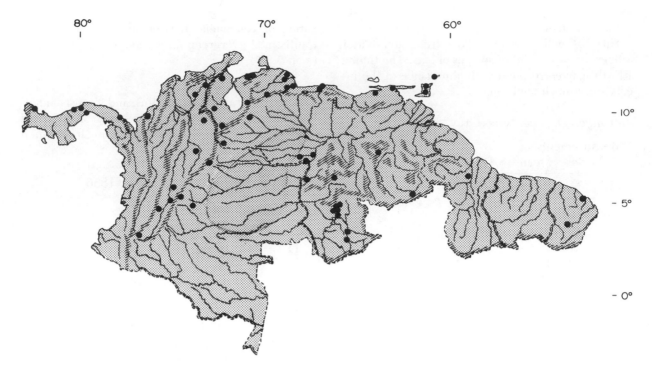

Map 5.23. Distribution of *Micronycteris megalotis*.

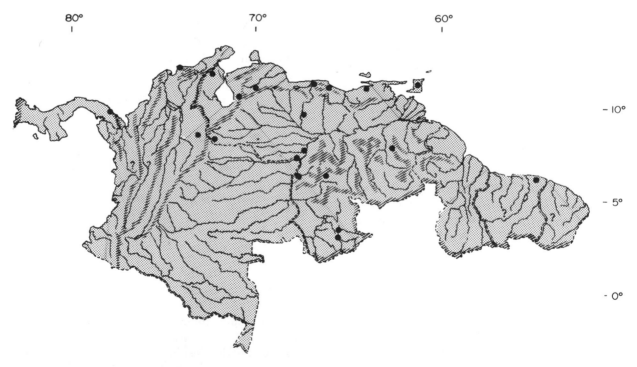

Map 5.24. Distribution of *Micronycteris minuta*.

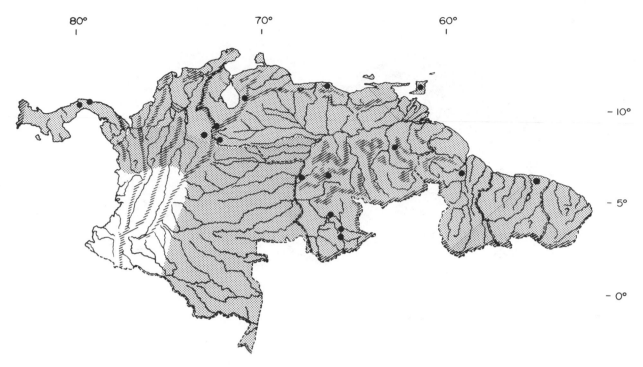

Map 5.25. Distribution of *Micronycteris nicefori*.

Micronycteris nicefori Sanborn, 1949

Measurements

	Sex	Mean	S.D.	N
TL	M	64.78	3.41	32
	F	68.31	2.34	36
HB	M	56.31	3.81	32
	F	58.89	2.93	36
T	M	8.47	1.87	32
	F	9.42	1.98	36
HF	M	12.91	0.93	32
	F	13.33	0.86	36
E	M	17.94	0.88	32
	F	18.39	0.99	36
FA	M	37.51	1.01	32
	F	39.01	0.90	36
Wta	M	7.80	0.56	24
	F	9.52	1.18	12

Location: Bolívar, Venezuela (USNMNH)

Description

This is the smallest of the species not exhibiting a connecting band between the ears. Males average 55 mm in head and body length, and females average 58 mm. Weight averages 7.9 g for males and 9.2 for females (see table 5.14). A reddish cast to the dorsal pelage is evident.

Chromosome number: $2n = 28$; FN = 52.

Range and Habitat

This species occurs from Nicaragua to northern Colombia, then eastward across Venezuela to Amazonian Brazil and south from Colombia into Amazo-

nian Peru. It is strongly associated with multistratal tropical evergreen forests and dry uplands (Handley 1976) (map 5.25).

Natural History

Groups of about twelve individuals roost in hollow trees in company with other species (Goodwin and Greenhall 1961).

Micronycteris schmidtorum Sanborn, 1935

Description

The species is immediately characterized by its relatively long tail, which averages about 16 mm. Mean measurements for males and females, respectively, are TL 64, 64.8; T 16, 16.2; HF 11; E 21, 23; FA 34, 35.8. Weights average 7.6 g for males and 7.2 g for females. The dorsal pelage is a very pale brown, and the venter is pale gray to almost white. Chromosome number: $2n = 38$; FN = 66.

Range and Habitat

This species occurs from southern Mexico to Venezuela. It is strongly associated with moist habitats when foraging, although it occurs in both multistratal tropical evergreen forest and dry thorn forest. It frequently roosts in tree holes (map 5.26).

Micronycteris sylvestris (Thomas, 1896)

Description

Total length ranges from 53 to 73 mm: T 9–14; HF 10–12; E 20–22 (Suriname, CMNH). Weight averages 7.2 g for males and 10.3 g for females (see

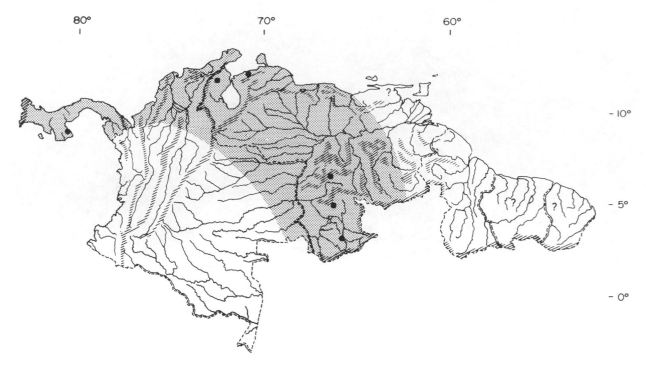

Map 5.26. Distribution of *Micronycteris schmidtorum*.

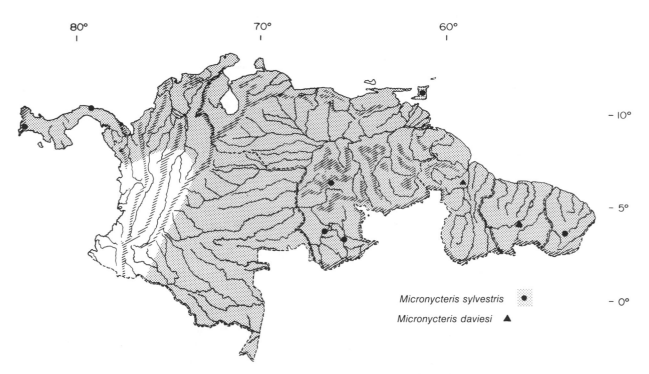

Micronycteris sylvestris

Micronycteris daviesi ▲

Map 5.27. Distribution of two species of *Micronycteris*.

table 5.14). The dorsum is dark brown with a frosted cast because the tips of the dorsal hairs are white. There is no band connecting the ears. The species may be distinguished from *M. nicefori* by its relatively longer tail. This species has been placed in its own genus *Glyphonycteris*, but here *Glyphonycteris* is considered at best a subgenus.
Chromosome number: $2n = 22$; FN = 40.

Range and Habitat

This species occurs from Veracruz, Mexico, south through the Isthmus to Peru and southeastern Brazil. It prefers low elevations and tropical evergreen forest habitat (map 5.27).

Natural History

The bats roost in hollow trees in groups of up to seventy-five and tolerate other species (Goodwin and Greenhall 1961).

Micronycteris (Barticonycteris) daviesi (Hill, 1964)

Description

Included by Honacki, Kinman, and Koeppl (1982) within *Micronycteris*, this species is certainly closely allied, but the dental formula is I 1/2, C 1/1, P 2/3, M 3/3. Head and body length ranges from 69 to 84 mm, and the forearm from 54 to 58 mm. Measurements of one specimen from Suriname (CMNH) were: TL 77; T 7; HF 16; E 31; Wt 8 g. The medium-sized ears are not connected by a band of skin. The dark brown dorsal pelage is long and lax; the venter is gray brown.
Chromosome number: $2n = 28$; FN = 52.

Range and Habitat

This lowland bat has a disjunct range. Specimens have been taken from Costa Rica, eastern Colombia, and the Guyanas (map 5.27). *Micronycteris pusilla* has been taken in extreme northwestern Brazil; the range of this species may yet prove to extend into Colombia.

Genus *Lonchorhina* Tomes, 1863
Sword-nosed Bat, Murciélago de Espada

Description

The dental formula is I 3/3, C 1/1, P 3/3, M 2/3 (according to Hernandez-Camacho and Cadena-G. 1978). These medium-sized bats have a forearm length exceeding 40 mm. Head and body length ranges from 52 to 60 mm, and the tail nearly equals the head and body in length. The nose leaf is exceedingly long, almost equaling the ears, which exceed 30 mm (fig. 5.9). The large ear, with a promi-

nent tragus, and the narrow, long nose leaf clearly distinguish this genus from any other New World bat. The fur is a light reddish brown, and the venter is not set off from the dorsum.

Distribution

The genus is broadly distributed from southern Veracruz, Mexico, to southeastern Brazil. These bats tend to roost in colonies, and *L. aurita* has been found in groups of over five hundred.

Lonchorhina aurita Tomes, 1863

Measurements

	Sex	Mean	S.D.	N
TL	M	112.58	3.30	36
	F	113.85	3.83	26
HB	M	60.00	2.78	36
	F	59.77	4.05	26
T	M	52.58	2.59	36
	F	54.08	4.05	26
HF	M	14.92	1.50	36
	F	14.77	1.82	26
E	M	32.58	1.40	36
	F	32.00	1.67	26
FA	M	50.94	1.09	39
	F	51.57	1.11	26
Wta	M	14.20	1.19	21
	F	15.20	1.51	9

Location: Northern Venezuela (USNMNH)

Description

This is the second largest species within the genus. Head and body length for males averages 60.9 mm, and females average 61.1 mm. Forearm length ranges from 48 to 55 mm (see table 5.14), and weight averages 14.3 g for males and 15 g for females. The bat has a medium brown dorsum with a gray brown venter. (See plate 5.)
Chromosome number: $2n = 32$; FN = 60.

Range and Habitat

This species is broadly distributed, from southern Veracruz, Mexico, to southeastern Brazil. It is strongly associated with moist habitats and is most frequently encountered in multistratal tropical forests. In Venezuela it was not taken at above 1,000 m elevation (map 5.28). (Handley 1976).

Natural History

This bat is strongly insectivorous, highly specialized for the aerial pursuit of insects. Gardner (1977a) records some plant material in its diet. It roosts in caves in colonies of twenty to twenty-five (Goodwin and Greenhall 1961).

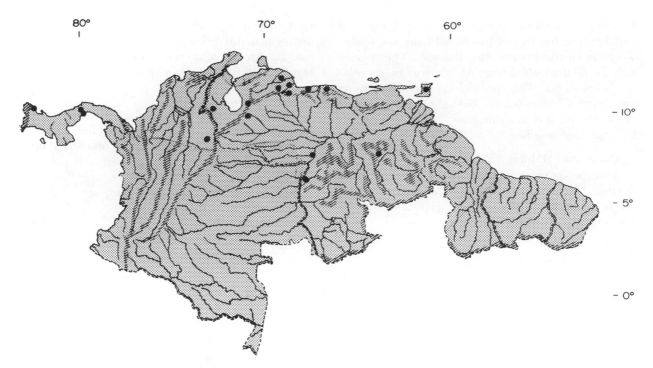

Map 5.28. Distribution of *Lonchorhina aurita*.

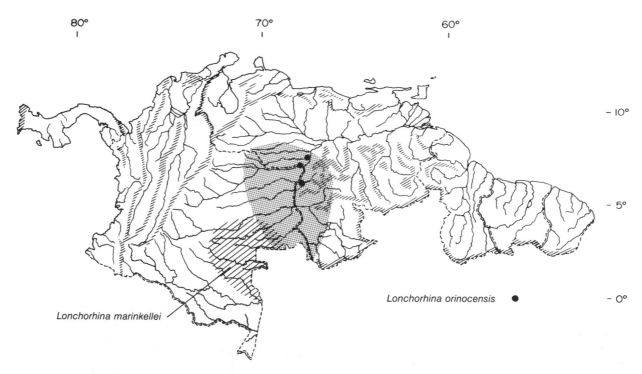

Lonchorhina orinocensis ●

Lonchorhina marinkellei

Map 5.29. Distribution of two species of *Lonchorhina* (see Linares and Ojasti 1971; Hernandez-Camacho and Cadena-G. 1978).

Lonchorhina fernandez Ochoa and
Ibañez, 1982

Description
This is the smallest species of the genus: TL
90.23; HF 9.0; E 19.0 (N = 33) (Ochoa-G. and
Ibañez 1982). The dorsum is brown and the venter
similar.

Range and Habitat
The species is known from the type locality, 40–50
km northeast of Puerto Ayacucho, TFA, Venezuela.

Lonchorhina orinocensis Linares and
Ojasti, 1971

Measurements

	Sex	Mean	S.D.	N
TL	M	102.80	3.37	25
	F	103.29	2.55	21
HB	M	52.56	2.50	25
	F	52.33	2.08	21
T	M	50.24	1.76	25
	F	50.95	2.31	21
HF	M	11.00	0.71	25
	F	11.05	0.67	21
E	M	30.88	0.97	25
	F	30.71	1.10	21
FA	M	42.13	1.69	25
	F	42.34	0.78	21
Wta	M	8.53	0.72	17
	F	8.97	0.64	10

Location: North-central Venezuela (USNMNH)

Description
This species is similar in appearance to *L. aurita* but smaller. Head and body length averages
52.3 mm for both males and females, and forearm
length ranges from 41 to 44 mm (see table 5.14).
Weight averages 8.7 g.

Range and Habitat
This species appears to be confined in the llanos
regions of western Venezuela and east-central Colombia. Strongly associated with savanna habitats, it
apparently roosts in rock crevices during the day
(map 5.29).

Lonchorhina marinkellei Hernandez-Camacho
and Cadena-G., 1978

Description
This species is similar to *L. aurita* but larger.
Forearm length averages 59 mm (see Hernandez-
Camacho and Cadena-G. 1978).

Range and Habitat
The species is known to occur in the extreme
southeast of Colombia, and the description is based
on a small number of specimens (map 5.29).

Genus *Macrophyllum* Gray, 1838
Long-legged Bat

Description
The dental formula is I 2/2, C 1/1, P 2/3, M 3/3.
In its superficial appearance, this bat is similar to
Micronycteris, but it is easily distinguished by its
tail, which extends to the edge of the uropatagium.
The hind feet are extremely long relative to the head
and body length, averaging some 14.5 mm (see table
5.14). The dorsum is sooty brown, and the venter is
paler brown.

Macrophyllum macrophyllum (Schinz, 1821)

Measurements

	Sex	Mean	S.D.	N
TL	M	86.00	4.69	4
	F	93.00	2.05	11
HB	M	44.50	3.11	4
	F	48.64	1.86	11
T	M	41.50	1.91	4
	F	44.36	1.12	11
HF	M	15.00	0.00	4
	F	15.09	0.30	11
E	M	18.00	0.82	4
	F	18.64	0.50	11
FA	M	34.23	0.05	4
	F	35.80	0.62	12
Wta	M	7.15	0.10	4
	F	7.15		2

Location: Northwestern Venezuela (USNMNH)

Description
Head and body length averages 44.7 mm for
males and 48.5 mm for females; weight averages 7 g
for males and 7.2 g for females (see table 5.14). The
dorsum is dark brown and the venter is paler.
Chromosome number: $2n = 32$; FN = 56

Range and Habitat
This species ranges from Tabasco, Mexico,
through the Isthmus and over most of northern
South America through Brazil to the south. It forages
near streams or in other moist areas and, though tolerant of deciduous scrub forest, is most abundant in
multistratal tropical evergreen forest (map 5.30).

Natural History
This species roosts singly or in very small groups
and has been found in caves. In Central America
there appears to be only a single breeding period
during the late dry season. It is largely insectivorous, and Gardner (1977a) suggests it specializes on
aquatic insects.

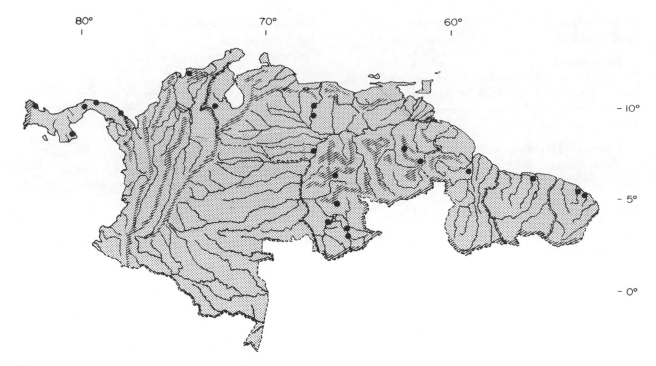

Map 5.30. Distribution of *Macrophyllum macrophyllum*.

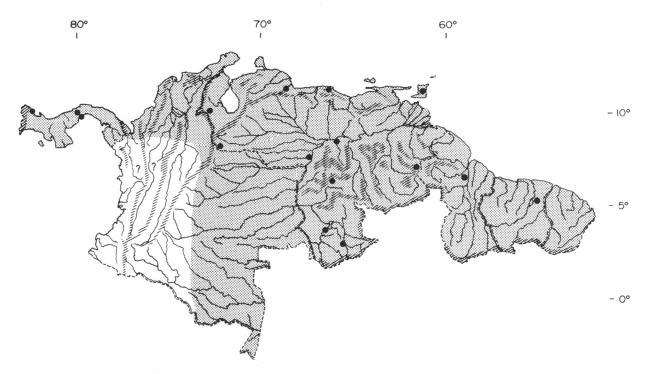

Map 5.31. Distribution of *Tonatia bidens*.

Genus *Tonatia* Gray, 1827
Round-eared Bat

Description
The dental formula is I 2/1, C 1/1, P 2/3, M 3/3. Within the subfamily Phyllostominae only three genera have the single lower incisor, *Tonatia*, *Mimon*, and *Chrotopterus*. These large-eared bats have an average nose leaf. The tail is short, but the uropatagium is retained. The dorsal pelage is dark to light brown, and the venter is slightly paler. Superficially, species of *Tonatia* resemble *Micronycteris*; they may be distinguished from *Mimon*, which has an extremely long nose leaf (see fig. 5.9). *Chrotopterus* cannot be confused with either *Mimon* or *Tonatia* because of its extremely large size.

Distribution
The genus *Tonatia* is distributed from Guatemala south through the Isthmus to extreme southeastern Brazil.

Natural History
This genus appears to be frugivorous, although insects are a significant part of its diet seasonally. Species of *Tonatia* typically bear a single young, but in some parts of the range there are two birth peaks, suggesting that individual females may conceive twice in a given year.

Tonatia bidens (Spix, 1823)

Measurements

	Sex	Mean	S.D.	N
TL	M	97.89	3.06	9
	F	95.00	3.70	7
HB	M	76.00	3.46	9
	F	75.43	5.65	7
T	M	21.10	4.12	10
	F	19.57	3.21	7
HF	M	16.80	0.79	10
	F	16.86	0.69	7
E	M	33.50	1.35	10
	F	31.14	2.12	7
FA	M	57.33	2.47	10
	F	57.04	1.02	7
Wta	M	26.25	3.31	4
	F	24.65	1.12	4

Location: Venezuela (USNMNH)

Description
This is the largest species of the genus *Tonatia*. Head and body length averages 76 mm for males and 75.4 mm for females. Average weight for males is 26.3 g and for females 24.7 g. The ears are rounded proximally, hence the common name round-eared bat. A light-colored median stripe occurs on the forehead in some populations. Upperparts vary from tawny to blackish brown, and underparts tend to be paler.
Chromosome number: $2n = 16$; FN = 20.

Range and Habitat
This species is widely distributed from southern Guatemala to southeastern Brazil. It has been taken in both moist sites and dry deciduous sites. All specimens taken in Venezuela were below 200 m elevation. Specimens have been found roosting with other species in hollow trees (Handley 1976) (map 5.31).

Tonatia brasiliense (Peters, 1866)

Measurements

	Sex	Mean	S.D.	N
TL	M	63.25	2.31	8
	F	60.60	3.21	5
HB	M	52.38	3.16	8
	F	51.00	4.18	5
T	M	10.88	2.59	8
	F	9.60	3.36	5
HF	M	11.88	0.83	8
	F	11.80	0.84	5
E	M	24.75	1.91	8
	F	24.40	0.89	5
FA	M	34.63	0.91	9
	F	35.34	1.69	5
Wta	M	9.94	0.53	7
	F	10.75		2

Location: TFA Venezuela (USNMNH)

Description
This is the smallest species of the genus in northern South America. Head and body length averages 54.8 mm for males and 55.7 mm for females. Weight averages 10.2 g for males and 10.6 g for females (see table 5.14). The species is easily distinguished from the larger *T. bidens*. *T. brasiliense* includes *T. minuta*, *T. nicaraguae*, and *T. venezuelae* (Honacki, Kinman, and Koeppl, 1982).
Chromosome number: $2n = 30$; FN = 56.

Range and Habitat
Distributed from southern Veracruz, Mexico, south to Peru and eastward across the northern Neotropics and northern Brazil (map 5.32). In northern Venezuela this species was common below 500 m elevation. It is strongly associated with streamside habitats and other moist areas but can range into deciduous forests. The preferred habitat appears to be multistratal evergreen forest, though it is broadly tolerant of man-made clearings (Handley 1976).

Natural History
In Trinidad this species has been seen roosting in abandoned termite nests (Goodwin and Greenhall 1961).

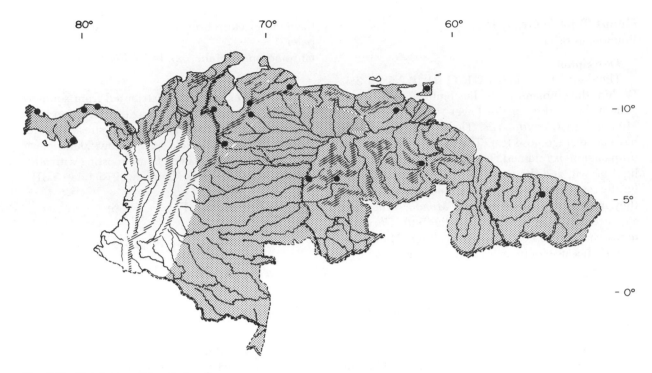

Map 5.32. Distribution of *Tonatia brasiliense*.

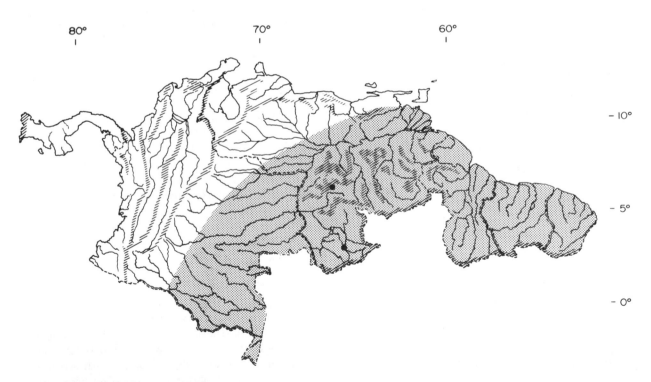

Map 5.33. Distribution of *Tonatia carrikeri*.

Tonatia carrikeri (J. A. Allen, 1910)

Description

This species resembles *T. bidens* in size but lacks the median white stripe on the head: TL 66–76; T 14–15; HF 12–16; E 22–25; FA 45–50 ($N = 7$; Colombia, CMNH). One male from Venezuela had a head and body length of 72 mm, a forearm of 44 mm, and a weight of 22 g.
Chromosome number: $2n = 26$; FN $= 46$.

Range and Habitat

Thus far this species has been described in southeastern Colombia and the extreme southern part of Amazon territory in Venezuela. The Venezuelan specimens were taken at elevations below 155 m in multistratal tropical evergreen forests near streams (map 5.33).

Tonatia schulzi Genoways and Williams, 1980

Description

This newly discovered species is intermediate in size: TL 68–78; T 11–13; HF 13–14; E 27–29 ($N = 2$; Suriname, CMNH).
Chromosome number: $2n = 28$; FN $= 36$ (Genoways and Williams 1980).

Range and Habitat

The species is known only from a restricted locality in Suriname, 38°48′ N, 56°08′ W (Genoways and Williams 1980).

Tonatia silvicola (d'Orbigny, 1836)

Measurements

	Sex	Mean	S.D.	N
TL	M	94.13	7.72	8
	F	91.62	6.29	13
HB	M	73.00	8.18	8
	F	69.46	6.41	13
T	M	21.13	3.80	8
	F	22.15	2.73	13
HF	M	17.38	0.74	8
	F	17.31	0.63	13
E	M	37.75	0.89	8
	F	37.85	2.23	13
FA	M	52.66	2.08	8
	F	51.83	1.73	12
Wta	M	31.16	5.20	7
	F	23.60	2.59	3

Location: TFA, Venezuela (USNMNH)

Description

Head and body length averages 73.3 mm for males and 71.5 mm for females, with weights averaging 30.6 g for males and 24.1 g for females. This bat is approximately the same size as *T. carrikeri* and

T. bidens. It lacks the median white stripe on the head characteristic of *T. bidens*, and it is distinguishable from *T. carrikeri* by its much larger ear. In *T. carrikeri*, the ear of one specimen was only 30 mm long, but in *T. silvicola* ear lengths average 36.6 mm for males and 37.2 mm for females (see table 5.14).
Chromosome number: $2n = 34$; FN $= 60$.

Range and Habitat

Most specimens obtained in Venezuela were found at below 460 m. Although the species is occasionally taken in deciduous forests near streams, most specimens were collected in multistratal tropical evergreen forests (Handley 1976) (map 5.34).

Natural History

This species occasionally roosts in termite nests. Fleming, Hooper, and Wilson (1972) recorded insects from the stomachs of twenty-two specimens taken in Panama.

Genus *Mimon* Gray, 1847
Gray's Spear-nosed Bat

Description

The dental formula is I 2/1, C 1/1, P 2/2, M 3/3. Although it shares the same dental formula as *Tonatia* and *Chrotopterus*, *Mimon* is easily distinguishable because of its much larger nose leaf. The fur is long and woolly. Color markings vary according to the species but are always some shade of brown.

Distribution

The genus *Mimon* is distributed from southern Veracruz, Mexico, to southern Brazil.

Natural History

Gardner (1977a) reports that the species of *Mimon* feed on fruits and insects.

Mimon bennettii (Gray, 1838)

Description

Total length ranges from 85 to 95 mm: T 20–25; HF 15–17; E 36–38 ($N = 4$; Suriname, CMNH). This bat typically has no striping on the back, which distinguishes it from *M. crenulatum*. *M. bennettii* has a pale brownish dorsum, and there are small whitish patches behind the ears. For this discussion *M. cozumelae* is considered a junior synonym of *M. bennettii*.
Chromosome number: $2n = 30$; FN $= 56$.

Range and Habitat

Mimon bennettii occurs from southern Veracruz, Mexico, to northern Colombia, then in the north coastal region of Venezuela to the Guyanas, and follows the coast of Brazil to southeastern Brazil (map 5.35).

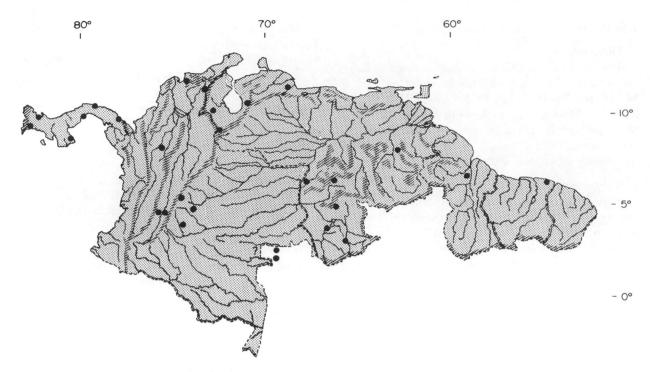

Map 5.34. Distribution of *Tonatia silvicola*.

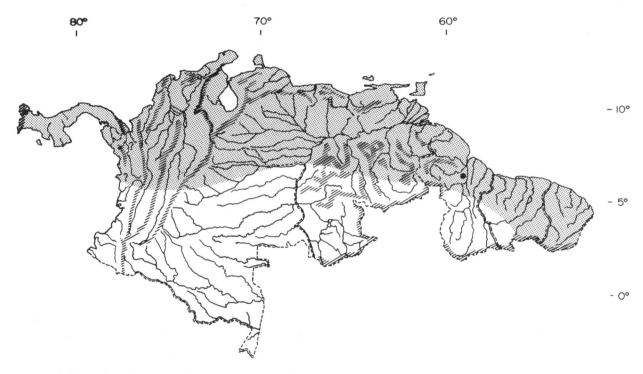

Map 5.35. Distribution of *Mimon bennettii* (total range from Koopman 1982).

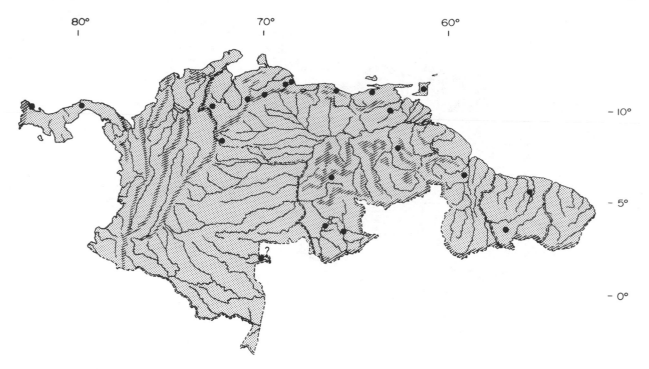

Map 5.36. Distribution of *Mimon crenulatum.*

Mimon crenulatum (E. Geoffroy, 1810)

Measurements

	Sex	Mean	S.D.	N
TL	M	83.76	3.35	25
	F	84.35	4.26	20
HB	M	57.24	2.49	25
	F	58.05	3.22	20
T	M	26.52	2.33	25
	F	26.30	1.75	20
HF	M	12.48	0.65	25
	F	12.25	0.64	20
E	M	25.88	1.39	25
	F	26.20	1.06	20
FA	M	49.44	1.43	24
	F	49.86	1.35	20
Wta	M	14.13	0.88	4

Location: Northern Venezuela (USNMNH)

Description

Head and body length of males averages 57.9 mm, while females are slightly larger at 58.5 mm. Weight averages 12.8 g for males and 12 g for females (see table 5.14). In this species there is generally a white line down the center of the back. The basic color of the dorsum is a bright mahogany brown, sometimes grading to blackish brown. The underparts are rusty to gray.
Chromosome number: $2n = 32$; FN = 60.

Range and Habitat

This species occurs from Campeche in Mexico south over most of the northern Neotropics, including northern Brazil (map 5.36). It was not taken at above 600 m in elevation in northern Venezuela. Although it ranges into dry deciduous forests, it prefers multistratal tropical evergreen forests. It frequently forages in natural openings or man-made fields, and it roosts in hollow tree trunks and buildings (Handley 1976).

Genus *Phyllostomus* Lacépède, 1799
Spear-nosed Bat

Description

The dental formula is I 2/2, C 1/1, P 2/2, M 3/3 (see fig. 5.11). These are medium to large bats with forearms exceeding 50 mm in length. The tail is short, and the interfemoral membrane is reduced in size. There is a glandular throat sac that is well developed in males but vestigial in females. The lower lip bears a V-shaped groove edged with small warts. The dorsal color ranges from light to dark brown.

Distribution

Species of this genus range from southern Veracruz, Mexico, to southern Brazil.

Natural History

The species *P. hastatus* is one of the largest New World bats. Although all species of this genus eat

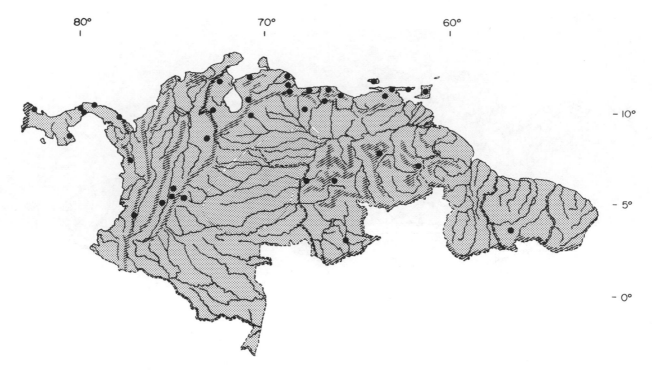

Map 5.37. Distribution of *Phyllostomus discolor*.

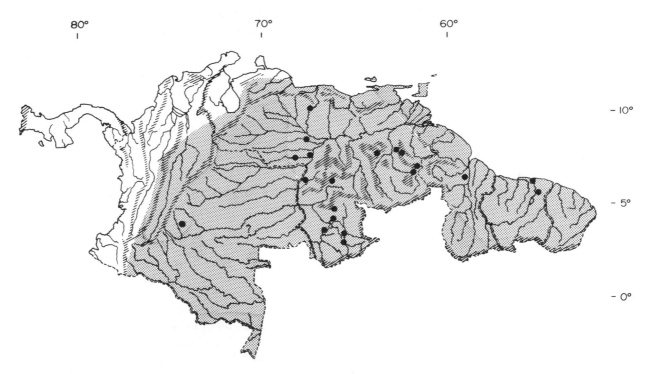

Map 5.38. Distribution of *Phyllostomus elongatus*.

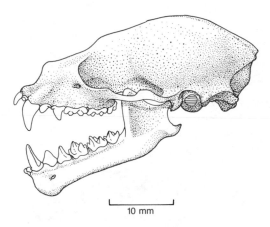

Figure 5.11. Skull of *Phyllostomus hastatus*.

fruit, the larger species are also carnivorous. *P. hastatus* preys on small rodents and other bats.

Phyllostomus discolor Wagner, 1843

Measurements

	Sex	Mean	S.D.	N
TL	M	99.01	4.16	68
	F	99.11	3.55	53
HB	M	85.00	3.69	68
	F	85.13	2.99	53
T	M	14.01	2.28	68
	F	13.98	2.04	53
HF	M	17.71	1.13	68
	F	17.68	0.98	53
E	M	23.91	0.86	68
	F	23.78	1.27	53
FA	M	61.44	1.63	96
	F	61.12	2.02	67
Wta	M	38.70	4.32	54
	F	35.45	4.24	20

Location: Northwestern Venezuela (USNMNH)

Description

This is one of the smallest species of the genus; forearm length averages about 61 mm. In males, head and body length averages 83.6 mm, and in females it is 84.2 mm. Weight averages 38.2 g for males and 35.4 g for females (table 5.14). The dorsal pelage is brown, but the white tips of the hairs convey a mottled effect.
Chromosome number: $2n = 32$; FN = 60.

Range and Habitat

P. discolor occurs from southern Mexico south across northern South America to northern Bolivia and southeastern Brazil. In Venezuela most specimens were taken at below 500 m elevation (map 5.37). The species is broadly tolerant of both dry and wet habitats, occurring in deciduous tropical forest and multistratal tropical evergreen forest.

Natural History

This species appears to be mainly frugivorous but consumes insects and pollen seasonally. It does not exhibit predatory behavior like that shown by *P. hastatus* (Gardner 1977a). The bats roost in hollow tree trunks in groups of twenty-five; in breeding roosts the sex ratio may be one male to twelve females.

Phyllostomus elongatus (E. Geoffroy, 1810)

Measurements

	Sex	Mean	S.D.	N
TL	M	102.65	5.08	46
	F	103.35	4.68	40
HB	M	79.49	5.26	45
	F	81.48	5.09	40
T	M	23.11	3.59	45
	F	21.88	2.69	40
HF	M	18.89	0.99	46
	F	18.18	1.28	40
E	M	30.46	1.43	46
	F	30.60	1.85	40
FA	M	66.30	1.56	46
	F	66.20	1.69	41
Wta	M	42.02	4.17	42
	F	39.35	3.82	14

Location: TFA, Venezuela (USNMNH)

Description

This bat is somewhat larger than *P. discolor:* forearms average 66 mm for males and females. Males average 80.4 mm in head and body length, while females average 82.5 mm. Weight for males and females averages 42 g and 40 g respectively (table 5.14). The dorsal and ventral pelage is a uniform dark brown.
Chromosome number: $2n = 32$; FN = 58.

Range and Habitat

P. elongatus is confined to South America to the east of the Andes, ranging across northern South America to southeastern Brazil (map 5.38). In Venezuela it is occasionally taken in dryer habitats near streams, but it strongly prefers multistratal tropical evergreen forests. In Venezuela most specimens occur below 350 m elevation (Handley 1976).

Natural History

This species roosts in tree cavities. Apparently it is strongly disposed to feed on fruits, but it may also take nectar and pollen (Gardner 1977a).

Phyllostomus latifolius (Thomas, 1901)

Description

This species is smaller than *P. elongatus:* TL 91–95; T 13–17; HF 14–17; E 27–29; FA 56–59;

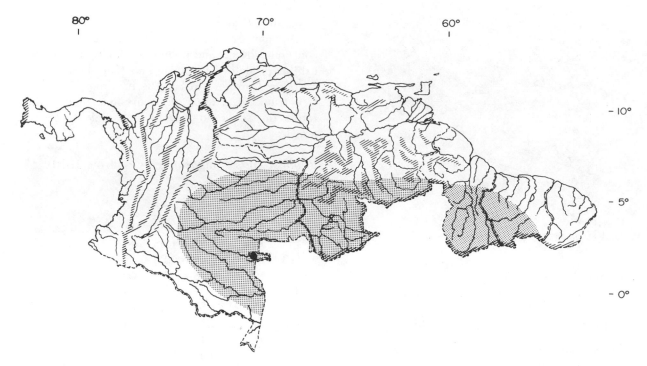

Map 5.39. Distribution of *Phyllostomus latifolius*.

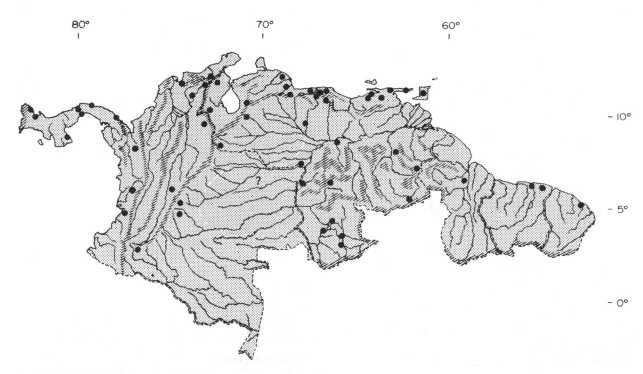

Map 5.40. Distribution of *Phyllostomus hastatus*.

Wt 25–30 g (N = 4; CMNH). The dorsum and venter are brown.
Chromosome number: $2n = 32$; FN = 58.

Range and Habitat

The species is confined to northwestern Brazil and adjacent portions of Guyana and Colombia (map 5.39). Little is known concerning this bat.

Phyllostomus hastatus (Pallas, 1767)
Vampiro de Lanza

Measurements

	Sex	Mean	S.D.	N
TL	M	130.63	7.32	40
	F	124.24	6.37	25
HB	M	111.55	6.79	40
	F	104.28	7.90	25
T	M	19.08	3.01	40
	F	19.96	4.33	25
HF	M	20.95	2.39	40
	F	21.28	1.72	25
E	M	31.54	1.70	39
	F	31.72	1.65	25
FA	M	83.91	1.98	47
	F	82.44	2.06	33
Wta	M	111.67	8.74	28
	F	90.25	9.75	10

Location: Northern Venezuela (USNMNH)

Description

This is one of the largest New World bats (see figs. 5.9 and 5.11 and table 5.14). The venter and dorsum are colored similarly, ranging from dark brown to reddish brown.
Chromosome number: $2n = 32$; FN = 58.

Range and Habitat

This species ranges from Honduras south through the Isthmus to southeastern Brazil (map 5.40). In Venezuela it usually occurs below 500 m elevation, but some specimens have been taken at as high as 1,394 m. It tolerates a variety of habitat types including deciduous forests, man-made clearings, and multistratal tropical evergreen forests (Handley 1976).

Natural History

This species roosts opportunistically in caves and buildings and under palm leaves, forming both small groups and colonies exceeding five hundred individuals. In addition to being frugivorous this species preys actively on lizards, rodents, bats, and other small vertebrates. Even within a colony, males will defend groups of females and form temporary harems of thirty females per male. A single young is born (McCracken and Bradbury 1977, 1981).

Genus *Phylloderma* Peters, 1865
Phylloderma stenops Peters, 1865

Measurements

	Sex	Mean	S.D.	N
TL	M	111.14	5.58	7
	F	112.80	6.97	10
HB	M	93.00	7.02	7
	F	91.80	7.30	10
T	M	18.14	2.67	7
	F	21.00	2.91	10
HF	M	21.71	1.11	7
	F	21.10	0.99	10
E	M	28.00	1.83	7
	F	28.00	1.41	10
FA	M	71.17	3.09	7
	F	70.90	1.61	10
Wta	M	47.65	7.27	6
	F	47.09	4.61	7

Location: TFA, Venezuela (USNMNH)

Description

There is a single species in South America, *P. stenops*. The dental formula is I 2/2, C 1/1, P 2/3, M 3/3. *Phylloderma* has one more lower premolar than does *Phyllostomus*. Head and body length ranges from 85 to 120 mm, and forearm length from 67 to 80 mm (see table 5.14). Males possess a glandular throat sac. The dorsum is brown, but the pale bases of the dorsal hairs are noticeable on the shoulders. The venter is gray, and the tips of the wing membranes may show depigmentation.

Range and Habitat

This species occurs from Honduras south to Brazil, but it appears to be absent from dry deciduous forest over much of southeastern Brazil. In Venezuela this species was taken below 206 m (map 5.41). It is strongly associated with multistratal tropical evergreen forests but is broadly tolerant of man-made clearings (Handley 1976).

Natural History

This mixed feeder will take fruit as well as insects. Females bear a single young.

Genus *Trachops* Gray, 1847
Trachops cirrhosus (Spix, 1823)
Fringe-lip Bat

Measurements

	Sex	Mean	S.D.	N
TL	M	98.84	5.09	38
	F	100.36	5.52	36
HB	M	79.21	6.36	38
	F	81.28	6.39	36
T	M	19.63	3.81	38
	F	19.08	3.03	36

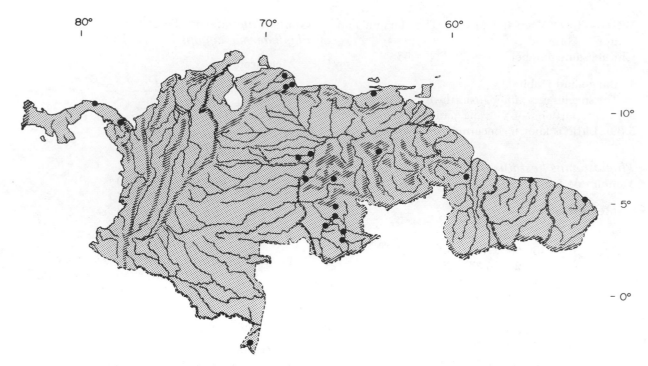

Map 5.41. Distribution of *Phylloderma stenops*.

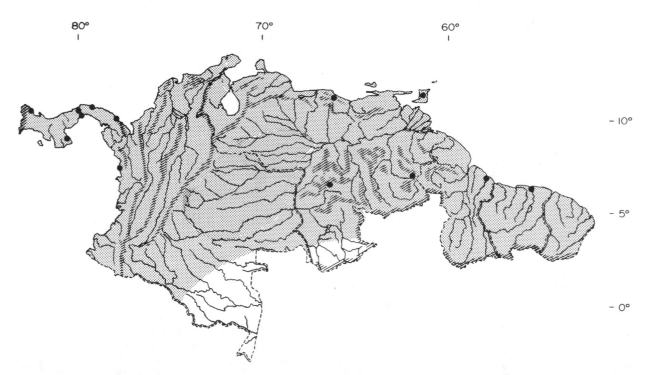

Map 5.42. Distribution of *Trachops cirrhosus*.

	Sex	Mean	S.D.	N
HF	M	20.39	0.92	38
	F	20.22	0.93	36
E	M	35.47	1.62	38
	F	35.53	1.61	36
FA	M	60.34	1.86	41
	F	60.82	1.40	39
Wta	M	33.74	3.97	31
	F	33.13	3.52	29

Location: TFA, Venezuela (USNMNH)

Description

The dentitiuon is I 2/2, C 1/1, P 2/3, M 3/3. The genus is typified by robust size. Head and body length averages 77.5 mm for males and 78.2 mm for females. Weight averages 34.2 g for males and 32.9 g for females (table 5.14). The lips, both upper and lower, are characteristic because of the great number of wartlike protrusions (see fig. 5.9). The upperparts vary from dark brown to cinnamon; the underparts are a dull brown, contrasting slightly with the dorsum.

Range and Habitat

This bat is distributed from southern Mexico south through the Isthmus and ranges broadly over the tropical portions of South America (map 5.42). In Venezuela it is found below 500 m elevation. It is strongly associated with tropical evergreen forest but occurs in regions of dry deciduous forest near moist habitats.

Natural History

This species tends to roost in caves and hollow trees. The colonies are small (fewer than six individuals). There is some evidence that the young associate with a parent for a considerable period. Although they eat insects, these bats are active predators and also feed on lizards, other bats, and frogs. In Panama, Tuttle, Taft, and Ryan (1982) report that individuals of this species may specialize on frogs. Indeed, some frog species have been under considerable selection to produce calls that render them less conspicuous to the ears of these predators (Tuttle and Ryan 1981).

Genus *Chrotopterus* Peters, 1865
Chrotopterus auritus (Peters, 1856)
Peters's Woolly False Vampire Bat

Measurements

	Sex	Mean	S.D.	N
TL	M	104.17	4.88	12
	F	109.83	5.85	6
HB	M	93.00	5.20	12
	F	99.80	2.49	5
T	M	11.17	2.08	12
	F	12.00	3.16	5
HF	M	24.83	1.64	12
	F	25.33	1.21	6
E	M	45.83	2.08	12
	F	48.00	0.89	6
FA	M	74.80	2.37	12
	F	77.87	1.07	6
Wta	M	60.23	5.05	12
	F	64.72	4.46	5

Location: TFA, Venezuela (USNMNH)

Description

The dental formula is I 2/1, C 1/1, P 2/3, M 3/3. This bat's large size and single lower incisor easily distinguish it from other phyllostomines. Forearm length ranges from 76 to 80 mm. A glandular throat pouch is conspicuous in the male, reminiscent of some species of *Phyllostomus*. The hair of the dorsum is long and soft, from light to dark brown. The venter tends to be more grayish.

Range and Habitat

C. auritus ranges from southern Mexico south through the Isthmus to southeastern Brazil (map 5.43) but appears to be absent over much of the Amazon region. In Venezuela, specimens were taken at below 500 m. The species is strongly associated with multistratal tropical evergreen forests.

Natural History

This bat feeds on insects and fruit but also preys actively on other vertebrates; Medellín (1988) reports heavy predation on small rodents in Mexico. A single young is born after a gestation exceeding 100 days. The bats roost in caves and hollow trees (Taddei 1976).

Genus *Vampyrum* Rafinesque, 1815
Vampyrum spectrum (Linnaeus, 1758)
False Vampire Bat, Vampiro Falso

Description

The dental fomula is I 2/2, C 1/1, P 2/3, M 3/3, easily separating this bat from *Chrotopterus* (see fig. 5.12). This is the largest bat in the New World tropics, easily distinguished from any other species. Measurements for males and females, respectively, are: TL 140, 158; HF 32, 38; E 46, 48; FA 105, 107; Wt 169, 199 g (USNM). It has no tail, and the interfemoral membrane is vastly reduced. The dorsum is reddish brown and the venter somewhat paler.

Range and Habitat

This species is distributed from southern Mexico to Peru, Bolivia, and southwestern Brazil. It appears to be absent over much of the Amazonian region of Brazil (map 5.44). It occurs in both dry deciduous forest and tropical evergreen forest but is strongly

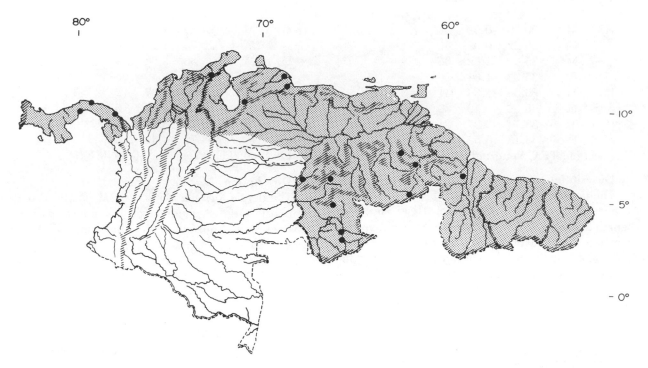

Map 5.43. Distribution of *Chrotopterus auritus*.

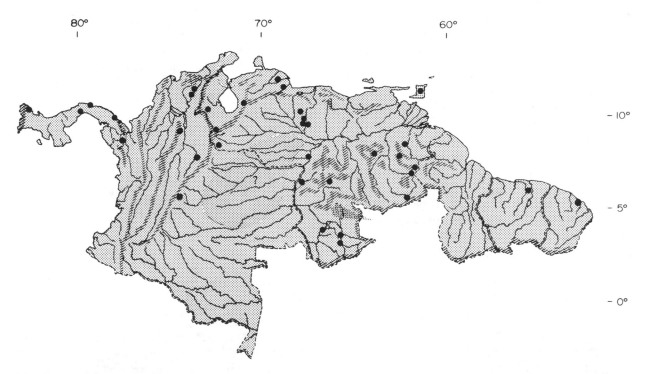

Map 5.44. Distribution of *Vampyrum spectrum*.

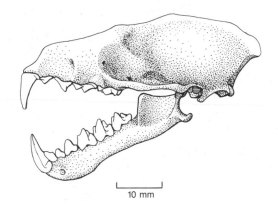

Figure 5.12. Skull of *Vampyrum spectrum*.

associated with moist habitats. It occurs at up to 1,032 m elevation in Venezuela.

Natural History

Our knowledge of the natural history of this interesting bat species has been greatly augmented by Vehrencamp, Stiles, and Bradbury (1977). It roosts in very small groups in hollow tree trunks and is apparently monogamous. Males have been observed to capture prey and bring it back to the roosting area, apparently to feed the female and dependent young. These bats are known to feed on fruit and small vertebrates, and there is reason to believe that individuals may specialize on certain types of vertebrates. In the Vehrencamp study one male actively preyed upon the groove-billed ani (*Crotophagus*); it attacked at night, killed by a bite to the brain, then transported the prey to the roosting area (Vehrencamp, Stiles, and Bradbury 1977; Navarro and Wilson 1982).

SUBFAMILY GLOSSOPHAGINAE
Long-tongued Bats

Diagnosis

For the most part, these bats are small with relatively short ears. The muzzle tends to be long, and the tongue, when extruded, is extremely long and has papillae on the tip. The tongue specialization and long rostrum correlate with nectar and pollen feeding. In conjunction with the tongue specialization, and to permit extrusion of the long tongue, in several genera the lower incisors are absent. Dental evolution in this subfamily is discussed by Phillips (1971). (See fig. 5.13 for anatomical details).

Distribution

Members of this subfamily are distributed from the southwestern United States through the Isthmus of Panama to southeastern Brazil.

Natural History

This subfamily of bats exhibits adaptations for feeding on pollen and nectar. Indeed, some species

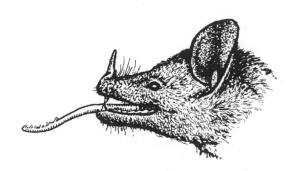

Figure 5.13. External features of the Glossophaginae: *Glossophaga longirostris*, demonstrating the long tongue and relatively long rostrum.

Table 5.7   Key to the Genera of Glossophaginae of Northern South America

1 Lower incisors present	2
1′ Lower incisors absent	5
2 Size large (head and body length greater than 75 mm); interfemoral membrane vastly reduced; tail rudimentary	*Leptonycteris*
2′ Not as above	3
3 Only two upper and lower molars per side	*Lionycteris*
3′ Three upper and lower molars per side	4
4 Upper incisors about the same length	*Glossophaga*
4′ Upper incisors unequal in length	*Lonchophylla*
5 Interfemoral membrane absent or vastly reduced; tail absent or rudimentary	*Anoura*
5′ Interfemoral membrane present; tail present but may be quite short	6
6 Only two upper and lower molars per side	*Lichonycteris*
6′ Three upper and lower molars per side	7
7 Interfemoral membrane somewhat abbreviated; first and second upper incisors separated by a distinct gap; pterygoids expanded at base, pterygoid wings in contact with the bullae	*Choeroniscus*
7′ Interfemoral membrane moderate to well developed; ptergoids normal	8
8 Upper molars lacking mesostyle	*Hylonycteris*
8′ Upper molars with mesostyle	*Scleronycteris*

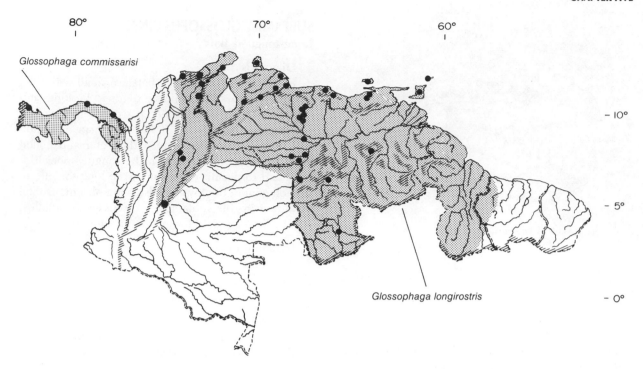

Map 5.45. Distribution of two species of *Glossophaga*.

of the genus *Leptonycteris* are important pollinators of certain plant species. On the other hand, pollen and nectar are not always available year round, and so insects and fruit are included in the diet. Adaptation for pollen and nectar feeding involves modifications of the teeth and tongue and is not equally developed among the genera within the subfamily: *Glossophaga* and *Leptonycteris* are far less specialized than are *Choeroniscus*, *Scleronycteris*, and *Anoura*. The importance of these bat species as plant pollinators and the degree of coevolution between certain plant species and bat species has yet to be explored, although an excellent start has been made in studies of the genus *Leptonycteris* (Howell 1974).

Identification of the Genera

Within the area covered by this book the genera can be identified only by reference to a combination of external and tooth characters (see table 5.7). Some genera are readily distinguished in the hand. As the group has specialized for nectar and pollen feeding, there has been a tendency to lose the lower incisors and reduces the molars to 2/2. Tooth number can vary from 34 in *Glossophaga* and *Lonchophylla* to a low of 26 in *Lichonycteris*. By inspecting the lower incisors one can immediately decide between two groups, those with 2 lower incisors per side (*Glossophaga*, *Lonchophylla*, *Lionycteris*, and *Leptonycteris*) and the other genera that have no lower incisors. Among the four genera with 2 lower incisors per side, *Leptonycteris* distinctively exhibits a rudimentary tail and a vastly reduced interfemoral

membrane. The genus *Lionycteris* has a low number of teeth (30), since it has reduced its molars to 2/2 per side.

The genera with no lower incisors are more difficult to separate, but *Anoura* is distinctive because of its loss of the interfemoral membrane and its rudimentary tail (see table 5.7).

Genus *Glossophaga* E. Geoffroy, 1818
Long-tongued Bat, Murciélago Nectario

Description

The dental formula is I 2/2, C 1/1, P 2/3, M 3/3 (see fig. 5.14). These are small bats, with head and body length ranging from 48 to 65 mm and forearm length from 32 to 42 mm. The interfemoral membrane is reduced but clearly visible, and the tail is less than half the length of the interfemoral membrane. Color varies from dark to reddish brown.

Distribution

The genus is distributed from Mexico to southeastern Brazil and extreme northeastern Argentina.

Natural History

These bats feed on insects, fruits, pollen, nectar, and flower parts.

Glossophaga commissarisi Gardner, 1962

Description

The lower incisors are very small. Head and body length ranges from 43 to 60 mm; HF 9–13; E 11–16; FA 33–36.

Range and Habitat

The species is found in western Mexico from Sinaloa south through Panama (map 5.45).

Glossophaga longirostris Miller, 1898

Measurements

	Sex	Mean	S.D.	N
TL	M	70.55	3.01	107
	F	71.50	3.40	56
HB	M	64.70	3.17	92
	F	65.53	3.23	43
T	M	5.72	1.20	92
	F	6.28	1.44	43
HF	M	12.91	0.61	107
	F	13.00	0.93	56
E	M	16.51	0.83	107
	F	16.41	0.89	56
FA	M	38.18	0.93	147
	F	38.57	0.97	70
Wta	M	13.07	1.32	103
	F	13.64	0.98	30

Location: Northwestern Venezuela (USNMNH)

Description

This is the largest of the three species that occur in northern South America. Head and body length averages 61.6 mm for males and 61.4 mm for females; weight averages 12.8 g for males and 12.9 g for females (see table 5.14). The dorsum is light brown.

Chromosome number: $2n = 32$; FN = 60.

Range and Habitat

This species is distributed in northeastern Colombia and across northern Venezuela to Guyana. In Venezuela most specimens were taken at below 500 m elevation (map 5.45). This bat is adapted to tropical dry deciduous forest habitats; it is the common nectar-feeding bat in the llanos.

Natural History

These bats are opportunistic roosters, having been found in caverns, rocks, and crevices as well as hollow trees. They roost in small colonies, often with other species (Goodwin and Greenhall 1961).

Glossophaga soricina (Pallas, 1766)

Measurements

	Sex	Mean	S.D.	N
TL	M	62.31	4.02	80
	F	64.99	4.10	69
HB	M	55.24	4.04	80
	F	57.12	3.86	68
T	M	7.08	1.63	80
	F	7.91	1.67	68

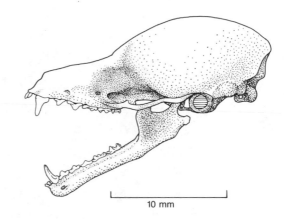

Figure 5.14. Skull of *Glossophaga soricina*.

	Sex	Mean	S.D.	N
HF	M	10.83	0.91	80
	F	10.96	0.78	69
E	M	13.99	1.43	80
	F	14.06	1.14	69
FA	M	34.36	0.86	108
	F	35.21	0.87	110
Wta	M	9.53	1.26	53
	F	10.05	1.09	35

Location: Northern Venezuela (USNMNH)

Description

Head and body length averages 53.9 mm for males and 54.9 mm for females; weight averages 9.5 g for males and 9.4 g for females (see table 5.14). The dorsum is dark brown to light brown (see plate 5). Chromosome number: $2n = 32$; FN = 60.

Range and Habitat

This species occurs from southern Sonora in western Mexico south through the Isthmus to northeastern Argentina and southeastern Brazil (map 5.46). It prefers moist situations and is rarely found in dry areas. It tolerates man-made clearings and croplands, but in undisturbed areas it prefers multistratal tropical evergreen forests. In northern Venezuela most specimens were caught below 500 m elevation, but it can range to 1,560 m (Handley 1976).

Natural History

This species forms maternity colonies in shelters such as caves and hollow trees. Several hundred females and their young can roost together, but Goodwin and Greenhall (1961) report colonies of twelve to sixteen in Trinidad. Normally a single young is born. The time of reproduction can be strongly seasonal in habitats with pronounced rainfall cycles, but females are polyestrous and can bear two or three young per year. Though fully capable of hovering in flight while taking nectar from flowers, this bat is also to some extent a foliage-gleaning in-

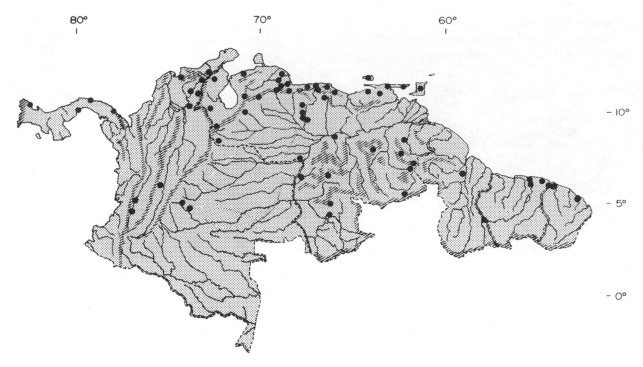

Map 5.46. Distribution of *Glossophaga soricina*.

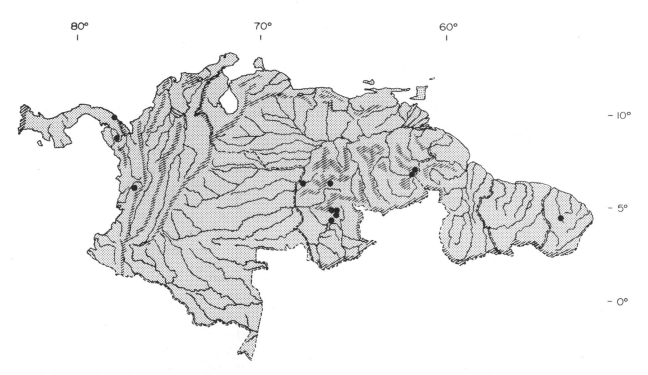

Map 5.47. Distribution of *Lionycteris spurrelli*.

sectivore. During the period of nectar production by *Agave*, individual bats will defend the plants against conspecifics to ensure an exclusive supply of nectar and pollen (Lemke 1984; Howell 1983).

Genus *Lionycteris* Thomas, 1913
Lionycteris spurrelli Thomas, 1913

Measurements

	Sex	Mean	S.D.	N
TL	M	62.68	2.44	28
	F	63.98	2.57	57
HB	M	54.57	3.10	28
	F	55.74	2.88	57
T	M	8.11	1.40	28
	F	8.25	2.06	57
HF	M	11.39	0.74	28
	F	11.39	0.84	57
E	M	13.25	0.93	28
	F	13.16	1.13	57
FA	M	34.47	0.74	32
	F	35.59	0.87	62
Wta	M	9.15	1.86	29
	F	9.11	1.15	49

Location: Bolívar, Venezuela (USNMNH)

Description
The dental formula is I 2/2, C 1/1, P 2/3, M 2/2. Head and body length averages 53.5 mm for males and 54.8 mm for females; weight averages 8.7 g for males and 8.9 g for females (see table 5.14). The tail extends about half the length of the well-developed interfemoral membrane. The coat is reddish brown. Chromosome number: $2n = 28$; FN = 50.

Range and Habitat
This species occurs from the extreme east of Panama southward to the east of the Andes and across northern South America, extending to west-central Brazil (map 5.47). It occurs from 135 to 1,400 m elevation in Venezuela. It prefers moist areas for foraging and is strongly associated with multistratal tropical evergreen forest (Handley 1976).

Natural History
These bats roost in caves and crevices. The diet has not been recorded but is probably similar to that of *Lonchophylla*.

Genus *Lonchophylla* Thomas, 1903

Description
The dental formula is I 2/2, C 1/1, P 2/3, M 3/3. Head and body length ranges from 45 to 60 mm and the short tail from 8 to 10 mm; weight ranges from 6 to 14 g. Species of this genus are graded in size, with *L. thomasi* the smallest, *L. mordax* intermediate, and *L. robusta* the largest. The uropatagium is well developed. The dorsum is rusty or dark brown, the venter somewhat paler.

Distribution
The genus is distributed from Nicaragua to Brazil. Some species show an extremely disjunct range.

Natural History
These bats frequently roost in caves. They are specialized for feeding on flowers and are strongly implicated in the pollination of night-blooming plants. In common with other members of the subfamily, they are not confined to pollen and nectar but feed on insects and fruit as well (Gardner 1977a).

Lonchophylla robusta Miller, 1912

Measurements

	Sex	Mean	S.D.	N
TL	M	76.64	2.38	11
	F	78.00	3.03	6
HB	M	67.64	2.73	11
	F	69.50	2.81	6
T	M	9.00	0.77	11
	F	8.50	1.05	6
HF	M	13.27	0.90	11
	F	13.33	0.52	6
E	M	16.73	0.65	11
	F	16.83	0.41	6
FA	M	42.64	1.16	11
	F	42.89	1.17	8
Wta	M	14.20	1.08	11
	F	14.00	0.50	3

Location: North-central Venezuela (USNMNH)

Description
This is the largest species of the genus. Head and body length averages 69.4 mm for males and 69.3 mm for females; Weight averages 14.3 g for males and 13.7 g for females (see table 5.14). The dorsum and venter are light brown. Chromosome number: $2n = 28$; FN = 50.

Range and Habitat
This species ranges from southern Nicaragua through the Isthmus to western Colombia and northwestern Venezuela (map 5.48). It tolerates elevations of 75 to 1,135 m in Venezuela and is strongly associated with multistratal tropical evergreen forests and moist areas (Handley 1976).

Lonchophylla mordax Thomas, 1903

Description
Head and body length ranges from 55 to 58 mm, placing this species between *L. robusta* and *L. thomasi*. The rostrum is considerably longer than in *L. thomasi*. The dorsum is reddish brown, the venter paler.

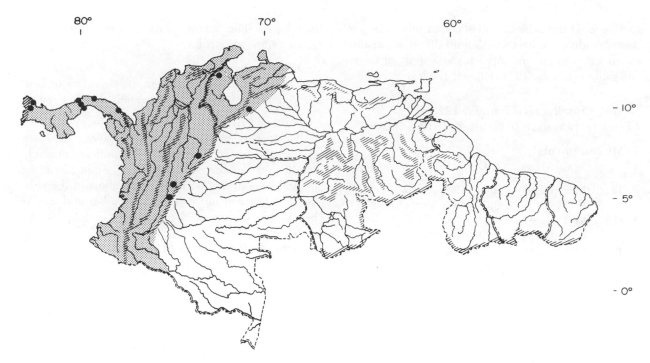

Map 5.48. Distribution of *Lonchophylla robusta*.

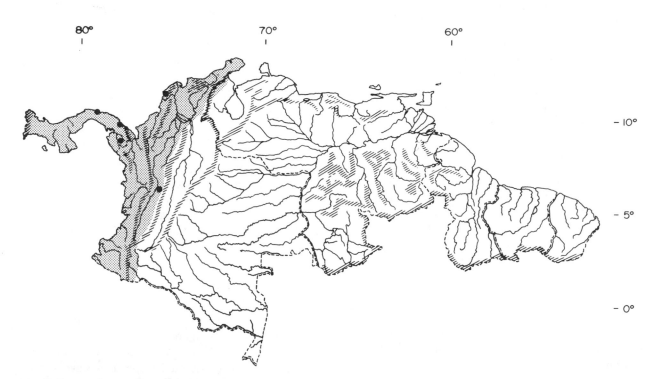

Map 5.49. Distribution of *Lonchophylla mordax*.

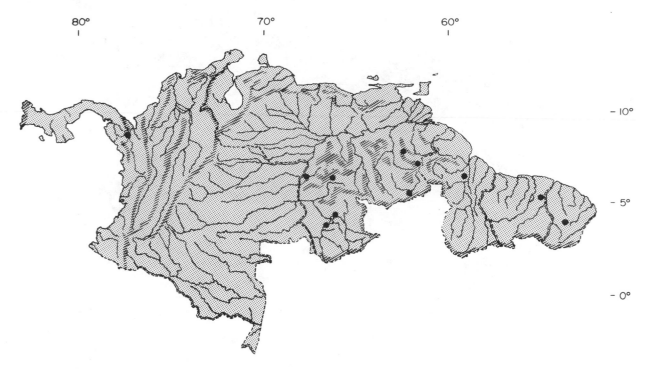

Map 5.50. Distribution of *Lonchophylla thomasi*.

Range and Habitat

This species, according to Koopman (1982), has a disjunct distribution. It is recorded from Costa Rica through the Isthmus of Panama and in western Colombia (map 5.49). Specimens referable to *L. mordax* are also described from east-central Brazil. No specimens have been recorded in areas between these two populations.

Lonchophylla thomasi J. A. Allen, 1904

Measurements

	Sex	Mean	S.D.	N
TL	M	58.00	2.24	5
	F	56.67	3.56	6
HB	M	50.20	2.68	5
	F	49.00	4.00	6
T	M	7.80	1.64	5
	F	7.67	1.63	6
HF	M	10.00	0.71	5
	F	10.17	0.75	6
E	M	15.00	1.41	5
	F	15.00	1.26	6
FA	M	31.86	0.90	5
	F	31.67	0.83	6
Wta	M	6.85	0.60	4
	F	6.10	0.72	5

Location: TFA, Venezuela (USNMNH)

Description

This is the smallest species of *Lonchophylla*. The dorsum and venter are light brown.
Chromosome number: $2n = 30–32$; FN = 34–38.

Range and Habitat

There is a disjunct distribution in northern South America (map 5.50), strongly associated with streams and moist areas. Although tolerant of man-made clearings, this species prefers multistratal tropical evergreen forests. All specimens in Venezuela were caught below 851 m elevation (Handley 1976).

Genus *Anoura* Gray, 1838
Tailless Bat

Description

The dental formula is I 2/0, C 1/1, P 3/3, M 3/3 (see fig. 5.15). There are six species of this genus; one, *A. latidens*, has recently been described by Handley from Venezuela and is similar in size to *A. caudifer*. Where three species co-occur, in general they are graded in size, ranging in head and body length from 50 to 90 mm and in forearm length

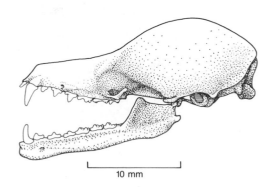

10 mm

Figure 5.15. Skull of *Anoura geoffroyi*.

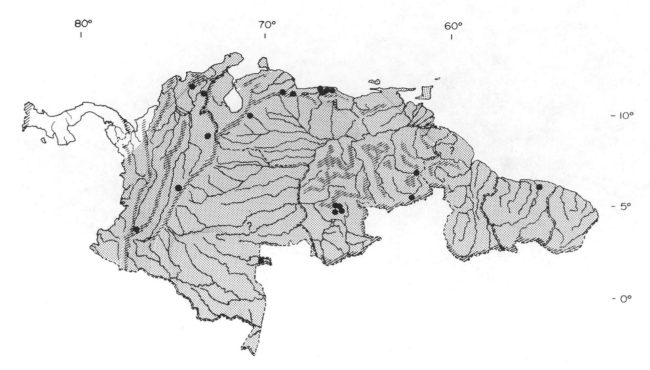

Map 5.51. Distribution of *Anoura caudifer*.

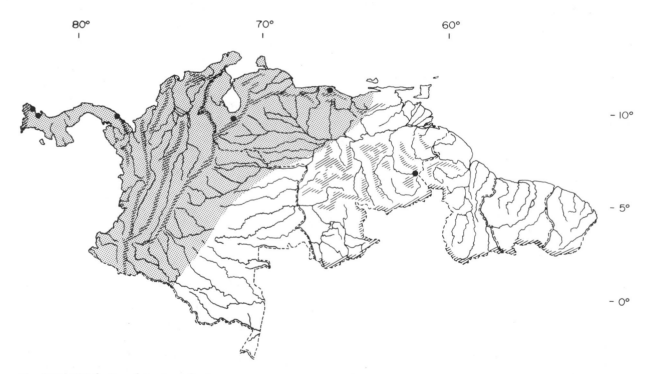

Map 5.52. Distribution of *Anoura cultrata*.

from 34 to 48 mm. The tail is rudimentary or absent, and the interfemoral membrane is greatly reduced.

Distribution

The genus is distributed from Sinaloa in western Mexico south through the Isthmus of Panama to southeastern Brazil. Species of *Anoura* appear to be absent from the central Amazonian region.

Natural History

Anoura feeds on fruit, nectar, pollen, and insects (Gardner 1977a).

Anoura caudifer (E. Geoffroy, 1818)

Measurements

	Sex	Mean	S.D.	N
TL	M	65.83	2.72	12
	F	65.38	3.46	16
HB	M	61.17	2.59	12
	F	60.75	3.24	16
T	M	4.67	0.65	12
	F	4.63	0.72	16
HF	M	12.08	0.29	12
	F	12.06	0.44	16
E	M	14.67	0.49	12
	F	14.44	0.73	16
FA	M	36.21	0.84	19
	F	36.29	0.88	24
Wta	M	10.31	0.94	12
	F	9.98	0.75	15

Location: North-central Venezuela (USNMNH)

Description

This is one of the smaller species of *Anoura*. The upperparts are dark brown, the venter paler brown. Chromosome number: $2n = 30$; FN = 56.

Range and Habitat

A. caudifer is distributed over northern South America; it is absent from the Amazonian region but broadly distributed over north-central Bolivia east to southeastern Brazil (map 5.51). In Venezuela it occurs from 500 to 1,500 m elevation. It is strongly associated with streams within multistratal tropical evergreen forest (Handley 1976).

Anoura cultrata Handley, 1960

Description

This is the largest species of the genus. The male is considerably larger than the female. Mean measurements for males and females, respectively, are: TL 79, 71; T 4, 4.5; HF 14.5, 14.5; E 16, 17; FA 42, 40; Wt 20.4, 14.9 g ($N = 4$; USNM). The dorsal pelage is dark brown to black. Chromosome number: $2n = 30$; FN = 54.

Range and Habitat

A. cultrata ranges from Costa Rica to northwestern Venezuela and in the western part of Colombia, south to Amazonian Peru (map 5.52). It occurs at modest elevations from 200 to 1,870 m in Venezuela, and its preferred habitat is moist premontane forest. It roosts in caves (Tamsitt and Nagorsen 1982).

Anoura geoffroyi Gray, 1838

Measurements

	Sex	Mean	S.D.	N
TL	M	65.86	4.51	21
	F	65.12	2.87	33
HB	M	65.86	4.51	21
	F	65.12	2.87	33
HF	M	13.14	0.85	21
	F	13.21	0.93	33
E	M	15.19	1.08	21
	F	15.52	1.23	33
FA	M	42.63	0.87	20
	F	42.86	1.19	32
Wta	M	15.08	1.34	10
	F	14.25	1.95	4

Location: Bolívar, Venezuela (USNMNH)

Description

This species is slightly larger than *A. caudifer*. Head and body length averages 69 mm for males and 68 mm for females; weight averages 15 g for males and 13 g for females. This species lacks a tail, which immediately sets it off from *A. caudifer* and *A. cultrata*, both of which have tiny tails, but in all species of the genus the interfemoral membrane is distinctively reduced. The dorsum is dull brown blending to gray on the shoulders, and the venter is gray. Chromosome number: $2n = 30$; FN = 56.

Range and Habitat

This species occurs from Sinaloa in western Mexico and Tamaulipas in eastern Mexico south through the Isthmus, across northern South America, through Peru and Bolivia to east-central Brazil (map 5.53). It appears to be absent from most of the Amazonian region. In Venezuela most specimens were taken at below 1,500 m. The species is broadly tolerant of man-made clearings and is strongly associated with moist areas and multistratal tropical evergreen forest (Handley 1976).

Natural History

A. geoffroyi has been implicated in the pollination of certain night-blooming plants, including *Eperua falcata*. This species roosts in caves in groups of up to fifty. Seasonally, sexes can be segregated or roost

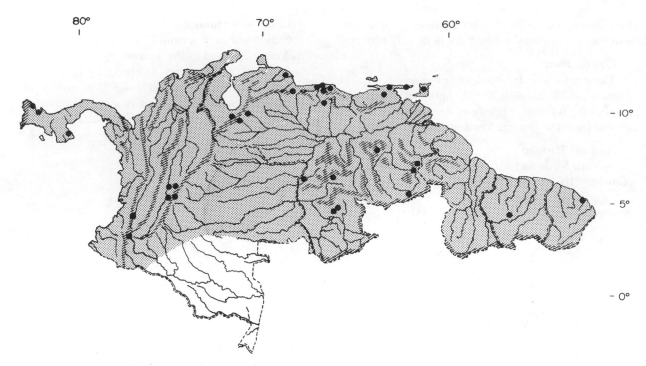

Map 5.53. Distribution of *Anoura geoffroyi*.

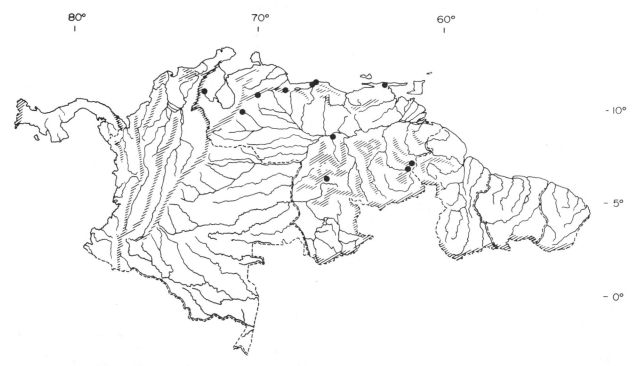

Map 5.54. Distribution of *Anoura latidens*.

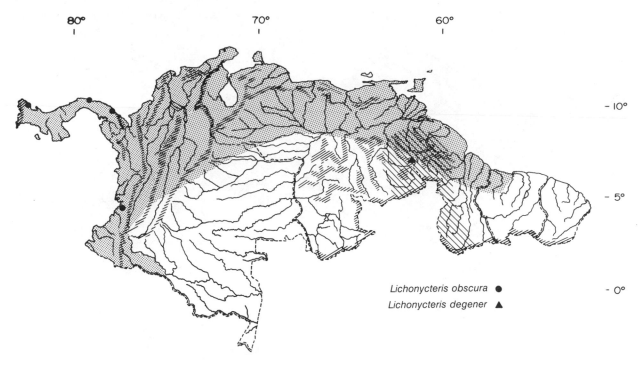

Map 5.55. Distribution of two species of *Lichonycteris* (total ranges from Koopman 1982).

in mixed-sex colonies. This bat is highly insectivorous at certain seasons of the year (Gardner 1977a; Wilson 1979).

Anoura latidens Handley, 1984

Measurements

	Sex	Mean	S.D.	N
TL	M	64.55	2.88	11
	F	67.31	4.00	16
HB	M	64.55	2.88	11
	F	67.31	4.00	16
HF	M	12.27	0.65	11
	F	12.31	0.70	16
E	M	13.91	1.38	11
	F	13.50	1.41	16
FA	M	41.81	0.85	15
	F	42.34	0.88	16
Wta	M	15.00	1.46	4
	F	13.35		2

Location: Northern Venezuela (USNMNH)

Description
This species is newly describd by Handley (1984). Males average 64.4 mm in head and body length, while females average 66.2 mm. Weight averages 14.5 g for males and 14.8 g for females.

Range and Habitat
Thus far this species has been found widely distributed in Venezuela (map 5.54). Strongly associated with moist areas and multistratal tropical evergreen forests, in Venezuela it tolerates elevations from 50 to 2,240 m.

Genus *Lichonycteris* Thomas, 1895
Lichonycteris obscura Thomas, 1895

Description
The dental formula is I 2/0, C 1/1, P 2/3, M 2/2. Head and body length averages 46 to 55 mm and forearm length 30 to 33.9. The upperparts are uniform dark brown and the underparts slightly darker (Hall 1981). *L. degener* is a synonym according to Gardner (1976), but Honacki, Kinman, and Koeppl (1982) retain the separation.
Chromosome number: $2n = 24$; FN = 44.

Range and Habitat
This species occurs from Honduras south through the Isthmus of Panama, through western Venezuela to northern Peru (map 5.55). It has been sporadically reported across northern Venezuela, Guyana, and Suriname in moist evergreen forests.

Natural History
This species visits flowers and probably feeds on nectar, pollen, and insects (Gardner 1977a).

Genus *Choeroniscus* Thomas, 1928
Hog-nosed Bat

Description
The dental formula is I 2/0, C 1/1, P 2/3, M 3/3. Head and body length ranges from 50 to 55 mm,

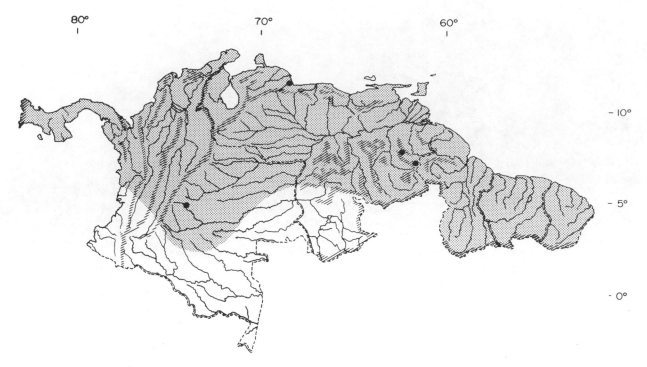

Map 5.56. Distribution of *Choeroniscus godmani.*

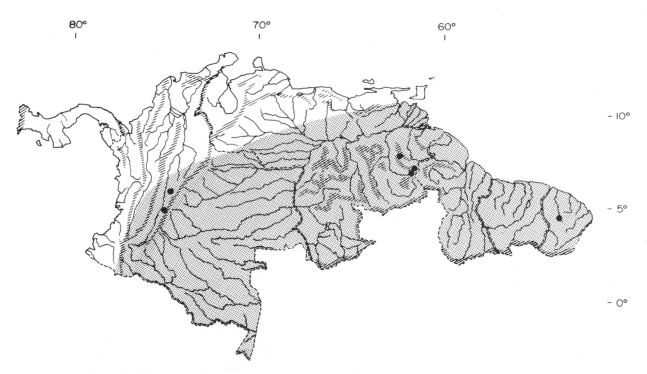

Map 5.57. Distribution of *Choeroniscus minor.*

with the tail averaging about 12 mm. Forearm length ranges from 32 to 38 mm. The pelage is usually dark to light brown above, with the venter paler to nearly the same color.

Distribution

Species of this genus are distributed from Sinaloa, Mexico, and south in western Mexico to Central America. In South America, species are distributed mostly in the northern parts. Some species tolerate rather arid habitats.

Choeroniscus godmani (Thomas, 1903)

Measurements

	Sex	Mean	S.D.	N
TL	M	59.75	4.57	4
	F	64.63	2.56	8
HB	M	53.75	3.30	4
	F	55.63	2.92	8
T	M	6.00	2.45	4
	F	9.00	1.31	8
HF	M	9.50	0.58	4
	F	9.88	0.99	8
E	M	8.25	5.56	4
	F	12.14	0.90	7
FA	M	31.87	0.71	3
	F	34.00	0.80	8
Wta	M	7.27	0.06	3
	F	7.75	0.91	8

Location: Bolívar, Venezuela (USNMNH)

Description

The muzzle is exceptionally long. This, together with the small nose leaf and relatively small ears, serves to identify the species. Coloration is uniformly dark brown on the dorsum and paler on the venter.
Chromosome number: $2n = 20$; FN $= 32$ or 36.

Range and Habitat

This species occurs from Sinaloa in western Mexico south through Central America and across northern South America (map 5.56). In Venezuela it ranges from 2 to 350 m elevation and is strongly associated with moist habitats and multistratal tropical evergreen forest. It may frequent orchards (Handley 1976).

Natural History

These bats roost in groups of twelve to twenty-four. They feed on fruit, pollen, nectar, and insects.

Choeroniscus minor (Peters, 1868)

Description

In spite of the name, this species is larger than *C. godmani.* Its measurements are TL 69–71; T 9; HF 9–10; E 12–13; Wt 10 g (Suriname, CMNH) (see also table 5.14). The dorsum and venter are light brown.

Range and Habitat

The species is confined to South America, including the southern parts of Venezuela, the Guyanas, southeastern Colombia, Ecuador, Peru, and northern Brazil (map 5.57). It is associated with moist areas in multistratal tropical evergreen forests (Handley 1976).

Choeroniscus intermedius (J. A. Allen and Chapman, 1893)

Description

Body proportions are approximately the same as for *C. minor.* Measurements of one female were TL 69; T 9; HF 10; E 13 (Suriname, CMNH). This very dark bat has an exceptionally long muzzle. Chromosome number: $2n = 20$; FN $= 36$.

Range and Habitat

This species occurs in the Guyanas and extreme eastern Venezuela and ranges south into northern Brazil (map 5.58).

Natural History

These bats roost in tree hollows in groups of up to eight (Goodwin and Greenhall 1961).

Choeroniscus periosus Handley, 1966

Description

This is a medium-sized species: TL 62; T 10; FA 41.2 (Handley 1966). The dorsum is brown and the venter lighter.

Range and Habitat

This species has been reported from the extreme southwest of Colombia (map 5.58).

Genus *Hylonycteris* Thomas, 1903
Hylonycteris underwoodi Thomas, 1903

Measurements

	Sex	Range	N
HB	M	49–66	6
	F	51–72	7
T	M	6–9	6
	F	7–11	7
FA	M	31.4–33.7	17
	F	31.5–35.9	7

Source: Hall (1981)

Description

The dental formula is I 2/0, C 1/1, P 2/3, M 3/3, and the molars are very slender. The species is similar to *Choeroniscus* except for skull characters. The

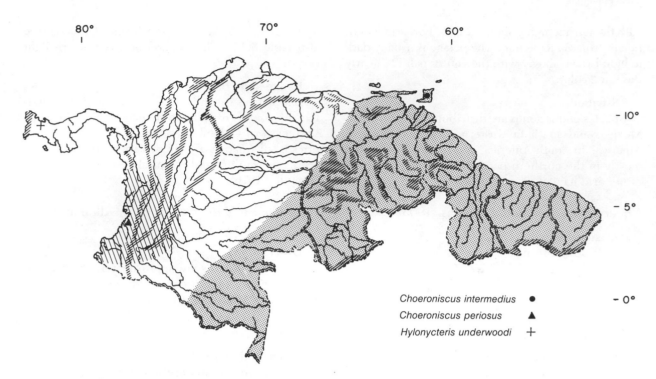

Map 5.58. Distribution of two species of *Choeroniscus* and *Hylonycteris underwoodi* (range of *Choeroniscus* from Koopman 1982).

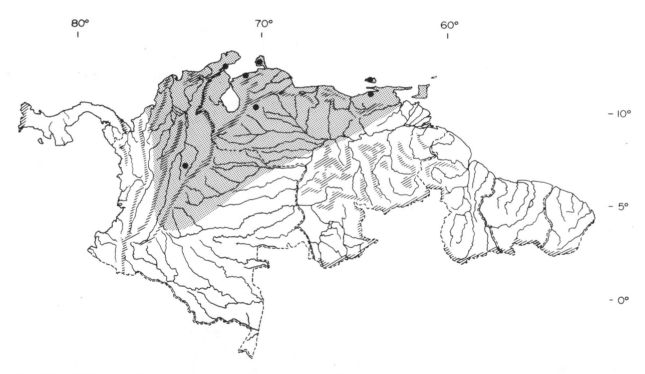

Map 5.59. Distribution of *Leponycteris curasoae* (this form may include *L. sanborni*; H. Arita, pers. comm.).

pterygoids are convex medially and not in contact with the bullae; the muzzle is long, and lower incisors are lacking. The dorsal varies from dark gray above and below to dark brown above with a pale brown venter.

Range and Habitat

The species ranges from western Mexico to western Panama and is not recorded from South America (map 5.58).

Natural History

This bat visits flowers, feeding on nectar, pollen, and insects (Gardner 1977a).

Genus *Leptonycteris* Lydekker, 1891
Leptonycteris curasoae Miller, 1900

Measurements

	Sex	Mean	S.D.	N
TL	M	82.22	3.86	65
	F	82.32	3.18	38
HB	M	82.22	3.86	65
	F	82.32	3.18	38
HF	M	16.74	0.85	65
	F	16.84	0.79	38
E	M	17.95	1.10	65
	F	17.95	0.70	38
FA	M	53.42	1.16	101
	F	53.40	0.91	43
Wta	M	26.80	2.34	65
	F	25.01	1.84	37

Location: Northwestern Venezuela (USNMNH)

Description

The dental formula is I 2/2, C 1/1, P 2/3, M 2/2. This bat is easily distinguished from all other species of the Glossophaginae in northern South America by its large size. It lacks a tail, a character it shares with the genus *Anoura*. The dorsum is dark brown with reddish brown on the shoulders; the venter is gray.

Range and Habitat

The genus *Leptonycteris* ranges from southern Texas and southern Arizona south through the Isthmus to northern South America. Only one species, *L. curasoae*, occurs in northern South America, where it is sporadically distributed along the Caribbean coast (map 5.59). Although it occasionally forages along streams and near ponds, it is strongly associated with dry deciduous tropical forests.

Natural History

A great deal of research has been done on the northern species of this genus (*L. sanborni*), which is an important pollinator of certain species of cactus

in the southwestern United States. The northern species forms nursery colonies in caves that can include several hundred individuals. The habits of *L. curasoae* are very poorly known.

Genus *Scleronycteris* Thomas, 1912
Scleronycteris ega Thomas, 1912

Description

The dental formula is I 2/0, C 1/1, P 2/3, M 3/3. Measurements of the type specimen are: HB 57; T 6; FA 35. The dorsum is dark brown, the venter light brown. This species is very poorly known.

Range and Habitat

This bat is known only from extreme southern Venezuela and adjacent portions of Brazil (map 5.60). In Venezuela the single specimen collected was caught in multistratal tropical evergreen forest at 135 m elevation.

SUBFAMILY CAROLLIINAE

Diagnosis

The muzzle is long and narrow but not as pronounced as in the Glossophaginae. The ears are relatively longer than in the Glossophaginae; when laid forward they reach almost to the tip of the nose. The naked pad on the chin with the large central O-shaped wart is diagnostic. No facial stripes are present. The tail is extremely reduced or absent, but the interfemoral membrane is still present, though much less extensive than in genera having longer tails. External features are shown in figure 5.16, and a key is given in table 5.8.

Distribution

These bats range from Sinaloa, Mexico, south through the Isthmus across South America to extreme southeastern Brazil and northeastern Argentina.

Table 5.8   Key to the Carolliinae

1	Short tail present (*Carollia*) 2
1'	Short tail absent (*Rhinophylla*) 5
2	Fur of dorsum long, forearm sparsely haired . . . . *C. subrufa* [a]
2'	Forearm well haired 3
3	Tail less than 7 mm . . . . *C. brevicauda*
3'	Tail greater than 7 mm 4
4	Head and body length less than 60 mm . . . . *C. castanea*
4'	Head and body length greater than 60 mm . . . *C. perspicillata*
5	Head and body length greater than 50 mm . . . . *R. alethina*
5'	Head and body length less than 50 mm . . . . *R. pumilio*

[a] Western Panama only.

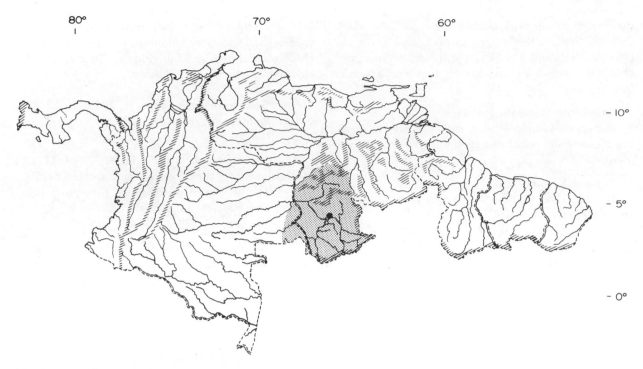

Map 5.60. Distribution of *Scleronycteris ega*.

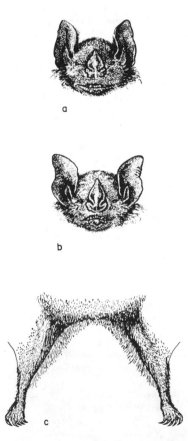

Figure 5.16. External features of the Carolliinae and Sturnirinae: (*a*) Face of *Sturnira lilium*. (*b*) Face of *Carollia perspicillata;* note the single large median wart on the lower lip. (*c*) Fringe of hairs and absence of uropatagium in *Sturnira lilium*.

Genus *Carollia* Gray, 1838
Short-tailed Bat, Leaf-nosed Bat, Murciélago Colicorto

Description
The dental formula is I 2/2, C 1/1, P 2/2, M 3/3 (see fig. 5.17). Head and body length ranges from 48 to 65 mm. The tail, which is lacking in the genus *Rhinophylla*, ranges from 3 to 14 mm. The dorsum is dark brown to reddish brown, and the venter is similar. The genus has been monographed by Pine (1972).

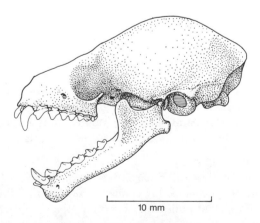

10 mm

Figure 5.17. Skull of *Carollia perspicillata*.

Distribution

The genus is distributed from southern Sinaloa, Mexico, to northeastern Argentina and southeastern Brazil.

Natural History

These bats are generalized feeders, eating fruits, flowers, and insects. They use a variety of roosting sites, including caves, hollow trees, and crevices.

Identification

Owen, Schmidley, and Davis (1984) have developed criteria for the field identification of the species of *Carollia*.

Carollia brevicauda (Schinz, 1821)

Measurements

	Sex	Mean	S.D.	N
TL	M	64.65	2.17	26
	F	64.58	1.72	26
HB	M	58.46	1.79	26
	F	58.46	1.53	26
T	M	6.19	1.02	26
	F	6.12	1.21	26
HF	M	14.00	0.00	26
	F	14.19	0.40	26
E	M	19.77	0.43	26
	F	19.65	0.56	26
FA	M	38.27	0.88	58
	F	38.55	0.81	58
Wta	M	13.77	0.74	26
	F	12.52	0.89	24

Location: North-central Venezuela (USNMNH)

Description

This species is characterized by an extremely short tail, averaging less than 6.7 mm, but *C. subrufa* also has a short tail. Although the head and body length is similar to that of *C. perspicillata*, the forearm is considerably shorter, averaging 38.5 mm for males and 38.6 mm for females (see table 5.14). The dorsum and venter are approximately the same shade of medium brown.
Chromosome number: $2n = 20-21$; FN = 36.

Range and Habitat

This species ranges from southern Veracruz, Mexico, south through the Isthmus over most of tropical South America to central Brazil (map 5.61). It is strongly associated with moist habitats for foraging and typically prefers multistratal tropical evergreen forests. It is broadly tolerant of a wide range of elevations. Roosts are found in caves, rock crevices, houses, and under the leaves of *Musa*. In northern Venezuela it occurs at from 24 to 2,147 m (Handley 1976).

Carollia castanea H. Allen, 1890

Measurements

	Sex	Mean	S.D.	N
TL	M	63.75	3.86	4
	F	62.60	5.13	5
HB	M	53.75	4.35	4
	F	53.80	6.38	5
T	M	10.00	0.82	4
	F	8.80	1.64	5
HF	M	13.50	1.00	4
	F	12.20	1.48	5
E	M	17.75	0.96	4
	F	18.60	0.89	5
FA	M	36.50	0.85	3
	F	35.73	0.88	4
Wta	M	14.55		2

Location: Northwestern Venezuela (USNMNH)

Description

This is the smallest of the three species occurring in northern South America (see table 5.14). It is similar in appearance to *C. brevicauda*.
Chromosome number: $2n = 20-22$; FN = 36-38.

Range and Habitat

This species is distributed from Honduras south through the Isthmus in the western portion of South America to Bolivia (map 5.62). It penetrates eastward in western Venezuela to the Guyanas but is absent over a vast portion of Brazil. In Venezuela it is strongly associated with multistratal tropical evergreen forests, and all specimens taken there were below 460 m elevation (Handley 1976).

Carollia perspicillata (Linnaeus, 1758)

Measurements

	Sex	Mean	S.D.	N
TL	M	66.40	5.22	209
	F	66.12	5.17	253
HB	M	56.14	6.12	207
	F	55.70	6.06	253
T	M	10.30	2.21	207
	F	10.43	2.23	253
HF	M	14.41	0.93	209
	F	14.49	1.06	253
E	M	20.81	1.23	208
	F	20.73	1.11	252
FA	M	40.77	1.06	220
	F	40.92	1.22	263
Wta	M	17.52	2.37	40
	F	17.21	1.78	12

Location: Northwestern Venezuela (USNMNH)

Description

This species has roughly the same head and body length and color as *Carollia brevicauda*, but the

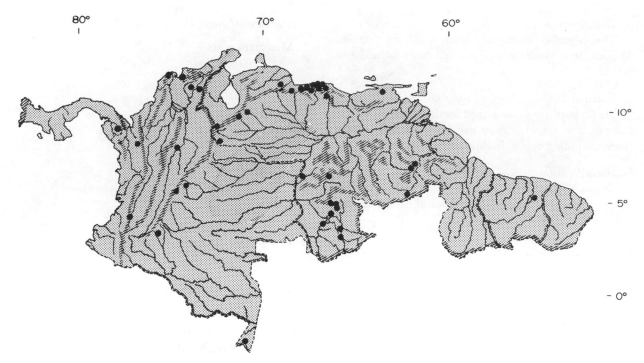

Map 5.61. Distribution of *Carollia brevicauda*.

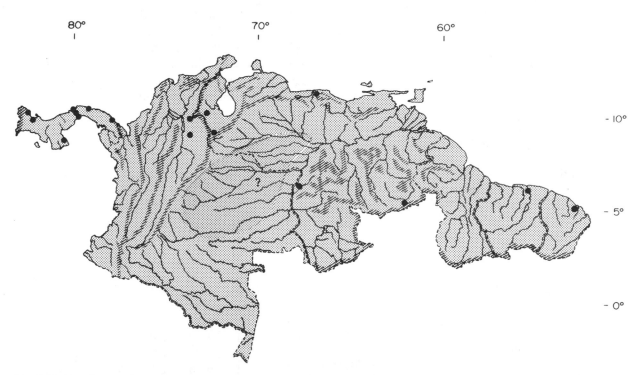

Map 5.62. Distribution of *Carollia castanea*.

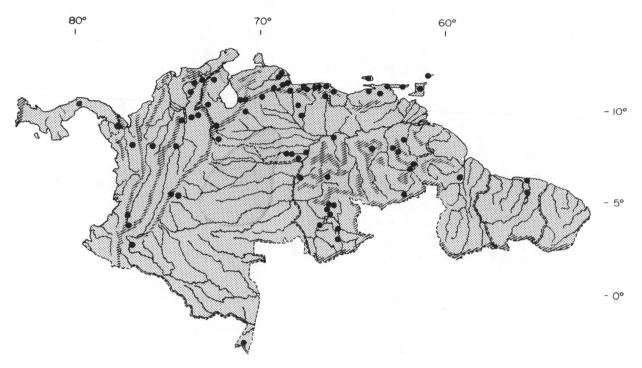

Map 5.63. Distribution of *Carollia perspicillata*.

forearms of males and females average 41 mm, and weights average 17 g for males and 16 g for females. The relatively longer tail, averaging approximately 10 mm, distinguishes *C. perspicillata* from *C. bre-vicauda* (see table 5.14).
Chromosome number: $2n = 20–21$; FN = 36.

Range and Habitat
This species is distributed from southern Ve-racruz south through the Isthmus, across northern South America and south through Amazonian Peru, Bolivia, and Brazil (map 5.63). In Venezuela it for-ages near moist areas, being taken most frequently in multistratal tropical evergreen forests. It has a wide altitudinal tolerance, ranging to 1,260 m eleva-tion in Venezuela (Handley 1976).

Natural History
This is one of the best-studied species of the genus. *Carollia* depends on the fruit of *Piper* for the major portion of its diet, but it also gleans foliage for insects. Males defend small groups of females in a harem breeding system. The bats roost in hollow trees, caves, crevices, and other moist places in colonies of up to one hundred. Over most parts of their range they have two birth peaks; a female thus produces two young annually (Fleming 1983). Inter-birth intervals range from 115 to 173 days, and ges-tation is approximately 115 to 120 days. The young are born in an advanced state with the eyes open. The newborn remains more or less continuously at-tached to the mother for the first fourteen days of life. Young weigh approximately 5 g at birth (Kleiman and Davis 1979).

Carollia subrufa (Hahn, 1905)

Description
Total length ranges from 68 to 73 mm; HF 12–15; E 17–21; FA 37.6–40.6. The dorsum is gray brown to pale gray. The dorsal fur is short, and the forearm is only sparsely haired, in contrast to *C. brevicauda*. These two species are easily confused in areas of sympatry (see Owen, Schmidley, and Davis 1984).

Range and Habitat
The species is distributed from western Mexico to Costa Rica and may reach Panama.

Genus *Rhinophylla* Peters, 1865

Description
The dental formula is I 2/2, C 1/1, P 2/2, M 3/3. This genus is similar to *Carollia* but is immediately distinguishable by its lack of a tail. The interfemoral membrane is small but still conspicuous, not nearly as reduced as that of *Sturnira*. The dorsum and ven-ter are medium brown to very dark brown.

Distribution
The genus is distributed solely in South America, ranging over the northern portion with *R. pumilio* extending into south-central Brazil.

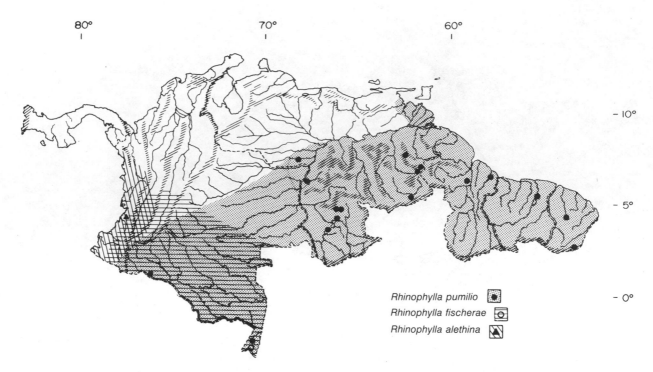

Map 5.64. Distribution of three species of *Rhinophylla* (ranges from Koopman 1982).

Natural History
These bats are believed to be primarily frugivorous.

Rhinophylla alethina Handley, 1966

Description
The dental formula is I 2/2, C 1/1, P 2/2, M 3/3. Head and body length ranges from 55 to 58 mm; HF 11; FA 35–37; Wt 12–16 g. The dorsum is blackish, shading to brownish black on the rump; the underparts are paler.

Range and Habitat
The species is known from Río Rapaso, Valle, Colombia; Pasco, Peru; and Belém, Brazil (map 5.64).

Rhinophylla fischerae Carter, 1966

Description
This small species has a head and body length ranging from 47 to 54 mm, and the forearm is 33 to 36 mm. The interfemoral membrane has a fringe of stiff hairs. The dorsal pelage varies from gray brown to reddish brown.
Chromosome number: $2n = 34$; FN = 56.

Range and Habitat
Known from 61 miles southeast of Pucallpa, Ucayali, Peru, the species may extend to southern Colombia. It was taken at 180 m elevation (map 5.64).

Rhinophylla pumilio Peters, 1865

Measurements

	Sex	Mean	S.D.	N
TL	M	46.67	1.21	6
	F	49.24	2.46	17
HB	M	46.67	1.21	6
	F	49.24	2.46	17
HF	M	10.83	0.41	6
	F	10.88	0.70	17
E	M	16.00	1.41	6
	F	15.47	1.23	17
FA	M	34.67	1.66	6
	F	34.92	0.74	17
Wta	M	9.77	1.06	6
	F	10.73	0.99	8

Location: Bolívar, Venezuela (USNMNH)

Description
Head and body length averages 48.3 mm for males and 50 mm for females; weight averages 9.4 g for males and 10.4 g for females (see table 5.14). The dorsum and venter are medium brown.
Chromosome number: $2n = 36$; FN = 62.

Range and Habitat
R. pumilio ranges across southern Colombia and Venezuela, the Guyanas, Amazonian Peru, and northern Brazil to the central east coast (map 5.64). In Venezuela this species was strongly associated

Table 5.9   Key to the Sturnirinae

1	High-elevation species occurring at from 1,500 to 2,500 m 2
1'	Low-elevation species, generally occurring below 1,000 m 4
2	Head and body length less than 60 mm . . . . *Sturnira erythromos*
2'	Head and body length greater than 60 mm 3
3	Head and body length less than 66 mm, only two lower molars . . . . *S. bidens*
3'	Head and body length greater than 66 mm, three lower molars . . . . *S. bogatensis*
4	Head and body length less than 65 mm . . . . *S. lilium*
4'	Head and body length greater than 65 mm 5
5	Head and body length less than 75 mm 6
5'	Head and body length greater than 75 mm 7
6	Ears greater than 20 mm . . . . *S. tildae*
6'	Ears less than 20 mm . . . . *S. ludovici*
7	Head and body length 85 to 90 mm . . . . *S. magna*
7'	Head and body length greater than 95 mm . . . . *S. aratathomasi*

with moist areas and multistratal tropical evergreen forests, ranging up to 1,400 m elevation.

SUBFAMILY STURNIRINAE

Diagnosis

This subfamily contains a single genus. The dental formula is I 2/2, C 1/1, P 2/2, M 3/2–3 (see fig. 5.18). These bats are tailless, and the interfemoral membrane is so reduced as to be inconspicuous, but the remnant fringe is well haired (see fig. 5.16). The ears are relatively short. Many species bear patches of stiff, yellowish hair on the shoulders that mark the position of shoulder glands. The dorsum varies from pinkish buff to dark brown, with the underparts usually considerably paler. There is little size dimorphism within this subfamily. A key is given in table 5.9, and keys for all species may be found in Davis (1980).

Distribution

The genus is distributed from southern Sinaloa in western Mexico and southern Tamaulipas in eastern Mexico south through the Isthmus to northern Argentina and Uruguay. Many species occur at modest to relatively high elevations.

Natural History

These bats are mainly frugivorous and possibly feed on pollen and nectar (Gardner 1977a). A single young is born annually.

Genus *Sturnira* Gray, 1842
Sturnira aratathomasi Peterson and Tamsitt, 1968
Yellow-shouldered Bat

Description

This large, rare species has a total length exceeding 100 mm, and the forearm may reach 60 mm. It may be distinguished from *S. magna* in that its inner

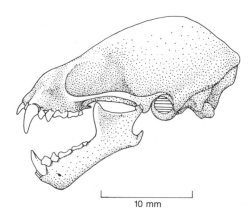

Figure 5.18. Skull of *Sturnira ludovici*.

upper incisors are long and centrally pointed and not in contact terminally (Tamsitt and Hauser 1985). The dorsal pelage is blackish brown.

Range and Habitat

The species is apparently confined to the Pacific side of the Andes from southern Colombia to Ecuador (map 5.65). It is known to feed on fruit (Thomas and McMurray 1974).

Sturnira bidens Thomas, 1915

Measurements

	Sex	Mean	S.D.	N
HB	M	64.3	2.9	3
	F	64.9	2.5	13
HF	M	14.0	0	3
	F	14.7	1.4	13
E	M	15.3	2.3	3
	F	16.3	1.5	13
FA	M	39.6	0.3	3
	F	40.6	0.5	13
Wta	M	17.2	0.5	3
	F	16.3		2

Location: Venezuela (USNMNH)

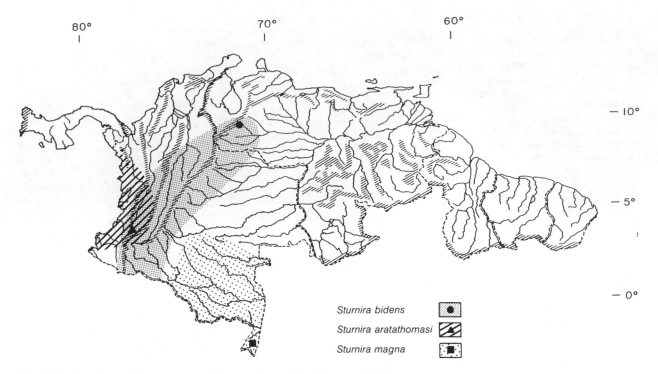

Map 5.65. Distribution of three species of *Sturnira* (ranges from Koopman 1982).

Sturnira bidens

Sturnira aratathomasi

Sturnira magna

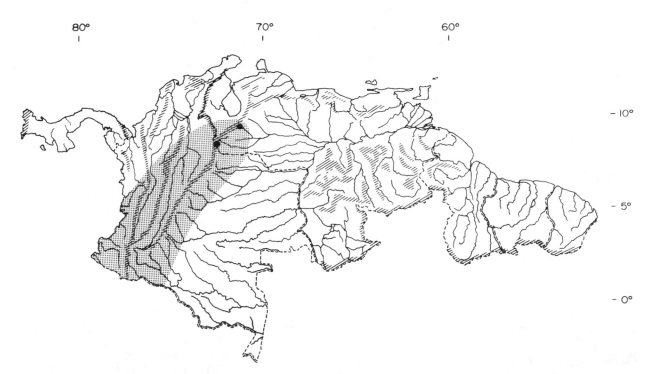

Map 5.66. Distribution of *Sturnira bogotensis*.

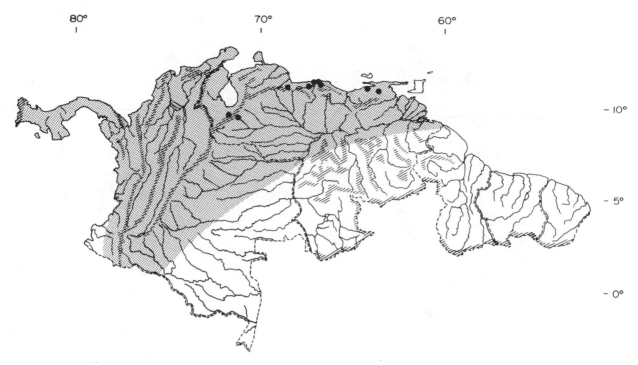

Map 5.67. Distribution of *Sturnira erythromos.*

Description

This bat differs from other species of *Sturnira* in possessing only two lower molars. The dorsum is a smoky gray deriving from the gray bases of the brown-tipped hairs; the venter is brown.
Chromosome number: $2n = 30$; FN = 56.

Range and Habitat

The species is distributed in the foothills of the Andes from western Venezuela south to Peru (map 5.65). In Venezuela it ranges between 2,550 and 2,640 m elevation and is strongly associated with cloud forest. It seems broadly tolerant of both moist and drier sites.

Sturnira bogotensis Shamel, 1927

Description

This species is larger than *S. bidens.* Its measurements are TL 65–67; HF 14–15.5; E 17–17.5; FA 43–43.3; Wt 19.6 g ($N = 4$; USNM). Forearms average approximately 43 mm. The coloration is similar to that of *S. bidens.*

Range and Habitat

This species co-occurs with *S. bidens* in the Andes from western Venezuela south to Bolivia (map 5.66). In Venezuela specimens were collected at between 2,107 and 2,640 m elevation. It is strongly associated with moist tropical evergreen forest or cloud forest (Handley 1976).

Sturnira erythromos (Tschudi, 1844)

Measurements

	Sex	Mean	S.D.	N
TL	M	56.45	3.38	29
	F	57.04	2.57	25
HB	M	56.45	3.38	29
	F	57.04	2.57	25
HF	M	13.07	1.00	29
	F	12.60	1.04	25
E	M	15.59	1.52	29
	F	15.56	1.00	25
FA	M	39.41	2.00	31
	F	38.70	0.92	28
Wta	M	14.42	2.02	5
	F	12.00		1

Location: Northern Venezuela (USNMNH)

Description

This is the smallest species in the genus; the sexes are nearly equal in size. It can be distinguished from *S. bogotensis* by its small size. The coloration is similar to that of *S. bidens,* with the dorsum smoky gray and the venter brown.

Range and Habitat

This species occurs in the premontane areas of the Andes in Venezuela, Colombia, and south to Bolivia (map 5.67). In Venezuela most specimens were taken at between 1,000 and 2,500 m elevation. It prefers moist habitats of multistratal evergreen forest or cloud forest.

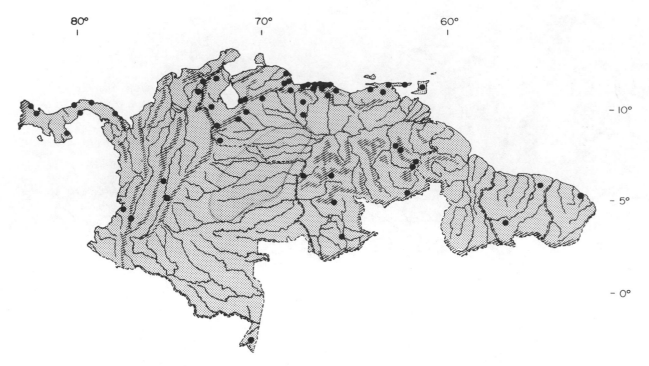

Map 5.68. Distribution of *Sturnira lilium*.

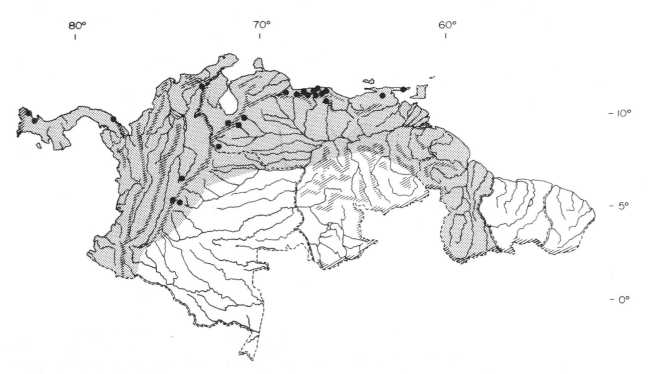

Map 5.69. Distribution of *Sturnira ludovici*.

Sturnira lilium (E. Geoffroy, 1810)

Measurements

	Sex	Mean	S.D.	N
TL	M	61.30	3.77	76
	F	60.61	3.88	89
HB	M	61.30	3.77	76
	F	60.61	3.88	89
HF	M	13.88	1.08	76
	F	13.62	1.01	89
E	M	16.66	1.11	76
	F	16.48	1.08	89
FA	M	41.03	1.37	117
	F	40.56	1.56	144
Wta	M	20.43	2.25	36
	F	18.37	1.95	23

Location: Northern Venezuela (USNMNH)

Description

There is little sexual size dimorphism (see plate 5). Yellow shoulders are very pronounced in males; the dorsum is brown, the venter lighter.

Range and Habitat

This species is widely distributed from southern Sonora in western Mexico and Tamaulipas in eastern Mexico south through the Isthmus, over all of South America to northeastern Argentina (map 5.68). In Venezuela it ranges from sea level to 1,982 m elevation. At higher elevations this species is replaced by *S. bogotensis* and *S. erythromos*.

Natural History

These bats roost in hollow trees and caves. This species is strongly frugivorous and at times will even forage on the ground for fallen fruit; it is occasionally taken in live traps set on the ground for rodents.

Sturnira ludovici Anthony, 1924

Measurements

	Sex	Mean	S.D.	N
TL	M	71.00	3.96	24
	F	71.04	4.45	25
HB	M	71.00	3.96	24
	F	71.04	4.45	25
HF	M	16.13	0.95	24
	F	15.80	0.96	25
E	M	18.46	1.77	24
	F	18.64	0.91	25
FA	M	46.69	1.47	31
	F	46.59	1.24	35
Wta	M	26.85	1.71	13
	F	24.84	1.93	5

Location: Northern Venezuela (USNMNH)

Description

This is one of the largest species within the genus; only *S. aratathomasi* and *S. magna* are larger. This species may be confused with *S. tildae*, but the ears of *S. ludovici* are relatively shorter, averaging about 19 mm. Basically this is a brown bat with the venter not contrasting, but the bases of the dorsal hairs are pale.

Range and Habitat

This species ranges from southern Sinaloa in western Mexico and southern Tamaulipas in eastern Mexico south through the Isthmus to northern South America. It extends to Ecuador and across Venezuela to Guyana (map 5.69). It ranges from sea level to 2,240 m elevation in Venezuela, but most specimens are taken at below 1,500 m. It forages in moist areas and mainly occurs in multistratal tropical evergreen forests, though it is occasionally taken in dry deciduous forests.

Remarks

Over much of the northern range of *Sturnira lilium*, *Sturnira ludovici* co-occurs. They are easily separable by size. In the southern part of South America, *S. ludovici* is replaced by the similar-sized *S. tildae* and co-occurs with *S. lilium*.

Sturnira magna de la Torre, 1966

Description

This species is similar to *S. aratathomasi*, but the inner upper incisors are broad and almost in contact terminally. The head and body length ranges from 85 to 90 mm and the forearm from 56 to 57 mm. The dorsal pelage is yellow brown to gray brown; the venter is pale yellow, brown, or gray, and the bases of the hairs are very pale.

Range and Habitat

The species is known from the Río Manito, Iquitos, Peru, and extreme southeastern Colombia (map 5.65). It has a wide altitudinal tolerance, ranging from 200 to 2,300 m in Peru. Moist lowland and premontane forests are preferred. It is assumed to be a frugivore (Tamsitt and Hauser 1985).

Sturnira tildae de la Torre, 1959

Measurements

	Sex	Mean	S.D.	N
TL	M	69.69	3.31	42
	F	69.63	3.18	65
HB	M	69.69	3.31	42
	F	69.63	3.18	65
HF	M	16.55	0.97	42
	F	16.40	1.42	65

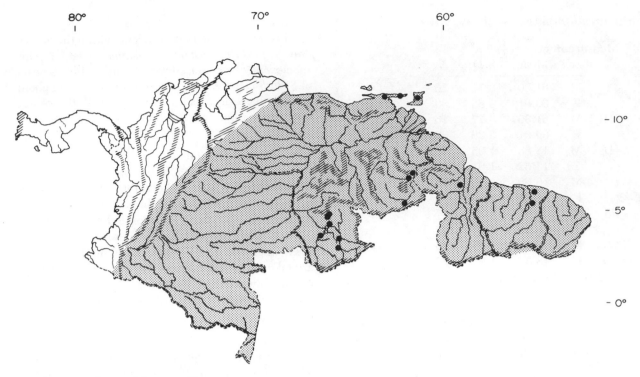

Map 5.70. Distribution of *Sturnira tildae*.

E	M	20.71	0.97	42
	F	20.43	0.88	65
FA	M	47.92	1.59	43
	F	47.69	1.13	65
Wta	M	26.28	3.27	42
	F	25.49	4.14	21

Location: TFA, Venezuela (USNMNH)

Description

Head and body length averages 69.6 mm. There is little sexual size dimorphism, but considerable variation in size among individuals can be noted (see Marinkelle and Cadena 1971). The ears are slightly larger in proportion to head and body length than in *S. ludovici*. The bases of the dorsal hairs are very light while the tips are brown, giving the general effect of a variable dorsal coloration of brown washed with white. The venter is pale brown.

Range and Habitat

This species is entirely South American and occurs in the southeastern portion of Colombia, southern Venezuela, the Guyanas, Amazonian Peru, and across much of Brazil (map 5.70). In Venezuela specimens were taken at below 1,165 m. It is strongly associated with moist habitats and multistratal tropical evergreen forests. This species of *Sturnira* replaces *S. ludovici* at low elevations over much of the South American continent.

SUBFAMILY STENODERMINAE

Diagnosis

There is a tendency within the subfamily to reduce the number of molars to two uppers and two lowers per side. These are medium-sized to small bats. The tail is rudimentary or absent, but the interfemoral membrane is present and supported by the calcar bone on the heel. Although there may be varying reduction in the interfemoral membrane, it

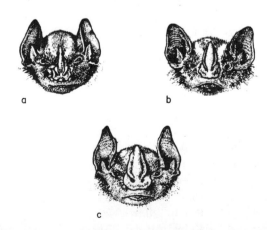

Figure 5.19. External features of the Stenoderminae: (*a*) Face of *Vampyrops helleri*; note white facial stripes. (*b*) Face of *Uroderma bilobatum*; note white facial stripes. (*c*) Face of *Vampyrodes caraccioli*; note faint facial stripes. (Redrawn with permission from Goodwin and Greenhall 1961.)

is not as pronounced as one finds in the genera *Anoura* and *Sturnira*. The muzzle is short, in some cases very short and broad, which apparently correlates with an increase in specialization for feeding on fruit. Many species of the subfamily have white facial stripes, generally two pairs from the nose to just above the eye and from the corner of the mouth to the ear. The dorsum frequently has a median white stripe. Facial stripes are absent in the genera *Pygoderma, Ametrida, Sphaeronycteris,* and *Centurio.* A key to the genera is given in table 5.10. Anatomical details are portrayed in figures 5.19 and 5.20.

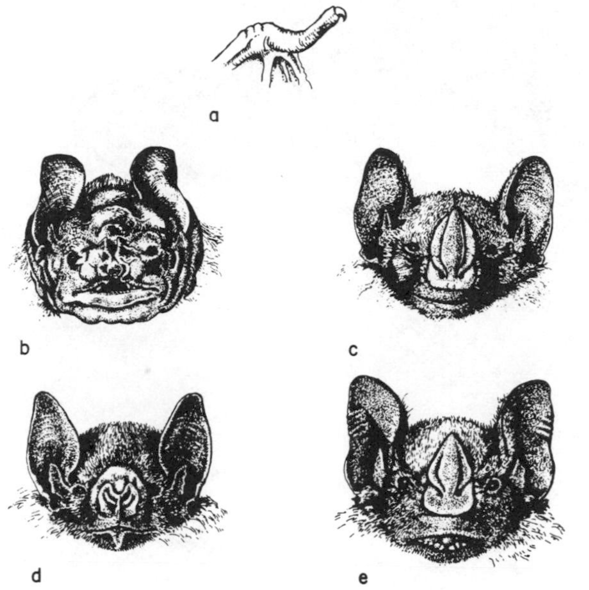

Genus *Uroderma* Peters, 1865
Tent-making Bat

Description

The dental formula is I 2/2, C 1/1, P 2/2, M 3/3 (fig. 5.21). Head and body length ranges from 60 to 62.4 mm. The white facial stripes are extremely pronounced. The basic color of the dorsum is gray brown to brown, and the venter is usually paler.

Figure 5.20. External features of the Stenoderminae and Desmodontinae: (*a*) Elongated thumb of *Desmodus rotundus*. (*b*) Face of *Centurio senex*. (*c*) Face of *Chiroderma villosum*. (*d*) Face of *Desmodus rotundus;* note reduced and modified nose leaf. (*e*) Face of *Artibeus jamaicensis*. (Redrawn with permission from Goodwin and Greenhall 1961.)

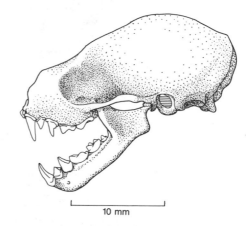

Figure 5.21. Skull of *Uroderma bilobatum*.

Table 5.10   Key to the Genera of the Stenoderminae

1	Face naked; nose leaf very reduced 2
1'	Face furred; nose leaf prominent 3
2	Face very wrinkled, fold of skin under chin .... *Centurio*
2'	Face not wrinkled; hornlike protuberance above nostrils of male; fold of skin under the chin prominent in male .... *Sphaeronycteris*
3	Interfemoral membrane furred on dorsal side; nasal bones absent .... *Chiroderma*
3'	Interfemoral membrane not furred 4
4	Fur whitish, wings yellow .... *Ectophylla alba*
4'	Fur colored 5
5	No striping on face or dorsum 6
5'	Stripes usually present on face and/or dorsum 7
6	Single white spot on each shoulder, dermal outgrowths on cheeks .... *Ametrida*
6'	Single white spot on each shoulder, no dermal outgrowths on face .... *Pygoderma*
7	Inner upper incisor slightly higher than outer incisor 8
7'	Inner upper incisor much higher than outer 10
8	Basal portion of nose leaf bilobate .... *Uroderma*
8'	Basal portion of nose leaf not bilobate 9
9	Inner upper incisor bifid .... *Artibeus*
9'	Inner upper incisor entire .... *Enchisthenes*
10	Upper incisor trilobate .... *Vampyrops*
10'	Not as above 11
11	Stripes on face and back prominent; incisors with cutting edge entire .... *Vampyrodes*
11'	No dorsal stripe .... *Vampyressa*

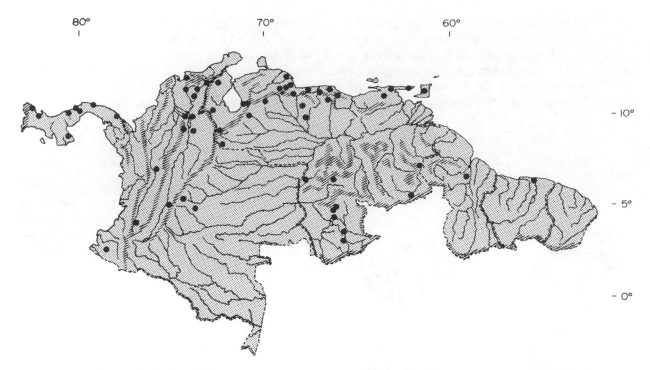

Map 5.71. Distribution of *Uroderma bilobatum*.

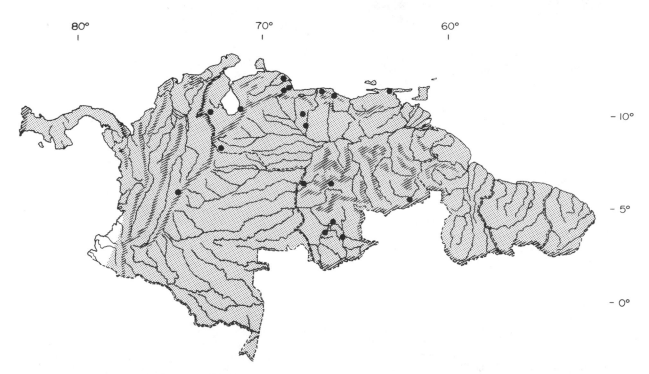

Map 5.72. Distribution of *Uroderma magnirostrum*.

Distribution

The genus occurs from southern Veracruz, Mexico, south to southeastern coastal Brazil.

Natural History

Species of this genus roost under palm fronds or banana leaves. They often bite through the ribs of fronds and cause the leaf to collapse on itself, thereby providing a shelter. For this reason they are called tent-making bats.

Uroderma bilobatum Peters, 1866

Measurements

	Sex	Mean	S.D.	N
TL	M	60.04	3.93	51
	F	60.02	4.13	89
HB	M	60.04	3.93	51
	F	60.02	4.13	89
HF	M	11.80	0.80	51
	F	12.19	1.08	89
E	M	17.69	1.14	51
	F	17.69	1.13	89
FA	M	41.03	0.99	70
	F	41.25	1.17	128
Wta	M	14.94	1.55	20
	F	17.71	1.44	15

Location: Northwestern Venezuela (USNMNH)

Description

Head and body length ranges from 54 to 61 mm. In Venezuela specimens averaged about 60.8 mm. The sexes are nearly identical in size. The facial stripes are very pronounced in *U. bilobatum*, and there is a white stripe down the midline of the back. The ear edge is pigmented yellowish white. Chromosome number: variable, $2n = 42, 44, 38$; FN $= 48, 50$.

Range and Habitat

This species occurs from southern Veracruz, Mexico, south through the Isthmus to southeastern Brazil (map 5.71). It is found widely over all tropical areas of South America. It generally occurs below 1,000 m elevation. It tolerates man-made clearings, and though it is strongly associated with multistratal tropical evergreen forest, it also occurs in drier situations (Handley 1976).

Natural History

These bats typically cut the surface of palm fronds with a series of bites, causing the leaf to bend in half and form a shelter. They may roost in small colonies (up to ten) of both sexes. They are strongly frugivorous but include insects in their diet (Goodwin and Greenhall 1961).

Uroderma magnirostrum Davis, 1968

Measurements

	Sex	Mean	S.D.	N
TL	M	60.06	3.03	36
	F	62.53	3.27	72
HB	M	60.06	3.03	36
	F	62.53	3.27	72
HF	M	12.42	1.42	36
	F	12.64	1.43	72
E	M	17.97	0.65	36
	F	17.73	0.83	71
FA	M	42.43	1.35	42
	F	43.44	1.50	81
Wta	M	15.90	1.95	34
	F	17.66	2.63	50

Location: TFA, Venezuela (USNMNH)

Description

This species is somewhat larger than *U. bilobatum*. Head and body length ranges from 58 to 65 mm (see table 5.14). The facial stripes are less conspicuous than in *U. bilobatum*. The ear tends to be uniformly colored in this species, whereas in *U. bilobatum* the ear is edged with yellowish white. Chromosome number: $2n = 36$; FN $= 60$ or 62.

Range and Habitat

This species ranges from Oaxaca, Mexico, south through the Isthmus to central Brazil (map 5.72). In Venezuela it ranges below 1,000 m; most specimens were taken at below 500 m. It is often associated with moist habitats. It makes use of open areas and man-made clearings and seems less tolerant of arid habitats than *U. bilobatum* (Handley 1976).

Natural History

Females form roosting colonies when they bear their young, and the sexes tend to roost separately during the rearing season. In Panama the young are born from February through April (Wilson 1979).

Genus *Vampyrops* (= *Platyrrhinus*) Peters, 1865

White-lined Bat

Description

The dental formula is I 2/2, C 1/1, P 2/2, M 3/3 (see fig. 5.22). Head and body length ranges from 48 to 98 mm. The basic color of the dorsum is dark brown to almost black, and the white or gray dorsal stripe is prominent, extending from the ears to the tail membrane (see plate 5). There are distinct white facial stripes. Chromosome number: all species show $2n = 30$; FN $= 56$.

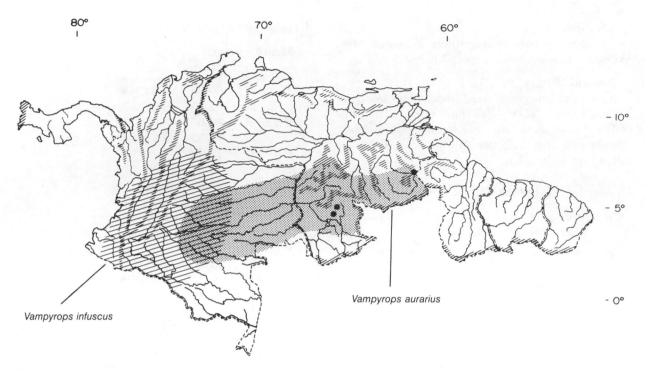

Map 5.73. Distribution of two species of *Vampyrops* (ranges from Koopman 1982).

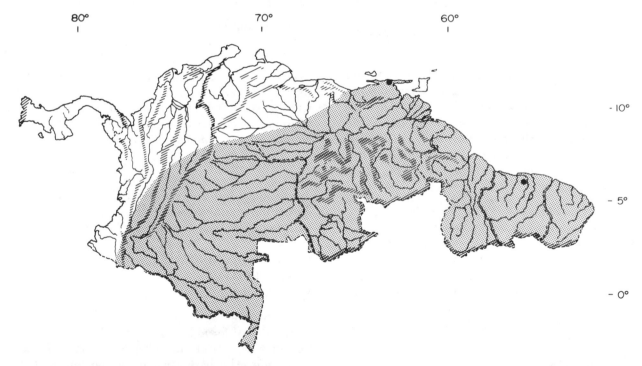

Map 5.74. Distribution of *Vampyrops brachycephalus*.

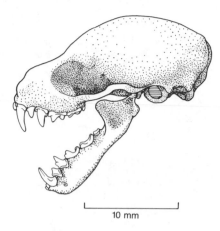

Figure 5.22. Skull of *Vampyrops helleri*.

Distribution

The genus is distributed from southern Mexico through the Isthmus to northeastern Argentina and Uruguay. Within its range it is divisible into a number of different species.

Natural History

White-lined bats are basically frugivorous. They roost in small groups of three to ten in leafy tangles, tree hollows, or caves. Reproduction usually coincides with the onset of the rainy season and varies locally.

Vampyrops aurarius Handley and Ferris, 1972

Measurements

	Sex	Mean	S.D.	N
TL	M	75.67	2.50	15
	F	77.62	2.79	13
HB	M	75.67	2.50	15
	F	77.62	2.79	13
HF	M	16.07	0.88	15
	F	16.77	0.60	13
E	M	22.00	0.65	15
	F	21.46	0.66	13
FA	M	52.03	1.22	15
	F	53.37	1.40	14
Wta	M	33.64	3.08	15
	F	34.68	2.77	5

Location: Bolívar, Venezuela (USNMNH)

Description

This is one of the larger species within the genus, exceeded in size only by *V. vittatus*. The dorsum and venter are dark brown; a conspicuous white line is found on the middorsum, and there is a white line from the nose passing over the eye.

Range and Habitat

The species has been described from central Colombia to the east across southern Venezuela (map 5.73). Most specimens are taken at above 1,000 m elevation in Venezuela, and the bats occur at 1,400 m. They are strongly associated with multistratal premontane evergreen tropical forest and do not utilize the low-altitude dry zone.

Vampyrops brachycephalus Rouk and Carter, 1972

Measurements

	Mean	S.D.	N
TL	62.67	4.15	3
HF	11.3	1.15	3
E	17.0	1.0	3
FA	39.8	0.46	3
Wta	15.5	1.54	3

Location: Northern Venezuela (USNMNH)

Description

This is one of the smaller species, only slightly larger than *V. helleri*, and the sexes are nearly the same size. Head and body length averaged 66 mm and weight 15.5 g for three specimens from Suriname (CMNH). The dorsum is light to medium brown, the venter pale brown. There is one bold line on the side of the face; the second line is fainter. The median dorsal stripe is conspicuous.

Range and Habitat

The species is confined to northern South America in the Amazonian portion of Brazil, the Guyanas, southern Venezuela, Colombia, and Amazonian Peru (map 5.74). Strongly associated with multistratal tropical evergreen forest and moist sites, in Venezuela it was taken at between 175 and 375 m (Handley 1976).

Vampyrops dorsalis (Thomas, 1900)

Description

Head and body length ranges from 69 to 80 mm and forearm length from 44.6 to 50.1 mm. One female specimen from Colombia had the following measurements: TL 80; HF 13; E 20; FA 47 (CMNH). This species is intermediate in size between *V. helleri* and *V. vittatus*. The middorsal stripe is prominent, as are the two stripes on each side of the face. The dorsum is dark brown and the venter gray brown.

Range and Habitat

This species occurs from the extreme eastern portion of Panama west of the Andes, and southward through Colombia to Bolivia (map 5.75).

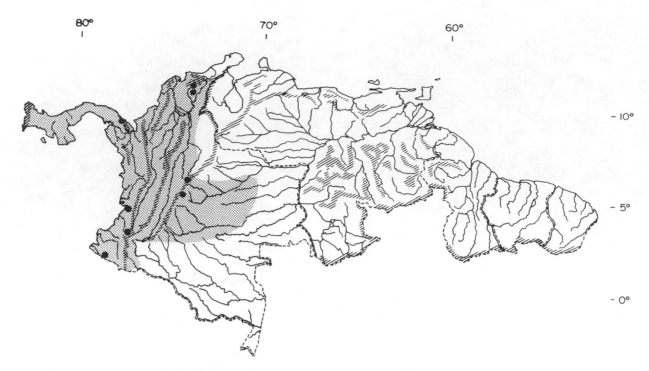

Map 5.75. Distribution of *Vampyrops dorsalis*.

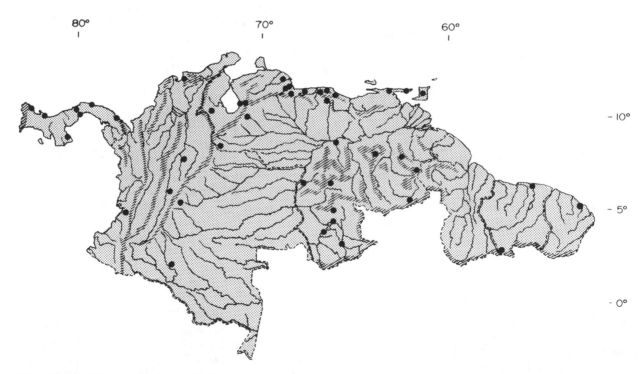

Map 5.76. Distribution of *Vampyrops helleri*.

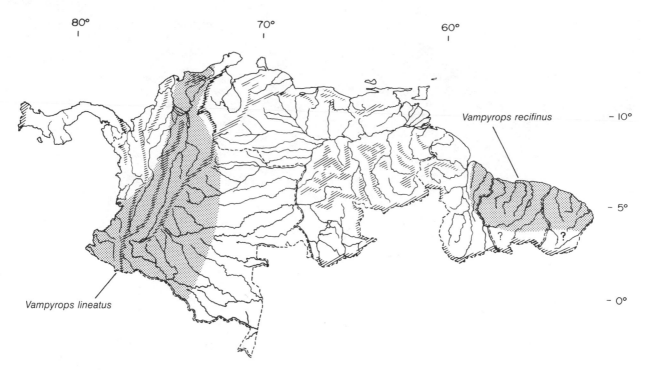

Map 5.77. Distribution of *Vampyrops recifinus* and *V. lineatus* (ranges from Koopman 1982).

Vampyrops helleri (Peters, 1867)

Measurements

	Sex	Mean	S.D.	N
TL	M	55.56	2.22	48
	F	56.35	2.22	68
HB	M	55.56	2.22	48
	F	56.35	2.22	68
HF	M	11.94	0.81	48
	F	12.35	0.81	68
E	M	17.15	0.92	48
	F	17.69	0.76	68
FA	M	37.67	1.09	63
	F	38.23	1.17	82
Wta	M	12.87	0.99	26
	F	13.99	0.93	22

Location: TFA, Venezuela (USNMNH)

Description

This is the smallest species in the genus, and males and females are nearly the same size. The dorsum is brown and the venter gray; a dorsal stripe and facial stripes are present.

Range and Habitat

The species ranges from southern Mexico through the Isthmus, broadly across northern South America and the eastern portion of Brazil (map 5.76). In Venezuela it is found below 1,000 m elevation and prefers moist habitats, but it can range into dry deciduous forest. It tolerates man-made clearings.

Natural History

These bats roost in pairs high in the crowns of trees. Evidence indicates that they are strongly frugivorous (Gardner 1977a). The species may be very abundant in favorable habitat but can be netted in numbers only if one sets nets in the canopy (Handley 1967).

Vampyrops infuscus Peters, 1881

Description

This large species is exceeded in size only by *V. vittatus*. Its measurements are TL 97–101; HF 15–16; E 21–23; FA 55–57 (CMNH). The dorsum is buff to light brown with the venter paler. The dorsal and facial stripes are faint.

Range and Habitat

The species is found in the Amazonian region of Peru north to Colombia (Gardner and Carter 1972). (See map 5.73.)

Vampyrops lineatus (E. Geoffroy, 1810)
Vampiro de Franjas Blancas

Description

This small, dark species has forearms measuring 41 to 48 mm. One male specimen from Colombia measured TL 41; HF 12; E 13; FA 41 (CMNH). The dorsal pelage varies from light to blackish brown with a prominent white stripe; the venter is a paler

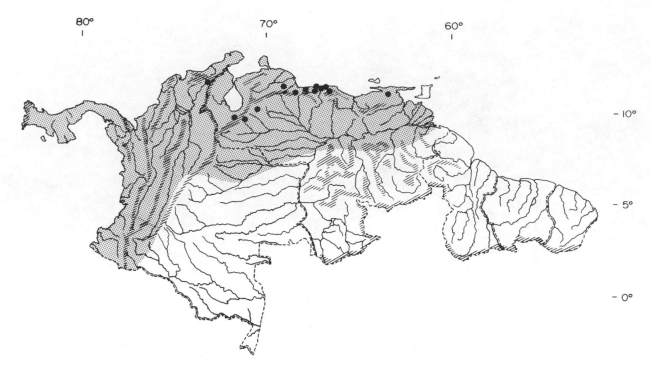

Map 5.78. Distribution of *Vampyrops umbratus*.

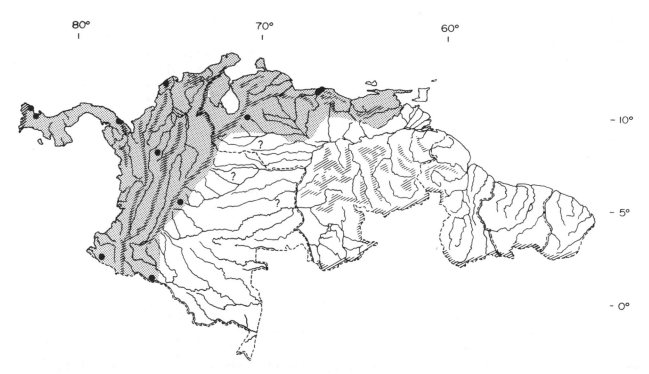

Map 5.79. Distribution of *Vampyrops vittatus*.

shade of brown. The four facial stripes are buff and indistinct. Includes *V. nigellus*.

Range and Habitat

The species ranges from Amazonian Colombia southeast of the Andes to northern Argentina and eastern Brazil (map 5.77).

Vampyrops recifinus Thomas, 1901

Description

This species may be conspecific with *V. lineatus*, which it resembles closely (see *V. lineatus*). The forearm measures from 40 to 42 mm.

Range and Habitat

The range extends from Guyana east through eastern Brazil (map 5.77).

Vampyrops umbratus Lyon, 1902

Measurements

	Sex	Mean	S.D.	N
TL	M	69.73	3.48	71
	F	71.58	4.28	72
HB	M	69.73	3.48	71
	F	71.58	4.28	72
HF	M	13.42	0.95	71
	F	13.75	1.00	72
E	M	18.63	1.23	71
	F	18.64	1.26	72
FA	M	44.23	1.21	79
	F	44.66	1.10	80
Wta	M	23.04	2.84	16
	F	25.72	2.48	12

Location: Northern Venezuela (USNMNH)

Description

The coloration and markings are similar to those of *V. helleri*, with dorsum brown and venter gray. The facial stripes and dorsal stripe are prominent.

Range and Habitat

This species is confined to the extreme northern part of South America, occurring in northern Colombia and Venezuela (map 5.78). In Venezuela it ranges from 395 to 2,550 m elevation. Most specimens are taken at above 1,000 m. It occurs in multistratal tropical evergreen forest as well as cloud forest and is strongly associated with moist areas.

Vampyrops vittatus (Peters, 1860)

Measurements

	Sex	Mean	S.D.	N
TL	M	86.8	4.9	5
	F	91		2
HF	M	17.4	1.14	5
	F	18		2

E	M	23.2	0.8	5
	F	24		2
FA	M	57.2	1.8	5
	F	59.8		2

Location: Northern Venezuela (USNMNH)

Description

This is the largest species of the genus. The dorsal pelage is a dark blackish brown. The lower facial stripes are poorly developed, but the upper facial stripe and dorsal stripe are buffy and contrast sharply with the basic dorsal color.

Range and Habitat

This species occurs from Costa Rica south through the Isthmus over much of South America, with the exception of eastern Brazil (map 5.79). In Venezuela it occurred up to 2,119 m elevation and was strongly associated with moist habitats and multistratal evergreen forest (Handley 1976).

Genus *Vampyrodes* Thomas, 1900
Vampyrodes caraccioli (Thomas, 1889)
Great Stripe-faced Bat

Measurements

	Sex	Mean	S.D.	N
TL	M	71.75	2.36	4
	F	74.50	2.47	14
HB	M	71.75	2.36	4
	F	74.50	2.47	14
HF	M	15.75	0.50	4
	F	16.14	0.77	14
E	M	21.25	0.50	5
	F	21.71	0.47	14
FA	M	50.88	1.24	4
	F	53.08	1.27	14
Wta	M	26.88	1.61	4
	F	29.81	3.56	7

Location: TFA, Venezuela (USNMNH)

Description

The dental formula is I 2/2, C 1/1, P 2/2, M 2/2; the reduction in number of molars distinguishes this genus from *Vampyrops* and *Uroderma*. The dorsal pelage varies from brownish to gray brown, and the venter is paler. There are four white facial stripes, and a white line extends from the head down the dorsal midline.
Chromosome number: $2n = 30$; FN = 56.

Range and Habitat

This species occurs from southern Veracruz, Mexico, south through the Isthmus and over most of northern South America to Amazonian Peru and northwestern Brazil (map 5.80). In northern Venezuela it occurs below 1,000 m elevation and is strongly associated with multistratal tropical evergreen forest.

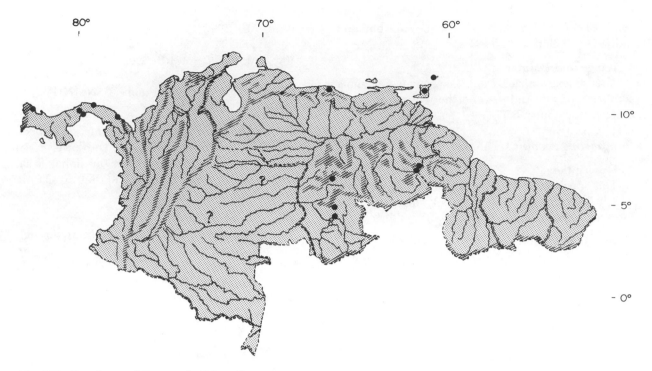

Map 5.80. Distribution of *Vampyrodes caraccioli.*

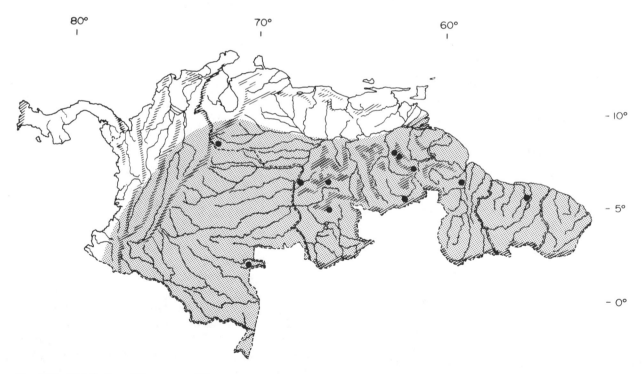

Map 5.81. Distribution of *Vampyressa bidens.*

Natural History

These bats may be found roosting in small groups (two to four) under shrub branches. They are frugivores (Goodwin and Greenhall 1961).

Genus *Vampyressa* Thomas, 1900
Yellow-eared Bat

Description

The dental formula is variable; I 2/2, C 1/1, P 2/2, M 2/2 or I 2/1, C 1/1, P 2/2, M 2/2. Head and body length ranges from 43 to 65 mm. The dorsum varies from smoky gray to pale brown or dark brown, and a dorsal stripe is usually not present. The white facial stripes are prominent in some species but lacking in others. The ears are typified by a yellow margin. The genus has been reviewed by Gardner (1977b) and Peterson (1968). For a key to the genus see table 5.10.

Distribution

The genus occurs from southern Mexico south through the Isthmus to northeastern Brazil and adjacent portions of Amazonian Peru. One species, *V. pusilla*, extends its range to southeastern Brazil but is absent from the Amazonian region.

Natural History

The species of this genus are believed to be largely frugivorous. Reproduction is seasonally timed by the onset of the rains (Gardner 1977a; Wilson 1979).

Vampyressa bidens (Dobson, 1878)

Measurements

	Sex	Mean	S.D.	N
TL	M	50.20	1.87	55
	F	52.11	2.17	44
HB	M	50.20	1.87	55
	F	52.11	2.17	44
HF	M	11.02	0.53	55
	F	11.25	0.53	44
E	M	17.20	0.58	55
	F	17.09	0.60	44
FA	M	35.39	0.76	55
	F	35.83	0.89	44
Wta	M	11.00	0.72	54
	F	12.33	1.06	18

Location: TFA, Venezuela (USNMNH)

Description

Only one pair of lower incisors is present. The dorsum is dark brown and the venter grayish brown (plate 5). In contrast to some species of the genus, a faint dorsal line is present, and there are four facial stripes.

Chromosome number: $2n = 26$; FN = 48.

Range and Habitat

This species is confined solely to South America and is found in the Amazonian region north of the Amazon River (map 5.81). In Venezuela most specimens were taken at below 500 m elevation, and the species was strongly associated with multistratal tropical evergreen forest (Handley 1976).

Vampyressa brocki Peterson, 1968

Description

This is a small species, similar in size to *V. pusilla*: TL 49–51; HF 9–10; E 13–15; Wt 7–8 g ($N = 4$; Suriname, CMNH). The dorsum is light brown and the venter gray, and there is no dorsal stripe. The facial stripes are conspicuous and demarcate a dark stripe from the nose leaf on each side of the eye to the ear.

Chromosome number: $2n = 24$; FN = 44.

Range and Habitat

The type specimen was taken in Guyana at 2°50′ N, 58°55′ W. Since the original description, specimens have been taken from Suriname and southeastern Colombia (Peterson 1968; Genoways and Williams 1979).

Vampyressa nymphaea Thomas, 1909

Description

Head and body length ranges from 55 to 60 mm and forearm length from 36 to 39 mm, and the hind foot is 13 mm. The dorsum is a smoky gray, the venter lighter. The white facial stripes are conspicuous. These bats are frequently referred to by the common name of yellow-eared bat because of the yellow margin of the ears.

Chromosome number: $2n = 26$; FN = 48.

Range and Habitat

The species is found from southern Nicaragua south through the Isthmus to extreme western Colombia (map 5.82). It has not been taken east of the Andes.

Vampyressa pusilla (Wagner, 1843)
Little Yellow-eared Bat

Measurements

	Sex	Mean	S.D.	N
TL	M	46.10	1.37	10
	F	46.86	1.85	21
HB	M	46.10	1.37	10
	F	46.86	1.85	21
HF	M	10.00	0.00	10
	F	10.10	0.30	21

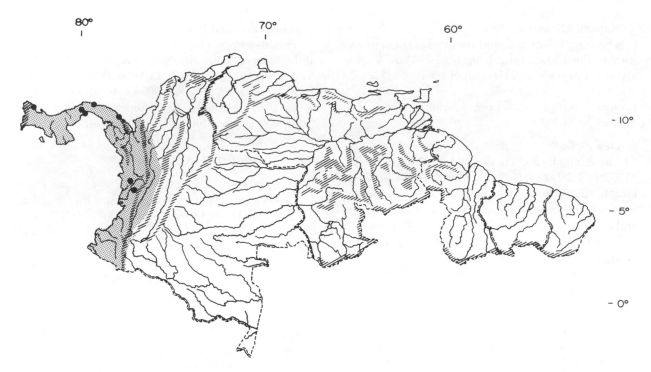

Map 5.82. Distribution of *Vampyressa nymphaea*.

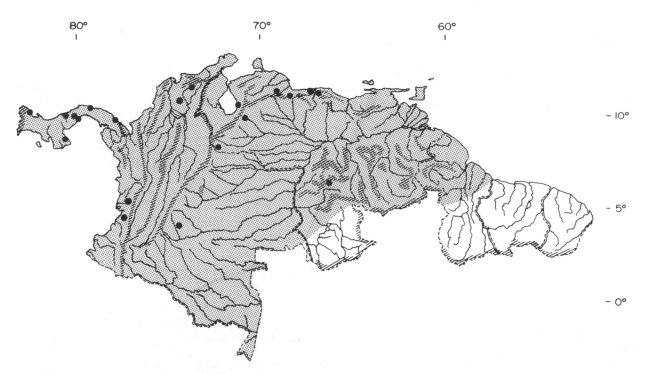

Map 5.83. Distribution of *Vampyressa pusilla*.

E	M	14.80	0.63	10
	F	14.76	0.54	21
FA	M	31.11	1.04	11
	F	31.31	0.73	24
Wta	M	7.72	0.55	10
	F	7.85		2

Location: North-central Venezuela (USNMNH)

Description

The dorsum is light brown, and the facial stripes are not conspicuous, though the upper facial stripe is usually more boldly marked than the lower. The venter is light gray, and the throat region may be yellowish.

Chromosome number: variable, $2n = 18, 24$; FN $= 20-22$.

Range and Habitat

The species is distributed from southern Mexico south through the Isthmus over most of northern South America, then in the Amazonian portion of Peru and Bolivia to southeastern Brazil. So far it has not been described from Amazonian Brazil (map 5.83). In Venezuela it occurs up to 1,537 m, but most specimens were taken at below 500 m. It is strongly associated with moist habitats and multistratal evergreen forest (Handley 1976).

Note

V. melissa occurs in the premontane forests of Peru and may extend into southwestern Colombia. A key to the species of Vampyressa is given in table 5.11.

Genus *Chiroderma* Peters, 1860
Large-eyed Bat

Description

The dental formula is I 2/2, C 1/1, P 2/2, M 2/2. The skull is unusual in that the nasal bones are absent. Head and body length ranges from 55 to 77 mm, and the forearm ranges from 37 to 53 mm. The upperparts are brown, and the underparts are usually lighter. The presence of facial stripes varies from species to species; the dorsal stripe may be faint or absent.

Chromosome number: all species show $2n = 26$; FN $= 48$.

Distribution

The genus is distributed from Sinaloa in western Mexico and southern Veracruz in eastern Mexico south through the Isthmus to Brazil.

Natural History

The species of *Chiroderma* are frugivorous. Reproduction is seasonal and timed by the onset of the rains.

Chiroderma salvini Dobson, 1878

Measurements

	Sex	Mean	S.D.	N
TL	M	72.13	4.45	8
	F	73.23	4.48	13
HB	M	72.13	4.45	8
	F	73.23	4.48	13
HF	M	13.63	1.06	8
	F	14.23	1.01	13
E	M	18.50	1.20	8
	F	19.23	1.17	13
FA	M	48.66	1.15	8
	F	48.19	0.84	13
Wta	M	25.90	1.54	4
	F	29.10	1.89	6

Location: Northern Venezuela (USNMNH)

Description

Head and body length averages 72.1 mm for males and 73.2 mm for females; weight averages 27.2 g for males and 29.1 g for females (see table 5.14). The facial and dorsal stripes are very conspicuous in this species.

Range and Habitat

The species is distributed from Sinaloa in western Mexico and southern Veracruz south through the Isthmus to northern Venezuela, western Colombia,

---

Table 5.11   Key to the Known Species and Subspecies of the Genus *Vampyressa*

1 One pair of lower incisors .... *V. bidens*
1′ Two pair of lower incisors 2
2 M₃ present .... *V. melissa*
2′ M₃ absent 3
3 M₂ approximately as long as wide, with high anterior and posterior cusps 4
3′ M₂ longer than wide, with lower anterior and posterior cusps 5
4 Forearm in full adults more than 34 mm .... *V. pusilla pusilla*
4′ Forearm in full adults less than 34 mm .... *V. pusilla thyone*
5 Forearm in full adults more than 34 mm, greatest length of skull more than 20 mm .... *V. nymphaea*
5′ Forearm in full adults less than 34 mm, greatest length of skull less than 20 mm .... *V. brocki*

Source: After Peterson (1968).

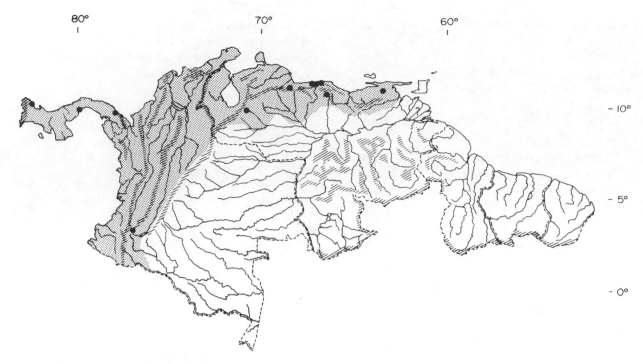

Map 5.84. Distribution of *Chiroderma salvini*.

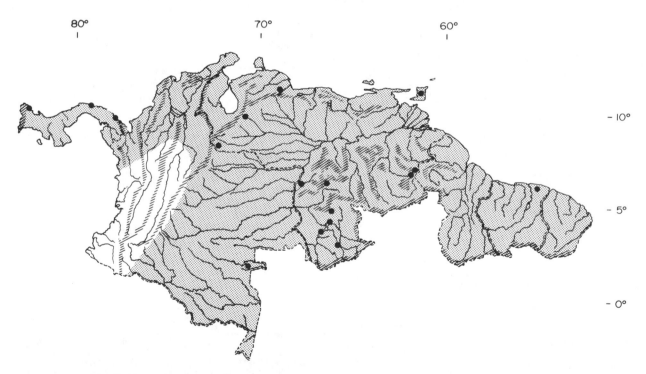

Map 5.85. Distribution of *Chiroderma trinitatum*.

and Amazonian Peru (map 5.84). It occurs over a range of elevations, from 611 to 2,240 m in Venezuela, and prefers moist habitats and multistratal tropical evergreen forest (Handley 1976).

Chiroderma trinitatum Goodwin, 1958

Measurements

	Sex	Mean	S.D.	N
TL	M	54.43	1.27	7
	F	56.59	2.80	39
HB	M	54.43	1.27	7
	F	56.69	2.80	39
HF	M	11.86	0.38	7
	F	12.00	0.76	39
E	M	17.57	0.53	7
	F	17.74	0.75	39
FA	M	38.38	1.28	6
	F	38.48	1.12	40
Wta	M	12.97	0.68	7
	F	13.98	1.27	19

Location: TFA, Venezuela (USNMNH)

Description
This is the smallest species of the genus. Males average 54.8 mm in total length and females 56.7 mm. Weight averages 13 g for males and 13.9 g for females. The basic dorsal color is brown, and facial and dorsal stripes are prominent.

Range and Habitat
The species is distributed from Panama to central Brazil. In Venezuela it occurred from 24 to 1,032 m elevation; most specimens were taken at below 500 m (map 5.85). It prefers moist habitats and multistratal evergreen tropical forest (Handley 1976).

Chiroderma villosum Peters, 1860

Measurements

	Sex	Mean	S.D.	N
TL	M	63.03	2.74	31
	F	67.61	2.66	168
HB	M	63.03	2.74	31
	F	67.61	2.66	168
HF	M	13.48	0.93	31
	F	13.76	1.01	168
E	M	19.29	0.94	31
	F	19.36	0.85	168
FA	M	43.77	1.40	33
	F	45.75	1.54	188
Wta	M	20.19	1.61	11
	F	22.67	2.10	75

Location: TFA, Venezuela (USNMNH)

Description
This species is considerably smaller than *C. salvini*. Total length averages 64.57 mm for males and 67.55 for females; weight averges 21 g for males and 22.9 g for females. The long, soft dorsal pelage is light brown and does not contrast with the venter. The facial stripes are faint or absent, which with its small size, serves to distinguish it from *C. salvini*.

Range and Habitat
This species occurs from southern Mexico south through the Isthmus, over much of northern South America to western Brazil (map 5.86). In Venezuela most specimens were taken at below 500 m. It is strongly associated with moist habitats and multistratal tropical evergreen forest (Handley 1976).

Genus *Ectophylla* H. Allen, 1892
Ectophylla (= *Mesophylla*) *macconelli* (Thomas, 1901)

Measurements

	Sex	Mean	S.D.	N
TL	M	42.73	1.49	15
	F	43.91	1.68	45
HB	M	42.73	1.49	15
	F	43.91	1.68	45
HF	M	10.27	0.80	15
	F	9.91	0.73	45
E	M	15.60	0.63	15
	F	15.47	0.63	45
FA	M	30.07	1.07	15
	F	30.94	0.69	45
Wta	M	6.33	0.36	15
	F	6.52	0.43	35

Location: TFA, Venezuela (USNMNH)

Description
The dental formula is I 2/2, C 1/1, P 2/2, M 2/2. Head and body length averages 42.8 mm for males and 44.1 mm for females; there is no external tail. Weight averages 6.4 g for males and 6.6 g for females. The dorsum is a dull brownish white, darkening to brown on the lower back. This species is markedly different from the white *E. alba*, the northern species that penetrates northwestern Panama. Chromosome number: $2n = 21-22$; FN = 20.

Range and Habitat
The species ranges from southern Costa Rica through the Isthmus over much of northern South America and northeastern Brazil (map 5.87). In Venezuela it is common below 500 m elevation. It prefers moist habitats and multistratal tropical evergreen forest (Handley 1976).

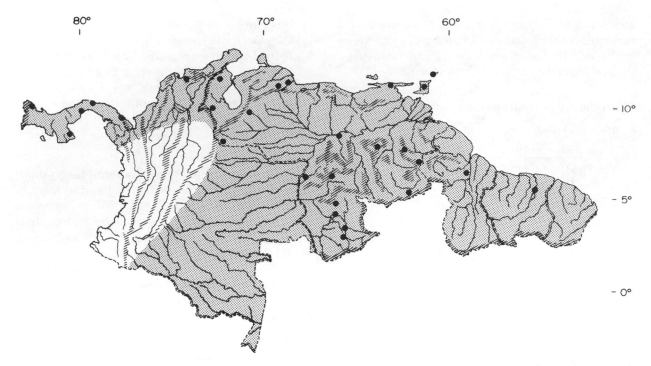

Map 5.86. Distribution of *Chiroderma villosum*.

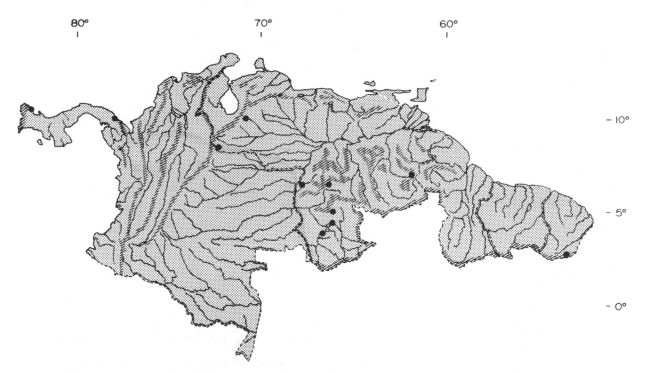

Map 5.87. Distribution of *Ectophylla macconnellii*.

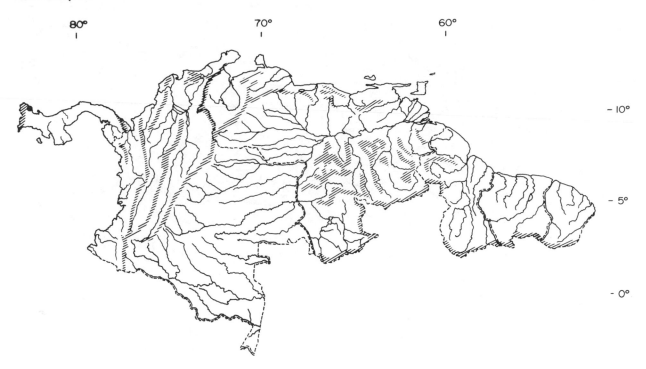

Map 5.88. Distribution of *Ectophylla alba*.

Natural History

This bat constructs a "tent" by biting through the midrib of *Heliconia* leaves. Roosting groups are small (two to four). The diet is believed to be mainly fruit.

Ectophylla alba H. Allen, 1892
White Bat, Murciélago Blanco

Description

The dental formula is I 2/2, C 1/1, P 2/2, M 2/2. Head and body length ranges from 35 to 48 mm and the forearm from 25 to 35 mm. Weight averages about 7.5 g. The fur of this distinctive bat is whitish, and the ears, nose leaf, and skin of the wings are yellow (plate 5). Its almost white coloring is shared only by diclidurine bats; this is the only white phyllostomid.

Range and Habitat

This species is found in Honduras, Nicaragua, Costa Rica, and northwestern Panama (map 5.88) (Hall 1981).

Genus *Artibeus* Leach, 1821
Fruit-eating Bat, Murciélago Frutero

Description

The dental formula is I 2/2, C 1/1, P 2/2, M 2–3/2–3 (fig. 5.23). Head and body length ranges from 53 to 100 mm and forearm length from 35 to 76

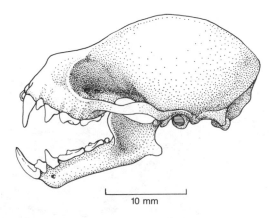

Figure 5.23. Skull of *Artibeus jamaicensis*.

mm. The species of this genus present an array of sizes. The interfemoral membrane is rather narrow. In general the dorsum is brownish to black, and the underparts are usually paler. The dorsal line, so characteristic of most stenodermine bats, is absent. Four whitish facial stripes may be present, but this varies from species to species.

Chromosome number: all species show $2n = 30–31$; FN = 56.

Distribution

Species of *Artibeus* range from Sinaloa in western Mexico and Tamaulipas in eastern Mexico south

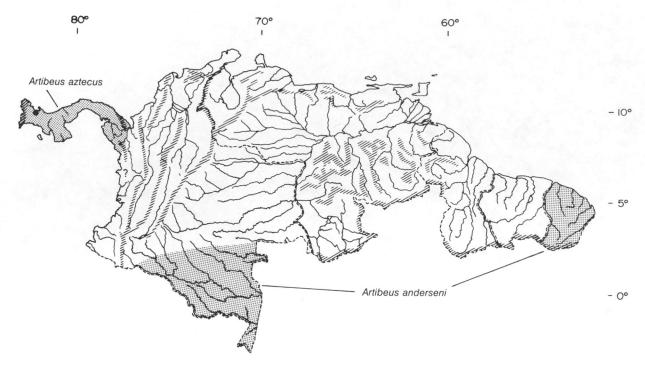

Map 5.89. Distribution of two species of *Artibeus* (based in part on Koopman 1982).

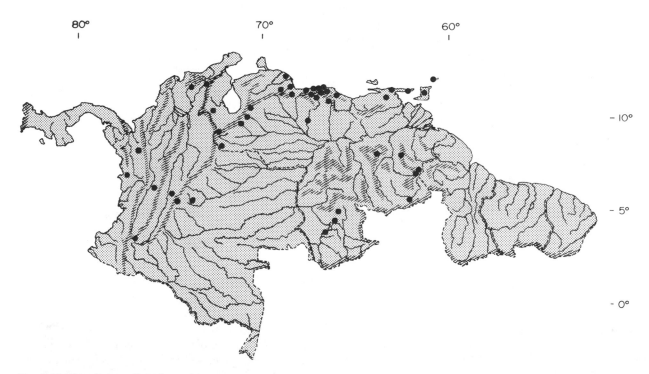

Map 5.90. Distribution of *Artibeus cinereus*.

over most of tropical South America to northern Argentina and southeastern Brazil.

Natural History

Artibeus contains some fourteen species (at least two more are yet to be described by Handley). Within any community they are usually graded in size, presumably reflecting specialization for different-sized foods. These bats are primarily frugivores, and one species (*A. jamaicensis*) has been studied in some detail (see species account). Most species exhibit a seasonal bimodal polyestrous pattern, with births occurring late in the dry season and in the midportion of the rainy season where rainfall is seasonal (Wilson 1979).

Artibeus amplus Handley, 1987

Measurements

	Sex	Mean	S.D.	N
TL	M	91.43	4.79	7
	F	87.46	4.63	13
HB	M	91.43	4.79	7
	F	87.46	4.63	13
HF	M	19.71	0.95	7
	F	19.54	1.05	13
E	M	23.14	2.12	7
	F	22.85	2.38	13
FA	M	69.65	2.59	8
	F	69.51	1.75	15
Wta	M	59.10		1

Location: TFA, Venezuela (USNMNH)

Description

This new species, described by Handley in 1987, is very similar to *A. literatus*. It is one of the largest species of the genus. Head and body length averages 91.5 mm for males and 90.3 mm for females; weight averages about 60 g. The skull is longer and narrower than in *A. jamaicensis*. The lower edge of the nose leaf is fused to the facial skin. The wings are never white tipped.

Range and Habitat

The species was diagnosed from specimens taken in Venezuela. It has been recorded from the "northern foothills of the Colombian Andes, . . . the Venezuelan Andes, . . . and the vicinity of Cerro Duida and the low southeastern mountains of Est. Bolívar in southeastern Venezuela. It probably occurs in adjacent parts of Guyana and Brazil as well" (Handley 1987, 164–65).

Note

Koopman (1982) suggests that *Artibeus andersoni* could extend to southeastern Colombia and the east-

ern Guyanas, but I cannot record a specimen (map 5.89).

Artibeus aztecus K. Anderson, 1906

Description

This small species shows variation in size and color over its entire range. Head and body length averages 61 to 71 mm and forearm length 40 to 48 mm. In Panama the dorsal pelage is very dark brown. Chromosome number: $2n = 30$; FN = 56.

Range and Habitat

The species occurs disjunctly from western Mexico to western Panama (map 5.89) but prefers higher altitudes and is associated with cloud forests (Webster and Knox Jones 1982a).

Artibeus cinereus (Gervais, 1856)

Measurements

	Sex	Mean	S.D.	N
TL	M	55.06	4.25	63
	F	54.95	4.73	60
HB	M	55.06	4.25	63
	F	54.95	4.73	60
HF	M	11.46	0.78	63
	F	11.68	0.68	60
E	M	16.67	1.05	63
	F	16.77	0.83	60
FA	M	39.90	1.16	82
	F	40.01	1.18	102
Wta	M	12.89	2.16	53
	F	13.08	1.46	36

Location: Northern Venezuela (USNMNH)

Description

This is one of the smaller species of the genus and includes *Artibeus andersoni*. Both dorsum and venter are medium brown, and the throat may be pale brown. The white facial stripes are prominent.

Range and Habitat

This species ranges from southern Veracruz, Mexico, through Panama and across most of northern South America to central Brazil. In Venezuela it ranges from 1,000 to 2,000 m elevation (map 5.90). It prefers multistratal tropical evergreen forest and in general occurs above 1,000 m. Where it co-occurs with *A. concolor,* the latter species generally replaces *A. cinereus* at lower elevations.

Natural History

These bats roost in small groups, usually in trees such as palms (Goodwin and Greenhall 1961).

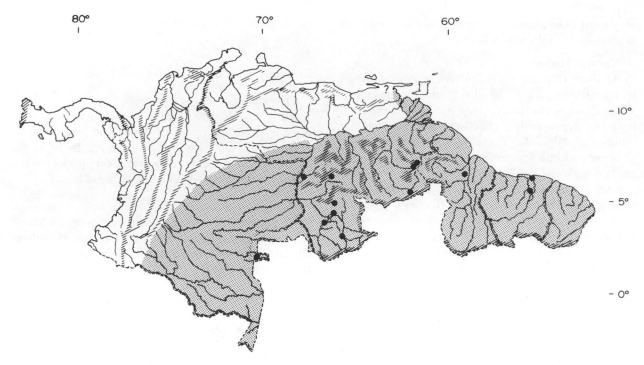

Map 5.91. Distribution of *Artibeus concolor*.

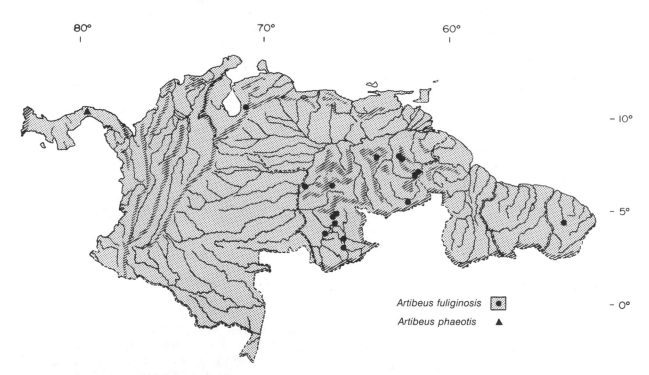

Artibeus fuliginosis ▣
Artibeus phaeotis ▲

Map 5.92. Distribution of two species of *Artibeus*.

Artibeus concolor Peters, 1865

Measurements

	Sex	Mean	S.D.	N
TL	M	61.63	2.62	8
	F	64.70	3.29	56
HB	M	61.63	2.62	8
	F	64.70	3.29	56
HF	M	11.50	1.20	8
	F	11.64	1.31	56
E	M	18.88	1.13	8
	F	19.04	1.08	56
FA	M	46.98	1.19	8
	F	48.40	1.66	63
Wta	M	18.33	2.26	7
	F	19.98	2.52	16

Location: TFA, Venezuela (USNMNH)

Description

This species is smaller than *A. jamaicensis* but larger than *A. cinereus;* females are considerably larger than males. The coloration is similar to that of *A. cinereus,* but the facial stripes are indistinct.

Range and Habitat

This species is found in southern Colombia, Venezuela, the Guyanas, and northern Brazil (map 5.91). In Venezuela specimens were caught in moist areas from 100 to 1,000 m elevation, but the vast majority were found below 500 m (Handley 1976).

Artibeus fuliginosis Gray, 1838

Measurements

	Sex	Mean	S.D.	N
TL	M	75.01	3.59	97
	F	77.05	3.54	44
HB	M	75.01	3.59	97
	F	77.05	3.54	44
HF	M	16.44	1.40	97
	F	17.18	1.17	44
E	M	23.07	1.13	97
	F	23.50	1.37	44
FA	M	58.89	1.83	101
	F	60.15	1.42	44
Wta	M	35.29	2.81	90
	F	36.48	4.48	23

Location: TFA, Venezuela (USNMNH)

Description

This species is very similar to *A. jamaicensis* in appearance but slightly smaller in some parts of its range (see Handley 1987). The dorsum is very dark brown, and the facial stripes are faint. Some authors consider this a subspecies of *A. jamaicensis.*

Range and Habitat

The species ranges south of the Orinoco in Venezuela and into the adjacent portions of the Guyanas and northern Brazil (map 5.92). Most specimens taken in Venezuela were caught below 500 m elevation. It is associated with moist habitats and multistratal tropical evergreen forest (Handley 1976).

Artibeus gnomus Handley, 1987

Measurements

	Sex	Mean	S.D.	N
TL	M	49.47	3.12	17
	F	51.15	2.48	13
HB	M	49.47	3.12	17
	F	51.15	1.48	13
HF	M	11.06	0.83	17
	F	11.08	0.64	13
E	M	17.35	0.93	17
	F	17.46	0.97	13
FA	M	37.60	1.06	18
	F	38.14	1.29	12
Wta	M	10.36	1.02	17
	F	10.62	0.84	5

Location: TFA, Venezuela (USNMNH)

Description

This is one of the smallest species of the genus (Handley 1987). The facial stripes and dorsal pelage are similar to those of *A. jamaicensis.*

Range and Habitat

The species has been taken in Venezuela, and I have seen similar specimens from Suriname (CMNH). Handley has the following comments: "Distribution—The Amazon Basin and bordering regions, from northern Amazonas Territory (14 km SSE Pto. Ayacucho) and northern Bolívar State (28 km SE El Manteco) in Venezuela and northern Guyana to Pará (Belém) and Mato Grosso (Serra do Roncador), Brazil and Loreto (Santa Rosa), Peru" (1987, 167).

Artibeus (= *Enchisthenes*) *hartii* Thomas, 1892

Measurements

	Sex	Mean	S.D.	N
TL	M	60.16	2.92	37
	F	61.09	3.52	32
HB	M	60.16	2.92	37
	F	61.09	3.52	32
HF	M	12.19	0.78	37
	F	12.09	0.93	32
E	M	15.92	1.36	36
	F	15.78	1.43	32
FA	M	39.26	1.24	38
	F	39.12	1.25	32
Wta	M	18.02	1.38	5
	F	17.50		1

Location: Northern Venezuela (USNMNH)

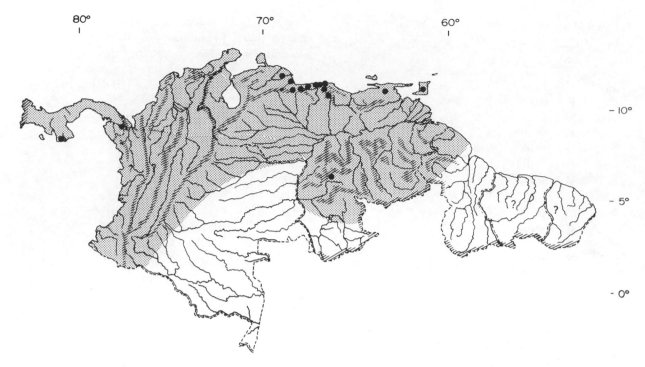

Map 5.93. Distribution of *Artibeus hartii*.

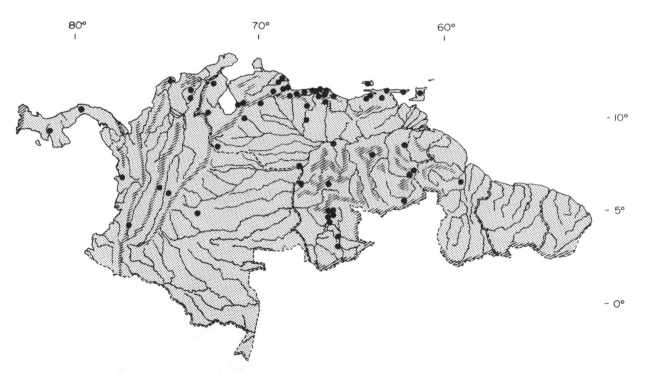

Map 5.94. Distribution of *Artibeus jamaicensis*.

Description

This species does not exhibit noticeable sexual size dimorphism. The dorsum is dark brown, the venter paler; it is exceptional for *Artibeus* in that it has a prominent white dorsal stripe. White facial stripes are present but faint. (See Owen 1988.)

Range and Habitat

The species is distributed from Nayarit in western Mexico and southern Tamaulipas in eastern Mexico south through Panama to northern Venezuela and western Colombia, south to Peru and Amazonian Bolivia (map 5.93). In Venezuela this species ranges from 2 to 2,250 m, but the vast majority of specimens were taken at between 1,000 and 2,250 m elevation (Handley 1976). It is strongly associated with moist habitats and multistratal tropical evergreen forest, but it does penetrate cloud forest, and some specimens were taken in dry deciduous forest.

Artibeus jamaicensis Leach, 1821

Measurements

	Sex	Mean	S.D.	N
TL	M	79.37	4.78	131
	F	80.86	5.28	110
HB	M	79.37	4.78	131
	F	80.86	5.28	110
HF	M	17.01	1.37	131
	F	17.29	1.24	110
E	M	21.65	1.36	130
	F	22.08	1.15	110
FA	M	58.18	1.71	226
	F	58.89	2.11	206
Wta	M	39.24	4.01	71
	F	40.56	4.72	43

Location: Northern Venezuela (USNMNH)

Description

Males average 80.1 mm in total length and females 81.8 mm; weight averages 40.4 g for males and 43.2 g for females (see table 5.14). The upperparts and venter are light brown to very dark brown; facial stripes may be faint. There is considerable geographic variation in size (see Handley 1987).

Range and Habitat

This is one of the most widespread species of the genus. It occurs from Sinaloa in western Mexico and southern Tamaulipas in eastern Mexico south through Panama to northern Argentina and southeastern Brazil (map 5.94). It tolerates a range of habitat types, occurring in both dry deciduous forest and multistratal tropical evergreen forest, and it even penetrates cloud forest. Although in northern Venezuela most specimens were taken below 1,500 m, it ranges to 2,135 m elevation (Handley 1976).

Natural History

This is one of the best-studied species of Neotropical bats, mainly owing to the efforts of Morrison (1978a,b,c,d, 1979). It is strongly frugivorous, feeding on *Ficus* (August 1981), and the timing of its reproduction is closely tied to maximum abundance of fruit (Fleming 1971). Breeding colonies of up to twenty-five involve harem defense by the male (Kunz, August, and Burnett 1983). The mating system is strongly polygynous. During lactation, females have a daytime roost, and when moving to forage they deposit their babies in crèches near the feeding tree. Bright moonlight strongly inhibits the flight of these bats. It is suspected that owls may be significant predators at night and snakes and the bat falcon, *Falco rufigularis*, during the day. When foraging the bats fly in small groups; if an individual is caught it will emit distress calls, inducing "mobbing" behavior by fellow flock members (August 1979). These bats occasionally construct tents by biting the midribs of large leaves, causing the leaves to fold over (Foster and Timm 1976). They also roost in hollow trees and well-lighted caves (Goodwin and Greenhall 1961).

Artibeus lituratus (Olfers, 1818)

Measurements

	Sex	Mean	S.D.	N
TL	M	88.82	6.34	39
	F	91.07	6.42	81
HB	M	88.82	6.34	39
	F	91.07	6.42	81
HF	M	19.77	1.20	39
	F	20.12	1.29	81
E	M	24.18	1.30	39
	F	24.37	1.05	81
FA	M	68.55	1.95	68
	F	69.20	2.13	127
Wta	M	64.50	5.24	28
	F	67.73	8.29	23

Location: Northwestern Venezuela (USNMNH)

Description

This is one of the largest species of *Artibeus*. The dorsum and venter are light brown, and the facial stripes are conspicuous.

Range and Habitat

This species occurs from Sinaloa in western Mexico and southern Tamaulipas in eastern Mexico south through Panama over most of the South American continent, to northern Argentina and southeastern Brazil (map 5.95). Together with *A. jamaicensis*, this is one of the most common species of fruit bats. In Venezuela it mainly occurs below 500 m, but some specimens may be taken at up to 2,000 m elevation.

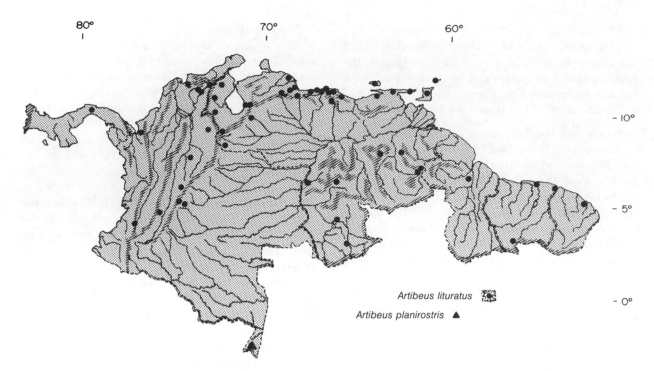

Map 5.95. Distribution of two species of *Artibeus*.

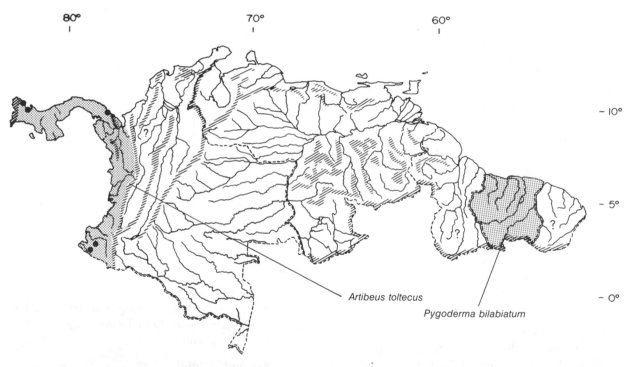

Map 5.96. Distribution of *Artibeus toltecus* and *Pygoderma bilabiatum* (*Pygoderma* based on Koopman 1982).

Although it prefers moist areas and multistratal evergreen forest, it tolerates man-made clearings.

Natural History

Handley (1967) noted that in Brazil this species forages high in the canopy. They roost in well-lighted caves and in palm trees. Colonies may be as large as twenty-five individuals (Goodwin and Greenhall 1961).

Artibeus phaeotis (Miller, 1902)

Description

This small species has a head and body length of 51 to 60 mm; forearm length is 35.2 to 41.8 mm. A single specimen from Veracruz measured TL 59; HF 10; E 17 (CMNH). The dorsum and venter are brown, and the facial stripes are prominent.

Range and Habitat

The species occurs from southern Sinaloa in western Mexico and southern Veracruz in eastern Mexico south through Panama, possibly to Colombia (map 5.92). The status of this species in Venezuela still needs to be clarified pending a future publication by Handley.

Natural History

Though primarily frugivorous, this bat will feed on insects and pollen. It constructs a "tent" as described for A. jamaicensis (Timm 1985).

Artibeus planirostris (Spix, 1823)

Description

This large Artibeus is similar to A. literatus in size and color but occurs only marginally within the area covered by this volume. It includes A. fattax and A. hercules (Koopman 1978).

Range and Habitat

The species is found in lowland tropical rain forest in Brazil, Peru, and Ecuador. Its exact limits in southern Colombia and Venezuela are poorly understood (map 5.95) (Koopman 1978; Albuja 1982).

Artibeus toltecus (Saussure, 1860)

Description

This bat is smaller than A. aztecus. Its measurements are TL 59.3; HF 10.7; E 17.7; Wt 15 g ($N = 7$; CMNH). In Panama and Colombia the dorsum is pale brown and the facial stripes are faint.

Range and Habitat

This small fruit bat ranges from Sinaloa in western Mexico and Tamaulipas in eastern Mexico south through Panama to northwestern Colombia (map

5.96). The specific status of this species with respect to A. phaeotis remains to be clarified (Webster and Knox Jones 1982b). (See Owen 1988.)

Genus Pygoderma Peters, 1863
Pygoderma bilabiatum (Wagner, 1843)

Description

The dental formula is I 2/2, C 1/1, P 2/2, M 2/2. Head and body length averages about 61 mm, and forearm length is about 39 mm in females and 36 mm in males. One female measured TL 75; HF 11.8; E 20.8; Wt 26 g. The dorsum is pale brown varying almost to black; the venter is grayish brown. There is a white spot on each shoulder near the wing, but this bat lacks facial stripes and a dorsal stripe. Males have large subdermal glands surrounding the eye (Meyers 1981). The fur of the chest is very thin or absent in the male.
Chromosome number: $2n = 30-31$; FN = 56.

Range and Habitat

This species has been described from southern Brazil, Bolivia, and Paraguay. Recently specimens have been collected in Suriname (map 5.96) (Webster and Owen 1984). This bat prefers tropical forest habitats.

Natural History

In Paraguay this species exhibits two marked birth peaks during the fall and winter. It is believed to feed on fruit pulp (Myers 1981).

Genus Ametrida Gray, 1847
Ametrida centurio Gray, 1847

Measurements

	Sex	Mean	S.D.	N
TL	M	41.34	2.10	59
	F	46.62	2.95	37
HB	M	41.34	2.10	59
	F	46.62	2.95	37
HF	M	10.64	0.58	59
	F	11.32	0.63	37
E	M	14.27	1.08	59
	F	15.08	1.21	37
FA	M	25.58	0.54	59
	F	31.88	0.70	36
Wta	M	7.99	0.70	15
	F	9.50		1

Location: Bolívar, Venezuela (USNMNH)

Description

The dental formula is I 2/2, C 1/1, P 2/2, M 3/3. Head and body length averages 42 mm for males and 47.9 mm for females. Males average 8 g, and females

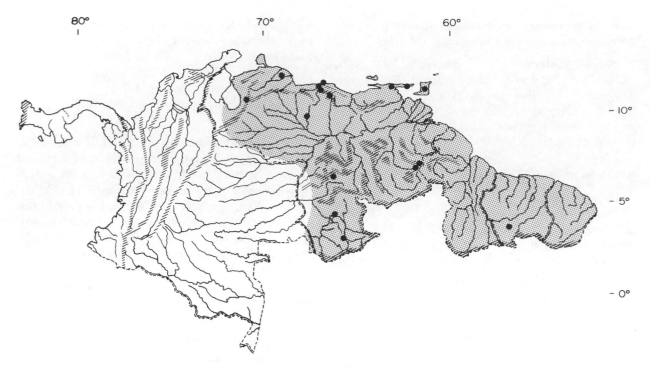

Map 5.97. Distribution of *Ametrida centurio*.

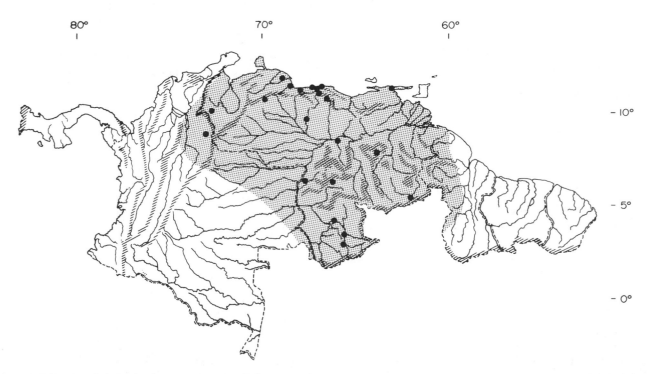

Map 5.98. Distribution of *Sphaeronycteris toxophyllum*.

average 12.6 g (see table 5.14). The pronounced sexual size dimorphism formerly led to the small males' being given the name *A. minor*. The rostrum is very short. There are some dermal outgrowths on the sides of the face, though they are less extensive than in *Centurio*, and the nose leaf is prominent. The dorsum is light to dark brown, and there is a single gray patch on each shoulder.
Chromosome number: $2n = 30-31$; FN = 56.

Range and Habitat

The species is distributed east of the Andes in Venezuela, across the Guyanas, and south into north-central Brazil (map 5.97). In Venezuela specimens were taken at up to 2,150 m elevation, but most were taken below 1,500. These bats are strongly associated with moist areas and multistratal tropical evergreen forest (Handley 1976). They are frugivores.

Genus *Sphaeronycteris* Peters, 1882
Sphaeronycteris toxophyllum Peters, 1882

Measurements

	Sex	Mean	S.D.	N
TL	M	56.00	3.67	20
	F	57.57	2.96	46
HB	M	56.00	3.67	20
	F	57.57	2.96	46
HF	M	12.00	0.92	20
	F	12.22	0.92	46
E	M	15.45	1.10	20
	F	16.00	1.28	46
FA	M	37.64	0.69	20
	F	40.06	0.67	48
Wta	M	17.10		2

Location: Northern Venezuela (USNMNH)

Description

The dental formula is I 2/2, C 1/1, P 2/2, M 3/3. The dorsum is cinnamon brown. There is some fleshy outgrowth on the side of the face, reminiscent of *Centurio*. The male has a hornlike growth on the forehead, but the structure is rudimentary in the female. A fold of skin is present under the chin, and in the male it can be rolled over the face like that described for *Centurio*.
Chromosome number: $2n = 28$; FN = 52.

Range and Habitat

This species occurs from Venezuela and eastern Colombia, east of the Andes, south to Amazonian Peru and northwestern Brazil (map 5.98). In Venezuela specimens were taken at up to 2,240 m. They may follow gallery forest into dry habitats but are usually associated with multistratal tropical evergreen forest, though they tolerate man-made clearings.

Genus *Centurio* Gray, 1842
Centurio senex Gray, 1842
Wrinkle-faced Bat

Description

The dental formula is I 2/2, C 1/1, P 2/2, M 2/2. The rostrum is very short (fig. 5.24). Head and body length averages about 55 mm, and forearm length ranges from 41 to 46.5 mm. Three specimens from Veracruz showed the following mean measurements: TL 65.3; HF 11.3; E 16 (CMNH). This distinctive bat has many folds of skin on its face, and glands are associated with them. There is a fold of skin that forms a pouchlike structure on the throat when retracted. This skin can be drawn over the face like a mask, and it has two spots that are generally devoid of hair, creating translucent areas so that the bat can distinguish light and dark while its face is covered. The nose leaf is reduced, but the distinctive appearance of the face renders this bat unmistakable. The dorsal color is brown to dark brown, and there is a white spot on each shoulder. The wing membrane between the second and third digits is usually striped horizontally (plate 5).
Chromosome number: $2n = 28$; FN = 52.

Range and Habitat

This bat occurs from Sinaloa in western Mexico and Tamaulipas in eastern Mexico south through the Isthmus, across the extreme northern part of South America. So far it is recorded only from Colombia and northern Venezuela (map 5.99). It is broadly tolerant of both moist and dry habitats.

Natural History

Little is known concerning the ecology and behavior of this extraordinary bat. Its diet is assumed to consist mainly of fruits, perhaps those with soft

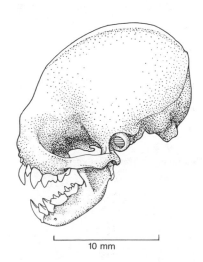

Figure 5.24. Skull of *Centurio senex*.

10 mm

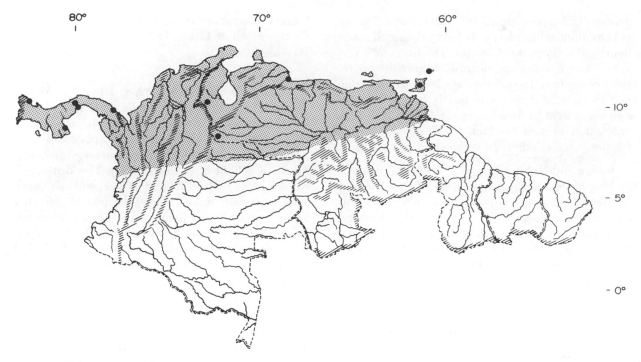

Map 5.99. Distribution of *Centurio senex*.

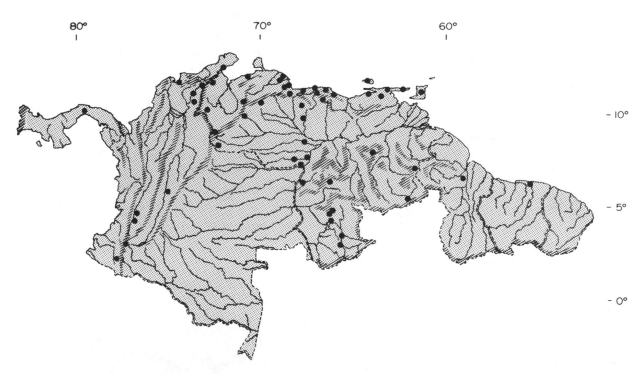

Map 5.100. Distribution of *Desmodus rotundus*.

pericarps, since the very short jaws suggest extreme specialization for biting into large, fleshy fruits (Snow, Knox Jones, and Webster 1980). The bats roost in groups of two to three, and the sexes may roost separately for part of the year (Goodwin and Greenhall 1961).

SUBFAMILY DESMODONTINAE (DESMODONTIDAE)

Diagnosis

This subfamily contains three species that range in head and body length from 65 to 90 mm. The dental formula for the three species is highly variable and will be discussed in the species accounts, but the incisors and canines are specialized for cutting, and the cutting edges form a V (fig. 5.25). These teeth are used to make incisions in their prey, for these are the true vampire bats. The premolars and molars are greatly reduced, indicating less selection for their retention, since these animals have specialized for their peculiar dietary habits. There is no tail. The nose leaf is reduced to U-shaped fleshy pads that surround each nostril. The dorsum is generally some shade of brown.

Distribution

Members of this subfamily occur from southern Sonora and southeastern Texas south across the continent of South America to northern Argentina, Uruguay, and northern Chile.

Natural History

The three species of this subfamily are specialized for feeding on the blood of warm-blooded vertebrates. The saliva contains an enzyme that retards coagulation, and they lap the blood after making an incision. They can walk on their feet and thumbs and usually crawl on the body of their prey before feeding. Vampire bats are of medical importance because they have been implicated in transmitting rabies.

Genus *Desmodus* Wied-Neuwied, 1826
Vampiro, Mordedor
Desmodus rotundus (E. Geoffroy, 1810)
Common Vampire Bat

Measurements

	Sex	Mean	S.D.	N
TL	M	74.51	4.59	71
	F	80.57	5.51	63
HB	M	74.51	4.59	71
	F	80.57	5.51	63
HF	M	16.93	1.30	71
	F	17.68	1.24	63
E	M	18.75	1.17	71
	F	19.33	0.92	63
FA	M	55.66	1.82	110
	F	59.65	2.04	113
Wta	M	27.89	3.17	49
	F	32.98	4.48	26

Location: Northern Venezuela (USNMNH)

Description

The dental formula is I 1/2, C 1/1, P 2/3, M 0/0 (fig. 5.25). Head and body length ranges from 75 to 90 mm and the forearm from 50 to 63 mm (see table 5.14). The weight is highly variable, since these bats become extremely distended after feeding. The thumb is rather long and has a distinct pad at the base (fig. 5.20). The dorsum is gray brown, and the underparts are somewhat paler.
Chromosome number: $2n = 28$; FN = 52.

Range and Habitat

The species is distributed from southern Sonora and Tamaulipas, Mexico, south through the Isthmus to northern Argentina and Uruguay. It rarely ranges above 1,000 m elevation (map 5.100). It tolerates a broad variety of cover types, being found in multistratal evergreen forest, man-made clearings, pastures, and deciduous tropical forest (Handley 1976).

Natural History

This species appears to feed almost exclusively on the blood of mammals. It is prone to feed on livestock and thus makes itself a pest. It is implicated in transmitting the rabies virus, and livestock losses may be considerable. The single young is born after a 110 day gestation period. These bats may roost in hollow trees, though they seem to prefer caves, and

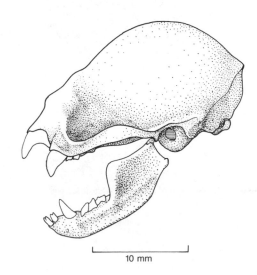

10 mm

Figure 5.25. Skull of *Desmodus rotundus*.

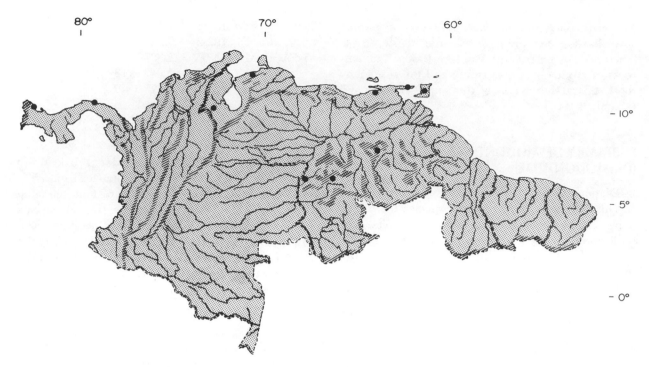

Map 5.101. Distribution of *Desmodus youngii*.

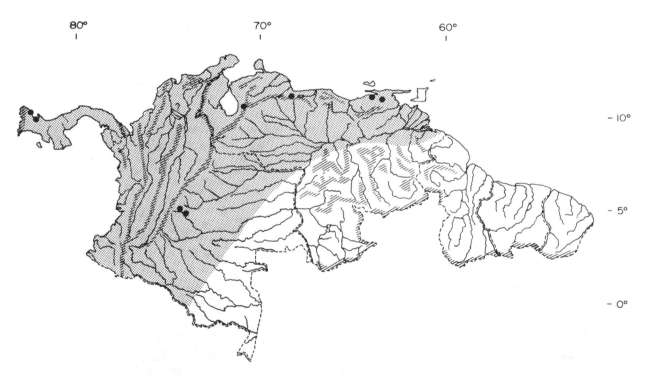

Map 5.102. Distribution of *Diphylla ecaudata*.

colonies may be as large as five hundred. Vampire bats often share roosting places with other species. The bats forage in small groups that may be families. They are quite able to run over surfaces, an adaptation they employ when selecting a place to bite on ungulates such as the domestic cow or white-tailed deer (Turner 1975; Schmidt and Manske 1973).

Desmodus (Diaemus) youngi Jentink, 1893

Measurements

	Sex	Mean	S.D.	N
TL	M	83.36	5.35	11
	F	84.00	2.94	4
HB	M	83.36	5.35	11
	F	84.00	2.94	4
HF	M	17.91	1.30	11
	F	19.50	1.29	4
E	M	18.18	1.33	11
	F	18.75	0.50	4
FA	M	51.65	1.39	11
	F	53.48	1.24	4
Wta	M	36.27	2.80	11
	F	32.20		1

Location: Venezuela (USNMNH)

Description

The dental formula is I 1/2, C 1/1, P 1/2, M 2/1. This species is similar in size to *D. rotundus*. Head and body length average approximately 83 mm, and weight averages about 35 g. The thumb is shorter than that of *D. rotundus*. The tips of the wings are white, which easily separates *D. youngi* from *D. rotundus*.
Chromosome number: $2n = 32$; FN = 60.

Range and Habitat

This bat has nearly the same distribution as *D. rotundus* but is absent from the west coast of Mexico. It is found in southern Tamaulipas and along the Gulf coast south to Central America, and from there it ranges throughout South America to southeastern Brazil (map 5.101). It prefers moist areas but forages in dry deciduous forest as well as in multistratal evergreen forest. In Venezuela specimens were taken at below 500 m elevation (Handley 1976).

Natural History

Like *D. rotundus*, this species feeds on the blood of higher vertebrates. Glands in the mouth produce a sharp odor in the male, but their exact function in communication is poorly understood. The bats nest in hollow trees, in colonies of up to thirty (Goodwin and Greenhall 1961).

Genus *Diphylla* Spix, 1823
Diphylla ecaudata Spix, 1823
Vampiro de Doble Escudo

Measurements

	Sex	Mean	S.D.	N
TL	M	73.17	6.31	6
	F	78.67	5.86	3
HB	M	73.17	6.31	6
	F	78.67	5.86	3
HF	M	15.67	1.75	6
	F	17.67	1.53	3
E	M	15.17	0.41	6
	F	15.67	0.58	3
FA	M	49.90	0.92	7
	F	51.80		2
Wta	M	25.98	1.94	5
	F	25.17	2.43	3

Location: Venezuela (USNMNH)

Description

The dental formula is I 2/2, C 1/1, P 1/2, M 2/2. The outer lower incisor is fan shaped and has seven tiny lobes; it may be important in grooming the fur. These bats are smaller than *Desmodus*. Head and body length averages 73.2 mm for males and 78.7 mm for females; weight averages 25.9 g for males and 25.2 g for females. The dorsum is medium brown, and the venter is nearly the same shade. Chromosome number: $2n = 32$; FN = 60.

Range and Habitat

This species ranges from the Gulf coast of Texas south in eastern Mexico to Central America, and thence to Colombia, Venezuela, Peru, Ecuador, Amazonian Bolivia, and Brazil south of the Amazon (map 5.102). In Venezuela it was collected at elevations up to 1,537 m. It tolerates a variety of habitats, including farms, multistratal evergreen forest, and dry deciduous forest (Handley 1976).

Natural History

This species apparently is specialized for feeding on the blood of birds and can be a serious pest to poultry farmers. It has been found roosting in caves, generally in colonies smaller than those typical of *D. rotundus*. It rests in small groups (one to three) in caves or hollow trees.

SUPERFAMILY VESPERTILIONOIDEA

These bats appear to be the most specialized for flight, based on the anatomy of the pectoral girdle. The greater tuberosity of the humerus is enlarged, and on the upstroke of the wingbeat it articulates

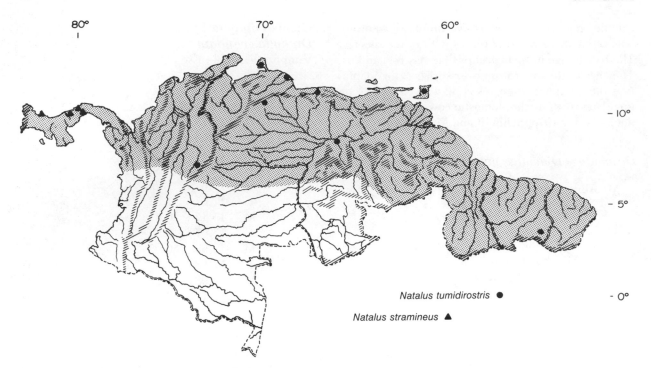

Map 5.103. Distribution of two species of *Natalus*.

with the scapula, thus mechanically terminating the upswing (Miller 1907).

FAMILY NATALIDAE
Genus *Natalus* Gray, 1838
Funnel-eared Bat

Diagnosis
This family includes a single genus, *Natalus*. The dental formula is I 2/3, C 1/1, P 3/3, M 3/3. The tail is very long and is completely enclosed in the interfemoral membrane, and it nearly equals or exceeds the head and body length (fig. 5.26). The thumb is short and is almost enveloped in the antebrachial membrane, and the third phalanx of the third finger remains cartilaginous even in the adult. Adult males have a glandlike structure in the center of the forehead. The ears have a peculiar funnel shape (see fig. 5.27).

Distribution
This family is confined to the Western Hemisphere. It is distributed from northern Mexico to northern South America and eastern Brazil and has speciated widely in the Antilles.

Natural History
These bats are aerial insectivores and find their prey through echolocation. They roost in caves or tunnels, sometimes with other cavernicolous bats, and the colonies may be very large.

Natalus stramineus Gray, 1838

Description
Head and body length is approximately 50 mm; the tail is longer than the head and body, averaging

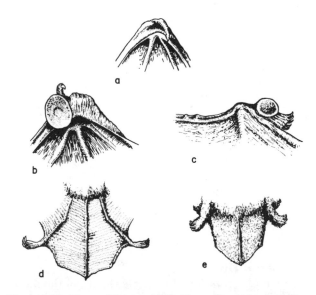

Figure 5.26. .External features of the Vespertilionoidea (part 1): (*a*) Reduced thumb of *Furipterus horrens*. (*b*) Disk on the underside of the thumb of *Thyroptera tricolor*. (*c*) Disk on the underside of the foot of *Thyroptera tricolor*. (*d*) Interfemoral membrane and extremely long tail of *Natalus tumidirostris*. (*e*) Hairy dorsal surface of the interfemoral membrane in *Lasiurus*. (Redrawn with permission from Goodwin and Greenhall 1961.)

about 55 mm, and forearms range from 35 to 45 mm. Two females from Colombia measured TL 92, 101; T 45, 53; HF 9, 16; E 13, 17; Wt. 4.3, 6.2 g. Over most of its range there are two color phases. The dorsum may be either buffy in the light phase or reddish brown in the dark phase. The underparts are always paler.

Range and Habitat

This species ranges from Sonora in western Mexico and Tamaulipas in eastern Mexico south to western Panama (map 5.103). Koopman (1982) gives the range as extending into eastern Brazil across northern South America, but there could be confusion about species identity in the Caribbean area. Either this species has an extremely disjunct range, or specimens from Brazil are in fact *N. tumidirostris*.

Natural History

Linares (1972) reports on the traditional use of caves as roosting sites as setting the stage for the evolution of local variation in measurements and pelage color.

Natalus tumidirostris Miller, 1900

Measurements

	Sex	Mean	S.D.	N
TL	M	101.62	3.84	65
	F	101.26	2.72	34
HB	M	49.12	3.48	65
	F	49.82	3.56	34
T	M	52.49	2.76	65
	F	51.44	2.29	34
HF	M	10.18	0.43	65
	F	10.21	0.59	34
E	M	15.23	0.68	65
	F	14.97	0.52	34
FA	M	40.24	0.81	63
	F	39.78	0.97	34
Wta	M	6.81	0.77	17
	F	6.33	0.24	14

Location: Northern Venezuela (USNMNH)

Description

The sexes are similar in size; total length averages about 101 mm. The tail averages 51.5 mm, head and body length 48.5 mm, and weight 6.2 g. The dorsum varies from cinnamon to dull buff.

Range and Habitat

The species is known from northern Colombia, Venezuela, Guyana, and Suriname. In Venezuela it ranges to 548 m elevation (map 5.103). It tolerates both dry and wet habitats, but most specimens are encountered in dry deciduous forest.

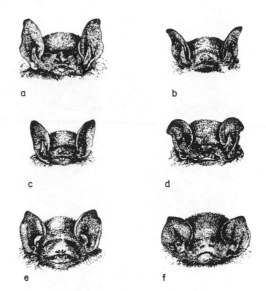

Figure 5.27. External features of the Vespertilionoidea (part 2): (*a*) Face of *Eptesicus brasiliensis*. (*b*) Face of *Myotis nigricans*. (*c*) Face of *Rhogeessa tumida*. (*d*) Face of *Lasiurus borealis*. (*e*) Face of *Natalus tumidirostris*. (*f*) Face of *Furipterus horrens*. (Redrawn with permission from Goodwin and Greenhall 1961.)

FAMILY FURIPTERIDAE
Smoky Bats

Diagnosis

The thumb is extremely reduced or absent. The claw, if present, is minute and functionless (see fig. 5.26). There are two genera, *Furipterus* and *Amorphochilus*.

Distribution

This family is confined to the New World tropics, ranging from Costa Rica across northern South America to eastern Brazil.

Genus *Furipterus* Bonaparte, 1837
Furipterus horrens (F. Cuvier, 1828)

Measurements

	Mean	S.D.
TL	62.0	2.76
T	24.5	2.42
HF	7.9	0.80
E	11.1	0.92
FA	35.3	1.16
Wta	3.08	0.68

Location: Venezuela (USNMNH [*N* = 8 males])

Description

In this small bat, head and body length averages 37.5 mm, the tail 24.5 mm, and the forearm 35.3 mm. The dorsum varies from brownish gray to almost black, and the venter is somewhat paler.

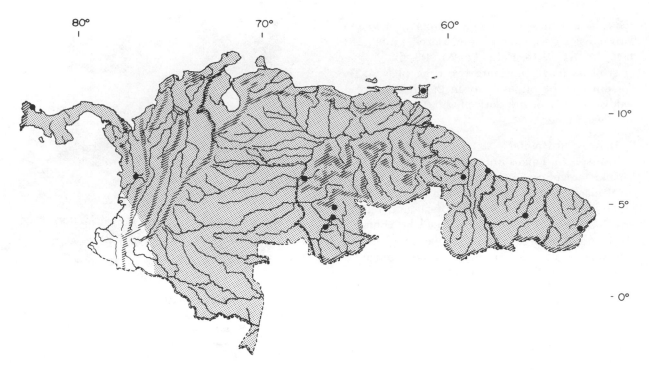

Map 5.104. Distribution of *Furipterus horrens*.

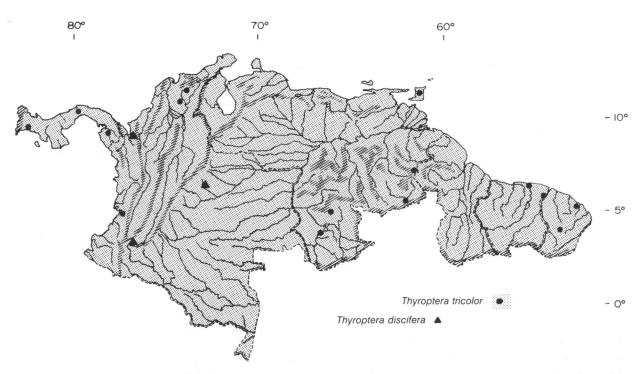

Thyroptera tricolor ▨
Thyroptera discifera ▲

Map 5.105. Distribution of two species of *Thyroptera*.

Range and Habitat

This species is distributed from Costa Rica south across northern South America to eastern Brazil (map 5.104). In northern Venezuela it was strongly associated with moist habitats. All specimens were taken at below 150 m elevation (Handley 1976).

Natural History

This insectivorous bat roosts in caves. Its natural history is very poorly known, and it is infrequently collected.

FAMILY THYROPTERIDAE

Disk-winged Bats, Murciélagos con Mamantones

Diagnosis

This distinctive family has suction disks at the bases of the thumbs and on the soles of the feet (see fig. 5.26). Such suction disks are possessed by only two other taxa of bats, the Myzopodidae confined to the Island of Madagascar and some vespertilionids including *Eudiscops* from Southeast Asia.

Distribution

This family is confined to the New World and distributed from Chiapas, Mexico, south through the Isthmus, across much of northern South America to eastern Brazil.

Genus *Thyroptera* Spix, 1823
Thyroptera discifera (Lichtenstein and Peters, 1855)

Description

The dental formula is I 2/3, C 1/1, P 3/3, M 3/3. Head and body length ranges from 37 to 47 mm, tail length from 24 to 33 mm, and forearm length from 31 to 35 mm. One male measured TL 73; T 36; HF 6; E 12; Wt 6 g. The calcar has a single cartilaginous projection extending into the posterior border of the interfemoral membrane. This species differs from *T. tricolor* in that the venter is only slightly paler than the light-brown dorsum.

Range and Habitat

This species is known from Nicaragua and from scattered localities in northern South America (map 5.105). The exact limits of its distribution are unknown.

Natural History

These bats have been found roosting communally under a dead banana leaf. Groups consist of adults of both sexes and young (Wilson 1978).

Thyroptera tricolor Spix, 1823
Murciélago Tricolor de Brasil

Measurements

	Sex	Mean	S.D.	N
TL	M	72.67	2.66	6
	F	72.50	2.38	4
HB	M	40.17	2.04	6
	F	40.25	2.06	4
T	M	32.50	2.17	6
	F	32.25	3.59	4
HF	M	6.67	0.52	6
	F	7.00	0.00	4
E	M	12.50	0.84	6
	F	12.50	0.58	4
FA	M	36.43	1.00	6
	F	37.63	0.82	4
Wta	M	4.68	0.40	6
	F	4.50	0.41	4

Location: Venezuela (USNMNH)

Description

This small bat has a total length of approximately 72.6 mm; the tail averages about 32.5 mm, and weight averages about 4.6 g. There is little difference in size between the sexes. The calcar has two cartilaginous projections extending onto the posterior border of the interfemoral membrane. The dorsum is reddish brown to almost black, but the roots of the hairs are pale. The venter contrasts strongly, being either white or yellowish (plate 5).

Range and Habitat

This species occurs from southern Mexico across northern South America to eastern Brazil (map 5.105). It is strongly associated with moist habitats. In northern Venezuela all specimens were collected below 850 m elevation (Handley 1976).

Natural History

This species is an aerial insectivore. The bats form small colonies, rarely exceeding nine individuals, that show stability over time although roosting sites are changed frequently. Their specialized roosting habits, inside rolled *Heliconia* leaves, may limit colony size. The suction disks allow them to cling to the smooth surface of leaves (Findley and Wilson 1974; Wilson and Findley 1977).

FAMILY VESPERTILIONIDAE

Diagnosis

These bats are highly specialized for flight. The trochiter of humerus is much larger than the trochin and has a surface of articulation with the scapula more than half as large as the glenoid fossa. The tail

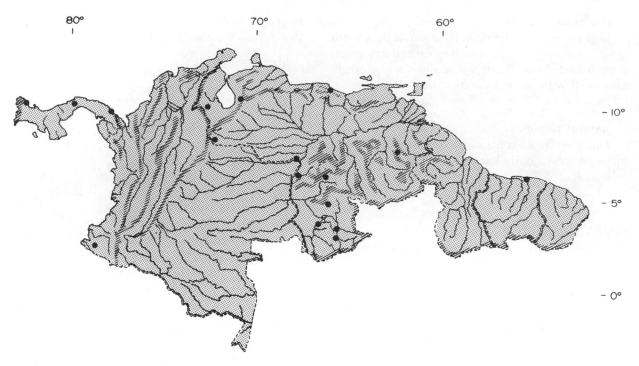

Map 5.106. Distribution of *Myotis albescens*.

Table 5.12   Key to the Genera of the Vespertilionidae
of Northern South America

1 Dorsal surface of interfemoral membrane furred at least
   halfway to the distal edge . . . . *Lasiurus*
1' Dorsal surface of interfemoral membrane naked 2
2 Only a single upper incisor on each side . . . . *Rhogeessa*
2' Two or more upper incisors per side 3
3 Only a single premolar per side 4
3' Two or more premolars per side . . . . *Myotis*
4 Ears 25% of head and body length . . . . *Histiotus*
4' Ears 12% or less of head and body length . . . . *Eptesicus*

is rather long and extends to the edge of a wide interfemoral membrane. These bats are of medium to
small size, with forearm length ranging from 90 to 24
mm. The nostrils and lips are relatively unmodified
(fig. 5.27), and there are no decorative facial folds.
The dental formula is highly variable and will be discussed under the genera. A key to the genera is
given in table 5.12.

Distribution
This family is worldwide in its distribution except
for Antarctica and New Zealand.

Natural History
Vespertilionid bats are usually specialized as aerial
insectivores, although some show specializations for
feeding on fish (e.g., *Pizonyx*). Bats of this family
roost in caves, hollow trees, or other sheltered areas.

These are the common bats of the north temperate
zone, but they also occupy the tropics. In the temperate zone they are specialized for winter hibernation. Echolocation in the capture of flying insects has
been well studied in certain species of this family
(Griffen 1958; Gould 1977).

Genus *Myotis* Kaup, 1829
Little Brown Bat

Description
The dental formula is usually I 2/3, C 1/1, P 3/3,
M 3/3 (fig. 5.28). Head and body length ranges from
35 to 80 mm, tail length from 40 to 60 mm, and fore-

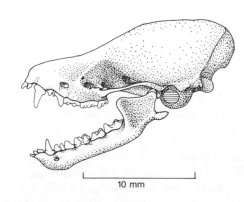

Figure 5.28. Skull of *Myotis nigricans*.

arm length from 29 to 68 mm. The dorsum is usually some shade of brown, and the underparts are somewhat paler. The Neotropical species were reviewed by LaVal (1973a).

Distribution

The genus is worldwide in its distribution except for the Arctic and Antarctic regions.

Myotis albescens (E. Geoffroy, 1806)

Measurements

	Sex	Mean	S.D.	N
TL	M	78.05	2.67	21
	F	80.38	4.28	39
HB	M	42.81	2.86	21
	F	44.89	3.41	38
T	M	35.24	3.49	21
	F	35.50	3.14	38
HF	M	9.00	0.77	21
	F	9.03	0.63	39
E	M	14.24	1.18	21
	F	14.13	1.20	39
FA	M	34.04	1.03	21
	F	34.61	1.07	39
Wta	M	5.44	0.67	21
	F	5.63	0.48	26

Location: TFA, Venezuela (USNMNH)

Description

This species is similar in size to *M. nigricans* but lighter in color. The dorsum is brown, but the dorsal hairs are black at the base and white at the tip, giving a frosted appearane. The venter is paler.

Range and Habitat

This species occurs from southern Veracruz, Mexico, south through the Isthmus and across most of South America to eastern Argentina (map 5.106). In contrast to *M. nigricans*, this species prefers lower elevations. In Venezuela specimens were all taken at below 155 m. This species tolerates both dry deciduous forest and humid multistratal tropical evergreen forest (Handley 1976).

Myotis keaysi J. A. Allen, 1914

Measurements

	Sex	Mean	S.D.	N
TL	M	85.00	5.74	7
	F	81.50	4.54	14
HB	M	48.71	5.02	7
	F	44.64	3.93	14
T	M	36.29	2.50	7
	F	36.86	1.66	14
HF	M	8.14	0.69	7
	F	8.36	0.50	14
E	M	12.71	0.95	7
	F	12.57	0.65	14
FA	M	35.99	0.95	7
	F	36.17	1.27	14
Wta	M	5.30	0.57	5
	F	5.25	0.37	4

Location: Northern Venezuela (USNMNH)

Description

This is one of the larger species of *Myotis*. Males tend to be slightly larger than females; head and body length averages 48.4 mm for males and 44.6 mm for females. The dorsum is a very dark brown, and the venter is paler.

Range and Habitat

The species is distributed from southern Veracruz, Mexico, through Panama to northern Venezuela and western Colombia, following the foothills of the Andes through Peru (map 5.107). This high-altitude bat usually occurs above 1,000 m elevation in Venezuela and can be found up to 2,000 m. It prefers moist habitats and montane tropical humid forest (Handley 1976).

Myotis nesopolus (= *larensis*) Miller, 1900

Measurements

	Sex	Mean	S.D.	N
TL	M	79.9	3.37	12
	F	80.3	2.25	6
T	M	37.6	2.02	12
	F	37.3	1.37	6
HF	M	7.7	0.49	12
	F	7.5	0.53	6
E	M	12.6	0.51	12
	F	13.2	0.41	6
FA	M	31.8	0.75	12
	F	32.5	0.65	6
Wta	M	3.50	0.56	12

Location: Venezuela (USNMNH)

Description

This is one of the smallest species in the genus occurring in northern South America. Head and body length averages 42.3 mm for males and 43 mm for females; the forearm averages 31.8 mm for males and 32 mm for females.

Range and Habitat

Known from the drier portions of northwestern Venezuela and adjacent Colombia (map 5.108), this bat is adapted to arid habitats. In Venezuela all specimens were caught below 55 m elevation in dry deciduous thorn forest (Handley 1976).

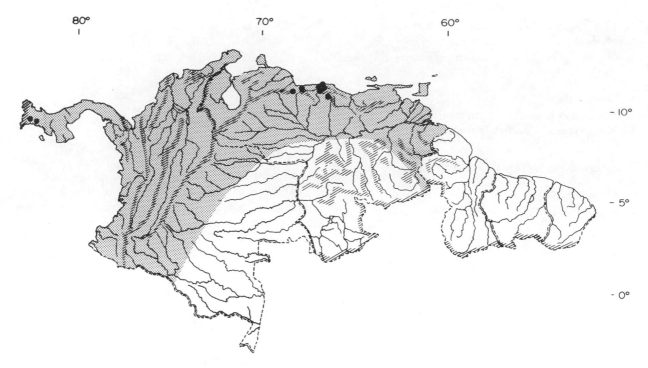

Map 5.107. Distribution of *Myotis keaysi*.

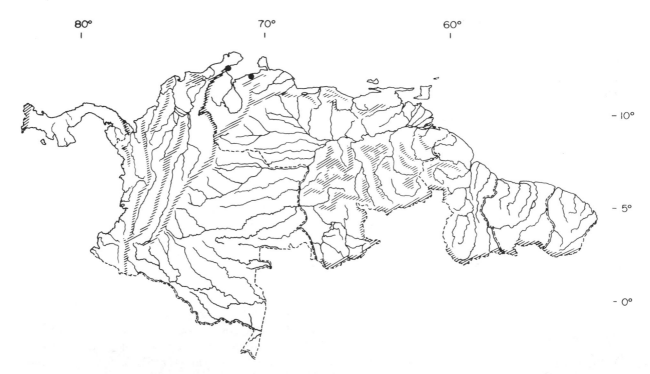

Map 5.108. Distribution of *Myotis nesopolus*.

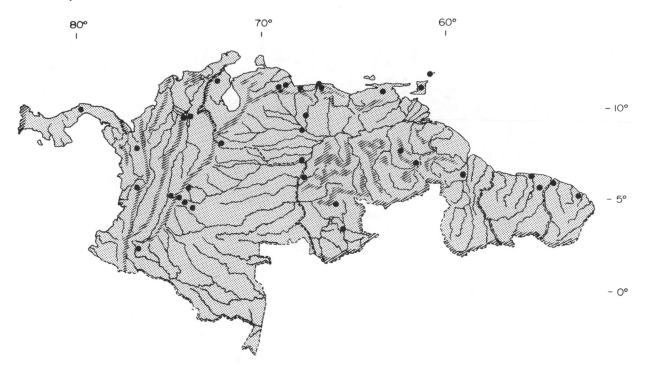

Map 5.109. Distribution of *Myotis nigricans*.

Natural History

Little is known concerning the biology of this bat. Clearly it is highly adapted for arid habitats, and it is probably a relict in its present range with the waning of arid habitats in northern South America since the close of the last glaciation.

Myotis nigricans (Schinz, 1821)

Measurements

	Sex	Mean	S.D.	N
TL	M	79.61	3.33	33
	F	77.60	2.53	42
HB	M	43.58	3.17	33
	F	41.05	1.97	42
T	M	36.03	2.97	33
	F	36.55	1.38	42
HF	M	8.03	0.59	33
	F	8.00	0.22	42
E	M	13.27	0.91	33
	F	12.38	0.62	42
FA	M	34.93	0.99	39
	F	33.41	0.78	42
Wta	M	4.91	0.40	27
	F	4.25		2

Location: Northern Venezuela (USNMNH)

Description

In northern Venezuela males are slightly larger than females. Head and body length averages 43.4 mm for males and 41.2 mm for females; forearm length averages 34.56 mm for males and 33.4 mm for females; weight ranges from 3 to 5.5 g (see table 5.14). The dorsal color varies geographically from light to dark brown, and the venter is nearly the same shade.

Range and Habitat

M. nigricans occurs from southern Mexico across most of South America to northern Argentina (map 5.109). Although it ranges to 2,240 m elevation in Venezuela, the vast majority of specimens were taken at below 1,200 m. It tolerates a wide range of vegetation types, occurring in dry deciduous forest as well as multistratal tropical evergreen forest, and is found near human habitations.

Natural History

This is one of the best studied of the tropical species of *Myotis*. These bats tend to roost in sheltered areas; females rearing young form separate groups. In Panama this species breeds in late December and early January. Gestation is approximately sixty days, and a birth peak occurs in February. The birth is followed by a postpartum estrus and a repeat of the cycle, with a second birth peak in April–May and another in August. Reproductive activity then declines, to be initiated again in December. Reproductive periods coincide with the time of maximum insect abundance. When feeding, females leave their young in large groups termed crèches. The young achieve adult size at approximately five to six weeks of age (Wilson 1971; Wilson and LaVal 1974).

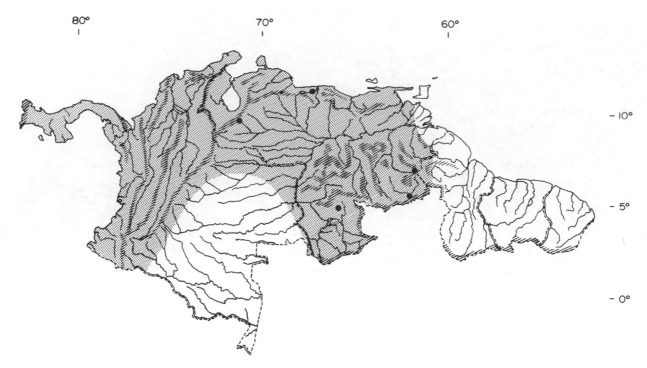

Map 5.110. Distribution of *Myotis oxyotus*.

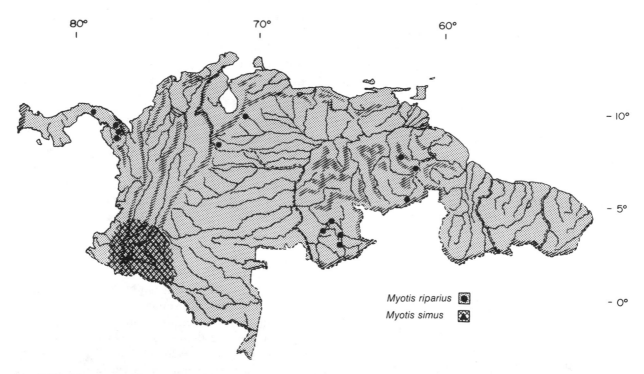

Myotis riparius
Myotis simus

Map 5.111. Distribution of two species of *Myotis*.

Myotis oxyotus (Peters, 1867)

Measurements

	Sex	Mean	S.D.	N
TL	M	84.00	2.94	4
	F	85.60	5.18	5
HB	M	45.50	3.42	4
	F	45.60	4.93	5
T	M	38.50	1.73	4
	F	40.00	1.58	5
HF	M	9.00	0.00	4
	F	8.40	0.89	5
E	M	14.00	0.82	4
	F	13.80	0.45	5
FA	M	36.93	1.05	4
	F	36.40	1.82	5
Wta	M	5.37	0.31	3
	F	5.60	0.88	4

Location: Venezuela (USNMNH)

Description

In this large species, head and body length averages 45.5 mm for males and 45.6 mm for females. The forearm averages about 36 mm, and weight generally exceeds 5 g. The dorsum is brown but has a slight frosted appearance in some specimens. The ventral hairs have brown bases with gray tips, creating a mottled effect.

Range and Habitat

The species occurs from Costa Rica south through Panama to Venezuela and in the west to Peru (map 5.110). These large bats are adapted to high elevations; in northern Venezuela specimens were taken at between 800 and 2,110 m elevation. They prefer multistratal montane tropical forest and wet habitats (Handley 1976).

Myotis riparius Handley, 1960

Measurements

	Sex	Mean	S.D.	N
TL	M	79.00	2.65	3
	F	81.38	4.79	13
HB	M	42.67	1.53	3
	F	42.85	2.67	13
T	M	36.33	4.04	3
	F	38.54	3.43	13
HF	M	8.00	0.00	3
	F	8.15	1.34	13
E	M	12.67	0.58	3
	F	12.92	0.76	13
FA	M	34.78	1.54	4
	F	35.33	1.10	14
Wta	M	4.33	0.31	3
	F	4.43	0.67	9

Location: Venezuela (USNMNH)

Description

Males and females are nearly the same size, averaging about 42.7 mm in head and body length, with forearms approximately 35 mm. The venter and dorsum are dark brown. The species may be confused with *M. simus* (see LaVal 1973a for a comparison).

Range and Habitat

This species is distributed from Honduras to South America, ranging widely over the continent; the southernmost distribution is in Uruguay (map 5.111). These bats generally range at lower elevations; although in northern Venezuela some specimens were taken at up to 1,000 m, 90% of all specimens collected were taken at below 200 m. Broadly tolerant of human activities, they may be found in croplands, dry deciduous forests, and multistratal tropical evergreen forest (Handley 1976).

Myotis simus Thomas, 1901

Description

This medium-sized species is similar in size and proportions to *M. riparius*, with forearm length about 35 mm. The dorsum is orange to cinnamon brown, and the fur is short and woolly.

Range and Habitat

The species is confined to the lowland rain forests of the Amazon basin; it reaches extreme southeastern Colombia (map 5.111).

Genus *Eptesicus* Rafinesque, 1820
Big Brown Bat, Murciélago con Orejas de Ratón

Description

The dental formula is I 2/2, C 1/1, P 1/1, M 3/3. The genus is distinguishable from *Myotis* by the reduction in premolar number (see fig. 5.29). Head and body length ranges from 35 to 75 mm and forearm length from 28 to 55 mm. In spite of the common name, there is a considerable size range among

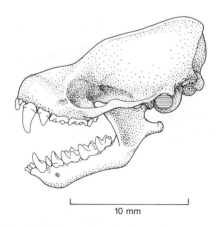

10 mm

Figure 5.29. Skull of *Eptesicus brasiliensis*.

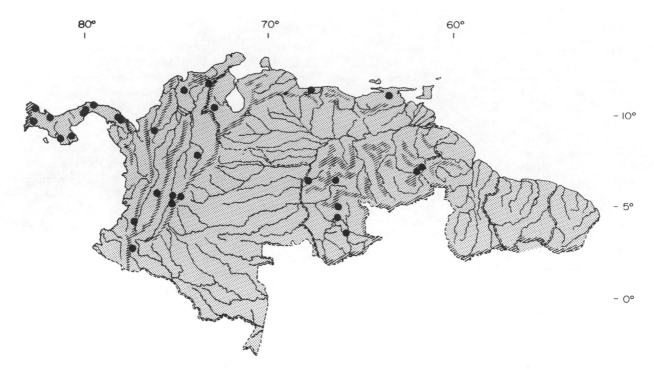

Map 5.112. Distribution of *Eptesicus brasiliensis*.

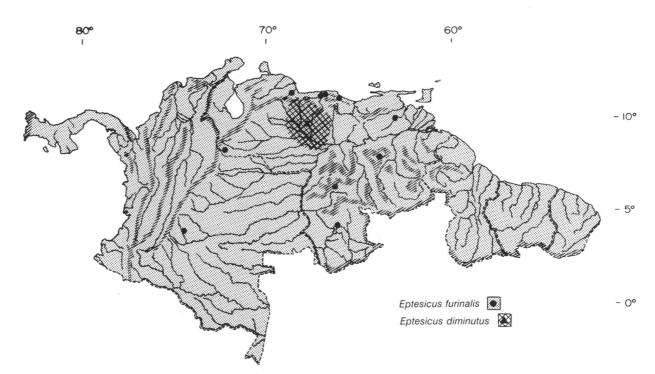

Eptesicus furinalis ⊡
Eptesicus diminutus ⊠

Map 5.113. Distribution of two species of *Eptesicus*.

the species of the genus. The dorsum is usually dark brown to almost black, and the venter is paler.

Distribution

This genus contains approximately thirty species and is worldwide in its distribution except for the Arctic and Antarctic regions.

Eptesicus brasiliensis (Desmarest, 1819)

Measurements

	Sex	Mean	S.D.	N
TL	M	94.78	3.27	9
	F	96.90	4.09	30
HB	M	54.00	4.00	9
	F	53.80	2.71	30
T	M	40.78	3.31	9
	F	43.10	2.84	30
HF	M	9.89	0.78	9
	F	9.83	0.75	30
E	M	14.78	1.20	9
	F	14.97	0.76	30
FA	M	40.04	0.84	9
	F	40.92	1.57	28
Wta	M	9.03	1.38	9
	F	9.89	1.06	22

Location: TFA, Venezuela (USNMNH)

Description

In this account *Eptesicus andinus* is considered a junior synonym of *E. brasiliensis*. This species is slightly larger than *E. furinalis* but considerably smaller than *E. fuscus*. Females are only slightly larger than males. The dorsum and venter are dark brown.

Range and Habitat

The species ranges through montane regions of southern Mexico, through the Isthmus, and across much of South America to Uruguay (map 5.112). In northern Venezuela these bats occur below 1,000 m elevation. They prefer moist habitats and multistratal tropical evergreen forest, though they will forage in man-made clearings (Handley 1976).

Eptesicus diminutus Osgood, 1915

Description

This is the smallest species of the genus in the northern Neotropics. Two specimens measured TL 91–91.9; T 37; HF 9; E 15; FA 35.5 (Venezuela, USNM). The dorsum is brown, as is the venter, but in some specimens the venter is washed with gray.

Range and Habitat

This species primarily occurs in Argentina and Brazil but was recently collected in Venezuela (map 5.113). Either this is an example of a very disjunct range or else we are dealing with a new species. In northern Venezuela this species was collected at 100 m elevation in a mixed habitat of grassland and deciduous tropical forest (Handley 1976).

Eptesicus furinalis (d'Orbigny, 1847)

Measurements

	Sex	Mean	S.D.	N
TL	M	92.50	2.62	8
	F	92.33	3.20	6
HB	M	52.00	2.62	8
	F	54.00	3.58	6
T	M	40.50	1.51	8
	F	38.33	2.34	6
HF	M	9.00	0.00	8
	F	9.33	0.82	6
E	M	13.25	1.28	8
	F	18.83	1.33	6
FA	M	38.35	1.24	8
	F	39.03	1.43	6
Wta	M	7.53	0.55	6
	F	8.08	0.50	4

Location: Venezuela (USNMNH)

Description

In this account *E. montosus* is considered a junior synonym of *E. furinalis*. It is smaller than *E. brasiliensis;* head and body length averages 52 mm and the forearm 38 mm. Females are only slightly larger than males in linear measurements but are about 1.5 g heavier; average weights range from 7.5 to 9 g. The dorsum and venter are dark brown.

Range and Habitat

This species occurs from Nayarit in western Mexico and Tamaulipas in eastern Mexico south through the Isthmus over most of South America to northern Argentina (map 5.113). It is generally found at higher altitudes up to 1,580 m and prefers moist habitats and montane tropical forest or multistratal evergreen forest.

Eptesicus fuscus (Beaubois, 1796)

Description

This is the largest species of the genus and is easily distinguishable from the other species by size alone. Measurements for two males averaged TL 117.5; T 48; HF 12.5; E 17; FA 56.6 (Venezuela, USNM). Head and body length averages 70 mm for males and 84 mm for females; weight may reach 26 g in females. The dorsum is dark brown with a slightly paler venter.

Range and Habitat

This bat occurs from Saskatchewan and Quebec in North America south through the Isthmus to the

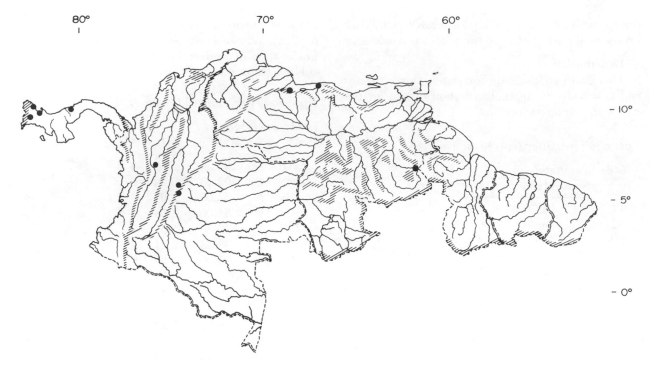

Map 5.114. Distribution of *Eptesicus fuscus*.

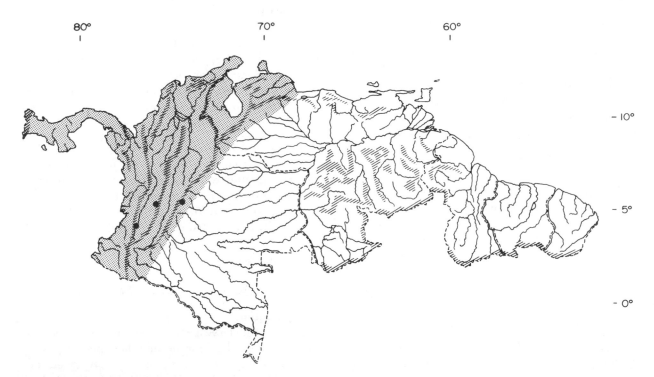

Map 5.115. Distribution of *Histiotus montanus*.

194

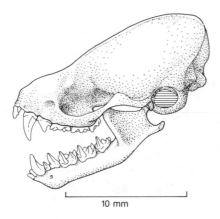

Figure 5.30. Skull of *Histiotus montanus*.

montane portions of Colombia and Venezuela (map 5.114). A high-elevation species in the tropical portion of its range, it was taken in northern Venezuela at from 1,260 to 1,524 m. It prefers moist habitats, cloud forest, and premontane tropical evergreen forest (Handley 1976).

Natural History
Although this is one of the best-studied bats in North America, little is known concerning the details of its biology in South America. This bat roosts during the day in small groups, usually in a hollow tree or a cave. In the northern part of its range it hibernates during the winter, sometimes with other species of bats but never in large groups of its own species. In the northern part of its range two or three young may be born, and they attain adult size within sixty days. These bats are strongly insectivorous, taking a wide variety of flying insects. In the north, beetles constitute up to 50% of their diet and they seldom take moths (Barbour and Davis 1969).

Genus *Histiotus* Gervais, 1856
Murciélago Orejudo

Description
Head and body length ranges from 54 to 70 mm and the forearm from 42 to 52 mm. This genus resembles *Eptesicus* in dental formula and shape of the skull (fig. 5.30), but it has extremely large ears compared with *Eptesicus*. For example, *E. fuscus* with a total length of 117 mm will have an ear approximately 17 mm long, whereas a comparable-sized *Histiotus* will have an ear 31 mm long. The upperparts of the body are generally light brown or grayish brown; the underparts are slightly paler.

Distribution
This genus is confined to South America, generally in the southern parts, but one species extends into the range covered by this volume.

Natural History
This aerial insectivore is found in small (three to seventeen) colonies inside natural cavities or buildings.

Histiotus montanus (Philippi and Landbeck, 1861)

Description
Mean measurements for three males were TL 108.7; T 49; HF 10.3; E 31; FA 45.6; for one female, TL 109; T 50; HF 12; FA 45.7 (Venezuela, USNM). The dorsum is medium brown and the venter somewhat paler.

Range and Habitat
The species ranges at high elevations in the Andean portions of Venezuela and Colombia. In Venezuela it occurred from 1,498 to 2,101 m (map 5.115). It is found in the montane portions of Venezuela, Colombia, and Peru and is widely distributed in temperate Argentina. It prefers moist habitats but is broadly tolerant of man-made clearings.

Genus *Rhogeessa* H. Allen, 1866
Little Yellow Bat

Description
The dental formula is I 1/3, C 1/1, P 1/2, M 3/3. This genus is distinguishable from *Myotis* and *Eptesicus* by its possession of a single upper incisor (see fig. 5.31). Head and body length ranges from 37 to 50 mm, and forearm length ranges from 25 to 34 mm. The color is yellowish brown or light brown dorsally and paler below. The genus was reviewed by LaVal (1973b).

Distribution
This New World genus has species ranging from Mexico to Brazil.

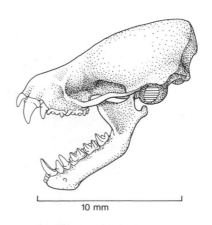

Figure 5.31. Skull of *Rhogeessa tumida*.

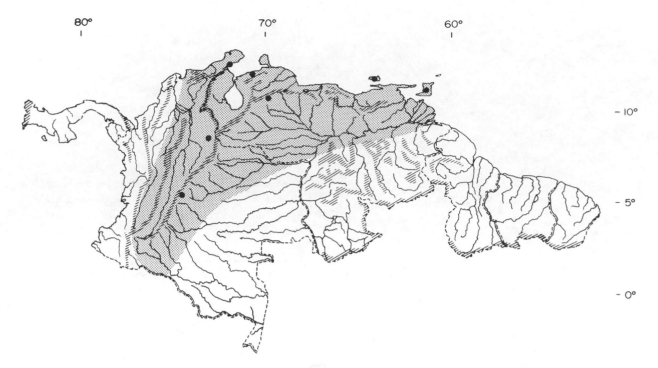

Map 5.116. Distribution of *Rhogeessa minutilla*.

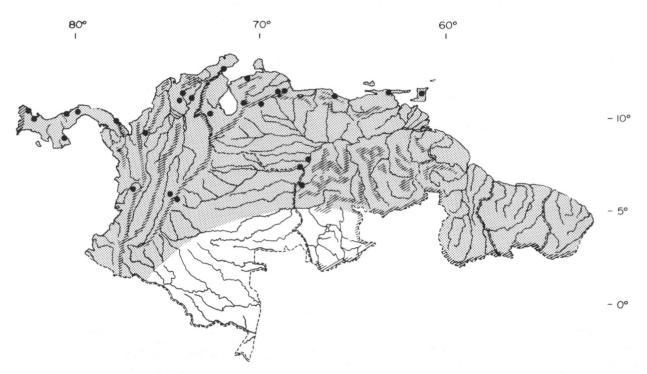

Map 5.117. Distribution of *Rhogeessa tumida*.

Rhogeessa minutilla Miller, 1897

Measurements

	Sex	Mean	S.D.	N
TL	M	72.18	3.94	17
	F	72.50	3.34	26
HB	M	40.59	2.72	17
	F	41.35	2.59	26
T	M	31.59	1.87	17
	F	31.15	2.19	26
HF	M	6.59	0.51	17
	F	6.77	0.51	26
E	M	14.53	0.80	17
	F	14.35	0.63	26
FA	M	27.29	0.84	22
	F	27.33	0.75	39
Wta	M	3.33	0.40	17
	F	3.40	0.38	25

Location: Northwestern Venezuela (USNMNH)

Description

The dorsum is reddish brown, the venter paler. The wings are very dark brown to almost black.

Range and Habitat

This species is confined to the more arid portions of northern Venezuela and adjacent parts of Colombia (map 5.116). It is strongly associated with dry habitats and dry deciduous tropical forest. These bats roost in hollow trees and houses (Handley 1976).

Rhogeessa tumida H. Allen, 1866

Measurements

	Sex	Mean	S.D.	N
TL	M	70.50	3.21	8
	F	71.29	3.05	14
HB	M	41.13	4.09	8
	F	42.36	2.53	14
T	M	29.38	2.62	8
	F	28.93	1.94	14
HF	M	6.29	0.49	7
	F	6.79	0.70	14
E	M	12.38	1.06	8
	F	12.79	1.19	14
FA	M	27.70	0.99	8
	F	28.42	0.88	15
Wta	M	3.47	0.50	6
	F	4.00	0.96	7

Location: Venezuela (USNMNH)

Description

This is an extremely small bat with an unadorned face (fig. 5.27). It is light yellow brown on the dorsum with a slighter paler venter.

Range and Habitat

This species occurs from southern Tamaulipas, Mexico, south through the Isthmus and across much of northern South America, extending to Central Brazil (map 5.117). In contrast to the preceding species, this form prefers moist habitats. It is broadly tolerant of man-made clearings and is the common species taken at below 570 m in appropriate habitat.

Natural History

These bats roost in hollow trees, and colonies may be large. They are aerial insectivores, and individuals appear to have established hunting routes. Females may bear up to two young (LaVal 1973b).

Genus Lasiurus Gray, 1831

Description

Species of this genus have a considerable range in size. Head and body length ranges from 50 to 90 mm, tail length from 40 to 75 mm, and the forearm from 37 to 57 mm. Weight ranges from 3 to 6 g. The distinguishing feature for this group is that the dorsal surface of the interfemoral membrane is well haired for at least half its length and usually for its entire length (see fig. 5.27).

Distribution

These bats are confined to the New World and have colonized the Galápagos and the Hawaiian Islands. They are distributed from Canada to Argentina.

Natural History

Bats of the genus Lasiurus have been studied in North America, but little is known concerning their habits in South America. In the north they exhibit seasonal migrations and are aerial insectivores. L. borealis has two or three young, thus departing from the general rule of one young characteristic for most of the Chiroptera.

Lasiurus borealis (Müller, 1776)
Red Bat

Measurements

	Sex	Mean	S.D.	N
TL	M	100.8	3.03	5
	F	102.5		2
T	M	9.3	0.48	5
	F	9.5		2
E	M	10.7	0.97	5
	F	11.5		2
FA	M	38.64	1.75	5
	F	40.5		2
Wta	M	6.43	0.97	4

Location: Venezuela (USNMNH)

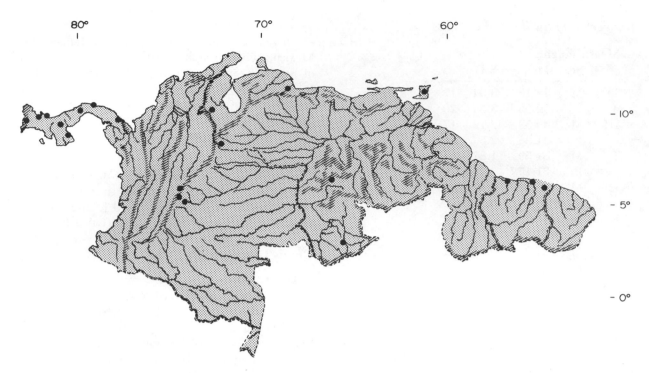

Map 5.118. Distribution of *Lasiurus borealis*.

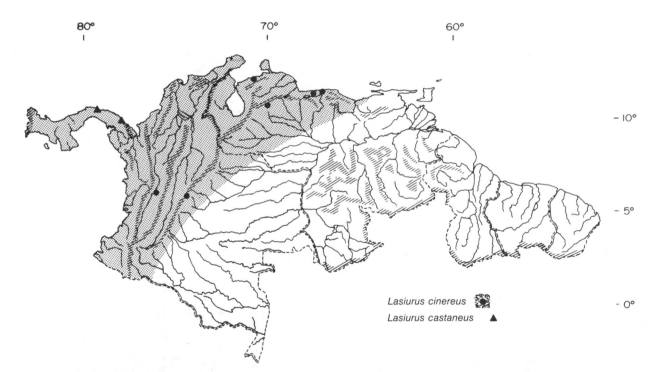

Lasiurus cinereus

Lasiurus castaneus ▲

Map 5.119. Distribution of two species of *Lasiurus*.

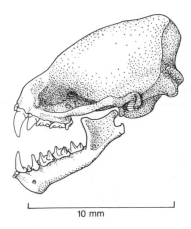

Figure 5.32. Skull of *Lasiurus borealis*.

Description

The dental formula is I 1/3, C 1/1, P 2/2, M 3/3 (see fig. 5.32). Head and body length averages 51.8 mm for males and 51.1 mm for females, and forearm length averages 38.6 mm for males and 40.5 for females. Weight exceeds 6 g. The dorsum is bright red or rust, and the venter is only slightly paler (see plate 5). Males tend to be brighter in color than females. There is usually a buffy patch on the front of each shoulder. The interfemoral membrane is furred on its entire upper surface.

Range and Habitat

This species ranges from southern Canada through the Isthmus to the southern tip of the South American continent in suitable habitats (Map 5.118). It is strongly associated with moist habitats and multistratal tropical evergreen forest. In Venezuela specimens were taken at below 155 m; at higher elevations it is replaced by *L. cinereus* (Handley 1976).

Natural History

Much is known concerning the biology of this bat in the northern parts of its range. In North America it tends to migrate and hibernates in the winter. When not hibernating it often roosts on the bark of trees. Its behavior in tropical climates is poorly understood. In North America approximately 26% of the diet of this aerial insectivore can consist of moths. Females may have up to five young, although three is the usual number (Barbour and Davis 1969). These bats are solitary and do not congregate in nurseries.

Lasiurus castaneus Handley, 1960

Description

The dental formula is I 1/3, C 1/1, P 2/2, M 3/3. Total length averages 112 mm, with tail 48 mm, hind foot 11 mm, and forearm 44.8 mm. The species is similar to *L. borealis*, with the same dental formula, but is distinguishable by its coloration. The dorsum is chestnut brown, and the venter dark brown.

Range and Habitat

The species has been taken in southern Panama and presumably extends into northern Colombia (map 5.119).

Lasiurus cinereus (Beauvois, 1796)
Hoary Bat

Description

The dental formula is I 1/3, C 1/1, P 2/2, M 3/3. This bat is distinguishable from all other species of *Lasiurus* by its size. Total length ranges from 134 to 140 mm, head and body length averages about 85 mm, and the forearm ranges from 46 to 55 mm. The dorsal pelage varies from yellowish brown to mahogany brown, but the tips of the hairs are white, giving a frosted or silver appearance. The underparts are whitish to yellow.

Range and Habitat

This species occurs from southern Canada south through the Isthmus in suitable habitat to southern Argentina (map 5.119). It is strongly associated with higher elevations in South America; in Venezuela it was taken at up to 1,456 m elevation. It prefers moist areas, though it has been taken in dry deciduous forest as well as multistratal tropical evergreen forest (Handley 1976).

Natural History

In the Northern Hemisphere this bat is migratory and hibernates. Little is known concerning its natural history in tropical regions. This species tends to be solitary. The female has an average litter size of two in the northern parts of its range. It is an aerial insectivore taking a wide variety of flying insects; moths seem to predominate in the diet of the temperate-zone races (Shump and Shump 1982).

Lasiurus egregius (Peters, 1871)
Big Red Bat

Description

This species is similar to *L. borealis* in coloration and dental characters, but it is distinguishable from all other red species of *Lasiurus* by its large size: total length is 127 mm and forearm 48 mm (the type).

Range and Habitat

The species occurs in extreme southeastern Panama, but the extent of its range into northwestern Colombia is unknown (map 5.120).

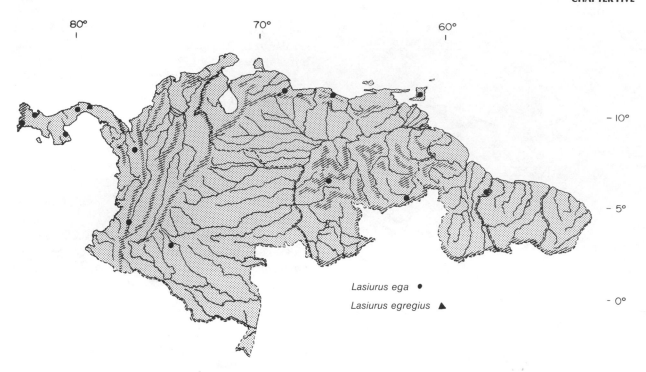

Map 5.120. Distribution of two species of *Lasiurus*.

Lasiurus ega (Gervais, 1856)
Southern Yellow Bat

Measurements

	Sex	Mean	S.D.	N
TL	M	114.44	3.75	9
	F	120.50	8.02	8
HB	M	61.11	2.62	9
	F	65.75	4.06	8
T	M	53.33	3.84	9
	F	54.75	5.20	8
HF	M	10.44	0.73	9
	F	10.88	0.64	8
	M	17.56	1.24	9
	F	17.75	1.75	8
FA	M	45.43	0.97	10
	F	46.15	2.51	8
Wta	M	11.87	1.42	6
	F	13.51	1.60	8

Location: Venezuela (USNMNH)

Description
The dental formula is I 1/3, C 1/1, P 1/2, M 3/3. Head and body length averages 61.1 mm for males and 65.8 mm for females. The dorsum is yellowish brown, which easily separates *L. ega* from *L. borealis* and *L. egregius*.

Range and Habitat
This species is distributed from southern California and Arizona south through Mexico and the Isthmus to northern Argentina (map 5.120). In Venezuela specimens were generally taken at below 500 m; the species was strongly associated with moist habitats and multistratal tropical evergreen forest (Handley 1976).

Natural History
Like other lasiurine bats, this species tends to roost individually on the bark of trees. Twins are born annually (Goodwin and Greenhall 1961).

FAMILY MOLOSSIDAE
Free-tailed Bats, Mastiff Bats, Murciélagos de Cola de Ratón, Murciélogos Moloso

Diagnosis
The family contains ten genera and about eighty species. Head and body length ranges from 40 to 130 mm, tail from 14 to 80 mm, and forearm from 27 to 85 mm. The fibula is well developed, supporting the lower leg. In these bats the tail extends beyond the edge of the interfemoral membrane, and this single character is diagnostic for the entire family, hence the common name "free-tailed bat" (see fig. 5.33).

Distribution
This family is worldwide in its distribution but is mainly confined to tropical regions.

Natural History
These aerial insectivores are swift, high fliers, and their wings are long and narrow. Freeman (1981) has made a thorough morphometric analysis of the

family. The jaw size covaries positively with body size, suggesting that there is great specialization for certain size classes of prey. Thus when several species co-occur they tend to present an array of sizes that probably reflects this specialization. Beetles seem to predominate in the diet. Most species nest in caves, tunnels, or hollow trees, and some are strongly colonial. In the northern parts of their range these bats may be seasonally migratory over moderate distances (see *Tadarida brasiliensis*).

Identification of the Genera

Within the area covered by this volume there are five genera of free-tailed bats: *Molossus, Molossops, Tadarida, Promops,* and *Eumops: Tadarida* includes *Nyctinimops* as a subgenus. The species span an array of sizes, with the forearm ranging from a minimum of 25 mm to a maximum of 80 mm. All the very large free-tailed bats with forearms longer than 60 mm belong to the genus *Eumops,* but this genus also includes some very small species such as *E. nanus.* Species of the genus *Molossops* are all quite small, and *Promops* and *Tadarida* are intermediate in size. Thus size alone does not discriminate among the genera, and one must pay close attention to dental characteristics and the shape of the ear. In all spe-

Table 5.13	Key to the Genera of the Molossidae of Northern South America
1	Single upper premolar per side 2
1′	Two upper premolars per side 3
2	Single lower incisor per side . . . . *Molossus*
2′	Two lower incisors per side; ears separate in the midline . . . . *Molossops*
3	Upper lip grooved and separable into two halves . . . . *Tadarida*
3′	Upper lip not divided 4
4	Ears almost reach the snout tip . . . . *Eumops*
4′	Ears fall far short of snout tip . . . . *Promops*

Note: Tadarida (= Nyctinomops).

cies of free-tailed bats the ears are very stout and somewhat flattened, projecting forward and laterally. Whether the ears are joined in the center of the forehead is an important diagnostic character. The peculiar shape of the ear has suggested that the ear itself may act as an airfoil and provide some lift to the head as the animal flies rapidly forward.

Although not readily visible in a live specimen, the number of premolars easily separates the genera into two groups. If the premolar formula is 1/2, you have in hand either *Molossops* or *Molossus.* If the premolar formula is 2/2, it could be *Tadarida, Eumops,* or *Promops.* An inspection of the lower incisors usually separates out *Molossus,* for which the incisor formula is 1/1; for all other genera the formula is 1/2 (but see *Molossops neglectus* and *M. temminckii*). Having eliminated *Molossus,* one should inspect the ears. If the ears are separate at the midline and do not connect with a fold of skin, the specimen belongs to the genus *Molossops.* If the ears are not separate but are joined or almost joined, one should inspect the upper lip. If the upper lip is grooved and separated into two parts as in a rodent, and if folds of skin from the ears almost meet at the midline, the specimen belongs to the genus *Tadarida* (see fig. 5.33). If there is no groove and the ears are completely fused across the forehead unambiguously, one should then look at the ears from the side. If they are very large and sculpted in a unique fashion, reaching almost to the end of the snout, the species belongs to the genus *Eumops* (see fig. 5.33). If the ears are not so sculpted and clearly do not reach to the end of the snout, the bat belongs to the genus *Promops* (see table 5.13).

Genus *Molossops* Peters, 1865
Malaga's Free-tailed Bat

Description

The dental formula is I 1/2, C 1/1, P 1/2, M 3/3 (but see *M. neglectus*). Head and body length ranges from 40 to 95 mm, and the tail shows a correspond-

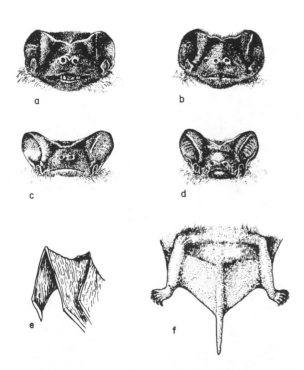

Figure 5.33. External features of the Molossidae: (*a*) Face of *Molossus ater.* (*b*) Face of *Promops centralis.* (*c*) Face of *Molossops greenhalli.* (*d*) Face of *Tadarida brasiliensis.* (*e*) Dorsal view of the third finger of *Molossus ater* showing reflexed second phalanx. (*f*) Dorsal view of *Molossus* showing tail length relative to the uropatagium. (Redrawn with permission from Goodwin and Greenhall 1961.)

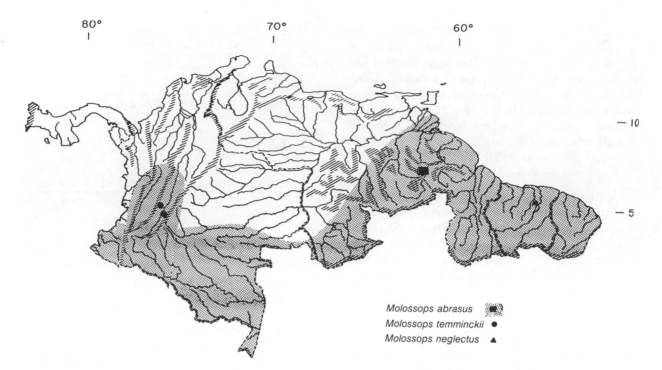

Map 5.121. Distribution of three species of *Molossops*.

Molossops abrasus ▨
Molossops temminckii ●
Molossops neglectus ▲

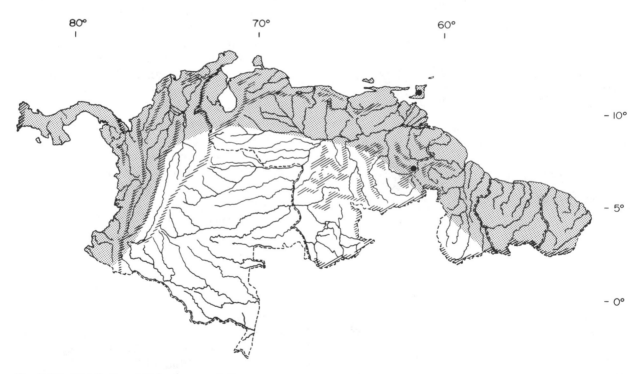

Map 5.122. Distribution of *Molossops greenhalli*.

ing range from 14 to 30 mm. The forearm averages 28–51 mm. The dorsum tends to be yellow brown to chocolate brown, and the underparts usually contrast, being gray or slate colored.

Distribution

The seven species within this genus are distributed from southern Mexico through Central America to northern Argentina.

Natural History

Species of *Molossops* are specialized for feeding on insects. Some are highly gregarious and may live in colonies of up to seventy-five. In forested habitats these bats tend to roost in hollow trees.

Molossops abrasus (Temminck, 1827)

Description

Within the range covered by this book *M. abrasus* is the largest species of the genus, with forearm averaging about 41.4 mm. Measurements of two females were TL 111.5; T 34; HF 13; E 18.5; FA 41.4; Wt 54.8 g (Venezuela USNM); one male measured TL 136; T 37; HF 14; E 19; FA 40 (CMNH). The dorsum is dark brown, and the venter is lighter. For this account *M. brachymeles* is included in *M. abrasus*.

Range and Habitat

This species is distributed in the Amazonian portions of Venezuela and Colombia, west through the Guyanas, and broadly over much of Amazonian Brazil and adjacent portions of Peru, Bolivia, and Paraguay. In northern Venezuela specimens were taken at below 150 m in association with mature tropical evergreen forest (map 5.121).

Molossops greenhalli (Goodwin, 1958)

Description

This species is only slightly smaller than *M. abrasus*. Measurements for two specimens were TL 94–105; T 34; HF 8–10; E 15–16; FA 35; Wt 18.9 g (Venezuela, USNMNH; Linares and Kiblinsky 1969). The dorsum is dark brown, the venter paler. Chromosome number: $2n = 34$.

Range and Habitat

This species appears to occupy the northern portions of the range covered by this book and is replaced by *M. abrasus* farther south (map 5.122). Its range extends from southern Nayarit in Mexico south along the Pacific coast to Honduras and thence through the Isthmus to the Caribbean coast of northern South America. It has been taken in the extreme northeastern portions of Brazil. In Venezuela it occurs at low elevations in association with multistratal

tropical evergreen forest (Handley 1976). It has been found roosting in colonies of fifty to seventy-five (Goodwin and Greenhall 1961).

Molossops planirostris (Peters, 1865)

Measurements

	Sex	Mean	S.D.	N
TL	M	89.19	2.88	16
	F	82.74	2.28	46
HB	M	61.38	2.36	16
	F	57.02	2.19	46
T	M	27.81	2.20	16
	F	25.72	1.67	46
HF	M	10.00	0.52	16
	F	9.64	0.53	47
E	M	16.38	0.62	16
	F	16.17	0.79	47
FA	M	33.64	0.42	16
	F	32.06	0.61	46
Wta	M	14.27	0.69	16
	F	11.38	0.86	18

Location: TFA, Venezuela (USNMNH)

Description

This bat is clearly separable by size alone from the two larger species of *Molossops*. The dorsum is pale brown, the venter yellow white. In this account *M. paranus* is included in *M. planirostris* (see Honacki, Kinman, and Koeppl 1982).

Range and Habitat

This species is distributed from eastern Panama through the Isthmus and across much of South America to the east of the Andes (map 5.123). It extends south to Paraguay and southeastern Brazil. In Venezuela it was taken at elevations below 150 m. It is strongly associated with moist habitats, apparently feeding on insects that are attracted to ponds or lakes. This species is associated with multistratal tropical evergreen forests and roosts in the hollows of rotting trees (Handley 1976).

Molossops (= *Neoplatymops*) *mattogrossensis* Vieira, 1942

Measurements

	Sex	Mean	S.D.	N
TL	M	78.90	4.48	10
	F	74.67	1.03	6
HB	M	53.50	3.63	10
	F	51.00	1.10	6
T	M	25.40	3.69	10
	F	23.67	1.03	6
HF	M	7.82	0.60	11
	F	7.83	0.41	6

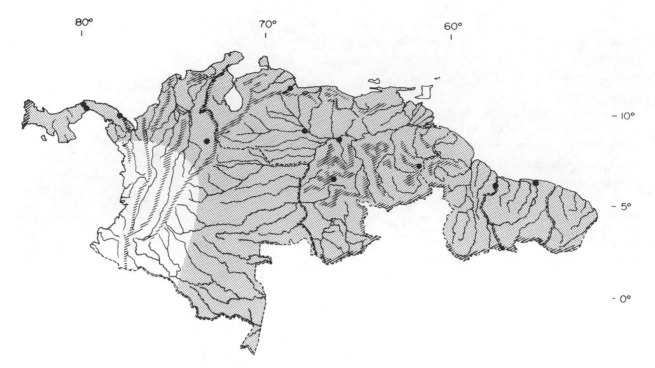

Map 5.123. Distribution of *Molossops planirostris*.

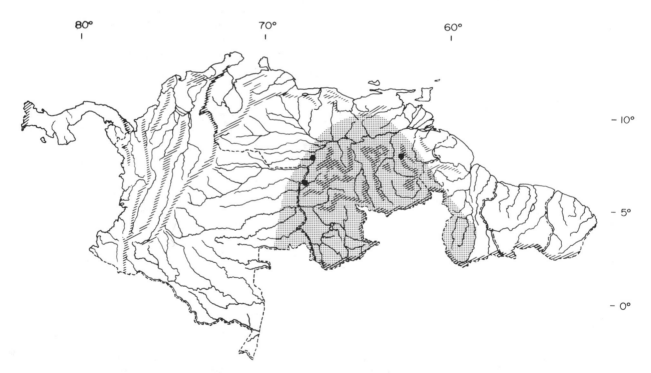

Map 5.124. Distribution of *Molossops mattogrossensis*.

E	M	13.55	0.69	11
	F	14.33	0.52	6
FA	M	29.81	0.84	9
	F	29.78	0.45	5
Wta	M	7.52	1.22	9
	F	7.15		2

Location: Venezuela (USNMNH)

Description

This is one of the smallest species of *Molossops* within the range covered by this volume. As with most species of this genus, the male is larger than the female. This bat is frequently placed in the genus *Neoplatymops* because of its very flattened skull (Willig and Knox Jones 1985). The upperparts are brown, but the white bases of the hairs yield a pale brown cast. The venter is white to gray, and the ears are dark brown.
Chromosome number: $2n = 48$; FN = 60.

Range and Habitat

This species occurs in the Amazonian portions of Venezuela, in Colombia, and south into Brazil (map 5.124). In Venezuela it was taken at below 200 m elevation in tropical evergreen forest. It has a habit of feeding over ponds and streams (Handley 1976).

Natural History

This bat is an aerial insectivore (Willig 1985). It roosts in rock crevices, often close to ground level.

Molossops neglectus Williams and Genoways, 1980

Description

The dental formula is I 1/1, C 1/1, P 1/2, M 3/3. The single lower incisor separates this species from other members of the genus except for *M. temminckii*, from which it may be distinguished by size. Total length averages 89 mm, tail 29 mm, and forearm 35 mm. The ears are distinctly pointed. The dorsum is brown and the venter paler.

Range and Habitat

The species is newly described from Suriname, 5°25′ N, 55°03′ W (Williams and Genoways 1980b) (map 5.121).

Molossops temminckii (Burmeister, 1854)

Description

The dental formula is I 1/1, C 1/1, P 1/2, M 3/3. This bat is similar to *M. mattogrossensis* in size and appearance; forearms range from 29.6 to 30.7 mm.

Range and Habitat

The species is broadly distributed in Amazonian Bolivia, Peru, and Brazil. It may extend to southeastern Colombia (map 5.121).

Genus *Tadarida* Rafinesque, 1814
Free-tailed Bat

Description

The dental formula is I 1/2, C 1/1, P 2/2, M 3/3. This genus contains approximately thirty-five species, which exhibit considerable variation in body size. Head and body length ranges from 45 to 100 mm, the tail from 20 to 60 mm, and the forearm from 27 to 65 mm. Males possess a small glandular throat sac. The dorsum is generally reddish brown to black, and the venter tends to be somewhat paler.

Distribution

This genus is distributed in the subtropical and tropical portions of both the Old World and the New World.

Tadarida brasiliensis (I. Geoffroy, 1824)
Murciélago de Cola de Ratón

Measurements

	Sex	Mean	S.D.	N
TL	M	96.33	2.89	3
	F	94.00	2.55	5
HB	M	63.33	2.31	3
	F	60.60	1.82	5
T	M	33.00	2.65	3
	F	33.40	0.89	5
HF	M	9.67	1.15	3
	F	10.60	0.55	5
E	M	20.00	0.00	3
	F	19.20	0.45	5
FA	M	44.57	0.46	3
	F	43.42	0.81	5
Wta	M	12.43	0.76	3
	F	11.04	0.54	5

Location: Venezuela (USNMNH)

Description

The dental formula is I 1/2, C 1/1, P 2/2, M 3/3 (see fig. 5.34). In this intermediate-sized species of the genus, head and body length averages 62.3 mm for males and 60.6 mm for females. The dorsal pelage ranges from dark gray to dark brown.

Range and Habitat

This species is distributed from California to Florida and southward through Mexico to southern Argentina and Chile (map 5.125). Its range then extends through Uruguay on the South Atlantic coast of Brazil. In Venezuela this tolerant species may range up to 2,107 m elevation. It utilizes a wide variety of habitats.

Natural History

This is one of the best-studied bats and one of the first in North America to be examined as a potential

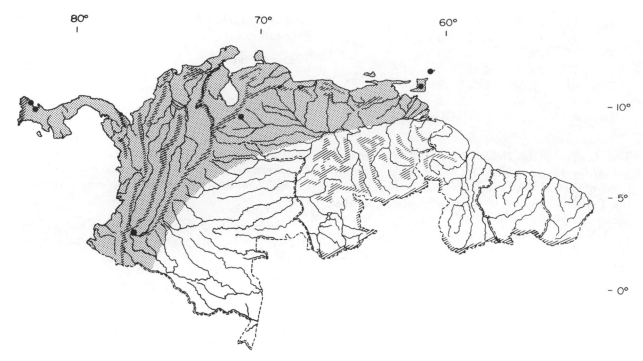

Map 5.125. Distribution of *Tadarida brasiliensis*.

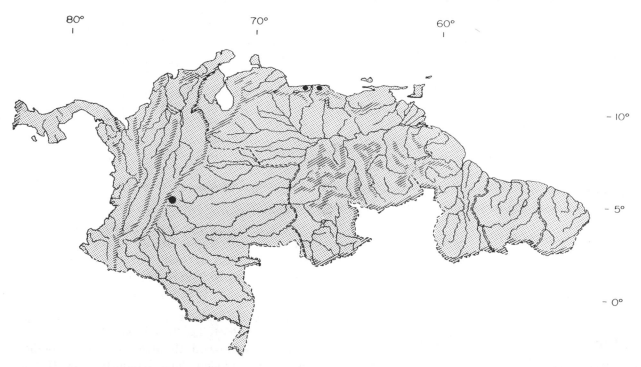

Map 5.126. Distribution of *Tadarida aurispinosa*.

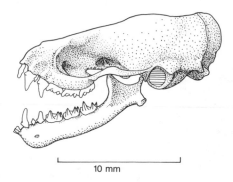

Figure 5.34. Skull of *Tadarida brasiliensis*.

vector of rabies. In North America this species shows seasonal migratory movements of several hundred kilometers; populations in southern Texas will migrate to Mexico for the winter (Davis, Herreid, and Short 1962). This species is highly adapted for feeding on small moths, and practically all its food is captured on the wing. The bats roost in caves in parts of their range but tolerate human dwellings as roosting sites. Davis, Herreid, and Short (1962) describe in detail the ecology of this species where it uses caves in Texas. When they form colonies in caves the bats may number in the hundreds of thousands. At Carlsbad Caverns the population has been estimated from as high as 8,700,000 in 1936 to a current low of approximately 250,000. Subpopulations of this species may exhibit seasonal torpor, but hibernation is not as profound as that shown by many of the north temperate vespertilionids. A single young is produced, and females may reproduce in the year following their own birth. During the rearing of the young the females form nursery colonies spatially separate from male groups; the young are not carried by the mothers when they feed but are left in the crèche.

Subgenus *Nyctinomops* Miller, 1902

Description

Use of *Nyctinomops* as a subgenus follows Koopman (1982). In species of this group the ears meet at the midline, though the point of intersection may be deeply notched, whereas in *T. brasiliensis* there is a slight gap in the fold of skin between the ears. This subgenus includes members distributed from the southwestern United States south across most of South America to northern Argentina.

Tadarida (Nyctinomops) aurispinosa (Peale, 1848)

Description

This is one of the largest species of the genus *Tadarida*, with forearm length ranging from 47.8 to 54.2 mm. Males are slightly larger than females. The dorsum is wood brown to russet, and the venter is somewhat paler.

Range and Habitat

This species occurs from Sinaloa in western Mexico and southern Tamaulipas in eastern Mexico south through the Isthmus and across most of northern South America, extending to Peru, Bolivia, and southeastern Brazil (see Ochoa-G. 1984a) (map 5.126).

Tadarida (Nyctinomops) laticaudata (E. Geoffroy, 1805)

Measurements

	Sex	Mean	S.D.	N
TL	M	96.63	4.37	81
	F	95.45	4.60	99
T	M	37.44	1.65	81
	F	37.51	2.31	99
HF	M	10.04	0.71	81
	F	10.52	0.72	96
E	M	19.48	0.96	81
	F	19.52	0.86	99
Wta	M	9.76	1.55	68
	F	9.73	1.80	49

Location: Venezuela (USNMNH)

Description

This account includes *T. europs*, *T. gracilis*, and *T. yucatanica*. This small species of *Tadarida* shows little sexual size dimorphism. The upperparts are deep brown, and the bases of the hairs are white. The venter is pinkish buff.

Range and Habitat

This species extends from southern Tamaulipas in Mexico south across northern South America to Argentina and southeastern Brazil (map 5.127). It was collected in Venezuela at elevations below 350 m. Although frequently foraging over streams and other moist sites, it is broadly tolerant of both dry deciduous forest and multistratal tropical evergreen forest (Handley 1976). It roosts in colonies of up to fifty.

Tadarida (Nyctinomops) macrotis (Gray, 1839)

Description

This is the largest species of *Tadarida* in the New World. Average measurements are TL 180; T 54; HF 12; E 28; FA 58–64. The tail of this large molossid extends a good 25 mm beyond the interfemoral membrane. The ears are large and join basally at the midline. The dorsum varies from reddish brown to black, and the basal portion of each hair is nearly white. The fur is shiny.

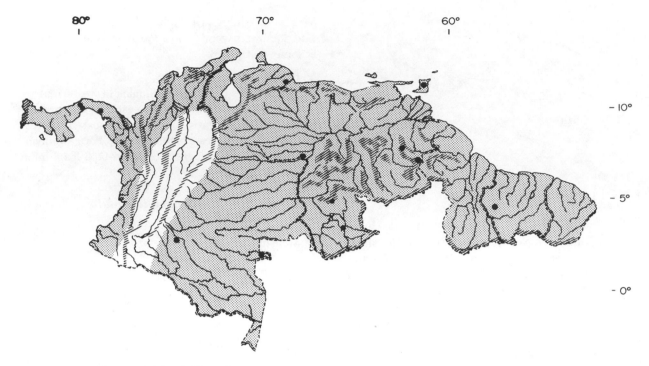

Map 5.127. Distribution of *Tadarida laticaudata*.

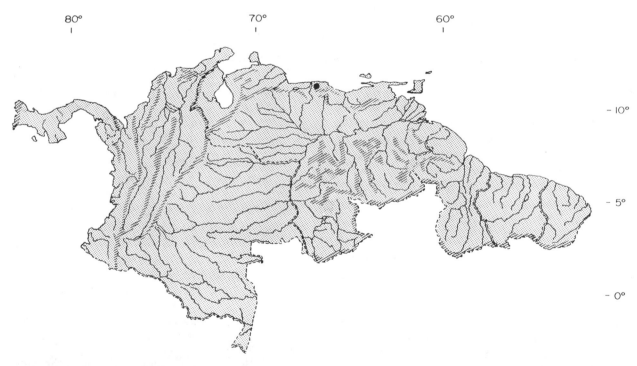

Map 5.128. Distribution of *Tadarida macrotis*.

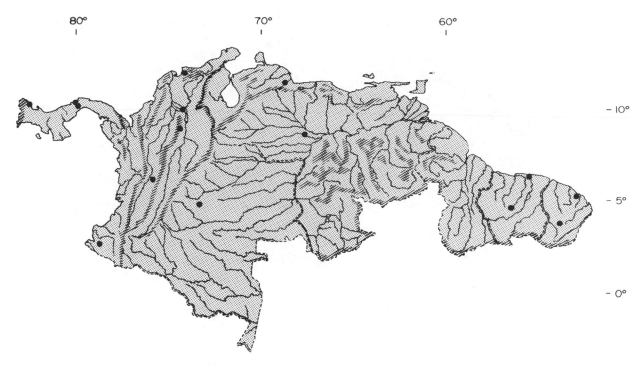

Map 5.129. Distribution of *Eumops auripendulus*.

Range and Habitat
This species extends from southern Utah and Colorado in the United States south through Mexico; it occurs intermittently throughout the Isthmus and then is broadly distributed across South America east of the Andes to Uruguay and northern Argentina (map 5.128). It prefers rocky outcrops and tolerates a variety of habitat types.

Natural History
This bat roosts in crevices in rocks, generally in small groups. Only a single young is produced each year, and the females form small nursery colonies.

Genus *Eumops* Miller, 1906
Bonneted Bat, Mastiff Bat

Description
The dental formula is I 1/2, C 1/1, P 2/2, M 3/3 (see fig. 5.35). There are eight species, showing an extreme range in size. Head and body length may vary between 40 and 130 mm, and the tail ranges from 35 to 80 mm. The large ears are rounded and usually connected across the head at the base. Males of some species have throat sacs that produce glandular secretions when in breeding condition. The genus has been monographed by Eger (1977).

Distribution
Species of this genus are distributed from southern California and Florida in the United States south through Central America to southern South America as far as northern Argentina.

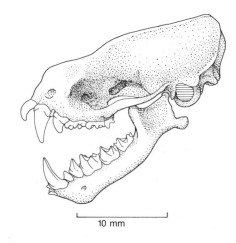

Figure 5.35. Skull of *Eumops glaucinus*.

Eumops auripendulus (Shaw, 1800)

Description
A medium-sized species of the genus, *Eumops auripendulus* has a forearm length from 56 to 63 mm. One female in Venezuela had a total length of 130.15 mm, of which 43 mm was the tail. A male from Trinidad measured TL 141; T 45; HF 15; E 23; Wt 32.5 g. The upperparts are dark reddish brown, the underparts slightly paler. The bases of the dorsal hairs are buffy.

Range and Habitat
This species is distributed from southern Veracruz, Mexico, south through the Isthmus and across South America to Paraguay and southeastern Brazil

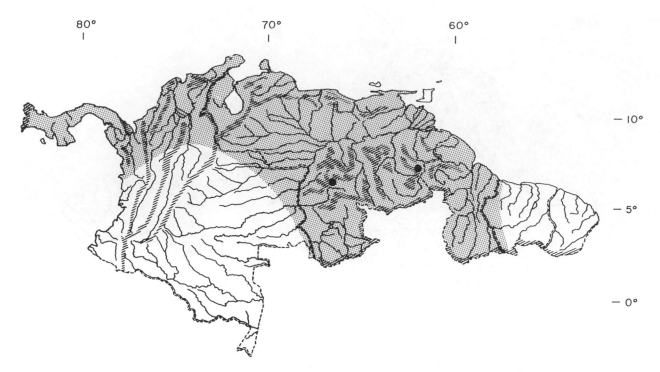

Map 5.130. Distribution of *Eumops hansae*.

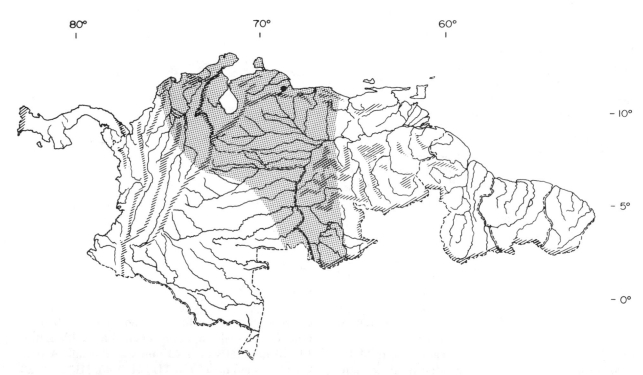

Map 5.131. Distribution of *Eumops dabbenei*.

(map 5.129). In northern Venezuela it prefers dry deciduous thorn forests with rocky outcroppings.

Eumops hansae Sanborn, 1932

Description

This small species of *Eumops* exhibits rather pronounced sexual size dimorphism. Males have forearm lengths from 40 to 41.6 mm, and females have forearm lengths of approximately 37 mm. Head and body length for a typical female is approximately 71 mm, and males have a head and body length of about 75 mm. The color is darker than that of *E. nanus*, but the species are similar in size. *E. amazonicus* is considered a junior synonym of *E. hansae*.

Range and Habitat

This species occurs from southern Costa Rica south through the Isthmus across northern South America to Guyana; it occurs sporadically in Amazonian Brazil (map 5.130). In Venezuela specimens were taken at below 155 m elevation in moist multistratal tropical evergreen forest.

Eumops dabbenei Thomas, 1914

Description

Because of its large size this species can be confused only with *E. perotis*. Measurements of a female taken in northern Venezuela were total length 177 mm; head and body length 120 mm; forearm 79 mm.

Range and Habitat

This species occurs from western Venezuela sporadically through central Brazil to northern Argentina (map 5.131). In northern Venezuela it was taken in dry deciduous thorn forest, near water (Handley 1976).

Eumops glaucinus (Wagner, 1843)

Measurements

	Sex	Mean	S.D.	N
TL	M	135.46	5.49	13
	F	134.88	4.98	41
HB	M	87.77	5.46	13
	F	86.44	3.69	41
T	M	47.69	3.79	13
	F	48.44	2.55	41
HF	M	14.38	1.04	13
	F	14.46	0.92	41
E	M	28.77	0.73	13
	F	27.85	0.80	40
FA	M	58.42	1.13	13
	F	58.37	1.02	41
Wta	M	33.83	2.09	13
	F	32.50	2.36	38

Location: TFA, Venezuela (USNMNH)

Description

This bat is smaller than *E. perotis* but larger than *E. auripendulus* and *E. maurus*. The sexes are nearly alike in size. The upperparts are dark cinnamon brown, the venter slightly paler (plate 5).

Range and Habitat

This species occurs from central Mexico and from Florida south through the Isthmus of Panama and the island of Cuba, and thence over most of South America to Paraguay and southeastern Brazil (map 5.132). In northern Venezuela it was taken at below 600 m elevation and generally was found foraging in the vicinity of streams, swamps, and lagoons. It is strongly associated with moist habitats and multistratal tropical evergreen forest. It roosts in trees (Handley 1976).

Eumops bonariensis (= E. nanus) (Peters, 1874)

Measurements

	Sex	Mean	S.D.	N
TL	M	89.00	4.24	4
	F	91.14	8.23	7
T	M	31.00	1.94	4
	F	30.71	1.80	7
HF	M	9.00	0.00	4
	F	8.43	0.79	7
E	M	18.00	1.15	4
	F	18.71	0.76	7
FA	M	37.72	0.24	5
	F	38.11	0.97	8
Wta	M	7.35	0.13	4
	F	6.95		2

Location: Venezuela (USNMNH)

Description

This is generally the smallest species of the genus taken over northern South America. *E. nanus* is included in *E. bonariensis* (Honacki, Kinman, and Koeppl 1982).

Range and Habitat

This species is found from southern Veracruz south through the Isthmus, across most of South America, to northern Argentina (map 5.133). In northern Venezuela it was found at extremely low elevations and was strongly associated with dry deciduous tropical forest.

Eumops maurus (Thomas, 1901)

Description

Forearm length ranges from 51 to 52.8 mm. *E. geijskesi* is a junior synonym of *E. maurus*.

Range and Habitat

This species is known from a very restricted area of Suriname and adjacent Guyana (map 5.133).

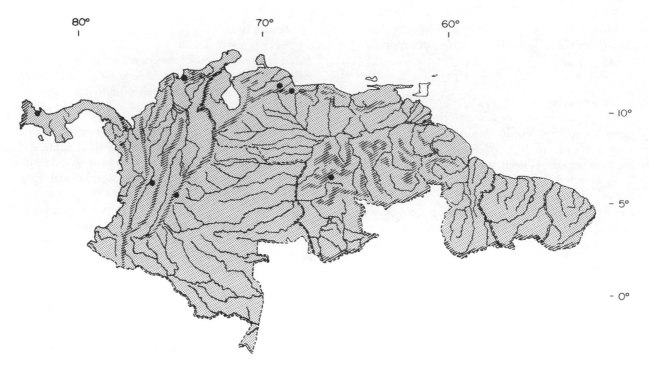

Map 5.132. Distribution of *Eumops glaucinus*.

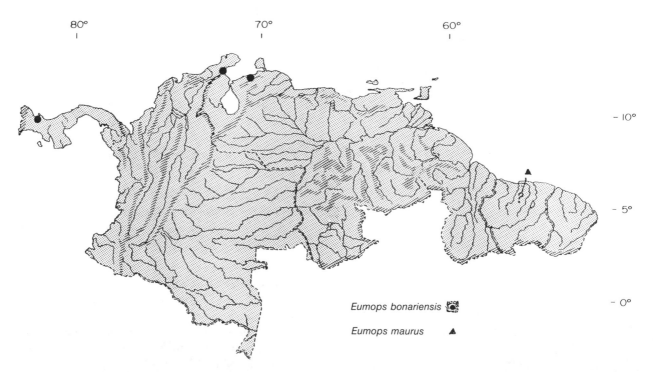

Eumops bonariensis 🔘

Eumops maurus ▲

Map 5.133. Distribution of two species of *Eumops*.

Eumops perotis (Schinz, 1821)
Moloso de Orejas Anchas

Description

This large bat has a forearm ranging from 72 to 82 mm in length. Average measurements are TL 171; T 55; HF 17; E 40. The dorsum ranges from dark gray to brownish gray, and the venter barely contrasts. The species *E. trumbulli* is considered a junior synonym of *E. perotis*.

Range and Habitat

This species ranges from central California south into western Mexico. It is sparsely distributed in Central America but is broadly distributed over most of South America south to northern Argentina (map 5.134).

Natural History

Like all members of the Molossidae, this large bat has very narrow wings. As a result, it has difficulty taking flight from flat surfaces and generally prefers crevices in rock outcroppings for daytime roosting sites, where it can make a vertical drop of several feet before launching into flight. This bat has been extensively studied in the southwestern United States by Vaughan (1959). Females form maternity colonies during the breeding season. A single young is born each year. The species, at least in the southern parts of the United States, feeds extensively upon Hymenoptera.

Genus *Molossus* E. Geoffroy, 1805

Description

The dental formula is I 1/1, C 1/1, P 1/2, M 3/3. Head and body length ranges from 50 to 95 mm, tail length from 20 to 70 mm, and the forearm from 33 to 60 mm. The general dorsal color is reddish brown to dark chestnut brown. It has been remarked that several species have two color phases: dark brown to almost black, and a lighter brown phase.

Distribution

This New World genus is distributed from northern Mexico south to northern Argentina and Uruguay.

Molossus ater E. Geoffroy, 1805
Black Mastiff Bat

Measurements

	Sex	Mean	S.D.	N
TL	M	124.88	4.31	68
	F	120.65	4.24	93
HB	M	78.82	4.66	68
	F	75.69	4.69	93
T	M	46.06	2.97	68
	F	44.98	3.15	94
HF	M	13.24	1.17	68
	F	13.00	1.18	95
E	M	16.57	1.51	68
	F	15.96	1.26	95
FA	M	48.19	1.39	76
	F	47.51	1.40	131
Wta	M	30.96	4.27	14
	F	27.95	3.06	28

Location: Northern Venezuela (USNMNH)

Description

This species is rather large for the genus. The upperparts are a rich russet to blackish; the underparts are only slightly paler.

Range and Habitat

Molossus ater occurs from Sinaloa in western Mexico and Nuevo León in eastern Mexico south through the Isthmus over much of South America to northern Argentina (map 5.135). In northern Venezuela this species was taken at below 1,180 m; most specimens were found below 500 m elevation. It is broadly tolerant of dry deciduous forest, man-made clearings, and multistratal tropical evergreen forest (Handley 1976).

Natural History

This species tends to roost in hollow trees or human dwellings and generally prefers moist areas. Colony size may reach fifty (Goodwin and Greenhall 1961).

Molossus aztecus Saussure, 1860

Measurements

	Sex	Mean	S.D.	N
TL	M	91.50	3.05	18
	F	88.47	3.03	32
HB	M	57.94	2.18	18
	F	55.16	2.83	32
T	M	33.56	2.04	18
	F	33.31	1.65	32
HF	M	11.11	0.90	18
	F	10.59	0.84	32
E	M	13.78	0.65	18
	F	13.94	0.62	32
FA	M	35.21	1.17	18
	F	34.87	0.68	32
Wta	M	16.11	3.51	8
	F	12.48	2.18	24

Location: TFA, Venezuela (USNMNH)

Description

This is one of the smallest species within the genus. The upperparts are dark brown to gray brown, and the venter is paler. This species is considered a subspecies of *M. molossus* by Hall (1981).

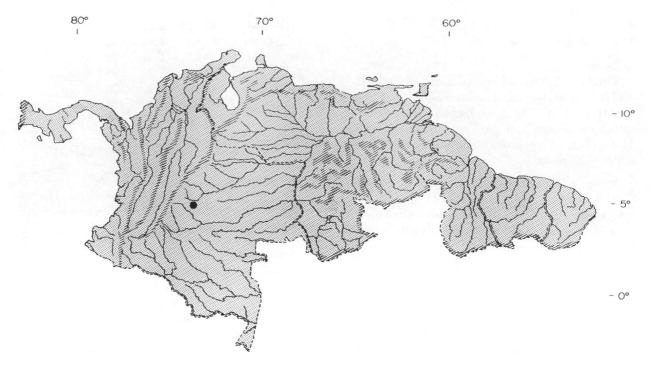

Map 5.134. Distribution of *Eumops perotis* (total range from Koopman 1982).

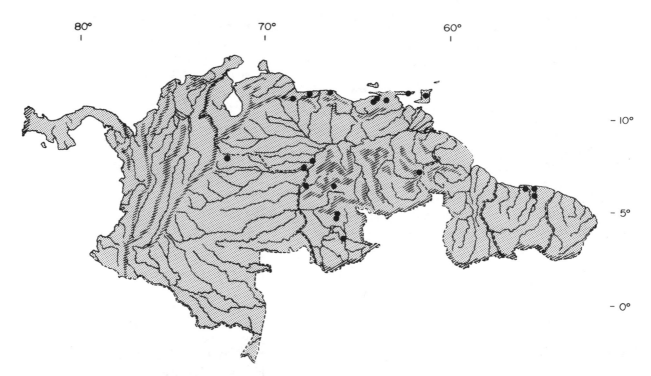

Map 5.135. Distribution of *Molossus ater*.

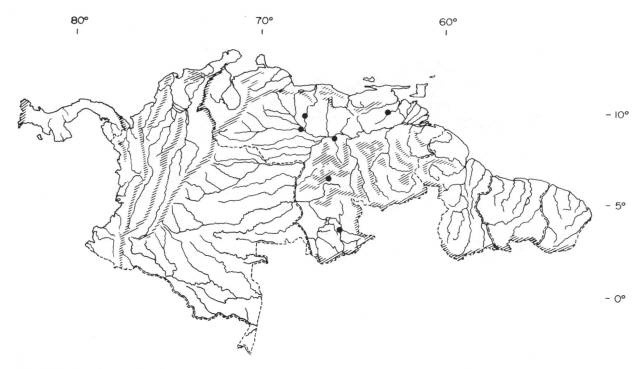

Map 5.136. Distribution of *Molossus aztecus*.

Range and Habitat

This species ranges from Sinaloa in western Mexico and Tamaulipas in eastern Mexico south to Panama and Venezuela (map 5.136). In northern Venezuela most specimens were taken at below 155 m elevation. The bats tend to roost in the hollows of trees and are strongly associated with moist habitats. Most specimens were taken in swampy areas or in association with multistratal tropical evergreen forest (Handley 1976).

Description

This species is intermediate in size. The forearm length in Venezuela averaged 41 mm for males and 39 mm for females. The dorsum is dark brown to reddish brown, the venter paler.

Range and Habitat

This species is distributed from Costa Rica south through the Isthmus across northern South America (map 5.137). In Venezuela specimens were taken at below 600 m and in association with dry deciduous tropical forest in coastal areas.

Molossus bondae J. A. Allen, 1904

Measurements

	Sex	Mean	S.D.	N
TL	M	111.50	6.45	4
	F	104.47	3.18	15
HB	M	71.00	3.92	4
	F	65.93	4.45	15
T	M	40.50	4.43	4
	F	38.53	3.38	15
HF	M	12.50	1.00	4
	F	11.20	0.68	15
E	M	14.25	0.96	4
	F	12.27	1.53	15
FA	M	41.05	1.94	4
	F	39.23	1.16	15
Wta	M	18.80	1.84	4
	F	17.00	1.31	3

Location: Venezuela (USNMNH)

Molossus molossus (Pallas, 1766)

Measurements

	Sex	Mean	S.D.	N
TL	M	101.70	4.26	43
	F	99.57	4.11	89
HB	M	65.37	3.97	43
	F	63.90	4.93	88
T	M	36.33	3.23	43
	F	35.73	2.69	88
HF	M	9.95	0.95	43
	F	9.56	1.02	89
E	M	12.47	1.53	43
	F	12.55	1.40	89
FA	M	38.95	0.94	44
	F	38.12	0.96	94
Wta	M	15.40	1.76	11
	F	13.29	1.61	20

Location: Northern Venezuela (USNMNH)

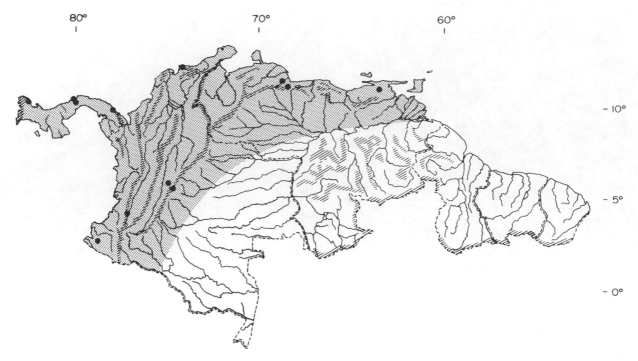

Map 5.137. Distribution of *Molossus bondae*.

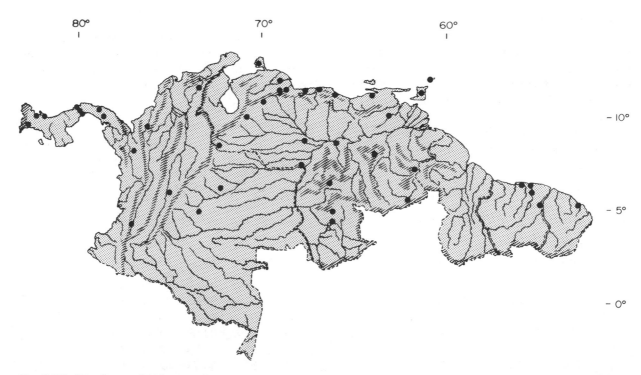

Map 5.138. Distribution of *Molossus molossus*.

80°   70°   60°

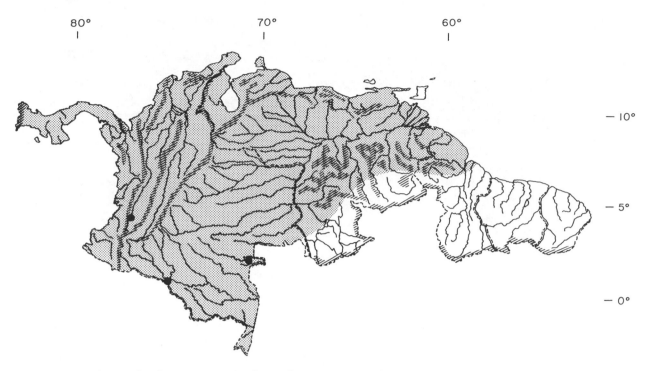

Map 5.139. Distribution of *Molossus pretiosus* (total range from Koopman 1982).

Description

The dorsum is gray brown to dark brown, the venter pale brown. This species includes *M. coibensis*. It may prove that *M. aztecus* is a subspecies of *M. molossus*.

Range and Habitat

This species ranges from Central America south across almost all of South America well into northern Argentina and Uruguay (map 5.138). It is the common small species of *Molossus* taken over an enormous range. In northern Venezuela most specimens were taken at below 900 m elevation. It is broadly tolerant of both dry deciduous forest and moist tropical multistratal evergreen forest. Although it forages near streams, it roosts in hollow trees and in human dwellings. In drier areas it frequently feeds near ponds and may be the dominant species taken in mist nets set near pond sites (Handley 1976).

Natural History

Recently Haussler, Moller and Schmidt (1981) conducted a captive study on the growth and development of this species. Reproduction takes place at the onset of the rainy season, and females form nursery colonies for rearing the young. Although the young hang as a compact group while the mothers leave the cave to forage, each female can identify her own young by its cry. The young begin thermoregulation at twenty days of age; they nurse until

sixty-five days of age, and the forearm reaches adult size at sixty days.

Molossus pretiosus Miller, 1902

Description

This bat is nearly the same size as *M. bondae*. Head and body length ranges from 63 to 65 mm, tail 42 to 45 mm, and forearm 41 to 46 mm. The dorsal pelage is very dark brown, almost black.

Range and Habitat

This species occurs from Nicaragua south through the Isthmus to Colombia, Venezuela, and Guyana (map 5.139).

Molossus sinaloae J. A. Allen, 1906

Description

This is the largest species of *Molossus* occurring in the range covered by this book. Its measurements are TL 115–48; T 38–54; E 12–17; FA 46–49. The dorsum is brown, the venter paler. This description includes *M. trinitatus*.

Range and Habitat

The species is distributed from southern Sinaloa in Mexico to the Yucatán peninsula, south through the Isthmus and across northern South America to Suriname (map 5.140). In northern Venezuela this species was taken at below 1,160 m. It is broadly tol-

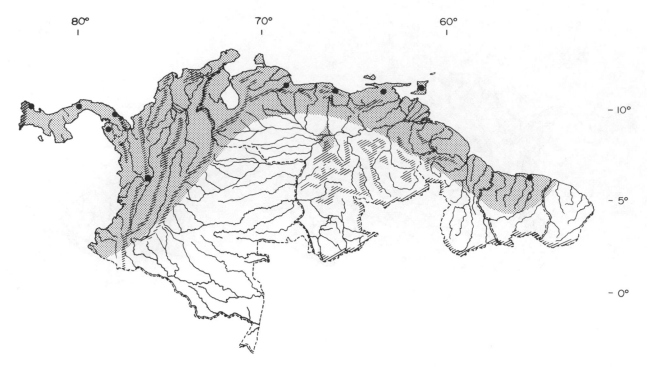

Map 5.140. Distribution of *Molossus sinaloae*.

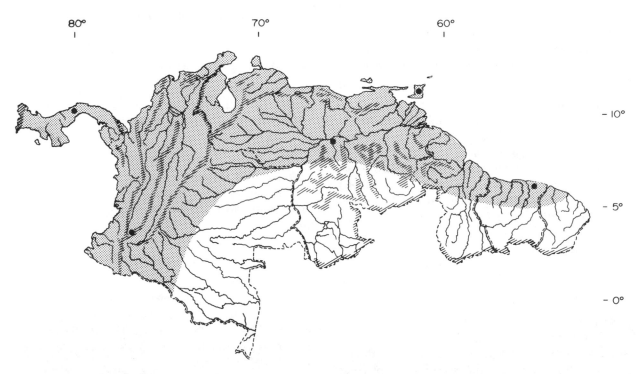

Map 5.141. Distribution of *Promops centralis*.

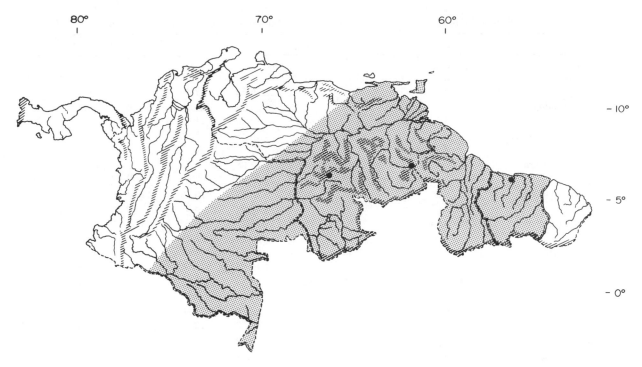

Map 5.142. Distribution of *Promops nasutus*.

erant of dry deciduous forest and premontane moist tropical forest (Handley 1976).

Genus *Promops* Gervais, 1855

Description
The dental formula is I 1/2, C 1/1, P 2/2, M 3/3. In these intermediate-sized molossid bats, head and body length ranges from 60 to 90 mm, tail length from 45 to 75 mm, and forearm from 43 to 63 mm. The dorsum is drab brown to glossy black, with the venter paler.

Distribution
Species of this genus are distributed from Central America south to Paraguay. There are two species within the range covered by this volume.

Promops centralis Thomas, 1915

Description
This is the largest species of the genus found in northern South America. Head and body length averages about 89 mm, and the forearm exceeds 50 mm. One female from Venezuela measured TL 139; T 50; HF 13; E 18; FA 52.5. The upperparts are dark brown to glossy black, the underparts slightly paler.

Range and Habitat
This species occurs from the Yucatán peninsula in Mexico south through the Isthmus across northern South America and in western Colombia and Peru to Paraguay (map 5.141). It is absent over much of Brazil. It roosts in colonies of up to six individuals under palm fronds (Goodwin and Greenhall 1961).

Promops nasutus (Spix, 1823)

Description
This bat is slightly smaller than *P. centralis:* TL 126–41; T 57–61; HF 12–13; E 13–17; FA 48.5; Wt 19.9 g (*N* = 2 females; Venezuela USNMNH)

Range and Habitat
This species is confined to the continent of South America, being found in the Amazonian portions of Venezuela, Colombia, Peru, and Bolivia and sporadically distributed from south-central Brazil to northern Argentina (map 5.142). Specimens in Venezuela were taken at below 1,000 m in wet habitat and multistratal tropical evergreen forest. Like other tropical molossids, it roosts in hollow trees.

(References follow on page 225)

Table 5.14   Average External Measurements (mm) and Weights (g) of Some Neotropical Chiroptera

	Sex	N	TL	HB	T	HF	E	FA	Wta
Emballonuridae									
Rhynchonycteris naso	m	46	56.48	41.09	15.39	7.18	13.20	36.87	3.80
	f	38	59.18	42.34	16.84	7.45	13.63	38.57	3.99
Saccopteryx bilineata	m	44	72.98	53.39	19.59	11.76	16.60	46.27	8.58
	f	99	74.93	55.33	19.59	11.65	16.24	47.74	8.82
Saccopteryx canescens	m	8	54.75	38.63	16.13	7.25	12.63	35.93	3.23
	f	8	61.63	44.00	17.63	7.25	13.75	38.68	3.47
Saccopteryx leptura	m	12	58.83	43.75	15.08	8.42	13.92	37.98	4.47
	f	21	61.14	45.05	15.90	8.33	14.33	39.48	5.71
Cormura brevirostris	m	6	67.83	54.00	13.83	10.33	14.50	45.73	8.62
	f	5	69.80	54.20	15.60	8.80	16.20	46.48	10.15
Peropteryx kappleri	m	14	71.93	58.21	13.71	10.86	18.07	48.30	7.73
	f	24	74.67	60.96	13.71	11.42	18.00	50.28	8.54
Peropteryx macrotis	m	47	60.49	47.21	13.28	8.57	15.30	41.94	5.14
	f	76	64.43	49.37	15.07	9.01	15.94	44.88	6.09
Peropteryx m. trinitatis	m	26	56.58	46.38	10.19	8.35	12.46	38.57	3.61
	f	11	62.18	51.09	11.09	8.64	12.45	41.81	4.42
Diclidurus albus	m	8	90.50	70.38	20.13	12.50	16.63	62.45	14.93
	f	14	91.93	71.43	20.50	12.57	16.93	64.38	16.25
Diclidurus ingens	f	3	103.67	80.67	23.00	13.00	20.67	71.43	—
Diclidurus isabellus	m	21	80.38	59.67	20.71	12.19	17.71	57.93	13.36
	f	6	81.33	59.33	22.00	12.83	17.67	59.10	15.63
Diclidurus scutatus	m	2	86.00	68.00	18.00	10.00	14.50	53.10	13.60
	f	10	83.40	65.10	18.30	10.40	13.80	56.22	12.94
Noctilionidae									
Noctilio albiventris	m	65	91.88	74.88	17.00	18.20	23.70	59.72	28.71
	f	146	88.26	71.55	16.71	17.14	23.03	58.64	22.99
Noctilio leporinus	m	26	127.42	96.88	30.54	33.65	28.00	85.12	67.72
	f	37	120.03	91.27	28.76	30.62	28.11	83.36	55.87
Mormoopidae									
Pteronotus davyi	m	55	78.69	56.44	22.25	11.75	17.87	46.52	9.38
	f	49	80.51	57.96	22.55	11.90	18.08	47.19	9.06
Pteronotus parnellii	m	127	95.13	71.77	23.44	15.06	23.69	59.95	20.42
	f	118	93.88	70.44	23.44	15.14	23.56	60.45	19.60
Pteronotus personatus	m	2	70.50	53.50	17.00	12.00	17.50	44.55	8.00
	f	2	75.00	58.00	17.00	11.00	16.00	45.00	6.90
Pteronotus suapurensis	m	18	87.89	64.33	23.56	12.94	18.94	51.13	12.62
	f	15	88.00	64.07	23.93	12.88	19.06	51.87	13.98
Mormoops megalophylla	m	46	94.57	68.07	26.50	12.17	15.35	55.41	17.21
	f	15	95.53	69.60	25.93	12.53	16.40	55.06	16.55
Phyllostomidae									
Phyllostominae									
Micronycteris brachyotis	m	1	68.00	57.00	11.00	11.00	17.00	38.60	10.60
	f	1	70.00	59.00	11.00	13.00	16.00	40.00	—
Micronycteris hirsuta	m	2	78.00	64.00	14.00	11.50	25.50	44.15	14.00
	f	3	78.00	61.00	17.00	13.33	27.00	43.60	12.43
Micronycteris megalotis	m	24	56.41	43.78	12.63	9.96	21.85	32.13	5.09
	f	68	58.32	44.68	13.64	10.07	21.69	33.25	5.71
Micronycteris microtis	m	23	57.74	45.35	12.39	10.26	22.04	33.64	5.97
	f	17	60.35	46.12	14.24	10.47	22.59	34.97	6.64
Micronycteris minuta	m	25	59.52	47.40	12.12	11.56	22.32	34.96	6.86
	f	25	61.24	49.08	12.12	11.58	21.88	35.11	7.20
Micronycteris nicefori	m	41	64.02	54.98	9.05	13.00	18.37	37.51	7.90
	f	39	67.49	58.03	9.46	13.38	18.49	38.84	9.20
Micronycteris schmidtorum	m	2	64.00	48.00	16.00	11.00	23.00	35.80	7.60
	f	5	64.80	48.60	16.20	11.00	21.00	34.60	7.23
Micronycteris sylvestris	m	2	61.50	47.50	14.00	12.50	21.50	37.95	7.25
	f	1	65.00	50.00	15.00	11.00	22.00	41.90	10.30

Table 5.14   (continued)

	Sex	N	TL	HB	T	HF	E	FA	Wta
Lonchorhina aurita	m	61	113.92	60.97	59.89	14.55	32.66	50.97	14.34
	f	40	115.03	61.10	53.93	14.48	32.18	51.48	15.07
Lonchorhina orinocensis	m	29	102.34	52.28	50.07	11.00	30.76	42.16	8.59
	f	24	103.08	52.38	50.71	11.00	30.71	42.37	8.97
Macrophyllum macrophyllum	m	4	86.60	44.70	41.90	14.53	18.83	35.10	7.08
	f	11	93.38	48.54	44.85	15.00	18.77	35.89	7.15
Tonatia bidens	m	9	97.89	76.00	21.10	16.80	33.50	57.33	26.25
	f	7	95.00	75.43	19.57	16.86	31.14	57.04	24.65
Tonatia brasiliense	m	24	65.50	54.83	10.67	11.96	24.58	36.15	10.18
	f	21	65.43	55.71	9.71	12.09	24.86	36.06	10.61
Tonatia carrikeri	m	1	92.00	72.00	20.00	17.00	30.00	44.00	22.10
Tonatia silvicola	m	15	92.80	73.33	19.47	16.93	36.60	52.56	30.64
	f	20	92.20	71.50	20.70	17.25	37.20	51.91	24.13
Mimon crenulatum	m	33	83.15	57.88	25.53	12.58	26.00	49.18	12.85
	f	32	83.94	58.50	25.44	12.47	26.41	49.49	12.03
Phyllostomus discolor	m	128	97.70	83.57	14.13	17.29	23.36	60.91	38.15
	f	107	98.23	84.24	13.99	17.06	23.18	60.86	35.35
Phyllostomus elongatus	m	52	103.17	80.37	22.76	18.85	30.35	66.35	41.08
	f	55	104.78	82.47	22.31	18.22	29.96	66.36	40.06
Phyllostomus hastatus	m	8	131.09	109.75	21.34	22.56	32.36	84.24	104.54
	f	12	126.92	107.78	19.14	22.49	31.86	82.82	87.55
Phylloderma stenops	m	9	113.71	—	17.71	21.24	27.67	70.92	50.63
	f								
Trachops cirrhosus	m	125	97.02	77.54	19.47	19.89	34.36	60.06	34.20
	f	122	97.15	78.16	18.98	19.80	34.65	60.84	32.86
Chrotopterus auritus	m	24	107.58	97.48	9.38	25.50	45.50	76.53	65.06
	f	9	111.67	103.43	10.00	25.67	47.56	79.47	68.92
Vampyrum spectrum	m	3	140.00	140.00	0.00	32.67	46.33	105.17	169.43
	f	1	158.00	158.00	0.00	38.00	48.00	107.40	199.70
Glossophaginae									
Glossophaga longirostris	m	236	68.30	61.62	6.47	12.30	15.90	37.78	12.81
	f	209	68.91	61.36	7.61	11.97	15.69	38.24	12.91
Glossophaga soricina	m	221	61.67	53.96	7.71	11.14	14.29	34.27	9.46
	f	218	62.82	54.94	7.91	11.14	14.52	34.97	9.35
Lionycteris spurrelli	m	48	61.52	53.50	8.02	11.19	13.35	34.59	8.66
	f	77	63.30	54.73	8.57	11.35	13.29	35.55	8.89
Lonchophylla robusta	m	13	77.31	68.38	8.92	13.23	16.92	42.75	14.34
	f	9	78.44	69.33	9.11	13.00	16.67	42.85	13.72
Lonchophylla thomasi	m	6	58.83	51.33	7.50	10.17	15.00	31.84	7.08
	f	7	56.86	49.14	7.71	10.14	14.86	31.73	5.93
Anoura caudifer	m	36	65.81	62.61	5.50	12.03	14.36	36.64	10.98
	f	31	64.55	61.65	4.72	12.06	14.48	36.39	10.00
Anoura cultrata	m	2	79.00	75.00	4.00	14.50	16.50	41.90	20.35
	f	2	71.00	66.50	4.50	14.50	17.00	40.50	14.95
Anoura geoffroyi	m	21	65.86	65.86	—	13.14	15.41	42.63	15.80
	f	33	65.12	65.12	—	13.21	15.52	42.86	14.50
Anoura latidens	m	45	64.44	64.44	—	12.93	15.33	42.91	14.53
	f	46	66.20	66.20	—	12.65	14.65	43.15	14.81
Choeronsicus godmani	m	4	59.75	53.75	6.00	9.50	8.25	31.87	7.27
	f	9	64.78	55.67	9.11	9.89	12.13	33.98	7.61
Choeroniscus minor	m	1	69.00	61.00	8.00	10.00	13.00	34.60	8.20
	f	2	70.50	61.00	9.50	11.00	12.50	35.30	8.30
Leptonycteris curasoae	m	97	81.35	81.35	—	16.85	17.47	53.37	26.33
	f	86	79.02	79.02	—	17.03	17.22	52.55	24.93
Scleronycteris ega	m	1	—	—	—	—	—	35.20	—
Carolliinae									
Carollia brevicauda	m	66	65.35	58.75	6.69	13.88	19.49	38.48	14.57
	f	83	65.19	58.54	6.65	13.96	19.37	38.63	12.93

Table 5.14   (continued)

	Sex	N	TL	HB	T	HF	E	FA	Wta
Carollia castanea	m	4	63.75	53.75	10.00	13.50	17.75	36.50	14.55
	f	5	62.60	53.80	8.80	12.20	18.60	35.73	—
Carollia perspicillata	m	547	68.23	58.21	10.04	14.39	20.79	40.91	17.04
	f	655	69.29	58.43	9.85	14.33	20.63	41.14	15.90
Rhinophylla pumilio	m	12	48.33	48.33	—	10.33	16.33	34.71	9.36
	f	26	50.04	50.04	—	10.77	15.81	35.14	10.41
Sturnirinae									
Sturnira bidens	m	3	64.33	64.33	—	14.00	15.33	39.63	17.20
	f	13	64.92	64.92	—	14.77	16.31	40.64	16.30
Sturnira bogotensis	m	2	67.00	67.00	—	15.50	17.50	43.05	19.60
	f	2	65.50	65.50	—	14.00	17.00	43.30	—
Sturnira erythromos	m	40	58.23	58.23	—	13.40	15.83	39.50	15.11
	f	32	58.34	58.34	—	12.94	15.78	38.91	15.07
Sturnira lilium	m	76	61.30	61.30	—	13.88	16.66	41.03	20.43
	f	89	60.61	60.61	—	13.62	16.48	40.56	18.87
Sturnira ludovici	m	90	70.80	70.80	—	16.39	19.27	47.21	26.77
	f	48	70.92	70.92	—	16.13	18.88	46.82	23.98
Sturnira tildae	m	67	69.66	69.66	—	16.63	20.37	47.78	26.16
	f	89	69.47	69.47	—	16.40	20.22	47.67	25.46
Stenoderminae									
Uroderma bilobatum	m	167	60.92	60.92	—	12.50	18.16	41.68	16.05
	f	215	60.73	60.73	—	12.60	18.14	41.73	17.19
Uroderma magnirostrum	m	52	61.37	61.37	—	12.38	17.90	42.54	15.92
	f	103	62.48	62.48	—	12.71	17.66	43.34	17.83
Vampyrops aurarius	m	17	76.06	76.06	—	16.06	21.88	51.92	33.64
	f	15	78.27	78.27	—	16.47	21.53	53.22	34.68
Vampyrops brachycephalus	m	3	62.67	62.67	—	11.33	17.00	39.80	15.50
Vampyrops helleri	m	159	57.05	57.05	—	11.85	17.19	37.54	13.01
	f	181	57.60	57.60	—	12.08	17.39	37.92	14.33
Vampyrops umbratus	m	81	69.85	69.85	—	13.52	18.67	44.20	23.35
	f	79	71.61	71.61	—	13.80	18.66	44.58	25.56
Vampyrops vittatus	m	5	86.80	86.80	—	17.40	23.20	57.24	—
	f	3	89.67	89.67	—	18.33	24.33	59.77	—
Vampyrodes caraccioli	m	4	71.75	71.75	—	15.75	21.25	51.06	26.88
	f	15	74.50	74.50	—	16.13	21.67	53.24	29.95
Vampyressa bidens	m	57	50.14	50.14	—	11.04	17.19	35.42	11.04
	f	48	52.10	52.10	—	11.23	17.08	35.81	12.21
Vampyressa pusilla	m	18	46.72	46.72	—	10.16	15.00	31.37	8.22
	f	62	47.76	47.76	—	10.31	14.95	31.86	9.43
Chiroderma salvini	m	9	72.11	72.11	—	13.78	18.78	48.67	27.22
	f	13	73.23	73.23	—	14.23	19.23	48.19	29.10
Chiroderma trinitatum	m	13	54.85	54.85	—	11.92	17.46	38.85	13.09
	f	46	56.70	56.70	—	11.91	17.70	38.48	13.98
Chiroderma villosum	m	46	64.57	64.57	—	13.37	19.15	44.21	21.00
	f	198	67.55	67.55	—	13.72	19.34	45.80	22.91
Ectophylla macconnelli	m	16	42.81	42.81	—	10.25	15.56	30.15	6.39
	f	47	44.06	44.06	—	9.91	15.45	30.97	6.56
Artibeus cinereus	m	109	54.01	54.01	—	11.46	16.77	39.74	12.62
	f	101	53.99	53.99	—	11.74	16.95	39.98	13.05
Artibeus concolor	m	18	59.61	59.61	—	12.33	18.17	46.93	18.21
	f	69	64.09	64.09	—	11.91	18.88	48.33	19.96
Artibeus fuliginosus	m	105	75.00	75.00	—	16.53	23.01	58.94	35.34
	f	50	77.24	77.24	—	17.26	23.48	60.16	36.61
Artibeus hartii	m	39	60.33	60.33	—	12.13	15.92	39.18	17.37
	f	36	61.00	61.00	—	12.00	15.78	39.18	17.26
Artibeus jamaicensis	m	294	80.11	80.11	—	17.57	22.23	59.57	40.41
	f	344	81.83	81.83	—	17.96	22.77	60.45	43.16

Table 5.14   (continued)

	Sex	N	TL	HB	T	HF	E	FA	Wta
Artibeus lituratus	m	161	89.43	89.43	—	19.96	23.77	68.82	62.94
	f	188	91.14	91.14	—	20.30	24.19	69.59	68.15
Artibeus gnomus	m	20	48.95	48.95	—	11.05	17.25	37.50	10.16
	f	23	49.78	49.78	—	11.00	17.26	37.45	10.49
Artibeus amplus	m	10	91.50	91.50	—	20.10	23.10	69.12	60.03
	f	16	90.31	90.31	—	19.75	23.13	69.43	—
Ametrida centurio	m	67	42.01	42.01	—	10.63	14.27	25.58	8.01
	f	69	47.94	47.94	—	10.94	15.23	31.87	12.60
Sphaeronycteris toxophyllum	m	31	56.55	56.55	—	11.55	15.84	37.66	14.91
	f	93	57.89	57.89	—	11.92	16.38	40.14	14.73
Centurio senex	f	5	55.40	55.40	—	14.80	17.00	42.22	18.70
Desmodontinae									
Desmodus rotundus	m	71	74.51	74.51	—	16.93	18.75	—	—
	f	63	80.57	80.57	—	17.68	19.33	—	—
Desmodus youngi	m	11	83.36	83.36	—	17.91	18.18	51.65	36.27
	f	4	84.00	84.00	—	19.50	18.75	53.48	32.20
Diphylla ecaudata	m	6	73.17	73.17	—	15.67	15.17	49.90	25.98
	f	3	78.67	78.67	—	17.67	15.67	51.80	25.17
Natalidae									
Natalus tumidirostris	m	89	100.69	48.73	52.96	10.06	15.42	39.43	6.23
	f	42	101.40	49.95	51.45	9.95	14.95	39.41	6.29
Furipteridae									
Furipterus horrens	m	6	62.00	37.50	24.50	7.92	11.08	35.28	3.08
Thyropteridae									
Thyroptera tricolor	m	6	72.67	40.17	32.50	6.67	12.50	36.43	4.68
	f	4	72.50	40.25	32.35	7.00	12.50	37.63	4.50
Vespertilionidae									
Myotis albescens	m	27	79.11	44.48	34.63	8.96	14.00	34.18	5.59
	f	48	81.08	45.64	35.47	9.00	14.02	34.71	5.72
Myotis keaysi	m	8	85.00	48.38	36.63	8.25	12.63	36.00	5.25
	f	14	81.50	44.64	36.86	8.36	12.57	36.17	5.25
Myotis nesopolus	m	12	79.92	42.33	37.58	7.67	12.58	31.80	3.50
	f	6	80.33	43.00	37.33	7.50	13.17	32.49	—
Myotis nigricans	m	45	78.98	43.38	35.60	8.02	13.02	34.56	4.66
	f	46	77.54	41.17	36.37	8.02	12.46	33.41	4.48
Myotis oxyotus	m	4	84.00	45.50	38.50	9.00	14.00	36.93	5.37
	f	5	85.60	45.60	40.00	8.40	13.80	36.40	5.60
Myotis riparius	m	3	79.00	42.67	36.33	8.00	12.67	34.78	4.33
	f	13	81.38	42.85	38.54	8.15	12.92	35.33	4.43
Eptesicus andinus	m	3	94.00	54.67	39.33	10.00	14.00	41.63	8.73
	f	10	100.30	57.40	42.90	10.20	14.30	43.74	10.19
Eptesicus brasiliensis	m	12	94.67	54.42	40.25	10.00	14.42	40.05	8.90
	f	32	97.53	54.53	43.00	9.81	14.94	40.93	9.89
Eptesicus diminutus	f	1	91.90	54.00	37.00	9.00	15.00	35.50	—
Eptesicus furinalis	m	8	92.50	52.00	40.50	9.00	13.25	38.35	7.53
	f	6	92.33	54.00	38.33	13.83	39.03	8.08	54.00
Eptesicus fuscus	m	2	117.50	69.50	48.00	12.50	17.00	56.63	—
	f	1	132.00	84.00	48.00	11.00	19.00	50.30	26.30
Eptesicus montosus	m	8	89.25	52.25	37.00	9.25	14.38	38.63	7.67
	f	21	92.19	54.86	37.33	9.86	14.62	39.16	9.08
Histiotus sp.	m	3	108.67	59.67	49.00	10.33	31.00	45.57	—
	f	1	109.00	59.00	50.00	12.00	—	45.70	—
Rhogeessa minutilla	m	24	72.17	40.54	31.63	6.71	14.38	27.33	3.51
	f	47	72.36	41.23	31.13	6.85	14.23	27.39	3.78
Rhogeessa tumida	m	8	70.50	41.13	29.39	6.29	12.38	27.70	3.47
	f	14	71.29	42.36	28.93	6.79	12.79	28.42	4.00
Lasiurus borealis	m	5	100.80	51.80	49.00	9.30	10.70	38.64	6.43
	f	2	102.50	51.50	51.00	9.50	11.50	40.50	—

Table 5.14   (continued)

	Sex	N	TL	HB	T	HF	E	FA	Wta
Lasiurus cinereus	f	3	138.00	77.00	61.00	12.33	17.00	54.33	19.70
Lasiurus ega	m	9	114.44	61.11	53.33	10.44	17.56	45.43	11.87
	f	8	120.50	65.75	54.75	10.88	17.75	46.15	13.51
Molossidae									
Molossops abrasus	f	2	111.50	77.50	34.00	13.00	18.50	41.40	54.80
Molossops greenhalli	m	1	105.00	80.00	25.00	10.00	16.00	35.00	18.90
Molossops paranus	f	2	87.50	61.00	26.50	8.50	16.00	32.00	11.65
Molossops planirostris	m	18	89.56	62.22	27.33	10.00	16.22	33.54	14.19
	f	52	83.02	57.48	25.54	9.64	16.02	32.02	11.33
Molossops mattogrossensis	m	10	78.90	53.50	25.40	7.82	13.55	29.81	7.52
	f	6	74.67	51.00	23.67	7.83	14.33	29.78	7.15
Tadarida brasiliensis	m	3	96.33	63.33	33.00	9.67	20.00	44.57	12.43
	f	5	94.00	60.60	33.40	10.60	19.20	43.42	11.04
Tadarida gracilis	m	81	96.63	59.19	37.44	10.04	19.48	41.97	9.76
	f	99	95.45	57.95	37.51	10.52	19.52	41.69	9.73
Tadarida laticaudata	m	1	108.00	67.00	41.00	11.00	23.00	45.00	14.60
	f	1	102.00	60.00	42.00	10.00	18.00	44.30	13.80
Eumops amazonicus	f	1	100.00	71.00	29.00	9.00	18.00	37.50	15.40
Eumops auripendulus	f	4	100.50	86.75	43.40	15.00	22.40	57.35	23.10
Eumops dabbenei	f	1	177.00	120.00	57.00	21.00	32.00	79.10	64.70
Eumops glaucinus	m	15	136.93	88.93	48.00	14.40	28.80	58.78	35.33
	f	49	136.14	88.00	48.14	14.41	27.77	58.56	33.00
Eumops nanus	m	4	89.00	58.00	31.00	9.00	18.00	37.72	7.35
	f	7	91.14	60.43	30.71	8.43	18.71	38.11	6.95
Molossus ater	m	107	125.21	79.36	45.85	13.69	16.56	48.29	32.10
	f	160	120.69	76.29	44.42	13.48	16.00	47.36	26.45
Molossus aztecus	m	26	92.65	59.73	32.92	10.73	13.58	35.27	15.61
	f	34	88.44	55.35	33.09	10.51	13.74	34.84	12.48
Molossus bondae	m	4	111.50	71.00	40.50	12.50	14.25	41.05	18.80
	f	15	104.47	65.93	38.53	11.20	12.27	39.23	17.00
Molossus molossus	m	78	102.14	65.73	36.41	10.26	13.26	39.38	15.85
	f	167	100.17	64.49	35.71	9.94	13.11	38.66	13.55
Molossus sinaloae	f	4	130.25	82.50	47.75	12.75	15.50	49.45	24.23
Promops centralis	f	1	139.00	89.00	50.00	13.00	18.00	52.50	—
Promops nasutus	f	2	133.50	74.50	59.00	12.50	15.00	48.40	19.90

Source: Data from USNMNH Venezuela collection, courtesy of K. Ralls and C. Handley, Jr.
Note: TL, total length; HB, head and body length; T, tail length; HF, hind foot length; E, ear length as measured from the notch to the tip; FA, forearm length; Wta, adult weight (mean weights and measurements).

Table 5.15   Key to the Subfamilies and Genera of Phyllostomidae

Subfamilies

1	Single upper incisor and upper canine enlarged and bladelike	Desmodontinae
1'	Upper incisor(s) and canine not enlarged and bladelike	2
2	Nose leaf rudimentary, without distinct upright process; tail present	Phyllonycterinae
2'	Nose leaf usually well developed; tail absent if nose leaf rudimentary	3

3   Tongue elongate, with conspicuous bristlelike papillae on anterodorsal surface; first upper premolar usually distinctly separated from canine and rarely in contact with second upper premolar (first upper premolar sometimes in contact with canine in *Monophyllus*, but distinctly separated from second upper premolar) .... Glossophaginae

3'   Tongue not elongate, lacking conspicuous bristlelike papillae; first upper premolar in contact with canine and usually with second upper premolar 4

4	Zygomatic arch incomplete	Carolliinae
4'	Zygomatic arch complete	5
5	Molars dilambdodont (distinct **W**-shaped pattern of lophs on occlusal surface)	Phyllostominae
5'	Molars lacking dilambdodont pattern	Stenoderminae

Phyllostominae

1	One lower incisor	2
1'	Two lower incisors	4
2	Two lower premolars	*Mimon*
2'	Three lower premolars (second small to minute)	3
3	Second lower premolar crowded to lingual side of tooth row, first and third lower premolars usually in contact	*Chrotopterus*
3'	Second lower premolar not crowded from tooth row, first and third lower premolars not in contact	*Tonatia*
4	Two lower premolars	*Phyllostomus*
4'	Three lower premolars (second sometimes crowded to lingual side of tooth row)	5
5	Rostrum as long as braincase	*Vampyrum*
5'	Rostrum shorter than braincase	6
6	Second lower premolar large, but smaller than first and third premolars	7
6'	Second lower premolar small to minute, much smaller than first and third premolars	8
7	Auditory bullae large, greatest diameter much exceeding distance between them	*Macrotis*
7'	Auditory bullae small, greatest diameter less than distance between them	*Micronycteris*
8	Second lower premolar displaced lingually from tooth row; first and second lower premolars in contact or nearly so	9
8'	Second lower premolar not displaced lingually from tooth row; first and second lower premolars usually not in contact	10
9	Greatest length of skull less than 20 mm	*Macrophyllum*
9'	Greatest length of skull more than 20 mm	*Trachops*
10	Dorsal profile of rostrum strongly convex; deep depression present between orbits	*Lonchorhina*
10'	Dorsal profile of rostrum not convex; no depression between orbits	*Phylloderma*

Glossophaginae

1	Permanent lower incisors lacking	2
1'	Two pairs of permanent lower incisors, usually well developed	8
2	Premolars 3/3	*Anoura*
2'	Premolars 2/3	3
3	Molars 2/2	*Lichonycteris*
3'	Molars 3/3	4

4   Pterygoids highly modified, expanded at base and inflated in appearance; pterygoid wings long and in contact, or nearly so, with auditory bullae 5

4'   Pterygoids normal, not expanded at base or inflated in appearance, pterygoid wings short and not in contact with auditory bullae .. 7

5   First and second upper incisors separated by distinct gap; upper premolars low, barely exceeding height of molars ... *Choeroniscus*

5'   First and second upper incisors in contact, or nearly so; upper premolars distinctly higher than molars 6

6   Rostrum distinctly longer than postrostral part of cranium; upper molars essentially equal in size, all with a distinct metastyle .... *Musonycteris*

6'   Rostrum about equal in length to postrostral part of cranium; third upper molar somewhat smaller than first two and lacking a distinct metastyle .... *Choeronycteris*

7   Upper molars lacking mesostyle; lower molars long and narrow; known only from Middle America .... *Hylonycteris*

7'   Mesostyle present on all upper molars; lower molars only moderately compressed; known only from Brazil and Venezuela .... *Scleronycteris*

8	Molars 2/2	*Leptonycteris*
8'	Molars 3/3	9
9	Zygomatic arch complete, first upper incisor not markedly enlarged and spatulate	10
9'	Zygomatic arch incomplete, first upper incisor enlarged and spatulate	11

10   Evident gap between upper premolars and between them and adjacent teeth; tail relatively long and extending beyond posterior border of uropatagium .... *Monophyllus*

10'   Upper premolars usually in contact and filling space between canine and first molar; tail short and not extending beyond posterior border of uropatagium .... *Glossophaga*

11   Rostrum elongate, longer than postrostral part of cranium; postcanine maxillary teeth reduced in size and with evident gaps between them .... *Platalina*

Table 5.15   *(continued)*

11'	Rostrum not elongate, no longer than postrostral part of cranium; postcanine maxillary teeth of normal size; last premolars and molars in contact of nearly so 12
12	First upper premolar smaller than second and laterally compressed . . . . *Lonchophylla*
12'	First upper premolar essentially same size as second, not laterally compressed (triangular in outline) . . . . *Lionycteris*

Carolliinae

1	Tail present; upper premolars essentially equal in size . . . . *Carollia*
1'	Tail absent; first upper premolar much smaller than second . . . . *Rhinophylla*

Stenoderminae (includes Sturnirinae)

1	Molars 2/2 2
1'	Molars 2/3 or 3/3 7
2	Upper dental arcade semicircular, rostrum less than half as long as braincase . . . . *Centurio*
2'	Upper dental arcade not semicircular, rostrum more than half as long as braincase 3
3	Rostrum inflated, nearly cuboid . . . . *Pygoderma*
3'	Rostrum not inflated or cuboid 4
4	Posterior margin of external nares with marked, lyre-shaped emargination . . . . *Chiroderma*
4'	Posterior margin of external nares lacking lyre-shaped emargination 5
5	Second upper molar markedly larger than first; upper premolars separated from each other and adjacent teeth by evident gaps . . . . *Ectophylla* (part)
5'	Second upper molar essentially equal in size to, or smaller than, first; no gaps between anterior upper cheek teeth 6
6	Posterior margin of external nares more or less straight; second upper molar much smaller than first and differing in form . . *Artibeus* (part)
6'	Posterior margin of external nares broadly V-shaped; second upper molar resembling first in size and form . . . . . . *Vampyressa* (part)
7	Molars 2/3 8
7'	Molars 3/3 12
8	Palate short, posterior border having deep U-shaped emargination that reaches level of first molar . . . . *Ariteus*
8'	Palate long, posterior border having shallow emargination that falls far short of level of tooth row 9
9	First upper incisor markedly bifid, less than twice size of second incisor . . . . *Artibeus* (part)
9'	First upper incisor not bifid or only weakly so, more than twice size of second incisor 10
10	Second upper molar noticeably larger than first; upper premolars separated from each other and from adjacent teeth by evident gaps . . . . *Ectophylla* (part)
10'	Second upper molar equal to or smaller than first, no gaps between anterior upper cheek teeth 11
11	Incisors 2/1 or 2/2; height of first incisor greater than height of first premolar; greatest length of skull less than 22 mm . . *Vampyressa* (part)
11'	Incisors 2/2; height of first incisor much less than height of first premolar; greatest length of skull more than 24 mm . . . . *Vampyrodes*
12	Upper dental arcade expanded laterally to form semicircular arc 13
12'	Upper dental arcade not expanded laterally, U-shaped in occlusal view 14
13	Orbital space wider than long; interorbital constriction less than 5 mm . . . . *Ametrida*
13'	Orbital space longer than wide; interorbital constriction more than 5 mm . . . . *Sphaeronycteris*
14	Palate short, posterior palatal emargination reaching level of first upper molar 15
14'	Palate of medium length or long, posterior border variously emarginate but never to level of tooth row 17
15	Palatal emargination broadly V-shaped . . . . *Phyllops*
15'	Palatal emargination deeply U-shaped 16
16	Well-developed V-shaped ridge from sagittal crest to anterior margin of orbits, forming deep rostral depression . . . . *Stenoderma*
16'	V-shaped ridge from sagittal crest to anterior margin of orbits lacking, rostrum normal . . . . *Ardops*
17	Upper molars distinctly grooved longitudinally, the first two subquadrate in outline and lacking well-developed cusps; first upper incisor approximately half as high as canine . . . . *Sturnira*
17'	Upper molars lacking longitudinal groove, the first two not subquadrate in outline and possessing well-developed cusps; first upper incisor much less than half as high as canine 18
18	First upper incisor less than twice size of second and resembling it in shape; upper incisors in contact and filling space between canines 19
18'	First upper incisor more than twice size of second and differing from it in shape; evident gaps present betweem upper incisors . . . 20
19	First upper incisor deeply bifid; M3, if present, minute and peglike . . . . *Artibeus* (part)
19'	First upper incisor not bifid; M3 relatively large and well developed . . . . *Enchisthenes*
20	Crowns of first upper incisors parallel, deeply bifid; lower incisors in contact . . . . *Uroderma*
20'	Crowns of first upper incisors converge distally, not deeply bifid; lower incisors separated by distinct gaps . . . . *Vampyrops*

Phyllonycterinae

1	Tail not extending beyond edge of uropatagium . . . . *Brachyphylla*
1'	Tail extending beyond edge of uropatagium 2
2	Zygomatic arch complete; second and third lower molars distinctly cuspidate . . . . *Erophylla*
2'	Zygomatic arch incomplete; second and third lower molars not distinctly cuspidate . . . . *Phyllonycteris*

Desmodontinae

1	First lower incisors in contact; interfemoral membrane with distinct fringe of moderately long hairs . . . . *Diphylla*
1'	First lower incisors not in contact; interfemoral membrane without fringe of hair 2
2	Lower incisors not bifid; wing white from middle of proximal phalanx to tip . . . . *Diaemus*
2'	Lower incisors bifid; wing usually pigmented to tip (if white tipped, white does not extend proximally to first phalanx) . . . *Desmodus*

Source: Modified from Knox Jones and Carter (1976).

Table 5.16   Key to the Species of *Micronycteris*

1	Ears connected by a high, notched band; P3 about same size as P4	subgenus *Micronycteris* 2
1'	Ears not connected by a band; P3 and P4 unequal in size	5
2	Forearm greater than 40 mm; skull length greater than 21 mm	3
3(2)	Interauricular band with slight notch; venter brown	*M. megalotis*
3'	Interauricular band with deep notch; venter white	4
4(3)	Calcar shorter than foot	*M. minuta*
4'	Calcar longer than foot	*M. schmidtorum*
5(1)	Fifth metacarpal shortest; I2 bifid; P4 straight	subgenus *Lampronycteris*, *M. brachyotis*
5'	Fourth metacarpal shortest; I2 unicuspid; P4 recurved or reduced	6
6(5)	Third metacarpal longest	7
6'	Fifth metacarpal longest	8
7(6)	Ears rounded; nose leaf blunt; P3 and P4 reduced	subgenus *Neonycteris*, *M. pusilla*
7'	Ears pointed, nose leaf pointed; P3 smaller than P4	subgenus *Trinycteris*, *M. nicefori*
8(6)	Single pair of upper incisors	subgenus *Barticonycteris*, *M. daviesi*
8'	Two pairs of upper incisors	subgenus *Glyphonycteris*, 9
9(8)	Forearm less than 42.5 mm	*M. sylvestris*
9'	Forearm greater than 42.5 mm	*M. behni*

Source: Medellin, Wilson, and Navarro L. (1985).

References

• References used in preparing distribution maps

• Aellen, V. 1970. Catalogue raisonné des chiroptères de la Colombie. *Rev. Suisse Zool.* 77(1):1–37.

Albuja V., L. 1982. *Murciélagos de Ecuador.* Quito: Escuela Politechnica Nacional.

Arata, A. A., and J. B. Vaughan. 1970. Analyses of the relative abundance and reproductive activity of bats in southwestern Colombia. *Caldasia* 10:517–28.

August, P. V. 1979. Distress calls in *Artibeus jamaicensis*. In *Vertebrate ecology in the northern Neotropics*, ed. J. F. Eisenberg, 151–59. Washington, D.C.: Smithsonian Institution Press.

———. 1981. Fig fruit consumption by *Artibeus jamaicensis* in the llanos of Venezuela. *Biotropica* (Reprod. Bot. Suppl.) 13:70–76.

Ayala, S. C., and A. D'Alessandro. 1973. Insect feeding of some Colombian fruit-eating bats. *J. Mammal.* 54:266–67.

Baker, R. J., H. H. Genoways, and P. A. Seyfarth. 1981. Additional chromosomal data for bats (Mammalia: Chiroptera) from Suriname. *Ann. Carnegie Mus.* 50(12):333–43.

Baker, R. J., M. W. Haiduk, L. W. Robbins, A. Cadena, and B. F. Koop. 1982. Chromosomal studies of bats and their implications. In *Mammalian biology in South America*, ed. M. A. Mares and H. H. Genoways, 303–44. Pymatuning Symposia in Ecology 6. Special Publications Series. Pittsburgh: Pymatuning Laboratory of Ecology, University of Pittsburgh.

Baker, R. J., J. Knox Jones, Jr., and D. C. Carter, eds. 1976. *Biology of bats of the New World family Phyllostomatidae, part 1.* Special Publications of the Museum 10. Lubbock: Texas Tech Press.

———. 1977. *Biology of bats of the New World family Phyllostomatidae, part 2.* Special Publications of the Museum 13. Lubbock: Texas Tech Press.

———. 1979. *Biology of bats of the New World family Phyllostomatidae, part 3.* Special Publications of the Museum 16. Lubbock: Texas Tech Press.

Barbour, R. W., and W. H. Davis. 1969. *Bats of America.* Lexington: University Press of Kentucky.

Barriga-Bonilla, E. 1965. Estudios mastozoológicos Colombianos. I. Chiroptera. *Caldasia* 9:241–68.

• Bergmans, W. 1979. A record from Surinam of the bat *Chiroderma trinitatum* Goodwin, 1958 (Mammalia, Chiroptera). *Zool. Med. Ed.* 54(22):313–17.

Bonaccorso, F. J. 1979. Foraging and reproductive ecology in a Panamanian bat community. *Bull. Florida State Mus. (Biol. Sci.)* 24(4):359–408.

Bradbury, J. W. 1977. Social organization and communication. In *Biology of bats*, vol. 3, ed. W. A. Wimsatt, 1–72. New York: Academic Press.

Bradbury, J. W., and L. Emmons. 1974. Social organization of some Trinidad bats. I. Emballonuridae. *Z. Tierpsychol.* 36:137–83.

Bradbury, J. W., and Vehrencamp, S. L. 1976a. Social organization and foraging in emballonurid bats. 1. Field studies. *Behav. Ecol. Sociobiol.* 1:337–81.

———. 1976b. Social organization and foraging in emballonurid bats. 2. A model for the determination of group size. *Behav. Ecol. Sociobiol.* 1:383–404.

———. 1977a. Social organization and foraging in

emballonurid bats. 3. Mating systems. *Behav. Ecol. Sociobiol.* 2:1–17.

———. 1977b. Social organization and foraging in emballonurid bats. 4. Parental investment patterns. *Behav. Ecol. Sociobiol.* 2:19–30.

• Brosset, A., and G. Dubost. 1967. Chiroptères de la Guyane française. *Mammalia* 31(4):583–94.

Brown, J. H. 1968. Activity patterns of some Neotropical bats. *J. Mammal.* 49:754–57.

• Cabrera, A. 1958. *Catálogo de los mamíferos de América del Sur.* Vol. 1. Buenos Aires: Museo Argentino de Ciencias Naturales "Bernardino Rivadavia," Zoología.

• Carter, D. C. 1966. A new species of *Rhinophylia* (Mammalia, Chiroptera, Phyllostomatidae) from South America. *Proc. Biol. Soc. Washington* 79:235–38.

Carvalho, C. T. 1961. Sobre los habitos alimentares de phillostomidos. *Rev. Biol. Trop.* 9:53–60.

Davis, R. B., C. F. Herreid, and H. Short. 1962. Mexican free-tailed bats in Texas. *Ecol. Monogr.* 32:311–46.

Davis, W. B. 1980. New *Sturnira* from Central and South America with a key to currently recognized species. *Occas. Pap. Mus. Texas Tech Univ.* 70:1–5.

• de la Torre, L. 1966. New bats of the genus *Sturnira* (Phyllostomidae) from the Amazonian lowlands of Peru and the Windward Islands, West Indies. *Proc. Biol. Soc. Washington* 79:267–72.

• de la Torre, L., and A. Schwartz. 1966. New species of *Sturnira* (Chiroptera: Phyllostomidae) from the Islands of Guadeloupe and Saint Vincent, Lesser Antilles. *Proc. Biol. Soc. Washington* 79:297–304.

Eger, J. L. 1977. Systematics of the genus *Eumops* (Chiroptera: Molossidae) *Royal Ontario Mus. Life Sci. Contrib.* 110:1–69.

Eisenberg, J. F. 1981. *The mammalian radiations: An analysis of trends in evolution, adaptation, and behavior.* Chicago: University of Chicago Press.

Eisenberg, J. F., and D. E. Wilson. 1978. Relative brain size and feeding strategies in the Chiroptera. *Evolution* 32(4):740–51.

Erkert, H. G. 1982. Ecological aspects of bat activity rhythms. In *Ecology of bats,* ed. T. Kunz, 201–42. New York: Plenum.

Fenton, M. B. 1982. Echolocation, insect hearing and feeding ecology of insectivorous bats. In *Ecology of bats,* ed. T. Kunz, 261–86. New York: Plenum.

Findley, J. S., and D. E. Wilson. 1974. Observations on the Neotropical disk-winged bat *Thyroptera tricolor* Spix. *J. Mammal.* 55:562–71.

———. 1982. Ecological significance of Chiropteran morphology. In *Ecology of bats,* ed. T. H. Kunz, 243–60. New York: Plenum.

Fleming, T. H. 1971. *Artibeus jamaicensis:* Delayed embryonic development in a Neotropical bat. *Science* 171:402–4.

———. 1982. Foraging strategies of plant visiting bats. In *Ecology of bats,* ed. T. Kunz, 287–326. New York: Plenum.

———. 1983. *Carollia perspicillata* (murciélago candelaro, lesser short-tailed fruit bat). In *Costa Rican natural history,* ed. D. H. Janzen, 457–58. Chicago: University of Chicago Press.

Fleming, T. H., E. T. Hooper, and D. E. Wilson. 1972. Three Central American bat communities: Structure: reproductive cycles, and movement patterns. *Ecology* 53(4):555–69.

Foster, M. S., and R. M. Timm. 1976. Tent making by *Artibeus jamaicensis* with comments on plants used by bats for tents. *Biotropica* 8:265–69.

Freeman, P. W. 1979. Specialized insectivory: Beetle-eating and moth-eating molossid bats. *J. Mammal.* 60:467–79.

———. 1981. *A multivariate study of the family Molossidae: Morphology, ecology, evolution.* Fieldiana-Zoology 7. Chicago: Field Museum of Natural History.

Fuzessery, Z. M., and G. D. Pollak. 1984. Neural mechanisms of sound localization in an echolocating bat. *Science* 225:725–28.

Gardner, A. L. 1976. The distributional status of some Peruvian mammals. *Occas. Pap. Mus. Zool. Louisiana State Univ.* 48:1–18.

———. 1977a. Feeding habits. In *Biology of bats of the New World family Phyllostomatidae, part 2,* ed. R. J. Baker, J. Knox Jones, and D. C. Carter, 293–350. Special Publications of the Museum 13. Lubbock: Texas Tech Press.

———. 1977b. Chromosomal variation in *Vampyressa* and a review of chromosomal evolution in the Phyllostomidae. *Syst. Zool.* 26:300–318.

• Gardner, A. L., and D. C. Carter. 1972. A review of the Peruvian species of *Vampyrops* (Chiroptera: Phyllostomatidae). *J. Mammal.* 53:72–82.

• Gardner, A. L., and J. P. O'Neill. 1969. The taxonomic status of *Sturnira bidens* with notes on its karyotype and life history. *Occas. Papers Mus. Zool. Louisiana State Univ.* 38:1–8.

Gardner, A. L., and J. L. Patton. 1972. A new species of *Philander* and *Mimon* from Peru. *Occas. Pap. Mus. Zool. Louisiana State Univ.* 43:1–12.

Genoways, H. H., and S. L. Williams. 1979. Records of bats (Mammalia: Chiroptera) from Suriname. *Ann. Carnegie Mus.* 48:323–35.

• ———. 1980. A new species of bat of the genus *To-*

natia (Mammalia: Phyllostomatidae). *Ann. Carnegie Mus.* 49(14):203–11.

• Genoways, H. H., S. L. Williams, and J. A. Groen. 1981. Noteworthy records of Surinamese mammals. *Ann. Carnegie Mus.* 50(11):319–32.

• Goodwin, G. G., and A. M. Greenhall. 1961. A review of the bats of Trinidad and Tobago. *Bull. Amer. Mus. Nat. Hist.* 122(3):187–302.

Gould, E. 1977. Echolocation and communication. In *Biology of bats of the New World family Phyllostomatidae, part 2,* ed. R. J. Baker, J. Knox Jones, and D. C. Carter, 247–80. Special Publications of the Museum 13. Lubbock: Texas Tech Press.

• Greenhall, A. M. 1959. Bats of Guiana. *J. British Guiana Mus. Zoo* 22:55–57.

Griffin, D. R. 1958. *Listening in the dark.* New Haven: Yale University Press.

Griffin, D. R., F. A. Webster, and C. R. Michael. 1960. The echolocation of flying insects by bats. *Anim. Behav.* 8:141–54.

• Hall, E. R. 1981. *The mammals of North America.* 2d ed., 2 vols. New York: John Wiley.

Handley, C. O., Jr. 1966. Descriptions of new bats (*Choeroniscus* and *Rhinophylla*) from Colombia. *Proc. Biol. Soc. Washington* 79:83–88.

———. 1967. Bats of the canopy of an Amazonian forest. *Atlas Simp. Biota. Amazonica* 5:211–15.

• ———. 1976. Mammals of the Smithsonian Venezuelan project. *Brigham Young Univ. Sci. Bull., Biol. Ser.* 20(5):1–90.

———. 1984. New species of mammals from northern South America: A long-tongued bat, genus *Anoura* Gray. *Proc. Biol. Soc. Washington* 97(3):513–21.

———. 1987. New species of mammals from northern South America: Fruit-eating bats, genus *Artibeus* Leach. In *Studies in Neotropical mammalogy: Essays in honor of Philip Hershkovitz,* ed. B. D. Patterson and R. M. Timm, 163–72. Fieldiana-Zoology, n.s. 39. Chicago: Field Museum of Natural History.

Haussler, U., E. Moller, and U. Schmidt. 1981. Zur haltung und Jugendentwicklung von *Molossus molossus.* Z. *Säugetierk.* 46:337–51.

• Herd, R. M. 1983. *Pteronotus parnellii. Mammal. Species* 209:1–5.

• Hernandez-Camacho, J., and A. Cadena-G. 1978. Notas para la revisión del género *Lonchorhina* (Chiroptera, Phyllostomidae). *Caldasia* 12(57):199–251.

• Hershkovitz, P. 1949. Mammals of northern Colombia, preliminary report no. 5: Bats (Chiroptera). *Proc. U.S. Nat. Mus.* 99:429–54.

• Hill, J. E. 1964. Notes on bats from British Guiana with the description of a new genus and species of Phyllostomidae. *Mammalia* 28(4):553–72.

Honacki, J. H., K. E. Kinman, and J. W. Koeppl, eds. 1982. *Mammalian species of the world.* Lawrence, Kans.: Allen Press and Association of Systematics Collections.

• Honeycutt, R. L., R. J. Baker, and H. H. Genoways. 1980. Chromosomal data for bats (Mammalia: Chiroptera) from Suriname. *Ann. Carnegie Mus.* 49(16):237–50.

Hood, C. S., and J. Knox Jones, Jr. 1984. *Noctilio leporinus. Mammal. Species* 216:1–7.

Hood, C. S., and J. Pitocchelli. 1983. *Noctilio albiventris. Mammal. Species* 197:1–5.

Howell, D. J. 1974. Bats and pollen: Physiological aspects of chiropterophily. *Comp. Biochem. Physiol.* 48A:263–76.

———. 1983. *Glossophaga soricina.* In *Costa Rican natural history,* ed. D. H. Janzen, 472–74. Chicago: University of Chicago Press.

Humphrey, S. R., and F. J. Bonaccorso. 1979. Population and community ecology. In *Biology of bats of the New World family Phyllostomatidae, part 3,* ed. R. J. Baker, J. Knox Jones, and D. C. Carter, 406–41. Special Publications of the Museum 16. Lubbock: Texas Tech Press.

Humphrey, S. R., Bonaccorso, F. J., and T. L. Zinn. 1983. Guild structure of surface-gleaning bats in Panama. *Ecology* 64(2):284–94.

• Husson, A. M. 1958. Notes on the Neotropical leafnosed bat *Sphaeronycteris toxophyllum* Peters. *Arch. Neerland. Zool.* 13 (suppl. 1):114–19.

• ———. 1962. *The bats of Suriname.* Zoologische Verhandelingen 58. Leiden: E. J. Brill.

• ———. 1978. *The mammals of Suriname.* Leiden: E. J. Brill.

• Ibañez, C. 1979. Nuevos datos sobre *Eumops dabbenei* Thomas, 1914. *Donana-Acta Vert.* 4(2):248–52.

Jepsen, G. L. 1970. Bat origins and evolution. In *Biology of bats,* vol. 1, ed. W. A. Wimsatt, 1–64. New York: Academic Press.

Kleiman, D. G., and T. M. Davis. 1979. Ontogeny and maternal care. In *Biology of bats of the New World family Phyllostomatidae, part 3,* ed. R. J. Baker, J. Knox Jones, and D. C. Carter, 387–402. Special Publications of the Museum 16. Lubbock: Texas Tech Press.

Knox Jones, J., and D. C. Carter. 1976. Annotated checklist, with keys to subfamilies and genera. In *Biology of bats of the New World family Phyllostomatidae, part 1,* ed. R. J. Baker, J. Knox Jones, Jr., and D. C. Carter, 31–37. Special Publications of the Museum 10. Lubbock: Texas Tech Press.

Koopman, K. 1978. Zoogeography of Peruvian bats. *Amer. Mus. Novit.* 2651:1–33.

• ———. Biogeography of the bats of South America.

In *Mammalian biology in South America*, ed. M. A. Mares and H. H. Genoways, 273–302. Pymatuning Symposia in Ecology 6. Special Publication Series. Pittsburgh: Pymatuning Laboratory of Ecology, University of Pittsburgh.

Kunz, T. H., ed. 1982. *Ecology of bats*. New York: Plenum.

Kunz, T. H., P. V. August, and C. D. Burnett. 1983. Harem social organization in cave roosting *Artibeus jamaicensis* (Chiroptera: Phyllostomidae). *Biotropica* 15(21):133–38.

LaVal, R. K. 1973a. A revision of the Neotropical bats of the genus *Myotis*. *Los Angeles Nat. Hist. Mus. Sci. Bull.* 15:1–53.

———. 1973b. *Systematics of the genus* Rhogeessa. Occasional Papers 19. Lawrence: Museum of Natural History, University of Kansas.

Lemke, T. C. 1984. Foraging ecology of the long-nosed bat, *Glossophaga soricina*. *Ecology* 65:538–48.

• Lemke, T. O., A. Cadena, R. H. Pine, and J. Hernandez-Camacho. 1982. Notes on opossums, bats, and rodents new to the fauna of Colombia. *Mammalia* 46(2):225–34.

• Lemke, T. O., and J. R. Tamsitt. 1979. *Anoura culturata* from Colombia. *Mammalia* 43:579–81.

Linares, O. J. 1972. Studies in the bat *Natalus stramineus* of Venezuelan caves, with special reference to variation and isolation. *Bol. Soc. Venez. Espeleol.* 3(3):231–33.

———. 1986. *Murciélagos de Venezuela*. Caracas: Cuardenas Lagoven, Premio Nacional de Perodismo.

Linares, O. J., and P. Kiblinski. 1969. Note on a new locality and karyotype of *Molossops greenhalli* from Venezuela. *J. Mammal.* 50(4):831–32.

Linares, O. J., and J. Ojasti. 1971. Una nueva especie de murciélago del genero *Lonchorhina* (Chiroptera: Phyllostomidae) del sur de Venezuela. *Nov. Cient., Ser. Zool.*, 36:1–8.

• McCarthy, T. J. 1983. Comments on the first *Tonatia carrikeri* from Colombia. *Lozania* (Bogotá) 40:1–5.

McCracken, G. F., and J. W. Bradbury. 1977. Paternity and genetic heterogeneity in the polygynous bat *Phyllostomus hastatas*. *Science* 198:303–6.

———. 1981. Social organization and kinship in the polygynous bat *Phyllostomus hastatus*. *Behav. Ecol. Sociobiol.* 8:11–34.

McNab, B. K. 1969. The economics of temperature regulation in Neotropical bats. *Comp. Biochem. Physiol.* 31:227–68.

———. 1982. Evolutionary alternatives in the physiological ecology of bats. In *Ecology of bats*, ed. T. H. Kunz, 151–200. New York: Plenum.

• Marinkelle, C. J., and A. Cadena. 1971. Remarks on *Sturnira tildae* in Colombia. *J. Mammal.* 52(1):235–37.

• ———. 1972. Notes on bats new to the fauna of Colombia. *Mammalia* 36:50–58.

Medellin, R. A. 1988. Prey of *Chrotopterus*. *J. Mammal.* 69:841–44.

Medellin, R. A., D. E. Wilson, and D. Navarro L. 1985. *Micronycteris brachyotis*. *Mammal. Species* 251:1–4.

Miller, G. S., Jr. 1907. The families and genera of bats. *U.S. Nat. Mus. Bull.* 57:1–282.

Morrison, D. W. 1978a. Lunar phobia in a Neotropical fruit bat, *Artibeus jamaicensis* (Chiroptera: Phyllostomidae). *Anim. Behav.* 26:852–55.

———. 1978b. Influence of habitat on the foraging distances of the fruit bat, *Artibeus jamaicensis*. *J. Mammal.* 59(3):622–24.

———. 1978c. On the optimal searching strategy for refuging predators. *Amer. Nat.* 112:925–34.

———. 1978d. Foraging ecology and energetics of the frugivorous bat *Artibeus jamaicensis*. *Ecology* 59(4):716–23.

———. 1979. Apparent male defense of tree hollows in the fruit bat, *Artibeus jamaicensis*. *J. Mammal.* 60(1):11–15.

Myers, P. 1978. Sexual dimorphism in size of vespertilionid bats. *Amer. Nat.* 112(986):701–11.

———. 1981. Observations on *Pygoderma bilabiatum*. *Z. Säugetierk.* 46:146–51.

Navarro L. D., and D. E. Wilson. 1982. *Vampyrum spectrum*. *Mammal. Species* 184:1–4.

Novick, A. 1977. Acoustic orientation. In *Biology of bats*, vol. 3, ed. W. A. Wimsatt, 74–287. New York: Academic Press.

Novick, A., and B. A. Dale. 1971. Foraging behavior in fishing bats and their insectivorous relatives. *J. Mammal.* 52:817–18.

• Ochoa-G., J. 1980. Lista y comentarios ecológicos de las especies de murciélagos (Mammalia-Chiroptera) en la ciudad de Maracay y el Parque Nacional "Henri Pittier" (Rancho Grande), Aragua Venezuela. Thesis, Universidad Central de Venezuela, Facultad de Agronomía, Instituto de Zoología Agricola, Maracay.

———. 1984a. Presencia de *Nyctinomops aurispinosa* en Venezuela. *Acta Cient. Venez.* 35:147–50.

———. 1984b. Nuevo hallazgo de *Peronymus leucopterus* en Venezuela. *Acta Cient. Venez.* 35:160–61.

• Ochoa-G., J., and C. Ibañez. 1982. Nuevo murciélago del género *Lonchorhina* (Chiroptera: Phyllostomidae). *Memo. Soc. Cienc. Nat. La Salle* 42(118):145–59.

• Ojasti, J., and O. J. Linares. 1971. Adiciones a la fauna de murciélagos de Venezuela con notas sobre las especies del género *Diclidurus*. *Acta Biol. Venez.* 7:421–41.

Owen, J. G., D. J. Schmidly, and W. B. Davis. 1984. A morphometric analysis of three species of *Carollia* from Middle America. *Mammalia* 48:85–93.

Owen, R. D. 1988. Phenetic analysis of the bat subfamily Stenodermatinae (Chiroptera, Phyllostomidae). *J. Mammal.* 69:795–810.

Paradiso, J. L. 1967. A review of the wrinkle-faced bats (*Centurio senex*) with a description of a new subspecies. *Mammalia* 31:595–604.

Peterson, R. L. 1968. A new bat of the genus *Vampyressa* from Guyana, South America, with a brief systematic review of the genus. *Royal Ontario Mus. Life Sci. Contrib.* 73:1–17.

• Peterson, R. L., and J. R. Tamsitt. 1968. A new species of bat of the genus *Sturnira* (family Phyllostomatidae) from northwestern South America. *Royal Ontario Mus. Life Sci. Contrib.* 12:1–8.

Phillips, C. J. 1971. *The dentition of glossophagine bats: Development, morphological characteristics, variation, pathology, and evolution.* Miscellaneous Publications 54. Lawrence: Museum of Natural History, University of Kansas.

Pine, R. H. 1972. *The bats of the genus* Carollia. Technical Monograph 8. College Station: Agricultural Experiment Station, Texas A&M University.

• Pirlot, P. 1965. Chiroptères de l'est du Venezuela et delta de l'Orenoque. *Mammalia* 29:375–89.

• ———. 1967. Nouvelle récolte de chiroptères dans l'ouest du Venezuela. *Mammalia* 31:260–74.

Ralls, K. 1976. Mammals in which females are larger than males. *Quart. Rev. Biol.* 51:245–76.

Sanborn, C. C. 1937. American bats of the subfamily Emballonurinae. *Field Mus. Nat. Hist. Zool. Ser.* 20(24):321–54.

• ———. 1941. Descriptions and records of Neotropical bats. In Papers on mammalogy, published in honor of Wilfred Hudson Osgood. *Field Mus. Nat. Hist. Zool. Ser.* 27:371–87.

Sazima, I. 1976. Observations on the feeding habits of phyllostomatid bats (*Carollia, Anoura,* and *Vampyrops*) in southeastern Brazil. *J. Mammal.* 57:381–82.

Schmidt, U., and U. Manske. 1973. Jugendentwicklung der Vampirfledermause (*Desmodus rotundus*). *Z. Säugetierk.* 38:14–33.

Shump, K. A., Jr., and A. U. Shump. 1982. *Lasiurus cinereus. Mammal. Species* 185:1–5.

Smith, J. D. 1972. *Systematics of the chiropteran family Mormoopidae.* Miscellaneous Publications 56. Lawrence: Museum of Natural History, University of Kansas.

• Smith, J. D., and H. H. Genoways. 1974. Bats of Margarita Island, Venezuela, with zoogeographic comments. *Bull. So. Calif. Acad. Sci.* 73(2):64–79.

Snow, J. L., J. Knox Jones, Jr., and W. D. Webster. 1980. *Centurio senex. Mammal. Species* 138:1–3.

Starrett, A. 1972. *Cyttarops alecto. Mammal. Species* 13:1–2.

Taddei, V. A. 1976. The reproduction of some Phyllostomatidae from the northwestern regions of the state of São Paulo. *Bol. Zool. Univ. São Paulo* 1:313–30.

Tamsitt, J. R., and C. Hauser. 1985. *Sturnira magna. Mammal. Species* 240:1–4.

Tamsitt, J. R., and D. Nagorsen. 1982. *Anoura culturata. Mammal. Species* 179:1–5.

• Tamsitt, J. R., and D. Valdivieso. 1963. Records and observations on Colombian bats. *J. Mammal.* 44:168–80.

• Tate, G. H. H. 1939. The mammals of the Guiana region. *Bull. Amer. Mus. Nat. Hist.* 76(5):151–229.

• ———. 1947. A list of the mammals collected at Rancho Grande, in a montane cloud forest of northern Venezuela. *Zoologica* 32:65–66.

Thomas, M. E., and D. N. McMurray. 1974. Observations on *Sturnira aratathomasi* from Colombia. *J. Mammal.* 55:834–36.

Timm, R. M. 1985. *Artibeus phaeotis. Mammal. Species* 235:1–6.

• Trainer, M., and J. L. Berthier. 1984. Trois chauves-souris nouvelles pour la Guyane française. *Mammalia* 48:303.

Turner, D. C. 1975. *The vampire bat.* Baltimore: Johns Hopkins University Press.

Tuttle, M. D. 1970. Distribution and zoogeography of Peruvian bats with comments on natural history. *Univ. Kansas Sci. Bull.* 49:45–86.

Tuttle, M. D., and M. Ryan. 1981. Bat predation and the evolution of frog vocalizations in the Neotropics. *Science* 214:677–78.

Tuttle, M. D., and D. Stevenson. 1982. Growth and survival of bats. In *Ecology of bats,* ed. T. H. Kunz, 105–50. New York: Plenum.

Tuttle, M. D., L. Taft, and M. Ryan. 1982. Evasive behavior of a frog in response to bat predation. *Anim. Behav.* 30:393–97.

• Valdivieso, D. 1964. La fauna quiróptera del Departamento de Cudinamarca, Colombia. *Rev. Biol. Trop.* 12:19–45.

Vaughan, T. 1959. *Functional morphology of three bats:* Eumops, Myotis, *and* Macrotus. Publications 12. Lawrence: Museum of Natural History, University of Kansas.

Vehrencamp, S. L., F. G. Stiles, and J. W. Bradbury.

1977. Observations on the foraging behavior and avian prey of the Neotropical carnivorous bat, *Vampyrum spectrum*. *J. Mammal.* 158:469–78.

Webster, W. D., and J. Knox Jones, Jr. 1982a. *Artibeus aztecus*. *Mammal. Species* 177:1–3.

———. 1982b. *Artibeus toltecus*. *Mammal. Species* 178:1–3.

• Webster, W. D., and W. B. McGillivray. 1984. Additional records of bats from French Guiana. *Mammalia* 48:463–65.

Webster, W. D., and R. D. Owen. 1984. *Pygoderma bilabiatum*. *Mammal. Species* 220:1–3.

Whitaker, J. O., and J. S. Findley. 1980. Foods eaten by some bats from Costa Rica and Panama. *J. Mammal.* 61:540–44.

• Williams, S. L., and H. H. Genoways. 1980a. Additional records of bats (Mammalia: Chiroptera) from Suriname. *Ann. Carnegie Mus.* 49(15):213–36.

• ———. 1980b. A new species of bat of the genus *Molossops* (Mammalia: Molossidae). *Ann. Carnegie Mus.* 49(25):487–98.

Willig, M. R. 1985. Ecology, reproductive biology, and systematics of *Neoplatymops mattogrossensis* (Chiroptera: Molossidae). *J. Mammal.* 66:618–28.

Willig, M. R., and J. Knox Jones, Jr. 1985. *Neoplatymops mattogrossensis*. *Mammal. Species* 224:1–3.

Wilson, D. E. 1971. The ecology of *Myotis nigricans* on Barro Colorado Island, Panama. *J. Zool.* 163:1–13.

———. 1973a. Reproduction in Neotropical bats. *Period. Biol.* 75:215–17.

———. 1973b. Bat faunas: A trophic comparison. *System. Zool.* 22:14–29.

———. 1978. *Thyroptera discifera*. *Mammal. Species* 104:1–3.

Wilson, D. E. 1979. Reproductive patterns. In *Biology of bats of the New World family Phyllostomatidae, part 3*, ed. R. J. Baker, J. Knox Jones, and D. C. Carter, 317–78. Special Publications of the Museum 16. Lubbock: Texas Tech Press.

Wilson, D. E., and J. S. Findley. 1977. *Thyroptera tricolor*. *Mammal. Species* 71:1–3.

Wilson, D. E., and R. K. LaVal. 1974. *Myotis nigricans*. *Mammal. Species* 39:1–3.

Wimsatt, W. A. 1970a. *Biology of bats*. Vol. 1. New York: Academic Press.

———. 1970b. *Biology of bats*. Vol. 2. New York: Academic Press.

———. 1977. *Biology of bats*. Vol. 3. New York: Academic Press.

6   Order Primates

Diagnosis

If we exclude the fossil Plesiadapiformes, the order Primates comprises a group of species that exhibit rather unspecialized physical characteristics. Except for the extremely terrestrially adapted forms, most have flexible digits and retain five fingers and five toes. Only in some brachiating forms is the thumb lost. The shoulder joint is freely movable, and the radius and ulna remain unfused. The orbits of the skull are directed forward, and a strong postorbital bar separates the eye socket from the temporal fossa. The braincase is relatively large, and in the evolutionary history of this order one can see a trend toward progressive enlargement of the cerebral hemispheres and cerebellum (Hill 1957, 1960).

Distribution

The Recent distribution of this order, exclusive of man (*Homo sapiens*), is in both the Old World and New World tropics except for Australasia. In the Old World some members of the genera *Macaca* and *Rhinopithecus* extend their distributions into the temperate zone, but in the Western Hemisphere distributions are confined more or less between 23° north and 24° south latitude.

History and Classification

The earliest representatives of the order Primates can be distinguished in the Paleocene, when the first fossils appear in North America. By the time of the Eocene, lemurlike primates are recognizable from North America and Europe. In some manner the early prosimians of the Eocene made their way to Asia and Africa, with one or two stocks transit-

The Hominidae are not considered in this book.

ing to Madagascar where they became isolated and underwent an adaptive radiation. The New World primates, the Ceboidea, had an independent origin from the Old World primates and became established on what was then the island continent of South America, where they underwent an extensive adaptive radiation. The earliest ceboid primates have been found in the Oligocene strata of South America. The Cercopithecoidea and the Hominoidea had an Old World origin and underwent an adaptive radiation roughly parallel to the radiations in South America. In the New World tropics there are currently only three families extant, the Callithricidae, the Cebidae, and the recently immigrant family Hominidae (Szalay and Delson 1979). Hill (1957, 1960, 1962) summarizes anatomical data for the New World primates.

Natural History of New World Primates

In the past the New World primates received less attention from naturalists than their Old World cousins. Carpenter (1934) pioneered the study of New World primates with his study of *Alouatta palliata* on Barro Colorado Island, Panama. After a hiatus of some twenty years the behavior and ecology of Neotropical primates began to be explored by Moynihan (1964, 1976), and subsequently a number of long-term studies have focused on primate community ecology (Hladik and Hladik 1969; Hladik et al. 1971; Klein and Klein 1975; Izawa 1976; Mittermeier and van Roosmalen 1981; Terborgh, 1983).

With the exception of *Aotus*, the New World primates are diurnal. All species are highly arboreal, but they exhibit a variety of locomotor adaptations (Erikson 1963). Dietary specializations among New World primates are diverse, ranging from insect and fruit feeding by the tamarins to fruit and foliage feeding by *Alouatta*.

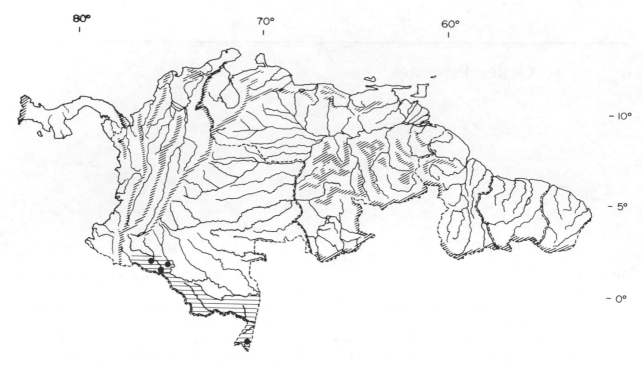

Map 6.1. Distribution of *Cebuella pygmaea*.

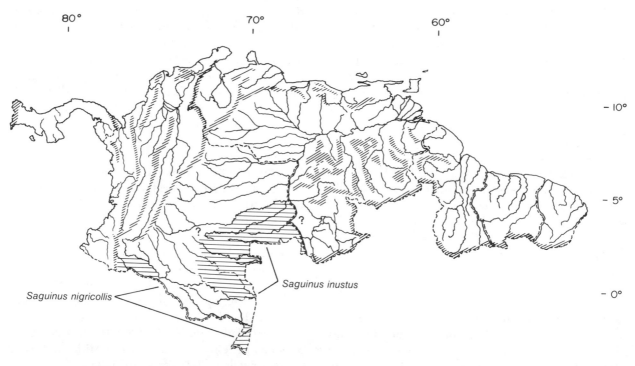

Saguinus nigricollis

Saguinus inustus

Map 6.2. Distribution of two species of *Saguinus* (from Hershkovitz 1977).

Primates typically give birth to a single young, but *Callithrix*, *Saguinus*, and *Cebuella* produce twins. Gestation is long relative to body size, and although the young are furred and have their eyes open at birth, there is a long period of dependency on the parents. Primates are sociable animals, and group sizes vary from nuclear families (four to seven) up to troops of mixed age and sex classes (twenty-four to fifty). The costs and benefits of group foraging have been analyzed by Terborgh (1983).

The communication system of New World primates involves the basic sensory inputs: tactile, visual, gustatory, auditory, and olfactory. The use of chemical signals is very important in the ceboids when contrasted with the cercopithecoids. Specialized glandular areas on the chest or on the genitalia are often used in marking, and chemical marking movements may be highly ritualized (Moynihan 1976; Eisenberg 1977; Epple 1973).

FAMILY CALLITHRICIDAE
Marmosets and Tamarins, Titis

Diagnosis
These small primates are highly adapted for an arboreal life, with nonprehensile tails. Rather than bearing nails at the ends of the digits, they have secondarily acquired clawlike nails. Head and body length ranges from 150 to 370 mm. The tail can be shorter or longer than the head and body. There are five Recent genera, three of which occur in the area delt with in this volume: *Saguinus*, *Cebuella*, and *Callimico*. The dental formula is I 2/2, C 1/1, P 3/3, M 2/2 (see fig. 6.1), with the exception of *Callimico*, which retains the third molar for a total of 36 teeth. On the basis of this characteristic and other skull features, *Callimico* has often been assigned to its own family (Hill 1957; Hershkovitz 1977). The standard reference for the family is Hershkovitz (1977).

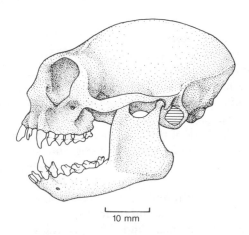

Figure 6.1. Skull of *Saguinus geoffroyi*, female.

Distribution
Species of this family do not occur north of Panama but range south to Paraguay. They are mainly confined to tropical forests and do not reach extremely high elevations.

Natural History
All species of this family are strictly diurnal. They have a varied diet including fruits, invertebrates, and in some cases small vertebrates. The genus *Cebuella*, in common with the southern genus *Callithrix*, is highly adapted for gouging tree bark to supplement its diet with sap. With the exception of the uniparous *Callimico*, all species characteristically give birth to two young. Field data indicate that strong pair-bonds exist between an adult male and an adult female during the rearing of the young. The male shares in the early rearing by transporting the young; the time at which this carrying behavior begins is variable from species to species, being as early as the first day in *Cebuella* and as late as fourteen days of age in some species of *Saguinus*. Although *Callimico* has a single young, it shows the same degree of male parental care. In both rearing patterns and patterns of affiliation there is a strong tendency toward a monogamous system. The animals shelter in tree cavities, in tangles of vines, or under the overhanging crowns of palms (see chapters in Kleiman 1977).

Group size varies from four to seven. Apparently only the adult pair reproduces in any family group (Kleiman 1976). The older offspring may assist the parents in rearing younger infants by carrying and by donating food (Hoage 1982). Communication involves an intricate combination of chemical, acoustic, and visual signals (Oppenheimer 1977). The ecology and behavior of this family have been summarized in a series of articles in Kleiman (1977).

Genus *Cebuella* Gray, 1866
Cebuella pygmaea (Spix, 1823)
Pygmy Marmoset, Titi Pigmeo

Description
In this smallest New World primate, unmistakable owing to its diminutive size, total length rarely exceeds 350 mm, and head and body length is approximately 200 mm. The tail is shorter than the head and body and is nonprehensile. A typical weight is 70 g. There are no conspicuous tufts on the head. The tricolored hairs give a definite salt-and-pepper pattern to the dorsal pelage (plate 6; fig. 6.2).

Range and Habitat
The pygmy marmoset is distributed from southern Colombia through Amazonian Peru and west into Brazil (map 6.1). Throughout its range it co-

exists with several species of the genus *Saguinus*. Its distribution is limited to multistratal tropical evergreen forests at lower elevations. It may be locally abundant, but it is dependent on certain tree species for sap.

Natural History

This strongly arboreal species rarely descends to the ground and typically shelters in tree cavities. The social structure appears to be based on a monogamous pair and their descendants. The male actively participates in rearing the young by carrying them from the day of their birth, transferring them to the female for nursing. Typically twins are born, and their development is relatively rapid. Solid food is taken from twenty to twenty-seven days of age. Earliest age at first conception for a female was eighteen months (Eisenberg 1977).

The animals forage actively under bark and on leaves, searching for various invertebrates. They remove bark from some trees by gouging out small pits with their incisors, then lick the sap at intervals. A tree so used may show hundreds of pits of varying ages that are visited regularly by a pygmy marmoset family (Ramirez, Freese, and Revilla C. 1977). Home ranges may center on such trees. The home range for a troop may be less than a hectare, and densities can reach 5.6 animals per hectare in favorable habitat (Castro and Soini 1977). Developmental studies and captive behavior have been studied by Christen (1974), and vocalizations have been analyzed in detail by Snowden and Cleveland (1980). The birdlike chittering and tweeting sounds of pygmy marmosets span a broad spectrum of frequencies. Parts of their vocal repertoire are out of the range of human hearing.

Genus *Saguinus* Hoffmannsegg, 1807
Tamarin, Pinche, Titi

Description

Approximately eleven species are assigned to this genus, and six occur within the area covered by this volume. These squirrel-sized monkeys exhibit a variety of color patterns, and tufts of hair often ornament their heads and faces. Head and body length rarely exceeds 370 mm, the tail is slightly longer than the head and body, and weight averages about 500 g. The heavily furred tail is nonprehensile. Clawlike nails are present on the digits. The canine teeth are rather large, in contrast to those of *Cebuella* (see figs. 6.1 and 6.2).

Distribution

The genus is distributed from eastern Brazil to Panama, but the distribution is discontinuous since no members of the Callithricidae have been re-

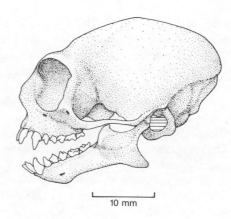

Figure 6.2. Skull of *Cebuella pygmaea*.

corded from east-central Colombia and northern Venezuela. In accordance with their arboreal adaptations, species of *Saguinus* occur only in forested areas. However, some seem to be adapted to second growth and may be found in very scrubby habitats (e.g., *Saguinus oedipus geoffroyi*).

Although *Saguinus* can co-occur with *Callimico* and with *Cebuella*, the pygmy marmoset, all distributions in northern South America appear to be allopatric, *Saguinus oedipus geoffroyi* occurs in Panama and extends into northwestern Colombia. *Saguinus leucopus* has a very limited range in the central valleys of north-central Colombia (see map 6.3). Over a vast part of eastern Colombia and Venezuela, no tamarins occur. The golden-handed tamarin, *Saguinus midas*, extends its range from northeastern Brazil to the Guyanas. Three species of tamarins are found in southeastern Colombia. *Saguinus nigricollis* occurs in southern Colombia with a disjunct distribution. *S. fuscicollis* occupies an intermediate position, effectively covering the range between the Río Putumayo and the Río Guejar. *Saguinus inustus* has a range to the northeast of that of *S. fuscicollis*, occupying the forested zone west of the Orinoco until the forest cover gives way to the grasslands of the llanos (see maps 6.2 and 6.3).

Natural History

The tamarins may be distinguished from the true marmosets (genera *Cebuella* and *Callithrix*) by their long canines and unspecialized incisors (see figs. 6.1 and 6.2). Although some species do feed on exudates, apparently the tamarins do not habitually gouge the bark of trees to obtain sap. Tamarins are generalized frugivore/insectivores and are strictly diurnal. They catch and kill small vertebrates, including lizards and birds. As in other members of the family Callithricidae, the typical social unit consists of a bonded pair and their offspring. Strong participation by the male and older juveniles in care of the young is characteristic. Two northern species

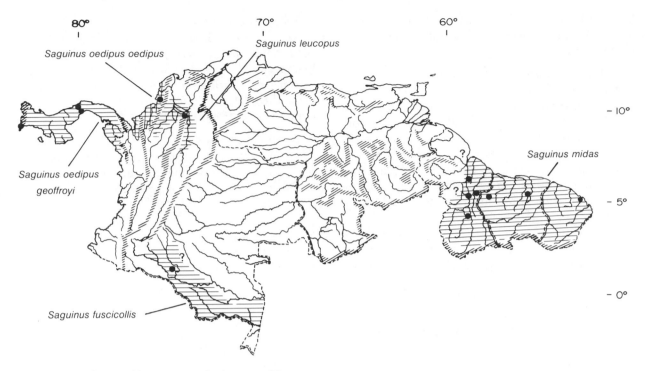

Map 6.3. Distribution of five species and subspecies of *Saguinus*.

have been the subject of long-term field studies, *Saguinus oedipus geoffroyi* by Dawson (1977) and *Saguinus o. oedipus* by Neyman (1977).

Identification of Tamarins

Following Hershkovitz (1977), it is possible to subdivide the tamarins into three groups: the hairy-faced tamarins, where the cheeks and sides of the face are haired; the mottled-faced tamarins, with sparse hair on the cheeks; and the bare-faced tamarins, with a nearly naked face. Within the geographical area under consideration, the hairy-faced tamarins include the white-mouthed tamarins (*Saguinus nigricollis* and *S. fuscicollis*) and the midas tamarins (*S. midas*). As the name implies, the hairy-faced tamarins have hair on their cheeks and chins. *Saguinus midas* may be easily distinguished from all others not only by its extreme eastern geographical distribution, but because the subspecies occurring within the range covered by this volume has reddish hands and feet. The white-mouthed tamarins include two species; *Saguinus fuscicollis* is easily distinguished from *S. nigricollis* by its salt-and-pepper saddle, contrasting strongly with the dark face and reddish arms and legs, whereas *S. nigricollis* has a uniformly colored back with some reddish on the hind legs, although this latter character varies between subspecies.

Within the mottled-faced tamarin group there is a single species, *Saguinus inustus*, which resembles the white-mouthed hairy-faced tamarins but does

not have a sharply contrasting white muzzle, although there is a white patch between the upper lip and the nostrils and in some races the cheeks may be whitish. It also does not have the salt-and-pepper saddle but is uniformly brown.

The bare-faced tamarins include three species, *Saguinus oedipus*, *S. geoffroyi*, and *S. leucopus*, all distributed in northwestern Colombia and Panama. Hershkovitz considers *S. geoffroyi* a subspecies of *Saguinus oedipus*, but there is reason to believe that *Saguinus geoffroyi* is more closely related to *S. leucopus*. *S. oedipus* (in the strict sense) is easily distinguished by its white front and long white crest. *S. geoffroy* has white on the head, but the hairs are very short. The neck of *S. geoffroyi* tends to be reddish, and the back is a salt-and-pepper patterning reminiscent of *S. fuscicollis*. *Saguinus leucopus* has no white on the head and is a uniform brown. In all three species the forearms and hands are white or pale brown (see plate 6).

Saguinus fuscicollis (Spix, 1823)
Saddle-backed Tamarin

Description

This average-sized tamarin is characterized by a white muzzle and a conspicuous black-and-white (salt-and-pepper) saddle on the middorsum (plate 6).

Range and Habitat

The species is distributed in the extreme southern part of Colombia and from there south into

Amazonian Peru and adjacent portions of Brazil (map 6.3). It is confined to multistratal tropical evergreen forests.

Natural History

The field data indicate that this marmoset lives in small groups, probably based on a nuclear family. At certain times of the year several families may congregate in fruiting trees, then separate into their original groupings. As one of the best-studied species in captivity, it has been the subject of numerous behavioral investigations by Epple (1973, 1975). The gestation period is approximately 140 days; the male begins to carry the young within ten days of their birth. The feeding ecology has been analyzed by Terborgh (1983). Insects and fruits form the important components of the diet, in common with other species of *Saguinus*.

In Peru *S. fuscicollis* forms mixed-species foraging groups with *S. imperator*. These associations are long term and involve subtle differences in feeding patterns, such as in vertical distribution while foraging. Apparently the mixed-species troop can more effectively defend the foraging area. Foraging together by sympatric species of *Saguinus* may be much more common than previously thought in areas where more than one species of *Saguinus* occurs (Terborgh 1983).

Saguinus inustus Schwartz, 1951

Description

Saguinus inustus is approximately the same size as *S. fuscicollis*. It does not have a distinctive white muzzle or the contrasting saddle and thereby is easily distinguished from the other two hairy-faced tamarins (plate 6).

Range and Habitat

The species occurs from southeastern Colombia into adjacent parts of northwestern Brazil (map 6.2).

Natural History

There are few natural history data concerning this species. Presumably its reproductive behavior and social structure conform to the pattern described for *Saguinus fuscicollis*.

Saguinus nigricollis (Spix, 1823)
Black-and-Red Tamarin

Description

Saguinus nigricollis is approximately the same size as *S. fuscicollis* but is distinguishable by the absence of the salt-and-pepper saddle on the dorsum. The muzzle is white, as in *S. fuscicollis* (plate 6).

Range and Habitat

This species is distributed in northwestern Brazil but extends into the extreme southeastern tip of Colombia and also crosses the Rio Putumayo into the extreme west of the southwestern portion of Colombia to the east of the Andes (map 6.2). It is confined to multistratal tropical evergreen forests.

Natural History

In captivity *Saguinus nigricollis* is very similar to *S. fuscicollis* in reproduction and behavior. There have been no long-term field studies of this species (Izawa 1976).

Saguinus midas (Linnaeus, 1758)
Red-handed Tamarin

Description

This tamarin is easily distinguishable within the range covered by this book by its characteristic red hands (plate 6).

Range and Habitat

The species is broadly distributed from Guyana east of the Río Essequibo through French Guiana and south into Brazil. It is broadly tolerant of both multistratal tropical evergreen forests and second-growth forests (map 6.3).

Natural History

This species has received attention in the recent publication by Mittermeier and van Roosmalen (1981). The red-handed tamarin occurs from lower montane rain forests to lowland rain forests. It utilizes edge habitats, especially where forest and savanna intersperse. It does not typically frequent the high emergent canopy, but rather forages in the lower crowns or the understory, eating both fruits and seeds. Seeds compose a large part of its diet, supplemented with insects. In its grouping tendencies and reproductive habits it does not appear to deviate from the typical tamarin pattern.

Saguinus geoffroyi (*S. oedipus geoffroyi*) (Reichenbach, 1862)
Geoffroy's Tamarin

Description

The color pattern is distinctive, with short whitish hairs on top of the head, reddish fur on the nape, salt-and-pepper pattern on the back, and white forearms and underparts (plate 6).

Range and Habitat

This is the most northern-ranging of the tamarins and formerly occurred throughout much of Panama, extending into the extreme northwest of Colombia.

It has been exterminated over much of its former range. In northern Colombia it is replaced by *S. o. oedipus* to the east of the first cordillera of the Andes (map 6.3). Broadly tolerant of second-growth habitats, it appears to occur in low numbers in mature multistratal tropical evergreen forests.

Natural History

This species was studied in the field by Dawson (1977). Its feeding habits and general ecology were described earlier by Hladik and Hladik (1969), and the behavior and vocal repertoire were described by Moynihan (1970) and by Muckenhirn (1967). In Panama *Saguinus geoffroyi* prefers second-growth habitats. It is strictly diurnal, foraging in midmorning and late afternoon. Its diet consists of 60% fruit, 30% insects, and 10% green plant material. Groups appear to use a stable home range, and the central focus of the social structure seems to be a monogamous pair. Neighboring groups may come together and share feeding trees. As young adults mature they may become peripheral and transfer from one group to another, especially if a vacancy occurs through the death of either member of the adult pair. In second-growth habitats of Panama, Geoffroy's tamarin does not typically use tree cavities for shelter but may shelter in tangles of vines. Home ranges average about 26 ha. Troop size averages about seven, which presumably includes the adult pair and two or three sets of offspring. Under favorable habitat conditions the animals can reach a density of twenty-three per km^2.

Saguinus oedipus (S. o. oedipus) (Linnaeus, 1758)
Cotton-topped Tamarin

Description

This average-sized tamarin is immediately recognizable by the long white hairs creating a crest on top of its head. It cannot be confused with any other species of tamarin (plate 6).

Range and Habitat

The species is confined to lowland tropical evergreen forests in north-central Colombia. Its current range has been vastly reduced through land clearing (map 6.3).

Natural History

This species has been studied extensively in captivity (Muckenhirn 1967) and has been studied in the field for several years by Neymen (1977). In its behavior, feeding habits, and social structure it does not differ from other species of tamarins. Typically feeding on fruits, insects, and small vertebrates, it

utilizes a home range of approximately 8–10 ha. Troop size is variable from three to thirteen. In the habitat where Neymen worked transfers of individuals from one troop to another were not uncommon, and she suggests that pair-bonds between an adult male and adult female may last only through a rearing cycle.

Saguinus leucopus (Gunther, 1877)

Description

This species is easily distinguished from *S. oedipus* by its lack of a long white crest on the head. No other tamarin occurs within its range (plate 6).

Range and Habitat

At present the species is confined to restricted forest patches in the Magdalena River valley of north-central Colombia. The range has been vastly reduced through land clearing; at one time this species was a typical denizen of the north-central lowland rain forests of Colombia (map 6.3).

Natural History

Green (1978) conducted field observations on this species. Troop size averaged about five, and in the forest patches where he worked density could approximate fifteen per km^2. From Green's observations, the behavior and social tendencies do not differ from the typical tamarin pattern.

Genus *Callimico* Miranda-Ribeiro, 1911
Callimico goeldii (Thomas, 1904)
Goeldi's Marmoset, Calimico

Description

This species differs from all other members of the family Callithricidae in possessing three molars. In the region under consideration it is the only all-black tamarin. The hair on its head is rather long, approximating a crest, but does not contrast with the basic black body color (plate 6).

Range and Habitat

This marmoset is widely distributed in the upper Amazon basin, but in the northern Neotropics it occurs only in the extreme southern part of Colombia (map 6.4).

Natural History

Goeldi's marmoset has been studied extensively in captivity by Heltne, Turner, and Wolhandler (1973) and in the field by Pook (Heltne, Wojeik, and Pook 1981; Pook and Pook 1981). It appears that *Callimico* prefers low-stature strata. It tends to forage in the understory of multistratal tropical evergreen forests, eating small fruits, berries, and insects. Group

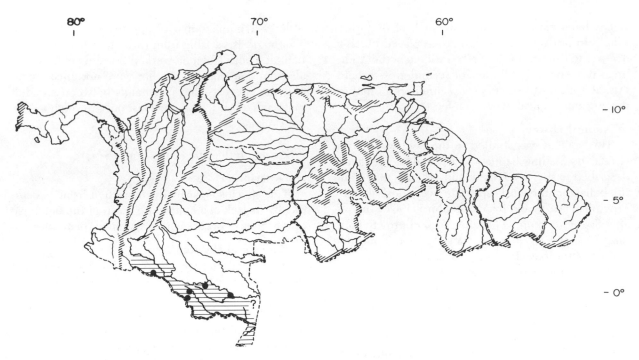

Map 6.4. Distribution of *Callimico goeldii*.

size can be up to eight. The female typically has one young, in marked contrast to other members of the family Callithricidae. The female transports the infant for the first two weeks of life, and thereafter the male participates. Gestation lies between 153 and 159 days. This species is never very abundant. Pook estimated a density of 0.25 groups per km². Home range for the group of eight studied by Pook was nearly 60 ha. Activity patterns indicate a range from crepuscular to diurnal. *Callimico* can occur not only with the pygmy marmoset but with other species of tamarins. While foraging, *Callimico* is occasionally seen in association with *S. fuscicollis*, following behind at a considerable distance.

FAMILY CEBIDAE

Diagnosis
The dental formula is I 1/1, C 2/2, P 3/3, M 3/3 (see fig. 6.3). The posterior molar may be missing in a small proportion of individuals within a larger population (15% for *Ateles*). There are eleven Recent genera, including some twenty-nine species. Five genera (*Cebus*, *Ateles*, *Alouatta*, *Brachyteles*, and *Lagothrix*) have prehensile tails. All species are small to medium in size.

Distribution
The species of this family occur from southern Veracruz, Mexico, to the gallery forests on the upper reaches of the Río Paraná in Argentina. All are adapted to multistratal tropical evergreen forests, but the genera *Alouatta* and *Cebus* have adapted to the semideciduous forests of the more xeric portions of northern and southern South America.

Natural History
All members of this family are diurnal except *Aotus*, which is the only nocturnal New World primate. Most species are adapted to a diet of fruit, seeds, small vertebrates, and invertebrates. *Alouatta* is exceptional in that it generally includes 50%

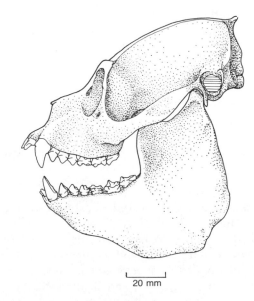

Figure 6.3. Skull of *Alouatta seniculus*.

foliage in its diet. The species of *Aotus*, *Callicebus*, and *Pithecia* show monogamous tendencies. All females of this family bear a single young.

Genus *Aotus* Illiger, 1811
Night Monkey, Mono de la Noche

Taxonomy

During the past thirty years most workers have recognized the single species *Aotus trivirgatus*, but recently cytogenetic research has indicated that the genus *Aotus* is divisible into populations having different chromosomal morphology (deBoer 1974). Up to nine karyotypes have been described, but it is unknown if hybridization is impossible among them all. Natural hybrids have been found in Colombia (Yunis, Torres de Caballero, and Ramirez 1977). Following Hershkovitz (1983), four species are recognized in the north.

Description

These small ceboid primates have nonprehensile tails. Head and body length ranges from 240 to 375 mm and tail length from 220 to 418 mm. Weight averages 700 g (Wright 1981). The dorsal color varies depending on the species, ranging from gray to red brown; the venter can be bright orange to white. There is usually a half-moon-shaped white patch above each eye, and black stripes are generally evident on the head, a median stripe and two lateral ones that come together at the top (plate 6).

All species of *Aotus* in northern South America belong to the gray-neck group. The pelage of the sides of the neck is grayish agouti to brownish agouti, corresponding to the color of the sides. The interscapular area may have a group of long hairs approximating a small crest. The fur of the pectoral area often separates along a distinct midventral line, where the chest gland is situated (Hershkovitz 1983).

Distribution

Aotus ranges from Panama to northern Argentina, but within this latitudinal range there are areas where it is completely absent. It is noticeably absent from the llanos of Venezuela and does not appear to be present in the Guyanas. The night monkey occurs in a variety of forest types and can occur at up to 3,200 m elevation in Colombia (map 6.5).

Natural History

Night monkeys are primarily frugivorous but take some insects. *Aotus* generally lives in small family groups based on a monogamous pair-bond. Apparently several family groups can share a large fruiting tree while foraging. In suitable old-growth habitat the animals may shelter in tree holes. In good habitats they may exist at densities of forty per km^2 (Wright 1981).

In Panama they are strictly nocturnal, generally emerging from their tree holes shortly after dark and returning at dawn. In northern Argentina they may be more diurnal. Home range for a family group appears to be rather small, perhaps less than 4 ha. The male transports the infant briefly during its first two weeks of life, but thereafter he carries it most of the time. The behavior patterns of *Aotus* have been described by Moynihan (1964) and Wright (1981).

When moving through the forest at night the animals produce "gulp" sounds, usually in bursts of two to five, perhaps as a contact note. Their alarm call, which has been termed a sneeze-grunt, sounds like a metallic click, usually produced in bursts of two or three, and serves as a warning signal (Moynihan 1964).

Aotus lemurinus

Description

The interscapular crest is absent. The chest is orange or buffy, but the color does not extend to the inner sides of the limbs.
Chromosome number: $2n = 55/56$.

Range and Habitat

The species ranges from Panama to Colombia west of the eastern cordillera of the Andes. In Venezuela it is confined to the Maracaibo basin (map 6.5).

Aotus brumbacki

Description

The interscapular area has a slight crest. The chest gland is long, and pectoral hairs separate into two groups on either side of the gland.
Chromosome number: $2n = 50$.

Range and Habitat

The species is confined to the tributaries and drainage basin of the Río Meta (map 6.5).

Aotus vociferans

Description

The interscapular hairs are organized in a whorl, distinguishing this species from *A. brumbacki*.
Chromosome number: $2n = 46$.

Range and Habitat

The species ranges east of the Andes in southern Colombia and in adjacent portions of Amazonian Brazil and Ecuador (map 6.5).

Aotus trivirgatus

Description

Aotus trivirgatus is similar to *A. lemurinus* in lacking the interscapular crest or whorl. It differs

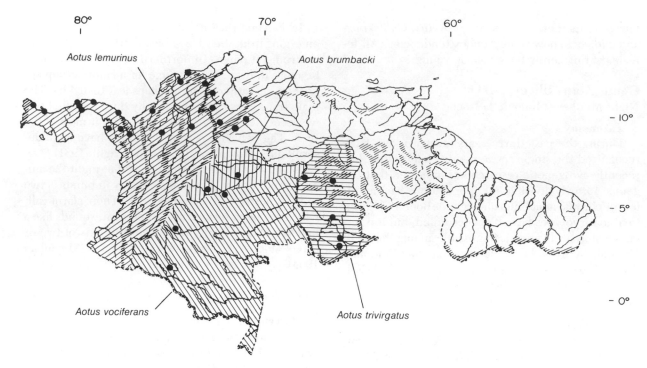

Map 6.5. Distribution of four species of *Aotus* (from Hershkovitz 1983).

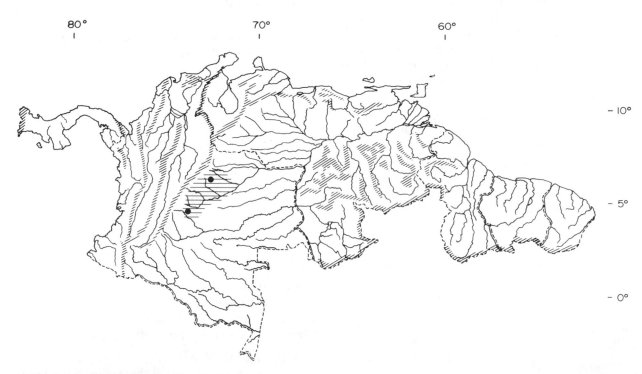

Map 6.6. Distribution of *Callicebus moloch*.

from *A. lemurinus* in that the orange or buff ventral color extends to the inner sides of the limbs.

Range and Habitat
The species ranges in Amazonian Venezuela and adjacent portions of Brazil (map 6.5).

Genus *Callicebus* Thomas, 1903
Titi Monkey, Sahui de Collar, Titi Zocayo

Description
This genus includes three species, two of which occur in the northern Neotropics. Titi monkeys are small, weighing less than a kilogram, with a head and body length of 287 to 390 mm. The tail exceeds the head and body length by about 20% (see plate 6).

Distribution
The genus is distributed from Colombia and Venezuela south over much of Brazil. The southern species, *C. personatus*, is allopatric with *C. torquatus* in Brazil, but in the upper Amazon basin *C. torquatus* can co-occur with the third species, *C. moloch*. In areas of macrosympatry the latter two species are usually segregated into different habitat types (Kinzey 1981).

Natural History
Titi monkeys typically travel in small groups, and social organization seems to be based on a bonded pair and their various-aged offspring. The ecology of the two northern species is quite different and will be discussed under the species accounts. Extensive captive studies have been carried out with *C. moloch* by Mason (1971). Field studies on *C. moloch* were done by Mason (1968) and Robinson (1977a,b,c, 1982). *C. torquatus* has been studied in the field by Defler (1983b) and by Kinzey and his associates (Kinzey 1977, 1981; Kinzey and Gentry 1978).

Callicebus rarely comes to the ground. *C. moloch* appears to forage in the understory, whereas *C. torquatus* seems to be more specialized for feeding in the crowns. A single young is born after a gestation period of approximately 160 days.

Callicebus moloch (Hoffmannsegg, 1807)
Dusky Titi Monkey, Orabazu

Description
This species is the smallest of the genus *Callicebus*. Mean head and body length for *C. moloch* is 334 mm, with the tail 429 mm long. There is virtually no sexual dimorphism in this species. It is reddish brown, without the distinctive facial striping so characteristic of *C. torquatus* (plate 6).

Range and Habitat
C. moloch is broadly distributed within the Amazonian portions of Brazil and Peru, but in the north it has a discontinuous distribution in Colombia east of the eastern cordillera of the Andes. In the more southern parts of its distribution, *C. moloch* is strongly associated with riverine habitats (map 6.6).

Natural History
Although *C. moloch* is basically a frugivore, at some times of year it may eat young leaves. The basic unit of social organization is the pair and various-aged offspring. The animals are strictly diurnal in their activity patterns and utilize a rather small home range; 3–5 ha seems average. They seek out a tangle of vines in a preferred sleeping tree when retiring at night. This species has a rather rich vocal repertoire. A bonded male and female will produce a duet calling sequence that has several elements, including bellows and pants. Countercalling between adjacent social groups is involved in territorial defense (Robinson 1977, 1979a,b).

Callicebus torquatus (Hofmannsegg, 1807)
Sahui de Collar

Description
This is the largest species within the genus, with an average head and body length of approximately 339 mm and a tail length of 460 mm. The pelage is dark brown to black, with white facial markings. In some parts of its range there may be a tendency toward a reduction in melanin deposits, resulting in a buff color, but this color variation is rare. Typically, the animals are dark and easily distinguished from *C. moloch* (plate 6).

Range and Habitat
This species is distributed in the forested parts of the Orinoco and Amazon drainages. *C. torquatus* is associated with tall, mature forests, generally in well-drained ground. Kinzey and Gentry (1978) noted that it often frequents forests adapted for white sand soils, which occur in nutrient-deficient areas. When in association with white sand soils and the so-called black-water forests, *C. torquatus* may stay near streams (map 6.7).

Natural History
This apparently monogamous species travels in small groups (Defler 1983b). Home ranges and day ranging seems to be greater than shown by *C. moloch*, perhaps because *C. torquatus* is adapted to black-water forests with a reduced carrying capacity. Long-term studies on *C. torquatus* indicate that a family's home range may approximate 29 ha and is vigorously defended against neighboring groups. *C. torquatus* eats only a small quantity of leaves; its diet is almost entirely fruit and seeds. The natural history of this species has been described in detail by Kinzey (1981) and Kinzey and Gentry (1978)

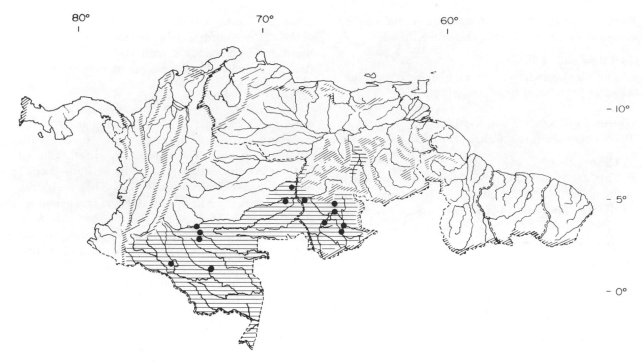

Map 6.7. Distribution of *Callicebus torquatus*.

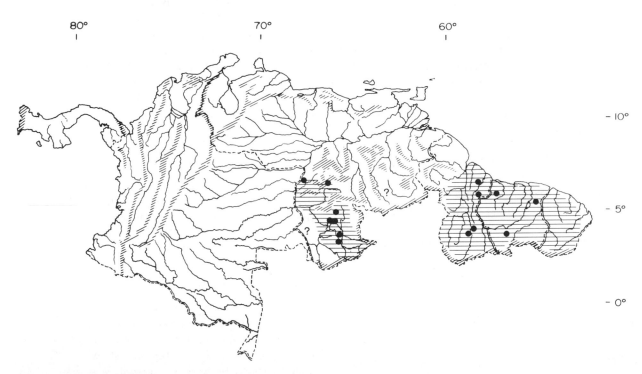

Map 6.8. Distribution of *Chiropotes satanas* (from Hershkovitz 1985).

Genus *Chiropotes* Lesson, 1840
Chiropotes satanas (Hoffmannsegg, 1807)
Bearded Saki, Cuxius

Description

Two species are recognized, but only one occurs in the northern Neotropics, *Chiropotes satanas* (Hershkovitz 1985). This medium-sized primate weighs from 2.7 to 3.2 kg. Head and body length ranges from 327 to 480 mm, with a tail length of 370 to 436. Males may slightly exceed females in size, but there is no conspicuous color dimorphism. Although the shoulders may be reddish, the coat is basically black, and there are no conspicuous white markings. The tail is heavily furred and nonprehensile. The hair on the male's head is organized into two rather rounded, dense masses overlying the enlarged temporal muscles, and there is a conspicuous beard. The female has a small beard and much less development of the temporal swellings on the head (plate 7).

Range and Habitat

The bearded saki is found from the Guyanas and southern Venezuela south on both sides of the Amazon to the mouth of the river. In the northern Neotropics as defined by this volume, the species occurs only in southern Venezuela and the Guyanas (map 6.8). The bearded saki generally frequents only mature rain forests.

Natural History

Only two recent field studies have been carried out (van Roosmalen, Mittermeier, and Milton 1981; Ayres 1981). *Chiropotes* tends to forage in the upper strata of mature rain forests. Typically this species is found in large bands of up to twenty-five. Several males are present in a band at any one time, and there is no tendency toward monogamy. In Suriname the diet consists primarily of fruits and seeds; van Roosmalen noted that the animals feed in no small measure on young seeds of unripe fruits from a wide variety of plant families; thus they are true seed predators. The feeding trees are widely dispersed, and a troop's daily movements may exceed 3 km. One troop in Suriname had a home range nearly 100 ha in extent. A single young is born after a gestation period of approximately five months (Hick 1973). Detailed analyses of communication patterns and vocalizations have yet to be performed for this species.

Genus *Cacajao* Lesson, 1840
Cacajao melanocephalus (Humboldt, 1812)
Black Uakari, Uakari Negro

Description

This genus is composed of two species, but only the black uakari, *C. melanocephalus*, occurs in the northern Neotropics. These are medium-sized primates, with adult males weighing 3.5 to 4.1 kg. Females are slightly smaller, weighing about 500 g less than males. The uakari cannot be mistaken for any other New World primate because it has an extremely short, nonprehensile tail, rarely exceeding one-third of the head and body length (plate 7). Whereas *C. calvus* has a red to pale yellow coat, the coat of *C. melanocephalus* tends to be black; however, there may be varying degrees of light brown on the hind limbs, tail, and lower back.

Range and Habitat

This species is confined to the Amazon basin and penetrates into the Amazonian portions of southern Venezuela and Colombia (map 6.9). In common with *Callicebus torquatus*, it appears that the genus *Cacajao* is adapted to black-water river systems.

Natural History

The limited field data suggest that *Cacajao* feeds on both leaves and fruits and also takes some invertebrates in the course of foraging. Most data concerning the foraging behavior of *Cacajao calvus* have been obtained from captive studies in a seminatural environment (Fontaine 1981). *C. melanocephalus* lives in rather large groups of fifteen to twenty-five animals; more than one adult-sized male may be present in a troop. The animals are strictly diurnal. The communication patterns of *C. calvus* have been analyzed for captives in a seminatural environment by Fontaine (1981), who noticed antiphonal vocalization bouts between members of the same group. A single young is born at a time, and the gestation period is imperfectly known, but is probably five and one-half months.

Genus *Pithecia* Desmarest, 1804
Saki Monkey, Parauacu

Description

According to the revision by Hershkovitz (1979) there are four species, two of which occur in northern South America. These medium-sized primates have rather long, thick hair and very well haired, nonprehensile tails. Head and body length averages 300 to 480 mm, with a tail length of 255 to 545 mm. Weight ranges from 1.4 to 2.25 kg. Size dimorphism is negligible, but in some species the sexes are strongly dimorphic in color pattern.

Distribution

The genus has a rather wide distribution in the Amazon basin.

Natural History

Saki monkeys are difficult to maintain in captivity and have received little attention from field-

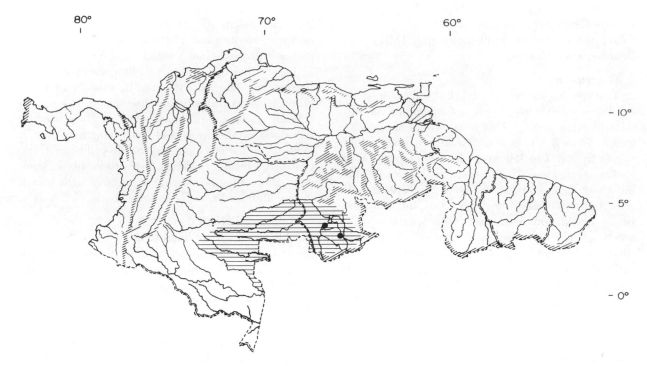

Map 6.9. Distribution of *Cacajao melanocephalus*.

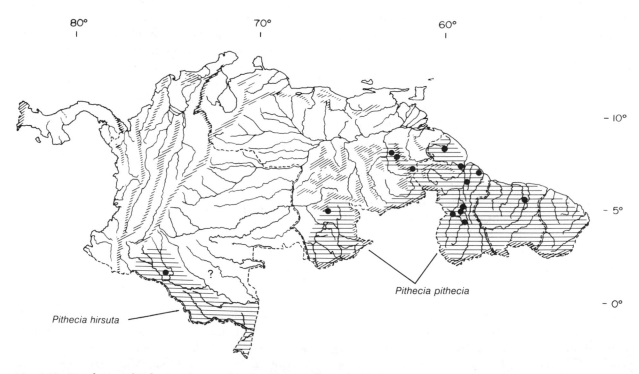

Map 6.10. Distribution of *Pithecia pithecia* and *Pithecia hirsuta* (from Hershkovitz 1979).

246

workers. The most recent summary of their natural history is contained in Buchanan, Mittermeier, and van Roosmalen (1981). Further remarks on natural history are included in the species accounts.

Pithecia pithecia (Linnaeus, 1766)
White-faced Saki

Description
In this species the sexes are strongly color dimorphic. The adult male is black except for a white face and throat. The female is basically an agouti brown on the dorsum with lighter underparts. White stripes extend from below the eye to the corner of the mouth. The female of *P. pithecia* is very similar to both the male and female of *P. hirsuta*, but the distinctive color of the male immediately identifies this species (plate 7).

Range and Habitat
The white-faced saki has its main center of distribution north of the Amazon and west through the Guyanas to Venezuela. It occurs sporadically in the state of Bolívar, Venezuela, but is broadly distributed through the Amazon portion of Venezuela. It is confined to mature multistratal tropical evergreen forests (map 6.10).

Natural History
The standard references for this species are Buchanan (1978) and Buchanan, Mittermeier, and van Roosmalen (1981). The animal is found from lowland forests to forests of moderate elevation. It rarely, if ever, enters seasonally flooded riverine forests. In Suriname it tends to forage in the understory or lower part of the canopy. It is primarily frugivorous but eats some leaves on occasion. Although this species will take insects in captivity, there are no precise data on insect consumption in the field. The saki tends to live in small family groups, apparently based on a monogamous pair. In adequate habitat population densities may reach seven per km^2. Day ranging has not been studied to any great extent; Buchanan estimates home ranges at 4–10 ha. A single young is born after a gestation of 163 to 176 days. Vocalizations have received a preliminary analysis by Buchanan (1978). Long, loud calls to promote group spacing are modified in form but are similar to those of the howler monkey though lower in amplitude.

Pithecia hirsuta Spix, 1823

Description
Hershkovitz (1979) separated this species from *P. monachus*. *P. hirsuta* is dark agouti brown with pale hands and feet, and its underparts and beard are blackish. The sexes are not strongly dimorphic in color or in size (plate 7).

Range and Habitat
This species occurs in the western Amazon basin, and in the area covered by this volume its distribution is confined to the extreme southern portions of Colombia. It typically prefers multistratal tropical evergreen forests (map 6.10).

Natural History
The natural history of this species is not as well known as that of *P. pithecia*. The field observations we have suggest that it lives in small family groups, feeding in the middle regions of the multistratal forest. It appears to be frugivorous, diurnal, and quite similar to *P. pithecia* in its general behavior patterns. No evidence of territorial defense has been noted (Happel 1982).

Genus *Cebus* Erxleben, 1777
Capuchin, Sapajou, Capuchino

Taxonomy
The species *Cebus olivaceus*, *C. albifrons*, and *C. capucinus* are all similar in body size, though they differ in color pattern. Over most of their range these three species are allopatric, but *Cebus albifrons* and *C. olivaceus* occur in sympatry in part of the Amazonian portion of Venezuela and again in Venezuela south of the eastern cordillera (see maps 6.12 and 6.13). This condition of near allopatry suggests that the three species are very similar in their mode of habitat exploitation; where they co-occur, they could possibly either be in the process of interbreeding or in strong competition. The question remains unclear. On the other hand, *Cebus apella* co-occurs over much of its range with either *C. olivaceus* or *C. albifrons*. The feeding ecology of *C. apella* reduces competition with the other species (see Terborgh 1983).

Description
These medium-sized primates have semiprehensile tails. Head and body length is 305 to 565 mm, the tail 300 to 560 mm. The adult male is larger and more robust than the adult female. Color patterns are variable (see species accounts and plate 7).

Distribution
There are four recognized species, and all occur in the northern Neotropics. *Cebus apella* is distributed from northern South America to northern Argentina. This species is sympatric with *C. olivaceus* and *C. albifrons* over many parts of its range. *Cebus capucinus* is confined to northwestern Colombia and then is distributed north to southern

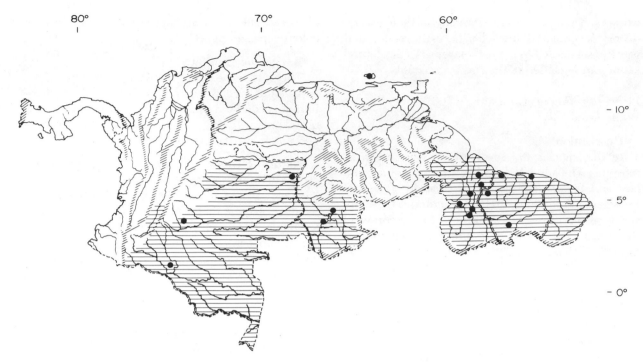

Map 6.11. Distribution of *Cebus apella*.

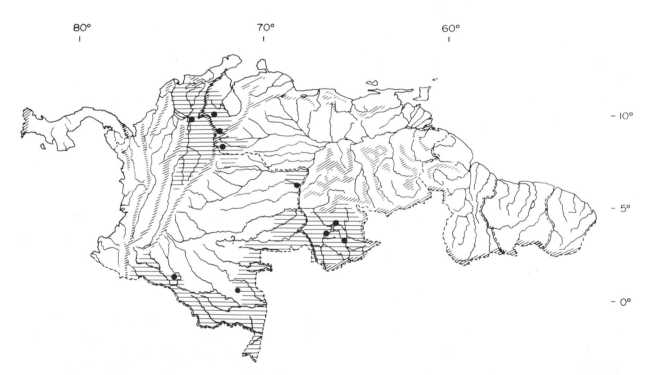

Map 6.12. Distribution of *Cebus albifrons*.

Nicaragua. Species of this genus have been able to penetrate and utilize dry deciduous forests. They appear to have an extremely broad tolerance for different habitat types and elevations.

Natural History

Capuchin monkeys typically live in moderate-sized bands, usually with a single adult male dominant over all younger males in the troop. They eat a variety of fruits and seeds and a modest amount of both vertebrates and invertebrates. In habitats supporting dry deciduous forests the troop can forage extensively on the ground, turning the leaf litter to find small vertebrates and invertebrates. Species of the genus are most versatile in their use of habitat and range of diet. Capuchin monkeys are often found with the squirrel monkey (*Saimiri*). It appears that *Cebus* is the focal species, with the squirrel monkey moving along in association. It is speculated that this association confers some antipredator benefit on the squirrel monkey but is of no special benefit to *Cebus*.

Cebus apella (Linnaeus, 1758)
Brown Capuchin

Description

This robust species averages about 2.5 kg in weight; males tend to be some 800 g heavier than females. The cap on top of the head is composed of short, dark, erect hairs that in males form ridges on either side of the crown, and it contrasts sharply with the light brown body (plate 7).

Range and Habitat

Cebus apella is broadly distributed from southern Colombia, Venezuela, and the Guyanas to northern Argentina. It occurs on Margarita Island, but this translocation from the range suggests it was originally introduced by Amerindians (map 6.11). It is broadly tolerant of a variety of forest types, from semideciduous to multistratal tropical rain forests.

Natural History

The diet includes fruit, palm nuts, seeds, and a considerable quantity of insects (Terborgh 1983). Group size ranges from five to forty, and population density varies from six to thirty-five per km². This species is strictly diurnal and, depending on carrying capacity, can show a home range of 25–40 ha. A single young is born after a gestation of approximately 160 days. Young males apparently do not become sexually mature until at least seven years of age. Females may conceive in their fourth year. The mating system is polygamous. Generall a single adult male is dominant over all other males in the troop. Communication patterns in this species have been described for captives. The standard review for the behavior of the genus is in Freese and Oppenheimer (1981).

Cebus albifrons (Humboldt, 1812)
Brown Pale-fronted Capuchin

Description

The weight averages 2.3 kg. The color pattern is brown dorsum with white circling the facial region and white on the upper forearms; the cap tends to be darker brown (plate 7).

Range and Habitat

The species occurs from southern Colombia and southern Venezuela south through the Amazon basin to northern Bolivia (map 6.12). It occupies a wide range of forest types, including dry deciduous tropical forests and multistratal evergreen forests.

Natural History

Like other species of this genus *Cebus albifrons* is omnivorous, taking fruit, seeds, vertebrates, and invertebrates. Group sizes range from seven to thirty, with fifteen as a mode. Generally there is only one adult male per troop, but instances of an adult male with a satellite subadult have been recorded. Densities in good habitat range from twenty-four to forty-five per km². The animals are diurnally active and generally have a rather large home range: 60–70 ha has been recorded. The foraging ecology has been analyzed by Terborgh (1983). In breeding biology, development, and vocalizations this species is similar to *C. capucinus*.

Cebus capucinus (Linnaeus, 1758)
White-throated Capuchin, Carita Blanca

Description

This capuchin is similar in size to *C. albifrons*, but its color is distinctive. The dorsum and hindquarters are black, the chest and forearms are a sharply contrasting white, and the pale face is surrounded by white with a black cap. This species is easily discriminated from all others (plate 7).

Range and Habitat

The species occurs from northwestern Colombia north to Honduras (map 6.13). It may be found in multistratal tropical evergreen forests to dry deciduous forests.

Natural History

The standard reference for the behavior and ecology of this species is Oppenheimer (1968). This diurnal primate has a broad diet, eating an enormous variety of fruits, flowers, and invertebrates. It forages over the entire vertical range of the forest, including the ground, and exhibits a versatility unmatched by any other New World primate. Group sizes average about fifteen. Usually only one adult male is present in a troop. They can exist at population densities of five to fifty per km². Home ranges

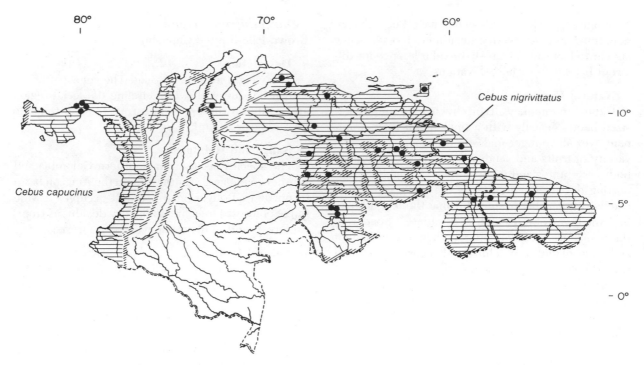

Map 6.13. Distribution of two species of *Cebus*.

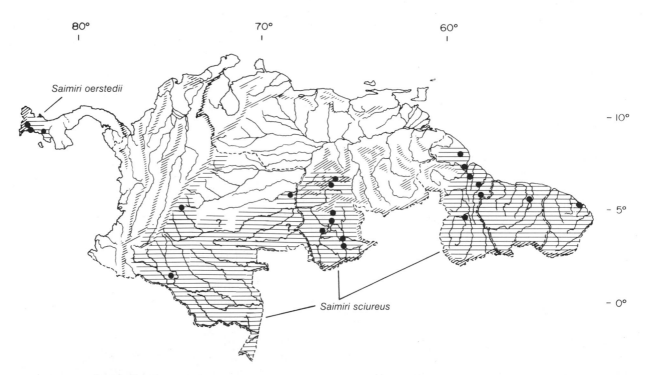

Map 6.14. Distribution of two species of *Saimiri*.

vary from 32 to 80 ha. Reproduction is similar to that described for *C. apella*. Communication patterns have been summarized by Freese and Oppenheimer (1981).

Cebus olivaceus (= *nigrivittatus*) Schomburgk, 1848
Weeper Capuchin, Cai Capuchino, Mico Común (Venezuela)

Description
This species is similar to *C. albifrons* in size and color: light brown on the dorsum, pale buff on the forearms and the hair around the face; but where *albifrons* is white, *C. olivaceus* tends to be buff (plate 7).

Range and Habitat
C. olivaceus is distributed in northern South America from northwestern Venezuela across through the Guyanas south to the Amazon (map 6.13).

Natural History
Diurnally active, this species is an omnivore that utilizes both the forest floor and various levels of the forest for feeding. Troops generally include a single adult male who is responsible for most of the breeding. Group foraging has been analyzed in detail by Robinson (1981, 1986). While foraging, the dominant male and dominant adult females tend to be at the center. Juvenile males tend to be at the front of the troop or toward the rear. A strict hierarchy among females and among males determines priority of access to preferred food items. Troop size ranges from ten to thirty-three. In the dry deciduous forests of northern Venezuela the home range may exceed 100 ha. In such a habitat home ranges of neighboring troops overlap considerably, and no territorial behavior is demonstrable. In reproduction, development, and age at sexual maturation this species is similar to *Cebus apella*.

Genus *Saimiri* Voigt, 1831
Squirrel Monkey, Titi (Venezuela), Mico de Chiero

Description
Authorities disagree on the number of species the genus *Saimiri* contains. Until the problem is resolved I will recognize two species, *S. sciureus* and *S. oerstedii*, which are allopatric. These small primates have long, nonprehensile tails. Head and body length of adult females averages about 280 mm, while that of males averages 300 mm; the tail is 350 to 430 mm. Adult females weigh from 500 to 750 g, while males range from 700 to 1,100 g. Color patterns are somewhat variable, but in general most of the body is a greenish yellow. The throat, face, and ears are usually white, and the muzzle is black, contrasting sharply. The venter and undersides of the limbs may be white or light yellow. Usually the tip of the tail is black (plate 7).

Distribution
S. oerstedii is found in Costa Rica and Panama. *S. sciureus* is found in the Amazonian portions of Brazil, Colombia, Ecuador, Peru, Bolivia, and Venezuela and in the Guyanas (map 6.14).

Saimiri oerstedii Reinhardt, 1872

Description
See the description for the genus.

Range and Habitat
The species is confined to Costa Rica and Panama. It is broadly tolerant of a variety of habitats and may be found in multistratal rain forests (map 6.14).

Natural History
In behavior and ecology this species is similar to *S. sciureus*. Play behavior and general ecology have been analyzed by Baldwin and Baldwin (1972).

Saimiri sciureus (Linnaeus, 1758)
Squirrel Monkey

Description
See the description for the genus.

Range and Habitat
As noted in the genus account, this species occurs widely throughout the Amazonian portions of South America (map 6.14). It is absent from the coastal forests of eastern Brazil. Broadly tolerant of a variety of habitat types, it can occur in multistratal evergreen tropical forests as well as mangroves and secondary forests.

Natural History
Saimiri primarily feeds on insects and fruit, though it occasionally takes some leafy material. Its foraging ecology has been analyzed by Terborgh (1983). Group size depends on the carrying capacity of the habitat. Groups as small as seven or eight have been noted in relict forest patches, and over one hundred were counted in one troop in the Amazon basin. Several adult males may occur in a troop. The ratio of adult females to adult males averages about four to one. Population density estimates in good habitats range from nineteen to thirty-one animals per km^2. They are diurnally active, and home-range size depends on the carrying capacity of the habitat. Home ranges of adjacent troops appear to overlap extensively. A single young is born after a gestation of approximately 165 days. Behavior patterns and vocalizations have been extensively stud-

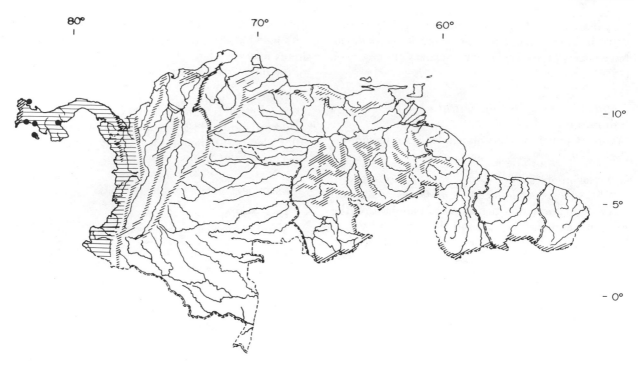

Map 6.15. Distribution of *Alouatta palliata*.

ied in the laboratory; Baldwin and Baldwin (1981) provide a useful overview of the material. Rosenbloom and Cooper (1968) summarize the anatomical, ecological, and captive maintenance data.

Ploog, Hopf, and Winter (1967) described the vocalization system. The basic vocalization repertoir can develop without imitative learning. Vocalizations can be classified into contact-seeking, fearful, aggressive, and warning calls. Within a single category there are many variants that express different intensities of motivation. A similar functional classification for New World primate vocalizations was developed by Eisenberg (1976).

Genus *Alouatta* Lacépède, 1799

Howler Monkey, Aullidor, Arguato

Description

Six species are recognized within this genus, and two occur in the region covered by this book: *A. palliata* and *A. seniculus*. Members of this genus are among the largest of the New World primates. Head and body length ranges from 559 to 915 mm, the prehensile tail from 585 to 920 mm. The weight of adult males ranges from 6.5 to 7.8 kg, and that of females from 4.5 to 6.6 kg. Color patterns are variable from species to species. *A. palliata* is black, *A. seniculus* is reddish. In *A. caraya* the males are black and the females and juveniles are buff, one of the few cases of color dimorphism among the New World primates. In the two species dealt with in this vol-

ume there is no color dimorphism, but they show the distinct size dimorphism typical of the genus.

Adult females of *A. palliata* weigh about 84% of the average adult male's weight; the value for *A. seniculus* is more extreme, with the female being only 69% of the male's average weight. Size dimorphism is less pronounced, with the female attaining 94% of the male's head and body length in *A. palliata* and 90% in *A. seniculus*. The hyoid bone is extremely enlarged and ossified and no doubt influences the low fundamental frequency of the long call (Thorington, Rudran, and Mack 1979).

Distribution

The genus is the most widely distributed among the New World primates, with a range extending from Vera Cruz, Mexico, to northern Argentina. The species tolerate a range of habitat types varying from semideciduous tropical forests to multistratal tropical evergreen forests (Eisenberg 1979).

Natural History

The howler monkeys are aptly named because they characteristically produce loud roaring bouts at daybreak. These roars have been implicated in intratroop spacing behavior, but they may also be directed at potential conspecific invaders of an occupied troop. Males roar toward alien males, and females roar when they encounter alien females. Both sexes can join in combined choruses. For a full account of the function of roaring in *A. seniculus*, see

Sekulic (1982). The howler monkey is organized into troops of both sexes, and the number of adult males in a troop varies from species to species. *A. palliata* seems to exhibit the greatest male-male tolerance and the highest incidence of multimale troops; *A. seniculus* tends to exhibit a unimale troop structure in many parts of its range. In favorable habitat, howler troops may have rather small home ranges, in part because they are the most folivorous New World primate and are not totally dependent on ripe fruit for nourishment. Up to 50% of the annual diet may be composed of young leaves. The foraging ecology of *A. palliata* has been amply documented by Milton (1980). A single young is born after a gestation period of 185 days.

Alouatta palliata (Gray, 1849)
Mantled Howler Monkey

Description

This species is the largest by weight within the genus; in Panama average weight for males was 7.8 kg. Mean head and body length for adult males on Barro Colorado Island is 506 mm; the comparable value for females is 475 mm. The tail exceeds the head and body length, and males average 665 mm while females average 651 mm. As can be seen from these figures, the female's tail is proportionately longer than the male's. The basic color is black (plate 8). The scrotal skin is white in the male, and the vulvar area is white in the female. A fringe of chestnut hairs runs laterally from the arm to the groin, hence the common name mantled howler. The development of the mantle varies widely throughout the range of the species, but some trace can usually be discerned even in those populations where it is reduced.

Range and Habitat

Alouatta palliata occurs from southern Veracruz, Mexico, through Panama and in Colombia west of the western cordillera. It is found in dry deciduous tropical forests as well as multistratal tropical evergreen forests (map 6.15).

Natural History

Howler monkeys eat a great deal of leafy material. Because they are not completely dependent upon fruit, their daily range and home range tend to be small for a primate of this size class. Howlers rarely descend to the ground, but they will do so in areas characterized by isolated tree islands, where they must cross open country to move from one grove of trees to the next. They are strictly diurnal and characteristically produce loud, roaring calls at dawn. Depending on the carrying capacity of the habitat and the stability of the troop, troop sizes can range from three to forty-four. Fourteen is average.

In some habitats *Alouatta palliata* can reach high densities; seventy animals per km² is not uncommon. In Costa Rica, where it co-occurs with *Ateles* and *Cebus*, it can represent 69% of the total primate biomass; in Colombia, where it co-occurs with three other primate species, it can represent 44% (Eisenberg 1979). *Alouatta palliata* has been the subject of long-term studies in both Costa Rica and Panama; see Carpenter (1934), Froehlich, Thorington, and Otis (1981), Milton (1978, 1980), and Glander (1975, 1981).

Alouatta seniculus (Linnaeus, 1766)
Red Howler Monkey, Guariba, Mono Colorado

Description

There is great variation in size over the range of this species. A population studied in Venezuela exhibited a mean head and body length of 523 mm for males and 468 mm for females (Thorington, Rudran, and Mack 1979). Average tail lengths for the two sexes were 630 mm and 590 mm. The basic body color is a chestnut brown, almost reddish brown in some parts of its range (plate 8).

Range and Habitat

The species is broadly distributed east of the western cordillera of the Andes in Colombia, south to Bolivia, and east to the Guyanas (map 6.16). It is found in a range of habitats from dry deciduous forest to multistratal tropical rain forests.

Natural History

This species has been intensively studied in Venezuela by Braza, Alvarez, and Azcarate (1983), Gaulin and Gaulin (1982), Neville (1972), Rudran (1979), Sekulic (1981), and Crockett (1984). In the llanos of Venezuela the animals depend on leaves for approximately 40% of their annual diet, and mean troop size there is 8.9. There is a pronounced tendency for only one adult male to be present in a given troop, but some troops have two adult males and the average for the Venezuela population was 1.65 males per troop.

At high population densities in Venezuela, alien males will actively attempt to enter a troop and depose the dominant male. During such times of social unrest infant mortality increases. If the alien male can establish himself within the troop, he usually kills infants less than six months old (Rudran 1979; Crockett and Sekulic 1984). Although the size of the home range varies depending on the quality of the habitat, there is usually a correlation between troop size and home-range size. In Venezuela the average home range is approximately 6 ha. In areas of good habitat in the llanos of Venezuela, howler populations can reach a density of 150 animals per km².

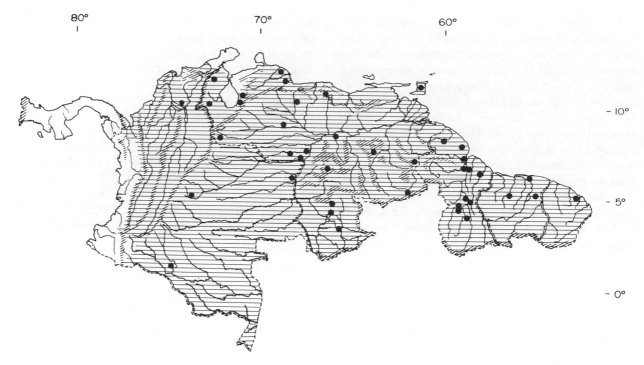

Map 6.16. Distribution of *Alouatta seniculus*.

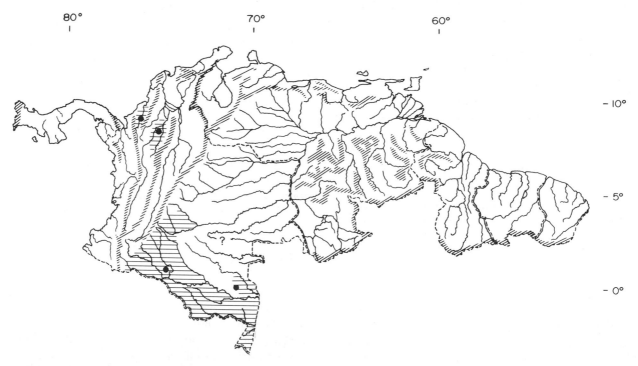

Map 6.17. Distribution of *Lagothrix lagothricha*.

The biology of this species is reviewed in Crockett and Eisenberg (1987).

Genus *Lagothrix* E. Geoffroy, 1812
Lagothrix lagothricha (Humboldt, 1812)
Woolly Monkey, Caparro, Barrigudo

Description

The genus *Lagothrix* contains two species, one of which occurs in the northern Neotropics. This is one of the largest New World primates by weight. Head and body length ranges from 558 to 686 mm, the prehensile tail 600 to 720 mm. Adult males may weigh over 10 kg and females 8 kg. Color varies from gray to brown, and the head is generally a darker brown than the body. The fur is dense and short, giving the animal a woolly appearance. The face is deeply pigmented (plate 8).

Range and Habitat

This monkey is distributed in the western part of the Amazon basin from Peru to southern Colombia. In the area covered by this volume the woolly monkey is known to occur only in southern Colombia to the east of the Andes and in several isolated areas in the Magdalena River basin (map 6.17).

Natural History

The woolly monkey has been little studied in the field but intensively studied in captivity (Williams 1968). From the field data we have available it appears that it feeds primarily on fruit and seeds supplemented by leaves (Ramirez 1980). Its consumption of leafy material, however, does not approach that of the howler monkey and is probably less than 20% of the diet. Troop size ranges between twelve and seventy. Depending on the carrying capacity, the home range can exceed 500 ha. Troops contain more than one adult male, but the exact nature of the dominance structure within a free-ranging troop has yet to be worked out. The vocalizations of the woolly monkey are extremely complicated; males' loud, long calls have a complex syllabic structure. There are similarities between the calls of the woolly monkey and the spider monkey (Eisenberg 1976). A single young is born after a gestation of seven and a half months.

Genus *Ateles* E. Geoffroy, 1806
Spider Monkey, Mono Araña, Marimonda

Taxonomy

Since the revision by Kellogg and Goldman (1944), the genus has not received a comprehensive taxonomic treatment. Although there has been a tendency to recognize only a single species within the genus, I firmly believe that the genus can be subdivided into at least five valid species. Some credence is offered to this viewpoint by the chromosomal studies of Heltne and Kunkel (1975) and Kunkel, Heltne, and Borgaonkar (1980), where karyotypic variants are demonstrable. The area of South America covered in this volume contains four valid species: *A. fusciceps*, *A. geoffroyi*, *A. paniscus*, and *A. belzebuth*.

Description

Spider monkeys are among the largest New World primates in linear measurements. Adult head and body length can range from 420 to 660 mm, and tail length ranges from 744 to 880 mm. A large adult male can weigh up to 11 kg. The limbs are long and slender, and the tail is extremely long and highly prehensile. Coat color is highly variable and will be discussed under the species accounts. The large size and long limbs immediately distinguish this genus from the other larger New World monkeys in the northern Neotropics, for example, the howler monkey and the woolly monkey (plate 8).

Distribution

The species of *Ateles* range from southern Veracruz, Mexico, to Bolivia. They are adapted to multistratal tropical evergreen forests, although in the drier parts of Costa Rica and Bolivia they are known to occupy semideciduous forests. The species of this genus tend to be allopatric.

Natural History

The spider monkeys are by and large frugivorous, although they may eat young leaves. Typically they forage in the middle to upper strata of the forest and rarely descend to the ground. When they do travel on the ground they walk bipedally. Troop size is variable, and there is a marked tendency for troops to fractionate into foraging subgroups, then reconstitute at preferred sleeping sites. The species is strongly diurnal, and a troop utilizes a rather large home range. Subgroups communicate through loud, long calls that are easily distinguished from the deeper, more monotonic roars of the howler monkey.

A single young is born after a gestation period of seven and a half months. The female carries the young until it is approximately ten months old, and thereafter she carries it at intervals until fourteen months of age. The most cohesive social unit within a troop of spider monkeys appears to be a female and her dependent young. The period of dependency is long, and the interbirth interval averages about thirty months. Spider monkeys show one of the lowest rates of recruitment of all New World monkeys and thus are slow to recover from catastrophic events and vulnerable to human predation. The standard references for behavior and ecology are

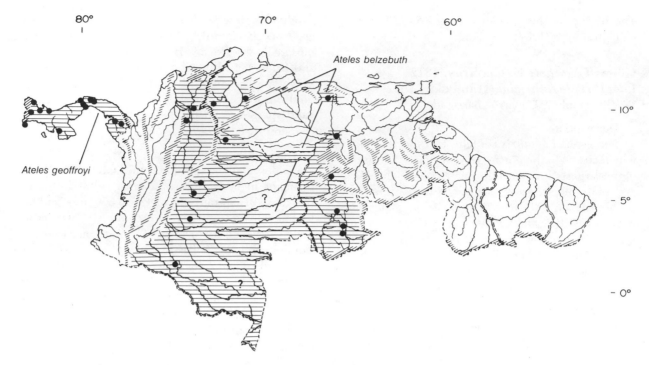

Map 6.18. Distribution of two species of *Ateles*.

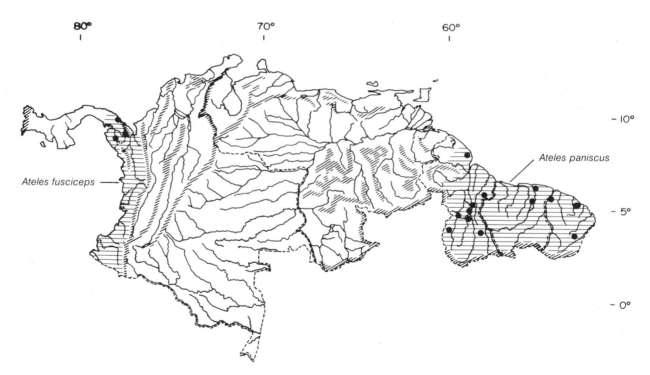

Map 6.19. Distribution of two species of *Ateles*.

Carpenter (1935), Dare (1974), Eisenberg and Keuhn (1966), Eisenberg (1973, 1976), and Klein (1972).

Ateles geoffroyi Kuhl, 1820
Black-handed Spider Monkey, Mono Colorado (Panama)

Description
This monkey is highly variable in color pattern over its range, from southern Mexico to Panama. Within the area covered by this volume the species is distinctive. All subspecies of *Ateles geoffroyi* have black forearms and black hands as well as black on the lower limbs, but they vary in dorsal color. In Panama the dorsum tends to be chestnut brown, with a black cape and black over the lumbar region (plate 8). In the Darién region of Panama, *Ateles geoffroyi* comes into contact with *Ateles fusciceps*.

Range and Habitat
This species is distributed from Veracruz to Panama. It generally is found in multistratal tropical evergreen forests but utilizes dry semideciduous forests in parts of Costa Rica and Mexico (map 6.18).

Natural History
In Panama *A. geoffroyi* is primarily frugivorous (Hladik and Hladik 1969). In suitable habitat home range may be between 1.1 and 1.15 km². A typical troop consists of fifteen to twenty individuals, but troops frequently fractionate into subgroups of one to five, which include females and infants or all males (Carpenter 1935). Males within a subgroup tend to be age graded, and the number of adult males per troop varies from one to three. The vocal repertoire is extremely rich; loud, long calls are used in subgroup intercommunication. A full description of vocalizations is included in Eisenberg (1976).

Ateles fusciceps Gray, 1866
Brown-headed Spider Monkey

Description
This species shows considerable variation in coat color over its entire range. A dark brown cap and brown back contrasting with a red venter are characteristic in the southern parts of its range, but in the area covered by this book the most frequent color is entirely black. The all-black form is described subspecifically as *Ateles fusciceps robustus* and extends from the Darién region of Panama west of the Andes through extreme western Colombia.

Range and Habitat
This species extends from southern Panama to Ecuador, west of the Andes. It typically inhabits multistratal tropical evergreen forests (map 6.19).

Natural History
There are few field data available for this species, but it has been intensively studied in captivity. Such data as we have indicate that it closely approximates *A. geoffroyi* in its basic natural history patterns. The standard reference for this species in captivity is Eisenberg (1976).

Ateles belzebuth E. Geoffroy, 1806
Long-haired Spider Monkey, Marimonda, Mono Frontino

Description
A large male may have a head and body length of 640 mm; males slightly exceed females in head and body length, but the female's tail is proportionately longer. The basic color is brown; the venter tends to be lighter, but the exact coloration is variable over the entire range of the species. The most distinguishing feature is a white diadem on the forehead (plate 8).

Range and Habitat
In the main, this species is distributed in the northeastern portion of the Amazon basin. Within the area covered by this book it occurs in southern Colombia, in Amazonian Venezuela, and in relict populations in the mountains of northern Venezuela (Mondolfi and Eisenberg 1979) (map 6.18). This species is replaced in the Guyanas by *A. paniscus*. It typically inhabits multistratal evergreen forests.

Natural History
This species received intensive field study by Klein (1972) and by Klein and Klein (1977). In ecology, behavior, and reproduction it resembles *A. geoffroyi*. Troops range from ten to thirty animals and frequently break into subgroups of one to seven. Home ranges vary from 2.6 to 3.9 km² (Klein and Klein 1977).

Ateles paniscus (Linnaeus, 1758)
Black Spider Monkey

Description
This species is almost indistinguishable from *A. fusciceps robustus* but may easily be identified in the field by its geographical location. In the north, *A. paniscus* is basically black with a pigmented face. In western Amazonas the face is pink (plate 8).

Range and Habitat
The species ranges from Guyana eastward and south to the Amazon River (map 6.19).

Natural History
In its behavior and ecology, this species strongly resembles *A. geoffroyi*. The species has been stud-

ied by Mittermeier and van Roosmalen (1981) and van Roosmalen (1980). Troop sizes range from ten to thirty. Home ranges may exceed 3.3 km². In Suriname, *Ateles paniscus* prefers mature rain forests and does not exploit edge habitats. It typically forages in the upper part of the canopy. Its diet largely consists of fruit, with leaves making up less than 6%. In common with other species of *Ateles*, this monkey is strongly persecuted by humans as a game animal. Over much of its range, *Ateles* thus has become rare or been exterminated in areas of prolonged human settlement.

References

• References used in preparing distribution maps

Ayres, J. M. 1981. Observacões sobre a ecologia e o comportamento dos cuxius (*Chiropotes albinasus* e *Chriopotes satanas*, Cebidae: Primates). Conselho Nacional de Desenvolvimento Cientifico e Tecnologico (CNPq) Instituto Nacional de Pesquisas da Amazonia (INPA) Fundacão Universidade do Amazonas (DUA).

Baldwin, J. D., and J. I. Baldwin. 1972. The ecology and behavior of squirrel monkeys (*Saimiri oerstedii*) in a natural forest in western Panama. *Folia Primatol.* 18:161–84.

———. 1981. The squirrel monkeys, genus *Saimiri*. In *Ecology and behavior of Neotropical primates*, ed. A. F. Coimbra-Filho and R. A. Mittermeier, 277–325. Rio de Janeiro: Academia Brasileira de Ciencias.

• Beebe, W. 1919. The higher vertebrates of British Guiana with special reference to the fauna of the Bartica district. 7. List of Amphibia, Reptilia, and Mammalia. *Zoologica* 2:205–38.

Braza, F., F. Alvarez, and T. Azcarate. 1983. Feeding habits of the red howler monkeys (*Alouatta seniculus*) in the llanos of Venezuela. *Mammalia* 47(2):205–14.

Buchanan, D. B. 1978. Communication and ecology of pithecine monkeys, with special reference to *Pithecia pithecia*. Ph.D. diss., Wayne State University.

Buchanan, D. B., R. A. Mittermeier, and M. G. M. van Roosmalen. 1981. The saki monkeys, genus *Pithecia*. In *Ecology and behavior of Neotropical primates*, ed. A. F. Coimbra-Filho and R. A. Mittermeier, 391–417. Rio de Janeiro: Academia Brasileira de Ciencias.

Carpenter, C. R. 1934. A field study of the behavior and social relations of howling monkeys. *Comp. Psychol. Monogr.* 10(2):1–168.

———. 1935. Behavior of red spider monkeys in Panama. *J. Mammal.*, 16:171–80.

Castro, R., and P. Soini. 1977. Field studies on *Saguinus mystax* and other callitrichids in Amazonian Peru. In *The biology and conservation of the Callitrichidae*, ed. D. G. Kleiman, 73–78. Washington, D.C.: Smithsonian Institution Press.

Christen, A. 1974. Fortpflanzungsbiologie und Verhalten bei *Cebuella pygmaea* und *Tamarin tamarin*. *Z. Tierpsychol.*, suppl. 14.

Crockett, C. M. 1984. Emigration by female red howler monkeys and the case for female competition. In *Female primates: Studies by women primatologists*, ed. M. Small, 159–73. New York: Alan R. Liss.

Crockett, C. M., and J. F. Eisenberg. 1987. Howlers: Variations in group size and demography. In *Primate societies*, ed. B. B. Smuts, D. L. Cheney, R. M. Seyfarth, R. W. Wrangham, and T. T. Struhsaker, 54–68. Chicago: University of Chicago Press.

Crockett, C. M., and R. Sekulic. 1984. Infanticide in red howler monkeys (*Alouatta seniculus*). In *Infanticide: Comparative and evolutionary perspectives*, ed. G. Hausfater and S. Hrdy. Hawthorne, N.Y.: Aldine.

Dare, R. 1974. The social behavior and ecology of spider monkeys, *Ateles geoffroyi*, on Barro Colorado Island, Panama. Ph.D. diss., University of Oregon.

Dawson, G. A. 1977. Composition and stability of social groups of the tamarin, *Saguinus oedipus geoffroyi*, in Panama: Ecological and behavioral implications. In *The biology and conservation of the Callitrichidae*, ed. D. G. Kleiman, 23–37. Washington, D.C.: Smithsonian Institution Press.

deBoer, L. B. M. 1974. Cytotaxonomy of the Platyrrhini. *Genen en Phaenen* 17:1–115.

• Defler, T. R. 1983a. A remote park in Colombia. *Oryx* 17:15–17.

———. 1983b. Some population characteristics of *Callicebus torquatus* in eastern Colombia. *Lozania* (Bogotá) 38:1–8.

Eisenberg, J. F. 1973. Reproduction in two species of spider monkeys, *Ateles fusciceps* and *A. geoffroyi*. *J. Mammal.* 54:955–57.

———. 1976. Communication mechanisms and social integration in the black spider monkey, *Ateles fusciceps robustus* and related species. *Smithsonian Contrib. Zool.* 213:1–108.

———. 1977. Comparative ecology and reproduction in New World primates. In *The biology and conseravtion of the Callitrichidae*, ed. D. G. Kleiman, 13–22. Washington, D.C.: Smithsonian Institution Press.

————. 1979. Habitat, economy and society: Some correlations and hypotheses for the Neotropical primates. In *Primate ecology and human origins*, ed. I. S. Bernstein and E. O. Smith, 215–62. New York: Garland Press.

Eisenberg, J. F., and R. E. Kuehn. 1966. The behavior of *Ateles geoffroyi* and related species. *Smithsonian Misc. Coll.* 151(8):1–63.

Epple, G. 1973. The role of pheromones in the social communication of marmoset monkeys. *J. Reprod. Fertil.*, suppl. 19:447–54.

————. 1975. The behavior of marmoset monkeys. In *Primate behavior*, ed. L. A. Rosenblum, 4: 195–239. New York: Academic Press.

Erikson, G. E. 1963. Brachiation in the New World monkeys. In *The primates*, ed. J. Napier and N. A. Barnicot, 135–64. Symposium 10. London: Zoological Society of London.

Fontaine, R. 1981. The uakaris, genus *Cacajao*. In *Ecology and behavior of Neotropical primates*, ed. A. F. Coimbra-Filho and R. A. Mittermeier, 443–93. Rio de Janeiro: Academia Brasileira de Ciencias.

Freese, C., and J. Oppenheimer. 1981. The capuchin monkeys, genus *Cebus*. In *Ecology and behavior of Neotropical primates*, ed. A. F. Coimbra-Filho and R. A. Mittermeier, 331–90. Rio de Janeiro: Academia Brasileira de Ciencias.

Froehlich, J. W., R. W. Thorington, Jr., and J. S. Otis. 1981. The demography of howler monkeys (*Alouatta palliata*) on Barro Colorado Island, Panama. *Int. J. Primatol.* 2(3):207–36.

Gaulin, S. J. C., and C. K. Gaulin. 1982. Behavioral ecology of *Alouatta seniculus* in Andean cloud forest. *Int. J. Primatol.* 3:1–32.

Glander, K. E. 1975. Habitat and resource utilization: An ecological view of social organization in the mantled howling monkey. Ph.D. diss., University of Chicago.

————. 1981. Reproduction and population growth in free-ranging mantled howling monkeys. *Amer. J. Phys. Anthropol.* 53:25–36.

Green, K. M. 1978. Neotropical primate censusing in northern Colombia. *Primates* 19:537–50.

• Hall, E. R. 1981. *The mammals of North America*. 2d ed., 2 vols. New York: John Wiley.

• Handley, C. O., Jr. 1976. Mammals of the Smithsonian Venezuelan project. *Brigham Young Univ. Sci. Bull., Biol. Ser.* 20(5):1–90.

Happel, R. E. 1982. Ecology of *Pithecia hirsuta* in Peru. *J. Human Evol.* 11:581–90.

Heltne, P. G., and L. Kunkel. 1975. Taxonomic notes on the pelage of *Ateles paniscus paniscus, A. p. chamek,* and *A. fusciceps rufiventris. J. Med. Primatol.* 4:83–102.

Heltne, P. F., D. C. Turner, and J. Wolhandler. 1973. Maternal and paternal periods in the development of infant *Callimico goeldii. Amer. J. Phys. Anthropol.* 3:555–60.

Heltne, P. G., J. F. Wojeik, and A. G. Pook. 1981. Goeldi's monkey, genus *Callimico*. In *Ecology and behavior of Neotropical primates*, ed. A. F. Coimbra-Filho and R. A. Mittermeier, 169–209. Rio de Janeiro: Academia Brasileira de Ciencias.

• Hernandez-Camacho, J., and R. W. Cooper. 1976. The nonhuman primates of Colombia. In *Neotropical primates: Field studies and conservation*, ed. R. W. Thorington, Jr., and P. G. Heltne, 35–69. Washington, D.C.: National Academy of Sciences.

• Hershkovitz, P. 1949. Mammals of northern Colombia, preliminary report no. 4: Monkeys (Primates), with taxonomic revisions of some forms. *Proc. U.S. Nat. Mus.* 98:323–427.

•————. 1977. *Living New World monkeys (Platyrrhini)*. Vol. 1. Chicago: University of Chicago Press.

•————. 1979. The species of sakis, genus *Pithecia* (Cebidae, Primates), with notes on sexual dichromatism. *Folia Primatol.* 31:1–22.

•————. 1983. Two new species of night monkeys, genus *Aotus* (Cebidae, Platyrrhini): A preliminary report on *Aotus* taxonomy. *Amer. J. Primatol.* 4:209–43.

————. 1985. *A preliminary review of the South American bearded saki monkeys genus* Chiropotes *(Cebidae; Platyrrhini), with a description of a new subspecies*. Fieldiana, Zoology, n.s., no. 27. Chicago: Field Museum of Natural History.

Hick, U. 1973. Wir sind umgezogen. Z. *Kolner Zoo* 16(4):127–45.

Hill, W. C. O. 1957. *Primates*. Vol. 3. Edinburgh: Edinburgh University Press.

————. 1960. *Primates*. Vol. 4. Edinburgh: Edinburgh University Press.

————. 1962. *Primates*. Vol. 5. Edinburgh: Edinburgh University Press.

Hladik, A., and C. M. Hladik. 1969. Rapports trophiques entre végétation et primates dans la forêt de Barro Colorado (Panama). *Terre et Vie* 23: 25–117.

Hladik, C. M., A. Hladik, J. Bousset, P. Valdebouze, G. Viroben, and J. Delort-Laval. 1971. Le régime alimentaire des primates de l'île de Barro-Colorado (Panama): Resultats des analyses quantitatives. *Folia Primatol.* 16:85–122.

Hoage, R. J. 1982. Social and physical maturation in captive lion tamarins, *Leontopithecus rosalia rosalia* (Primates: Callitrichidae). *Smithsonian Contrib. Zool.* 354:1–56.

• Husson, A. M. 1978. *The mammals of Suriname.* Leiden: E. J. Brill.

Izawa, K. 1976. Group sizes and compositions of monkeys in the upper Amazon basin. *Primates* 17:367–99.

• Kellogg, R., and E. A. Goldman. 1944. Review of the spider monkeys. *Proc. U.S. Nat. Mus.* 96 (3186):1–45.

Kinzey, W. G. 1977. Diet and feeding behavior of *Callicebus torquatus.* In *Primate ecology,* ed. T. H. Clutton-Brock, 127–51. London: Academic Press.

———. 1981. The titi monkeys, genus *Callicebus.* In *Ecology and behavior of Neotropical primates,* ed. A. F. Coimbra-Filho and R. A. Mittermeier, 241–76. Rio de Janeiro: Academia Brasileira de Ciencias.

Kinzey, W. G., and A. H. Gentry. 1978. Habitat utilization in two species of *Callicebus.* In *Primate ecology: Problem oriented field studies,* ed. R. W. Sussman, 89–100. New York: John Wiley.

Kleiman, D. G. 1976. Monogamy in mammals. *Quart. Rev. Biol.* 52:39–69.

———. ed. 1977. *The biology and conservation of the Callitrichidae.* Washington, D.C.: Smithsonian Institution Press.

Klein, L. L. 1972. Ecology and social organization of the spider monkey *Ateles belzebuth.* Ph.D. diss., University of California, Berkeley.

Klein, L. L., and D. B. Klein. 1975. Social and ecological contrasts between four taxa of Neotropical primates. In *Socioecology and psychology of primates,* ed. R. H. Tuttle, 59–85. The Hague: Mouton.

———. 1977. Feeding behavior of the Colombian spider monkey. In *Primate ecology,* ed. T. H. Clutton-Brock, 153–82. London: Academic Press.

Kunkel, L. M., P. G. Heltne, and D. S. Borgaonkar. 1980. Chromosomal variation and zoogeography in *Ateles. Int. J. Primatol.* 1(3):223–32.

Mason, W. A. 1968. Use of space by *Callicebus* groups. In *Primates: Studies in adaptation and variability,* ed. P. C. Jay, 200–216. New York: Holt, Rinehart and Winston.

———. 1971. Field and laboratory studies of social organization in *Saimiri* and *Callicebus. Primate Behav.* 2:107–38.

Milton, K. 1978. Behavioral adaptations to leaf eating in the mantled howler monkey. In *The ecology of arboreal folivores,* ed. G. G. Montgomery, 535–50. Washington, D.C.: Smithsonian Institution Press.

———. 1980. *The foraging strategy of howler monkeys.* New York: Colombia University Press.

Mittermeier, R. A., and M. G. M. van Roosmalen. 1981. Preliminary observations on habitat utilization and diet in eight Surinam monkeys. *Folia Primatol.* 36:1–39.

• Mondolfi, E., and J. F. Eisenberg. 1979. New records for *Ateles belzebuth hybridus* in northern Venezuela. In *Vertebrate ecology in the northern Neotropics,* ed. J. F. Eisenberg, 93–96. Washington, D.C.: Smithsonian Institution Press.

Moynihan, M. 1964. Some behavior patterns of platyrrhine monkeys. 1. The night monkey (*Aotus trivirgatus*). *Smithsonian Misc. Coll.* 146(5):1–84.

———. 1970. Some behavior patterns of platyrrhine monkeys. 2. *Saguinus geoffroyi* and some other tamarins. *Smithsonian Contrib. Zool.* 28:1–77.

———. 1976. *The New World primates.* Princeton: Princeton University Press.

Muckenhirn, N. A. 1967. The behavior and vocal repertoire of *Saguinus oedipus.* Master's thesis, University of Maryland.

• Muckenhirn, N. A., B. Mortensen, S. Vessey, C. E. Fraser, and B. Singh. 1976. *Report on a primate survey in Guyana.* Washington, D.C.: Panamerican Health Organization.

Neville, M. K. 1972. The population structure of red howler monkeys, *Alouatta seniculus* in Trinidad and Venezuela. *Folia Primatol.* 17:56–86.

Neyman, P. F. 1977. Aspects of the ecology and social organization of free-ranging cotton-top tamarins (*Saguinus oedipus*) and the conservation status of the species. In *The biology and conservation of the Callitrichidae,* ed. D. G. Kleiman, 39–71. Washington, D.C.: Smithsonian Institution Press.

Oppenheimer, J. R. 1968. Behavior, ecology of the white-faced monkey *Cebus capucinus* on Barro Colorado Island, Panama. Ph.D. diss., University of Illinois.

———. 1977. Communication in New World monkeys. In *How animals communicate,* ed. T. A. Sebeok, 851–89. Bloomington: Indiana University Press.

Ploog, D., S. Hopf, and P. Winter. 1967. Ontogenese des Verhaltens von Totenkopf-Affen (*Saimiri sciureus*). *Psychol. Fosch.* 31:1–41.

Pook, A. G., and G. Pook. 1981. A field study of the socioecology of Goeldi's monkey *Callimico goeldii* in northern Bolivia. *Folia Primatol.* 35:288–312.

Ramirez, M. F. 1980. Grouping patterns of the woolly monkey, *Lagothrix lagothricha* at the Manu National Park, Peru. *Amer. J. Phys. Anthropol.* 52:269.

Ramirez, M. F., C. H. Freese, and J. Revilla C. 1977. Feeding ecology of the pygmy marmoset,

Cebuella pygmaea, in northeastern Peru. In *The biology and conservation of the Callitrichidae*, ed. D. G. Kleiman, 91–104. Washington, D.C.: Smithsonian Institution Press.

Robinson, J. 1977. The vocal regulation of spacing in the titi monkey, *Callicebus moloch*. Ph.D. diss., University of North Carolina.

———. 1979a. An analysis of the organization of vocal communication in the titi monkey *Callicebus moloch*. *Z. Tierpsychol.* 49:381–405.

———. 1979b. Vocal regulation of use of space by groups of titi monkeys *Callicebus moloch*. *Behav. Ecol. Sociobiol.* 5:1–15.

———. 1981. Spatial structure in foraging groups of wedge-capped capuchin monkeys *Cebus nigrivittatus*. *Anim. Behav.* 29:1036–56.

———. 1982. Vocal systems regulating within-group spacing. In *Primate communication*, ed. C. T. Snowdon, C. H. Brown, and M. Petersen, 94–116. Cambridge: Cambridge University Press.

———. 1986. Seasonal variation in use of time and space by the wedge-capped capuchin monkey *Cebus olivaceus*: Implications for foraging theory. *Smithsonian Contrib. Zool.* 431:1–60.

Rosenblum, L. A., and R. W. Cooper, eds. 1968. *The squirrel monkey*. New York: Academic Press.

Rudran, R. 1979. The demography and social mobility of a red howler (*Alouatta seniculus*) population in Venezuela. In *Vertebrate ecology in the northern Neotropics*, ed. J. F. Eisenberg, 107–26. Washington, D.C.: Smithsonian Institution Press.

Sekulic, R. 1981. The significance of howling in the red howler monkeys *Alouatta seniculus*. Ph.D. diss., University of Maryland.

———. 1982. Daily and seasonal patterns of roaring and spacing in four red howler *Alouatta seniculus* troops. *Folia Primatol.* 39:22–48.

Snowdon, C. T., and J. Cleveland. 1980. Individual recognition of contact calls by pygmy marmosets. *Anim. Behav.* 28:717–27.

Szalay, F. S., and E. Delson. 1979. *The evolutionary history of primates*. New York: Academic Press.

• Tate, G. H. H. 1939. The mammals of the Guiana region. *Bull. Amer. Mus. Nat. Hist.* 76(5):151–229.

Terborgh, J. 1983. *Five New World primates: A study in comparative ecology*. Princeton: Princeton University Press.

Thorington, R. W., Jr., R. Rudran, and D. Mack. 1979. Sexual dimorphism of *Alouatta seniculus* and observation on capture techniques. In *Vertebrate ecology in the northern Neotropics*, ed. J. F. Eisenberg, 97–106. Washington, D.C.: Smithsonian Institution Press.

van Roosmalen, M. G. M. 1980. Habitat preferences, diet, feeding strategy and social organization of the black spider monkey (*Ateles p. paniscus* Linnaeus, 1758) in Suriname. Report, Rijksinstituut voor Natuurbeheer, Arnhem (the Netherlands).

van Roosmalen, M. G. M., R. A. Mittermeier, and K. Milton. 1981. The bearded sakis, genus *Chiropotes*. In *Ecology and behavior of Neotropical primates*, ed. A. F. Coimbra-Filho and R. A. Mittermeier, 419–42. Rio de Janeiro: Academia Brasileira de Ciencias.

Williams, L. 1968. *Man and monkey*. Philadelphia: J. B. Lippincott.

Wright, P. 1981. The night monkeys, genus *Aotus*. In *Ecology and behavior of Neotropical primates*, ed. A. F. Coimbra-Filho and R. A. Mittermeier, 211–40. Rio de Janeiro: Academia Brasileira de Ciencias.

Yunis, E., O. M. Torres de Caballero, and C. Ramirez. 1977. Genus *Aotus* Q- and G-band karyotypes and natural hybrids. *Folia Primatol.* 27:165–77.

7

Order Carnivora (Fissipedia)

Diagnosis

The order Carnivora includes most extant terrestrial mammals specialized for predation on other vertebrates. Many members are omnivores, and most are terrestrial or scansorial, except the semiaquatic otters. Dental formulas are somewhat variable, with reduction in tooth number pronounced in the Felidae, and will be presented under the family accounts. The canine is conical and prominent. The modern Carnivora are usually defined based on the major carnassial shearing effect of the molariform teeth; in modern carnivores the upper fourth premolar shears against the lower first molar. The digestive system shows no extreme modifications, and the cecum is usually small. The brain is relatively large and the braincase somewhat inflated, with the tympanic bullae enlarged and hemispheric. Of the eight extant families, five occur in South America.

I treat the Pinnipedia separately, since I consider them monophyletic. Although clearly derived from the Carnivora, they are sufficiently distinct to warrant ordinal status.

Distribution

The Recent distribution includes all continents except Antarctica and Australia; however, the dog (*Canis familiaris*) was introduced to Australia by aboriginal man about 4,000 to 7,000 B.P., possibly later than its introduction into the Western Hemisphere. Members of this order tolerate a variety of habitats, having extended to the extreme climatic conditions of the Arctic and the alpine zones of high mountains.

History and Classification

In Paleocene times the suborder Creodonta first appeared and radiated into the Eocene. Some creodonts persisted until the early Pliocene, but in general the original suborder was replaced by the suborder Fissipedia, which first appeared in the middle Paleocene as the family Miacidae, known from North America, Europe, and Asia. The modern families of carnivores are recognizable in the early Oligocene. Members of the order Carnivora were originally absent from South America and Australia; the carnivore niches there were filled by a parallel radiation within the order Marsupialia, resulting in the Dasyuridae (Australia) and the Borhyaenidae and Didelphidae (South America). The first true carnivores to enter South America appeared in the Miocene as an early raccoonlike form (Linares 1981). The remaining carnivores currently found there entered the continent by the Panamanian land bridge during the late Pliocene. A standard reference for the biology of the Carnivora is Ewer (1973).

FAMILY CANIDAE

Taxonomy

Cerdocyon has been placed in *Dusicyon* by Clutton-Brock, Corbet, and Hills (1976) and treated as a subgenus of *Canis* by Van Gelder (1978). *Atelocynus* has been included in *Dusicyon* by Clutton-Brock, Corbet, and Hills (1976) and considered a subgenus of *Canis* by Van Gelder (1978). Van Gelder raises interesting points concerning the criteria for determining generic rank. In recognition of the fossil canid data presented by Berta (1984, 1986), I retain the generic names *Atelocynus* and *Cerdocyon*.

Diagnosis

The dental formula is I 3/3, C 1/1, P 4/4, M 2/3, but some variation is shown in the number of molars. In the genus *Speothos*, molar number varies 1–2/2 (fig. 7.1). The canines are long and prominent. The molars have small crushing surfaces, and the lower first molar and upper third premolar provide the carnassial shearing surfaces. The rostrum is long. Members of this family are adapted for a cur-

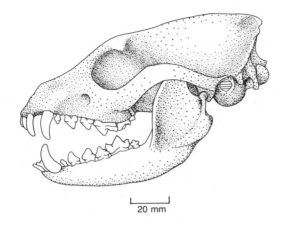

20 mm

Figure 7.1. Skull of *Speothos venaticus*.

sorial gait; they walk on their toes, and some reduction in toe number is demonstrable. There are
generally five toes on the forefoot and four on the
hind foot, but on the forefoot one toe is reduced to a
dewclaw. The claws are nonretractile.

Distribution

Species of this family occur on all continents except Antarctica. They were introduced to Australia
by early man.

Natural History

The natural history of the Canidae has been reviewed by Kleiman and Eisenberg (1973). Since canids are adapted for running, they can pursue prey
with a seemingly effortless gait, and some species
are adapted for taking animals larger than themselves. Their dietary habits are broad; fruit, invertebrates, and small vertebrates, as well as larger
vertebrates, are opportunistically taken. The basic
social grouping for most species of this family is a
mated pair and their various-aged offspring. Many
species exhibit a strong tendency for monogamy, although continuous contact by adult pair members is
variable. Some species are more solitary in their
habits and show a weaker pair-bond (e.g., the maned
wolf, *Chrysocyon*).

Most species of this family can excavate burrows
in the ground as shelters or for rearing the young.
They are opportunistic, however, and may take ad-
vantage of burrows constructed by other species,
enlarging and in various ways modifying them for
den sites. When not persecuted by humans, they
may be active into the early morning and late afternoon daylight hours as well as at night. A rich vocal
repertoire includes contact calls, greeting cries, and
chorusing that may serve to space adjacent social
groups. South American canids have been reviewed
by Langguth (1975).

Genus *Atelocynus* Cabrera, 1940
Atelocynus microtis (Sclater, 1883)
Small-eared Dog, Zorro de Orejas Cortas

Description
This genus contains a single species, *A. microtis*,
a medium-sized canid with an extremely long rostrum and very small external ears. Head and body
length ranges from 700 to 1,000 mm and the tail
from 250 to 350 mm; the hind foot is about 140 mm
and the ear 56 mm. The color is basically dark brown,
but white hairs scattered on the dorsal pelage give a
grizzled appearance and dark hairs in the midline
give the impression of a dorsal stripe. The species is
immediately distinguishable from all other canids in
the northern Neotropics by ear size, body size, and
coat color (plate 9).

Range and Habitat
Atelocynus is known from scattered localities in
the Amazon basin. It also occurs in southeastern Colombia (Hershkovitz 1961; Berta 1986) (map 7.1).

Natural History
The fragmentary information we have indicates
that the small-eared dog inhabits multistratal tropical evergreen forests and gallery forests. When
seen, it is usually solitary. Although pairs are tolerant in captivity, there is no strong expression of contact behavior, and one gains the impression that this
species is one of the least gregarious of the South
American canids. The vocal repertoire is limited, in
contrast to the rich variety of contact notes shown by
the closely related bush dog, *Speothos* (Hershkovitz
1961; pers. obs.).

Genus *Canis* Linnaeus, 1758
Canis latrans Say, 1823
Coyote

Description
This species is variable in size and coloration.
Head and body length ranges from 700 to 950 mm,
the tail ranges from 250 to 400 mm, and the ear is
large, measuring 100 to 120 mm from the notch.
Weight ranges from 8.2 to 13.6 kg. The dorsum varies from gray to yellow brown; black-tipped guard
hairs lend a salt-and-pepper effect to the pelage, and
the tail is often black tipped. The throat and venter
are white to pale gray.

Range and Habitat
The species is broadly distributed from Alaska
south across North America to western Panama. A
recent colonizer in Panama, it prefers open and edge
habitats. Land clearing for cattle grazing has allowed
it to extend its range (Mendez 1981; Vaughan 1983).

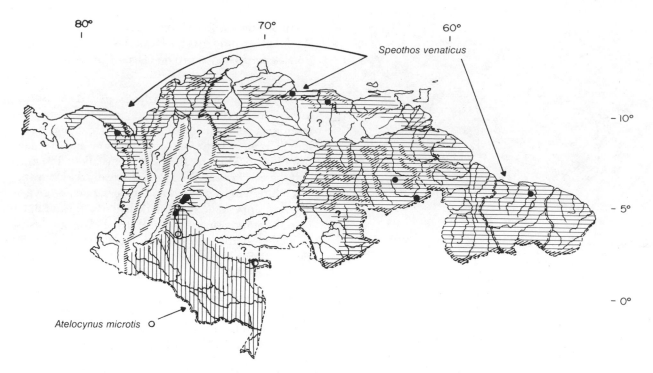

Map 7.1. Distribution of *Atelocynus microtis* and *Speothos venaticus*.

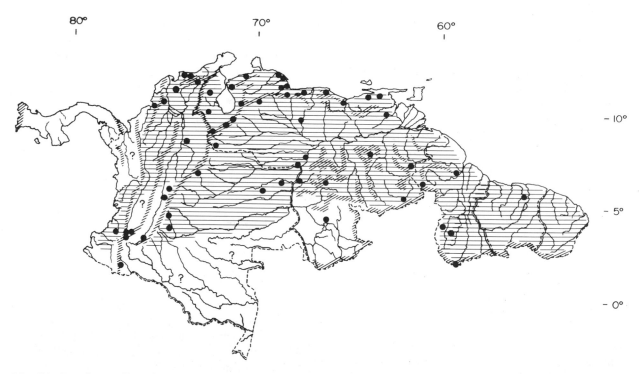

Map 7.2. Distribution of *Cerdocyon thous*.

Natural History

Coyotes take a variety of prey sizes, from young deer to mice. The social system is based on a mated pair and their offspring but is quite flexible. The animals may hunt in packs or singly. In the north they are seasonal breeders, mating in February. After a gestation of fifty-eight to sixty-three days, five to seven young are born in a burrow. At eight to twelve weeks of age the young begin to accompany the parents while foraging. Females can breed at one year of age, but the young do not usually establish a home range and pair until they are two years old. Coyotes vocalize singly or in a chorus by producing a series of sharp barks ending in a drawn-out moan (Bekoff 1978).

Genus *Cerdocyon* Hamilton-Smith, 1839
Cerdocyon thous (Linnaeus, 1766)
Crab-eating Fox, Zorro de Monte

Description

I use *Cerdocyon* in conformity with Berta (1984). This medium-sized canid is similar in body proportions to the temperate-zone red fox (*Vulpes fulva*). Head and body length averages 650 mm and the tail 300 mm. The coat color is somewhat variable but basically is gray on the dorsum, darkening to black in the midline. Specimens from the moist forests of northwestern Venezuela are very dark, while those coming from the llanos show a reduction in melanism and look more silver gray. They are distinguishable from the gray fox, *Urocyon*, in that *Urocyon* is reddish on the throat or upper forelimbs, while red markings are absent from the dorsal pelage of *Cerdocyon* (plate 9).

Range and Habitat

The crab-eating fox is the most common canid over much of northern South America. Its distribution extends south of the Amazon to Paraguay, and it is broadly tolerant of a variety of habitats but appears to be most common in savanna areas or gallery forests (map 7.2). See also Berta (1982).

Natural History

This species has been subjected to intensive study both in the field (Montgomery and Lubin 1978; Brady 1979; Schaller 1983; Bisbal and Ojasti 1980) and in captivity (Brady 1978; Biben 1983). The animals exploit a variety of food resources, including invertebrates, vertebrates, and fruit. The diet may show seasonal shifts with the abundance of various prey items (Bisbal and Ojasti 1980). The land crab may form a significant portion of its diet during the dry season in the llanos, when the crab is vulnerable to predation (Brady 1979). These foxes are significant predators on small rodents during cyclic highs in rodent populations.

They live in monogamous pairs that occupy exclusive territories. Spacing among neighbors is accomplished through a long call and urine marking. The home range of a pair varies from 0.6 to 0.9 km^2 in the llanos of Venezuela (Brady 1979; Sunquist, Sunquist, and Daneke, n.d.). Gestation rarely exceeds fifty-three days, and a pair potentially can reproduce in any month of the year. In practice, however, the seasonality of rainfall may limit the rearing phase to a single period in the annual cycle, so one litter per year tends to be the rule. Litter size is potentially high, up to five. The young begin to hunt with a parent at about six weeks of age and may remain in the parental home range for over a year. Partners may locate one another by a characteristic high-pitched whistling call (Brady 1979, 1981).

Genus *Speothos* Lund, 1839
Speothos venaticus (Lund, 1842)
Bush Dog, Zorro Vinagre

Description

The dental formula differs from the basic canid pattern, being I 3/3, C 1/1, P 4/4, M 1–2/2. This small canid is easily distinguishable from all other South American species by its extremely short ears, legs, and tail. Head and body length ranges from 600 to 750 mm, while the tail averages 130 mm and the weight 6 kg. The dorsal coat is reddish brown, and the legs show varying amounts of black (plate 9).

Range and Habitat

Never abundant, the bush dog occurs widely in northern South America, having been recorded from Panama, Colombia, Peru, Amazonian Brazil, Venezuela, and the Guyanas (Linares 1968). It appears to tolerate a variety of habitat types including gallery forests and multistratal tropical evergreen forests. It is generally found at lower elevations (map 7.1).

Natural History

This species has been only sporadically observed in the field but has been the subject of intensive captive studies (Biben 1982a,b). The animals are highly carnivorous and attack small mammals and birds (Deutsch 1983). Some of their prey, such as larger caviomorph rodents (the paca or capybara), may exceed them in weight.

All evidence from captivity indicates a strong pair-bond between an adult male and female. Young maintained in captivity with their parents after attaining maturity do not reproduce. Apparently the dominant female suppresses estrus in her daughters (Porton, Kleiman, and Rodden 1987). Accounts of pack hunting by this species surely refer to extended family groups. The male provisions the female and young throughout the rearing phase. The female can

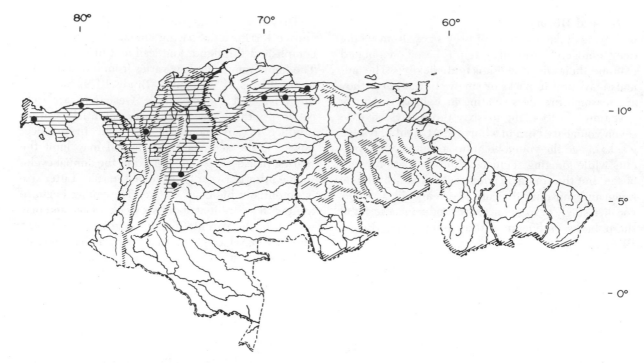

Map 7.3. Distribution of *Urocyon cinereoargenteus*.

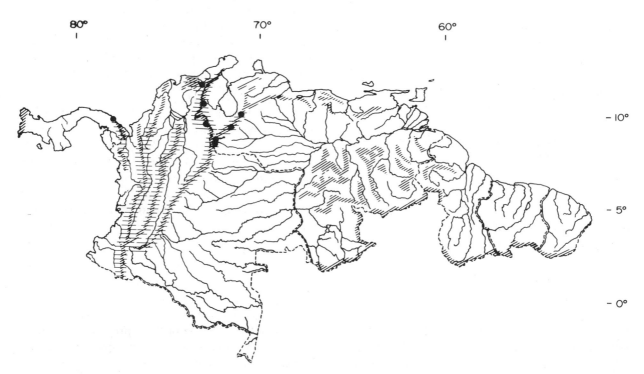

Map 7.4. Distribution of *Tremarctos ornatus*.

have a rather large litter, so the male's participation is probably essential for successful rearing. The gestation period is approximately sixty-five days, and three to five is the usual litter size. The female is polyestrous and can conceive at least twice in the same year (Porton, Kleiman, and Rodden 1987). Older offspring and adults show great social tolerance toward the young, and food is shared with a minimum of quarreling (Biben 1983). The animals are extremely vocal and use contact notes while moving about (Brady 1981). They urine mark in a rather unusual fashion, standing on the forelegs and micturating on a spot higher than would be possible on all fours (Kleiman 1972).

Genus *Urocyon* Baird, 1858
Urocyon cinereoargenteus (Schreber, 1775)
Gray Fox, Zorro Gris

Description
This fox is approximately the same size as the crab-eating fox. Total length ranges from 800 to 1,125 mm, and the tail ranges from 275 to 443 mm. The muzzle is delicate and pointed and the ears are large. The silver-gray dorsal coat makes it superficially resemble the crab-eating fox, but reddish markings over the ears, on the shoulders, and between the dorsum and venter and the lack of a black dorsal stripe clearly separate it from *Cerdocyon* (plate 9).

Range and Habitat
The gray fox occurs from the southeastern United States through Central America to northern South America. It was apparently a late entrant into South America but has become established in montane areas of Colombia and Venezuela (Bisbal 1982). Because of its preference for higher elevations, it does not come into appreciable contact with the crab-eating fox, although their ranges interdigitate in northern Venezuela (map 7.3).

Natural History
Most research on the gray fox has been concentrated in North America, but I infer that the behavior of the southern races is approximately the same. The gray fox prefers forested habitats and is broadly tolerant of a range of forest types. It feeds on fruits, invertebrates, and small vertebrates. Gestation is approximately 63 days, with a litter size of three to seven. In the southeastern United States gray foxes tend to form a pair-bond during the rearing season, and a mated pair generally has an exclusive home range. The male participates in rearing the young by provisioning the female and pups, conforming to the standard canid pattern (Lord 1961).

Comment on *Chrysocyon*
The maned wolf is suspected to occur in Colombia (Dietz 1985), but there is no "hard" evidence to support this claim.

FAMILY URSIDAE

Diagnosis
These large to medium-sized carnivores have a dental formula of I 3/3, C 1/1, P 4/4, M 2/3. They are powerfully built, with short ears and tail, and weigh up to 500 kg. There is no reduction in toe number, with five on each foot, and the animals have a plantigrade posture, bearing weight on the full sole of the foot (plate 9). Bears are immediately distinguishable from all other carnivores.

Distribution
The Recent distribution of this family includes all the major continents except Africa, Australia, and Antarctica.

Genus *Tremarctos* Gervais, 1855
Tremarctos ornatus (F. Cuvier, 1825)
Spectacled Bear, Oso

Description
In this medium-sized bear, weight may reach 200 kg. The basic color pattern is uniform dark brown to black, the face distinctively marked with varying degrees of white stripes. In their most complete form these white markings encircle the eyes, hence the vernacular name spectacled bear (plate 9).

Range and Habitat
The genus *Tremarctos* once was distributed from southern California to the southeastern United States and through Central America. Since the Pleistocene it exists only as a relict in South America and is confined to premontane and montane habitats in the Andes and adjacent foothills. The present range extends from Panama through Peru to Bolivia (map 7.4).

Natural History
The spectacled bear is a generalized omnivore, eating a variety of fruits, nuts, small vertebrates, and invertebrates. In Peru it may feed heavily on bromeliad hearts when fruit is unavailable. When climbing trees to reach fruits, it pulls small branches toward itself, creating a platform that will support its weight. This is not a true "nest." Daily resting places may be in cavities at the bases of trees or in rock caves. When it occurs near human populations, it may raid crops such as maize (Peyton 1980).

In foraging the bears tend to move as solitary in-

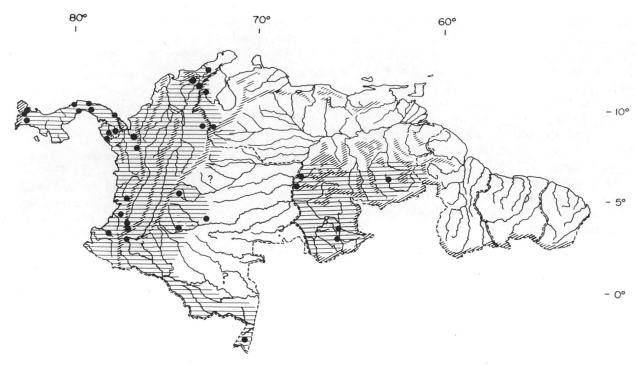

Map 7.5. Distribution of *Bassaricyon gabbii*.

dividuals except for females accompanied by young. The litter size averages two, born after seven months' gestation. This is not a true gestation period, however; a delayed implantation phase occurs, as in the temperate-zone bears of the Northern Hemisphere. The true gestation is probably about sixty-five days (Mondolfi 1971; Dathe 1967).

FAMILY PROCYONIDAE

Diagnosis

The molars of this family tend to be broad and do not exhibit the shearing edge so characteristic of canids and felids. In this feature Procyonidae resemble the bear family. The dental formula is usually I 3/3, C 1/1, P 4/4, M 2/2, except in the genus *Potos*, where the number of premolars is 3/4 (fig. 7.2). There is no toe reduction, and members of the family are either plantigrade or semiplantigrade. They are medium to small, with a total length ranging from 600 to 1,350 mm.

Distribution

If one excludes the Asiatic lesser panda, *Ailurus*, and the giant panda, *Ailuropoda*, from the procyonids, then the current distribution of the family is confined to the New World. Raccoons and their allies are distributed from Canada to Argentina.

They are adapted to a variety of habitats, but they generally occur only where there is tree cover.

Natural History

Procyonids are omnivores. Most are nocturnally adapted. The exception is the coati (*Nasua*), which is not only diurnal but also highly social. All members of the family show a pronounced arboreal ability.

Genus *Bassaricyon* J. A. Allen, 1876
Bassaricyon gabbii J. A. Allen, 1876
Olingo

Systematics

In the north this genus is represented by a single species, *B. gabbii*. *B. beddardi* has an incorrect provenience (R. Wetzel, pers. comm.). Two other species have been described with extremely restricted ranges in Costa Rica and Panama. Further study may indicate that *B. lasius* and *B. pauli* are conspecific with *B. gabbii*. *B. alleni* has been described from Ecuador.

Description

This genus is typified by its rather small ears and extremely long tail. Adult head and body length ranges from 370 to 420 mm; tail length is 380 to 432 mm; the ear averages 27 mm. The tail exceeds the head and body in length and often exhibits banding, though the pattern and degree of the banding

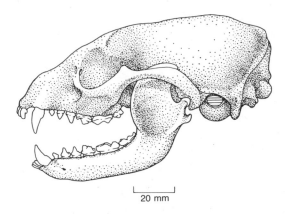

20 mm

Figure 7.2. Skull of *Procyon cancrivorus*.

are variable over the range of the species and in some specimens it is very indistinct. The basic color of the dorsum is honey brown. The tail banding and the lack of prehensility distinguish the olingo from the kinkajou (plate 10).

Range and Habitat

The genus *Bassaricyon* ranges from southern Nicaragua to the Amazon basin. It is absent from the llanos of Colombia and Venezuela and appears to be confined to areas of multistratal tropical evergreen forests below 2,000 m (map 7.5). Except for one record from Bolívar, Venezuela (E. Mondolfi, pers. comm.), it is absent from the eastern portion of northern South America.

Natural History

The olingo is a nocturnal, arboreal form that has been little studied in the wild; what we know of its behavior derives from captive studies. The standard reference is Poglayen-Neuwall and Poglayen-Neuwall (1965). Its diet includes fruits, invertebrates, and small vertebrates. There is some suggestion that it is more carnivorous than the kinkajou (*Potos*). Adults seem to forage singly, and it is believed to be less social than *Potos*; like-sexed adults are incompatible in captivity. The female bears a single young after seventy-three to seventy-four days' gestation. The olingo shelters in hollow trees.

Genus *Bassariscus* Coues, 1887
Cacomistle, Ring-tailed Cat, Bassarise

Description

Members of this genus are typified by a long, slender body, an elongated muzzle, and partially retractile claws on the five toes of each foot. Head and body length ranges from 306 to 380 mm, and weight ranges from 900 to 1,200 g. The tail is approximately

the same length as the head and body. The dorsum is light buff to dark brown, with the underparts contrasting white. The tail is moderately bushy and banded alternately with black and white, and the eyes are ringed with dark brown or black (plate 10).

Distribution

The genus *Bassariscus* occurs from the southwestern United States to Panama. There are two species, *B. astutus* and *B. sumichrasti*; the latter extends to the region covered in this volume.

Bassariscus sumichrasti (Saussure, 1860)
Central American Cacomistle, Guayanoche

Description

One specimen from Guatemala measured HB 397; T 433; HF 82. The basic dorsal coloration is gray brown and the venter ranges from tan to cream. The tail is distinctly banded.

Range and Habitat

This species occurs from southern Veracruz, Mexico, to extreme western Panama in moist forest (map 7.6).

Natural History

B. sumichrasti has not been well studied, but the more northern species, *B. astutus*, has been investigated in some detail (Trapp 1978). This species is nocturnal and solitary in its habits. It feeds on small vertebrates, fruit, and invertebrates. Three young are born after a fifty-four-day gestation period (Poglayen-Neuwall 1973; Poglayen-Neuwall and Poglayen-Neuwall 1980). *B. sumichrasti* in Veracruz, Mexico, is highly nocturnal and arboreal. It forages alone and emits a loud, high-pitched long call that apparently serves a spacing function (Coates-Estrada and Estrada 1986). We must wait for further studies to confirm if, in its behavior and ecology, *B. sumichrasti* resembles the better-known northern species.

Genus *Procyon* Storr, 1780
Raccoon, Mapache

Description

This genus contains six species, but only two, *P. lotor* and *P. cancrivorus*, are widely distributed, the other four being confined to islands. Although the muzzle is pointed, the raccoon's head is quite broad. The soles of the feet are naked and the toes long, and the forepaws are capable of considerable dexterity. The dorsal pelage is some shade of brown, with black-tipped guard hairs producing a grizzled effect in *P. lotor*. The tail is much shorter than the head and body and is banded alternately with dark brown

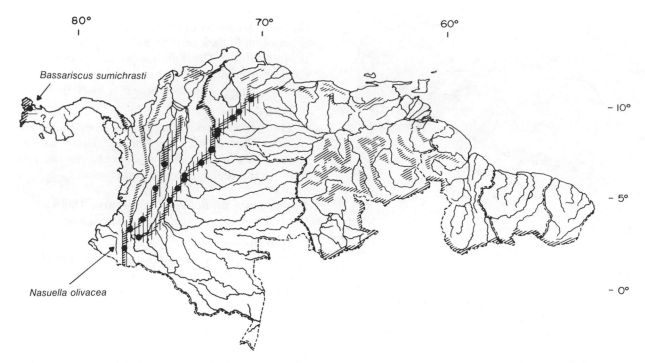

Map 7.6. Distribution of *Bassariscus sumichrasti* and *Nasuella olivacea.*

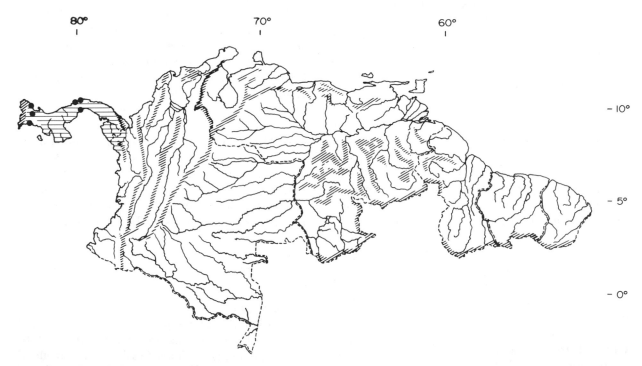

Map 7.7. Distribution of *Procyon lotor.*

and contrasting gray to yellow-orange bands. Generally a black mask surrounds the eyes, contrasting with the fur of the face (plate 10).

Distribution

The genus *Procyon* is widely distributed from southern Canada to Argentina. It prefers forested habitats and is often found in the vicinity of streams or ponds.

Natural History

Raccoons tend to forage singly or in groups composed of females and their young. They are nocturnal and feed upon a wide range of fruits, vertebrates, and invertebrates. They exhibit considerable manual dexterity and seek crustaceans and frogs by feeling under the surface of the water. The sense of touch is highly developed (Kaufmann 1982; Stains 1956).

Procyon lotor (Linnaeus, 1758)

Description

Head and body length averages 500 mm, and the tail averages 300 mm. Males are considerably larger than females. *P. lotor* is easily distinguished from *P. cancrivorus* because the pelage is grizzled in *P. lotor* and the soft underfur is dense. The underfur is sparse in *P. cancrivorus,* giving the animal a thinner appearance (plate 10).

Range and Habitat

P. lotor ranges from southern Canada to western Panama, where it is replaced by *P. cancrivorus* (map 7.7).

Natural History

The raccoon is a nocturnal omnivore that forages both arboreally and terrestrially. It generally seeks shelter in a hollow tree, or it may opportunistically use burrows made by other mammals such as armadillos. Home ranges vary depending on the locality; in Georgia in the United States, home ranges averaged 65 ha for males and 39 ha for females. Territoriality is not shown (Lotze 1979). Three to seven young are born after a sixty-three-day gestation period. Generally only one litter is produced each year (Kaufmann 1982).

Procyon cancrivorus (F. Cuvier, 1798)
Crab-eating Raccoon, Osito Lavador

Description

Head and body length may reach 900 mm, and the tail 350 mm (plate 10). Three specimens from Colombia exhibited the following ranges: HB 705–904; T 251–331; HF 130–40; E 56–61. The pelage

is short and underfur is absent. The dorsum lacks the grizzled appearance characteristic of *P. lotor* and is a more uniform brown.

Range and Habitat

P. cancrivorus is distributed from southern Costa Rica to Argentina. Its range overlaps with that of *P. lotor* in Costa Rica and Panama (map 7.8). It appears to be rare or absent in eastern Brazil.

Natural History

The crab-eating raccoon is nocturnal and a generalized omnivore. Very little is known about the ecology of this species; limited information is available from captive studies. It is frequently found in association with streams, lagoons, or lakes. The animals tend to forage alone except for the female-offspring unit. The average litter consists of three young. The developmental schedule is similar to that of *P. lotor* (Löhmer 1976).

Genus *Nasua* Storr, 1780
Nasua nasua (Linnaeus, 1766)
Coati, Cusumbo, Pizote

Description

This genus contains two species. One is an island form from off the coast of Yucatán; the other is the common coati, *N. nasua.* Although some workers separate the Central American coati as *N. narica,* I follow Honacki, Kinman, and Koeppl (1982). I believe that the genus *Nasua* is badly in need of revision. The tail is nearly equal to the head and body in length: head and body length ranges from 430 to 700 mm and the tail from 420 to 680 mm. The ears are short, the rostrum is narrow, and the snout is mobile. The dorsal pelage is variable, ranging from pale brown to reddish; in adult males, the shoulder region may show yellow or white hairs. The tail is banded with alternate yellow and brown markings, and the eyes are bordered by a mask that varies from reddish to brown (plate 10).

Range and Habitat

The genus is distributed from the southwestern United States to Argentina. Coatis appear to be absent from the llanos of Venezuela. They are broadly tolerant of a variety of habitat types, exploiting dry deciduous forests to multistratal tropical evergreen forests (map 7.9).

Natural History

The coati is diurnal and, unlike most members of the Procyonidae, highly social. Several females form permanent bands that forage together with their young. The males join female bands during the breeding season, but at other times of the year males

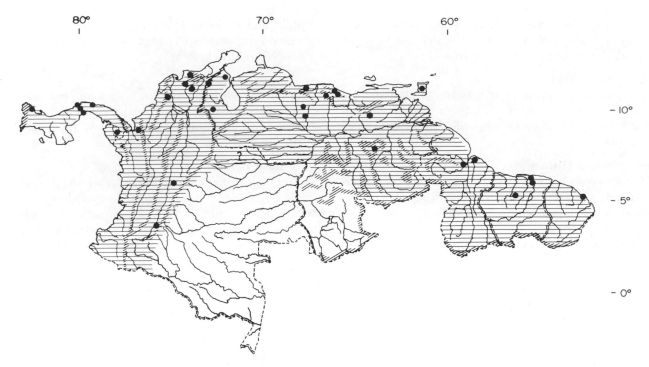

Map 7.8. Distribution of *Procyon cancrivorus*.

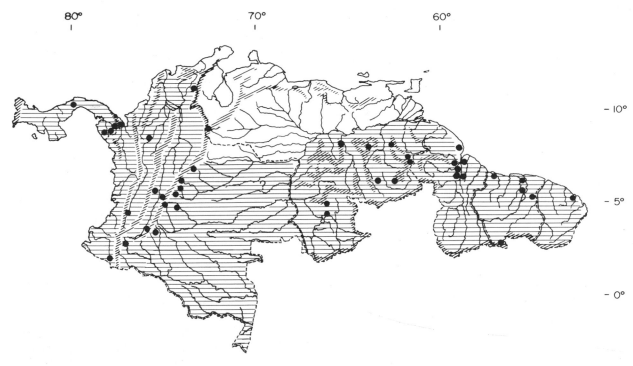

Map 7.9. Distribution of *Nasua nasua*.

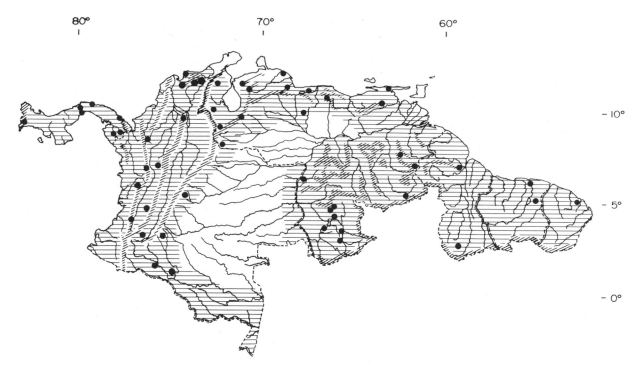

Map 7.10. Distribution of *Potos flavus*.

forage alone. Copulation may take place arboreally or terrestrially. In good habitats the animals have a regular home-range usage pattern and exclude neighboring bands from core areas. Home ranges may be up to 2 km². The animals sleep in trees at night, and the young are born in an arboreal nest after a seventy-seven-day gestation period. Litter size ranges from one to five. During the early part of life the young remain in a nest. When they are two to three weeks of age, they begin to accompany their mother and join a foraging unit with other females and their offspring. The animals forage both arboreally and terrestrially. Their diet includes fruit, vertebrates, and invertebrates. They prey actively on soil arthropods, using their sensitive snouts to probe the forest litter (Kaufmann 1962; Russell 1981; Sunquist and Montgomery 1973).

Genus *Nasuella* Hollister, 1915
Nasuella olivacea (Gray, 1865)
Mountain Coati, Coati Oliva

Description
This small coati is half the size of *N. nasua*. Head and body length ranges from 360 to 390 mm, and the tail ranges from 200 to 240 mm. The dorsum is gray brown. In body proportions and tail banding pattern *N. olivacea* resembles the common coati, but it is easily distinguished by its small size (plate 10).

Range and Habitat
Nasuella is found in montane habitats from north-

ern Colombia and Venezuela to Peru. It is a high-altitude specialist, preferring forested habitats at elevations over 2,000 m (Aagard 1982) (map 7.6).

Natural History
The mountain coati has not been the subject of detailed study, and little is known concerning its ecology and behavior. What fragmentary data are available suggest that its natural history is similar to that of the lowland coati, and confirmation awaits further fieldwork.

Genus *Potos* E. Geoffroy and G. Cuvier, 1795
Potos flavus (Schreber, 1774)
Kinkajou, Cuchicuchi

Description
The kinkajou is unique among the Procyonidae in having a prehensile tail. Head and body length averages 500 mm; tail length is about 450 mm. The range of measurements for four specimens from Colombia was HB 433–54; T 424–68; HF 87–98; E 40–42. Weight averages about 3 kg. The head is rounded and the muzzle short but pointed. The teeth are low crowned. The basic dorsal pelage is honey brown, and the fur is short and dense (plate 10).

Range and Habitat
The kinkajou is distributed from Veracruz, Mexico, to southern Brazil. It is associated with forested habitats, in keeping with its strong arboreal adaptations (map 7.10).

Natural History

Kinkajous specialize on fruits and arboreal vertebrates as their primary food items. Charles-Dominique et al. (1981) monitored their foraging behavior and found that they are important dispersers of the seeds of *Ficus*, *Virola*, and *Inga*, which are ingested with the fruit pulp and pass unharmed through the digestive tract. Often several kinkajous may be found eating in the same tree. In good habitat, densities can reach fifty-nine per km² (Walker and Cant 1977). A single young is born after a rather long gestation of 112 to 120 days; social bonds between mother and offspring may persist for a considerable time. This species has been studied in captivity (Poglayen-Neuwall 1962, 1966).

FAMILY MUSTELIDAE
Weasels and Their Allies

Diagnosis

These small to medium-sized carnivores generally show a reduction in molar number. Dental formula for most South American forms is I 3/3, C 1/1, P 3/3, M 1/2. The otters (*Lutra*) do not show the reduction in premolar number, and their formula is 4/3 (see figs. 7.3 and 7.4). The body tends to be elongated, with relatively short legs and usually a long tail. The family is classically divided into five subfamilies, three of which occur in northern South America: the subfamilies Mustelinae (the weasels), Mephitinae (the skunks), and Lutrinae (the otters).

Distribution

Species of the weasel family occur on all continents except Antarctica and Australia. The family had a northern origin, since mustelids first appear in the Oligocene in North America and Europe. Species of the Mustelidae entered South America in the late Pliocene and rapidly occupied the small-carnivore ecological niches.

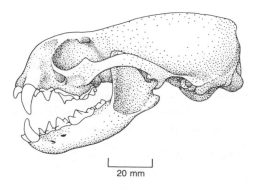

Figure 7.3. Skull of *Lutra* sp.

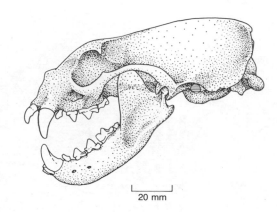

Figure 7.4. Skull of *Eira barbara*.

Genus *Mustela* Linnaeus, 1758

Description

Species of the genus *Mustela* exhibit a wide range of head and body lengths (114 to 228 mm). The tail is usually less than half the head and body length. Females are generally much smaller than males, but not in *M. felipei*. The soles of the feet are usually furred, but two South American species, *M. africana* and *M. felipei*, have naked soles. The dorsum is brown and the venter white or cream. The tail tip is often black.

Distribution

The genus is widely distributed in the temperate zones of North America, Europe, and Asia. Three species occur in South America: *M. africana*, *M. felipei*, and *M. frenata* (Izor and de la Torre 1978).

Mustela felipei Izor and de la Torre, 1978

Description

Head and body length ranges from 217 to 220 mm, and the tail averages 107 mm. The baculum is trifid, a character shared with *M. africana* (Izor and Peterson 1985). The face, dorsum, and tail are dark brown, and the venter is pale yellow from the chin to the groin. The plantar surfaces are naked.

Range and Habitat

The species is known from only two localities in the Cordillera Central of Colombia: San Augustin, Huila, and Popayán, Cauca, between 1,700 and 2,700 m (Izor and de la Torre 1978).

Mustela frenata Lichtenstein, 1831
Long-tailed Weasel

Description

Head and body length ranges from 180 to 220 mm, and the tail is approximately 50% to 60% of the head and body length. The plantar surfaces are furred.

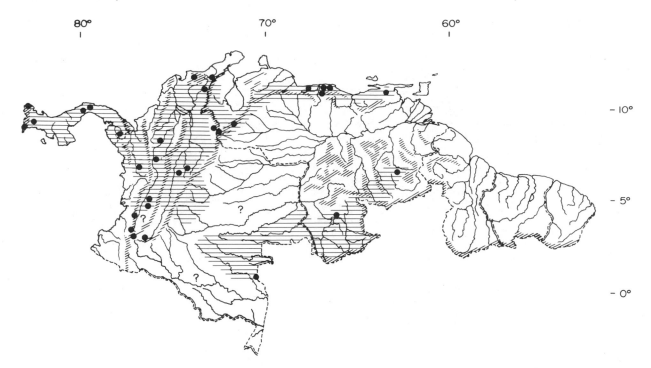

Map 7.11. Distribution of *Mustela frenata*.

The dorsum is brownish, the venter is cream to white, and the distal third of the tail is usually black. In the northern part of the range the dorsal coat changes to white in winter, but in the southern parts, including South America, there is no seasonal change (plate 11).

Range and Habitat
This species is distributed from southern British Columbia in Canada to northern South America. It has been taken in scattered localities of Colombia and Venezuela (map 7.11). It is associated with the premontane forests of the Andes and with the forests of the Guyana highlands in southern Venezuela. It prefers wooded habitats and tolerates a wide range of elevations (Hall 1951).

Natural History
The long-tailed weasel is a specialist predator on ground-nesting birds and rodents. It does not typically forage in trees, although it can climb. This species has been studied in some detail in the North American part of its range; little is known concerning the natural history of the southern subspecies. In the north temperate zone, the long-tailed weasel mates in the summer and exhibits delayed implantation so that the young are born the following spring. The actual gestation time, however, is twenty-three to twenty-four days when the time of delay is subtracted. Litter size may be high, up to seven in North America. Although the mother and young may forage as a social unit during the weaning phase, adults are typically solitary except during the breeding season (Hall 1951). Seasonality of breeding in the tropics has not been investigated.

Genus *Galictis* Bell, 1826
Galictis vittata (Schreber, 1776)
Grison, Huron

Description
This genus contains three species (*G. cuja*, *G. allamandi*, and *G. vittata*) distributed from southern Veracruz, Mexico, to Argentina. Most authorities consider the species *G. allamandi* to be a junior synonym of *G. vittata* (Honacki, Kinman, and Koeppl 1982). The genus *Grison* is a junior synonym for *Galictis*. Head and body length ranges from 475 to 550 mm; the tail is short, usually less than 150 mm. Animals may weigh up to 3.2 kg. The limbs, throat, and venter are black, and the black extends to the face. A white stripe from the forehead to the shoulders separates the venter from the dorsum, which has a grizzled salt-and-pepper pattern (plate 11).

Range and Habitat
Galictis vittata occurs from southern Veracruz, Mexico, to Brazil, where it is replaced by *G. cuja* (Krumbiegel 1942). It usually occurs below 1,200 m but is broadly tolerant of the vegetation cover, being found in both dry deciduous forests and multistratal tropical rain forests (map 7.12).

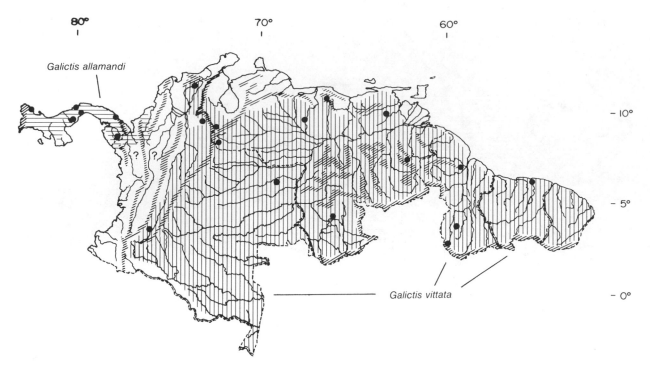

Map 7.12. Distribution of two species of *Galictis*.

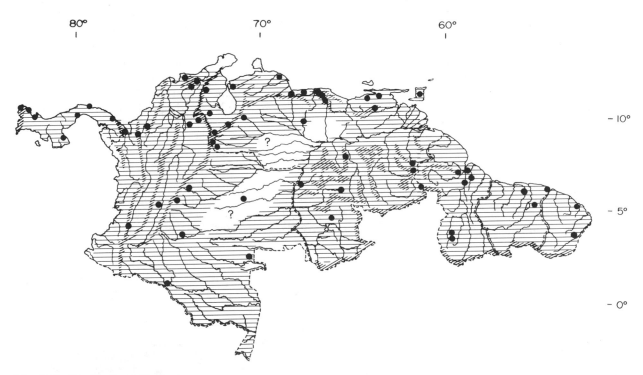

Map 7.13. Distribution of *Eira barbara*.

Natural History

Grisons are active in the early morning and late afternoon and at night. Although they can climb, they generally forage terrestrially (Kaufmann and Kaufmann 1965). They can eat a wide range of foods but are primary predators on reptiles, small birds, and small mammals. One radio-collared adult female had a home range of 4.2 km² in the llanos of Venezuela (Sunquist, Sunquist, and Daneke, n.d.). Grisons frequently shelter in burrows, often abandoned armadillo burrows. They may be seen in groups of three or four, probably a mother and her young. The gestation period is thirty-nine days, and litter size averages two (Miles Roberts, pers. comm.).

Genus *Eira* H. Smith, 1842
Eira barbara (Linnaeus, 1758)
Tayra, Cabeza de Viejo, Eirá

Description

Head and body length ranges from 600 to 680 mm, tail length is 380 to 470 mm, and weight may exceed 5 kg. The body is long and slender. Tayras are extremely able climbers, as evinced by the strong claws on the forefeet and hind feet. Coat color is variable over the range, but in general the animals are brown to black. In Panama they tend to be all black, but in other parts of their range the head and throat may be yellow (plate 11).

Range and Habitat

The tayra occurs from southern Veracruz, Mexico, to northern Argentina (Krumbiegel 1942). It prefers forested habitats but tolerates dry deciduous forests as well as multistratal tropical evergreen forests. It generally occurs at elevations below 1,200 m (map 7.13).

Natural History

Tayras are active both during the day and at night but are primarily diurnal, foraging on the ground and in the trees. They eat a wide range of foods including fruit, small vertebrates, and invertebrates. Often termed a frugivore, this large mustelid can be an active predator (Galef, Mittermeier, and Bailey 1976), but Defler (1980) does not believe tayras prey significantly on primates. They are usually encountered as solitaries or as family groups. A radio-collared female in the llanos of Venezuela had a home range of 9 km² (Sunquist, Sunquist, and Daneke, n.d.). The gestation period is believed to equal seventy days (Vaughn 1974; Poglayen-Neuwall 1975), and the average litter size is two. The young are playful and active during their early development. A wide range of vocalizations are employed during social interactions (Kaufmann and Kaufmann

1965). The species' behavior in captivity has been studied by Poglayan-Neuwall (1978).

Genus *Conepatus* Gray, 1837
Hog-nosed Skunk, Mapurite

Description

Head and body length ranges from 300 to 490 mm and the tail from 160 to 210 mm; weight may reach 4 kg. The muzzle is bare, and the nose pad is large and flat. The claws are well developed. Hog-nosed skunks have a typical pelage pattern, black on the sides and venter with either a strongly contrasting white dorsum or two white dorsal stripes. The species in northern South America is characterized by the white striping.

Distribution

The genus *Conepatus* is distributed from the southwestern United States to northern Argentina. It is broadly tolerant of many habitat types (Kipp 1965).

Conepatus semistriatus (Boddaert, 1784)

Description

Two white dorsal stripes contrasting with a black body immediately distinguish this species (plate 11).

Range and Habitat

Conepatus semistriatus is distributed from the Yucatán, Mexico, through northern Colombia into northern Venezuela (map 7.14).

Natural History

The hog-nosed skunk is nocturnal and forages for invertebrates and small vertebrates by digging in the soil. Home ranges of 18 to 53 ha have been recorded in Venezuela (Sunquist, Sunquist, and Daneke, n.d.). It uses the burrows of armadillos as shelters but can also construct its own. Four to five young are born after a gestation period of approximately sixty days. Aside from the female-young unit, hog-nosed skunks tend to forage singly. Strong-smelling secretions from the anal glands are employed in self-defense (Mondolfi 1973).

Comment on *Mephitis*

The hooded skunk, *Mephitis macroura*, has recently been recorded from Costa Rica. It may be expanding its range but thus far remains unrecorded from Panama (Janzen and Hallwachs 1982).

Genus *Lutra* Brunnich, 1771
Freshwater Otter, Perro de Agua

Description

The dental formula is I 3/3, C 1/1, P 4/3, M 1/2 (see fig. 7.3). A cylindrical body form, very short

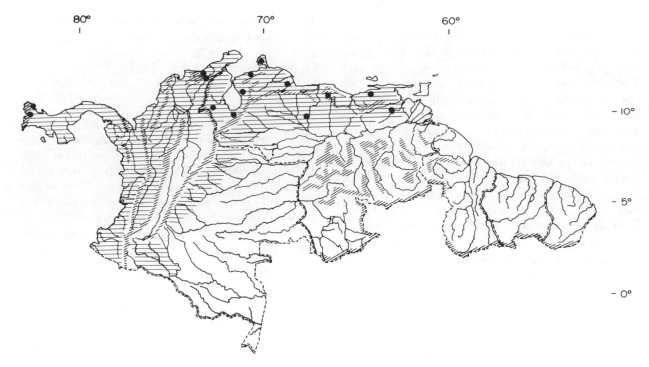

Map 7.14. Distribution of *Conepatus semistriatus*.

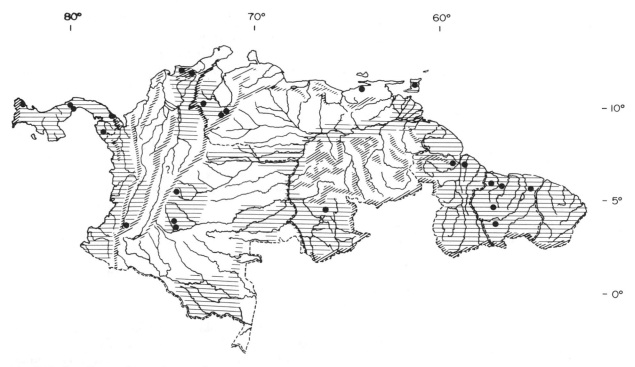

Map 7.15. Distribution of *Lutra longicaudis*.

ears, webbed feet, and stout vibrissae characterize this aquatic carnivore. The tail is thick at the base, tapering to a point, and the body contours are smooth. There are approximately twelve species worldwide.

Distribution

The genus is distributed in freshwater and coastal areas of North and South America, most of Africa, Europe, Asia, and parts of Southeast Asia. One species occurs in northern South America.

Lutra longicaudis (Olfers, 1818)

Description

This medium-sized otter has a head and body length ranging from 400 to 700 mm; the tail ranges from 390 to 500 mm. Weight is generally less than 12 kg (plate 11).

Range and Habitat

Lutra longicaudis is distributed from coastal Mexico to Brazil. It occurs in most freshwater streams, but it has been severely persecuted for its pelt and thus is locally extinct over many parts of its former range (Duplaix-Hall 1977) (map 7.15).

Natural History

This otter specializes in feeding on fish and crustaceans. Aquatically adapted, it hunts in rivers, streams, and ponds but shelters in a terrestrial burrow that it constructs itself. The southern river otter has been little studied compared with the northern species, *L. canadensis.* Litter size in the latter species is two or three, born after a true gestation of approximately fifty-six days. The northern river otter shows a delayed implantation, but this has not been confirmed for the southern species (Melquist and Hornocker 1983). Aside from the female-young unit, these otters tend to forage and dwell alone (Mondolfi 1970).

Genus *Pteronura* Gray, 1837
Pteronura brasiliensis (Gmelin, 1788)
Giant River Otter, "Lobo" del Río Grande

Description

Head and body length ranges from 1 to 1.5 m, the tail is approximately 700 mm, and weight can be up to 34 kg. Males average about 5 kg heavier than females. This is the largest freshwater otter and is clearly distinguishable from *Lutra* by its size and by its dark dorsal pelage with a white to cream throat patch (plate 11).

Range and Habitat

P. brasiliensis formerly inhabited most of the freshwater streams of South America from Vene-

zuela to Argentina. It has been severely persecuted for its pelt and thus is locally extinct over many areas of its former range (map 7.16).

Natural History

The giant river otter appears to forage mainly during the day, taking a wide variety of fishes. Although it forages in the water, it has resting places along streams, where it clears out the vegetation by biting off small shrubs to create an open area for sunning. The home range may exceed 12 km². This otter is highly sociable and appears to live in monogamous pairs. The gestation period is sixty-five to seventy days, and the litter size ranges from one to five. Bonds between parents and young appear to be strong; parents and young stay together for many months before the family unit begins to break up. Groups of five to eight are not uncommon. The vocal repertoire is extensive, including nine distinguishable calls. The standard reference on the natural history of this species is Duplaix (1980).

FAMILY FELIDAE
Cats, Gatos

Diagnosis

The tooth formula is extremely modified in this family; most notably, the number of molars is reduced. The dental formula typically is I 3/3, C 1/1, P 2–3/2, M 1/1 (fig. 7.5). The auditory bulla is inflated and subdivided into two chambers. Five toes are retained on the forefoot and four on the hind foot; the claws are retractile except in the African cheetah (*Acinonyx*). The eyes are directed forward, and the rostrum is rather short.

Distribution

The Recent distribution includes all continents except Australia and Antarctica. Members of the

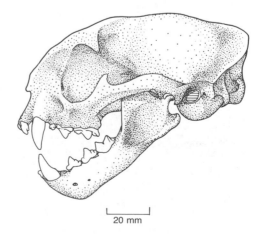

20 mm

Figure 7.5. Skull of *Felis pardalis.*

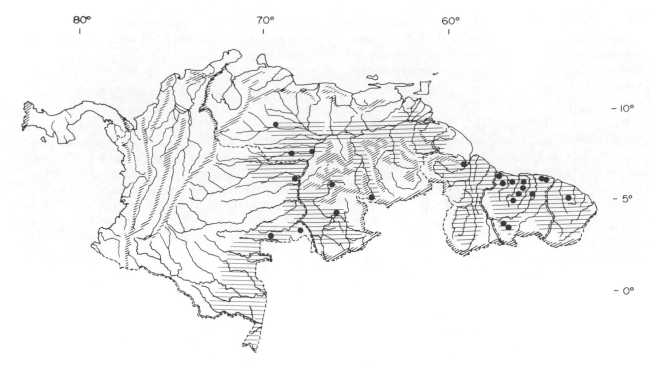

Map 7.16. Distribution of *Pteronura brasiliensis*.

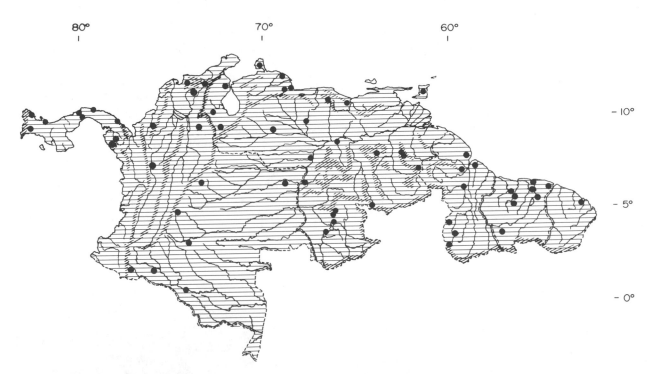

Map 7.17. Distribution of *Felis pardalis*.

family Felidae are distinguishable in the Oligocene, and at this point one could recognize two divergent groups, the true cats and the saber-toothed cats. The latter are frequently placed in their own subfamily, the Machairodontinae. The earliest true cats appear in the Oligocene of North America, Eurasia, and later in Africa and are placed in the subfamily Nimravinae. Members of the Felidae entered South America in the Pliocene and rapidly replaced the large carnivorous birds (Phorusrhachidae) that occupied the medium-to-large cursorial carnivore niche (Vuilleumier 1985).

Natural History

Species of this family are adapted for catching vertebrate prey. The largest can kill prey that exceed their own body weight. Although not especially adapted for long-distance running like the Canidae, the felids are built for maximum speed over a short distance. They retain a strong arboreal ability, much more pronounced in some species than in others. The cats are among the most carnivorous of the order Carnivora, eating a minimum of fruits and herbaceous material. The Felidae may be considered more specialized than the Canidae. With the exception of the African lion, most cats do not form complex social groupings. Indeed, the most common social unit is a mother and her semidependent offspring. Although rock shelters and burrows may be opportunistically used, the Felidae are not adapted for constructing their own burrows. The larger species in this family often have distinctive long-distance calls that make identification possible even when the animal is not visible to the fieldworker (Kleiman and Eisenberg 1973; Leyhausen 1979).

Genus *Felis* Linnaeus, 1758
Smaller Cats, Gatos

Description

This genus contains roughly thirty-five species and can be subdivided into thirteen subgenera. Four species in the subgenera *Leopardus* and *Herpailurus* occur in northern South America. These are the small cats easily recognized by their species and family characters (see plate 12).

Distribution

The genus *Felis* as defined in this volume has representatives in the Americas, Eurasia, and Africa.

Felis (Leopardus) pardalis Linnaeus, 1758
Ocelot, Mano Gordo

Description

This medium-sized cat has a head and body length ranging from 700 mm to 900 mm, tail length from 280 to 400 mm, and hind foot from 130 to 180 mm. Females are slightly smaller. The upperparts are grayish to cinnamon, with black markings forming streaks on the neck or elongated spots on the body. The tail exhibits incomplete banding. Spotting and banding extend to the dorsal surfaces of the limbs and onto the venter, whose basic color is white with occasional black spots (plate 12).

Range and Habitat

The ocelot is found from southern Texas in the United States to Argentina. It tolerates a variety of habitat types, occurring in areas of dry deciduous forests as well as in multistratal evergreen forests. It is seldom found far from continuous trees or dense shrub cover. It typically occurs at elevations below 1,200 m (map 7.17).

Natural History

An extremely secretive animal, the ocelot is seldom seen, but its presence is betrayed by characteristic tracks. Prey consists of larger caviomorph rodents, iguanas, and small rodents (Enders 1935; Mondolfi 1986). It tends to move and hunt alone. In Venezuela, adult females occupy an exclusive territory of about 3 km^2; an adult male's territory overlaps the territories of several females and may exceed 10 km^2. Ocelots tend to defecate at selected sites in their home ranges, and feces may accumulate at such latrines for several months during the dry season (Sunquist, Sunquist, and Daneke, n.d.; Ludlow 1986). One or two young are born after a gestation period of approximately seventy-five days and are reared by the mother in a secluded place. The male apparently does not take part in parental care. Retaining strong arboreal abilities, these cats frequently sleep in trees during the day.

Felis (Leopardus) tigrina Schreber, 1775
Little Spotted Cat

Description

This is one of the smallest wild cats in the New World; head and body length ranges from 340 to 560 mm and tail length from 198 to 220 mm. The basic ground color of the upperparts is light fulvous, overlaid with a spotted pattern and two pairs of strong black stripes on the sides of the neck. In certain parts of its range in northern Venezuela the little spotted cat exhibits a strong melanistic tendency, and all-black individuals are frequently caught. The spotted pattern against the basic black dorsal pelage is indistinctly visible when the light is appropriate (plate 12).

Range and Habitat

The little spotted cat is distributed from Costa Rica across northern South America to Brazil. It is

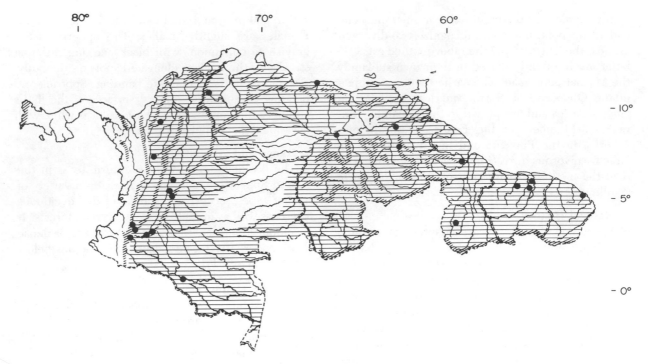

Map 7.18. Distribution of *Felis tigrina*.

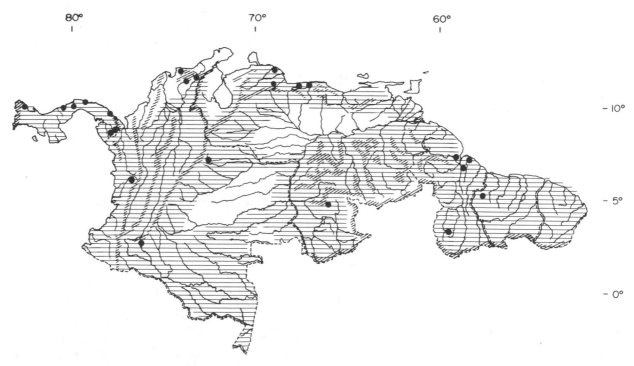

Map 7.19. Distribution of *Felis wiedii*.

frequently found in premontane forests, although it also occurs in lowland forests (map 7.18).

Natural History

One of the least known of the spotted cats, this shy, solitary species is frequently overlooked because it is seldom encountered and because its tracks are almost identical in size to those of the domestic cat. Its diet includes birds and small rodents (Gardner 1971; Leyhausen and Falkena 1966). One or two young are born after a gestation period of approximately seventy-five days.

Felis (Leopardus) wiedii Schinz, 1821
Margay Cat

Description

In this small cat, head and body length ranges from 530 to 700 mm, the tail from 331 to 510 mm, and the hind foot from 89 to 132 mm. The tail is proportionately longer in the margay than in the ocelot, and the margay tends to be smaller. Young margays, however, are markedly similar in appearance to young ocelots. The margay has the same basic ground color of the dorsum, from grayish to cinnamon, strongly marked with bands and spots on the dorsum and incomplete black stripes in a transverse pattern on the tail, and the venter is white (plate 12).

Range and Habitat

The margay cat prefers densely forested areas and ranges from southern Texas to southern Brazil. It is rarely found above 1,200 m elevation (map 7.19).

Natural History

The margay is adapted to an arboreal life. Unlike all other cats, it can pronate and supinate its hind foot; thus, when it descends from a tree the hind foot rotates around the ankle so that the animal can hang vertically, much like a squirrel (Leyhausen 1963). Its proportionately longer tail apparently is also an adaptation for arboreality, serving as a counterweight in springing or in maintaining balance while traversing limbs. Konecny (n.d.) reports that it forages arboreally. It preys actively on rodents and small birds (Mondolfi 1986). The margay has a single young after a seventy-five-day gestation period, one of the lowest reproductive potentials of the smaller cats. The young is proportionately large at birth, so postnatal development is rather rapid. As far as is known, the female assumes sole responsibility for rearing the offspring. This cat is seldom seen in the wild, and what information we do have concerning its natural history is derived almost entirely from observations on captives.

Felis (Herpailurus) yagouaroundi
E. Geoffroy, 1803
Jaguarundi, Gato Moro

Description

The first premolar is very reduced or occasionally absent. This medium-sized cat has a slender build and a rather long neck; head and body length ranges from 760 to 820 mm, the tail from 320 to 506 mm, and the hind foot from 120 to 156 mm. The head tends to be long in proportion to the body, lacking the rounded contours so characteristic of the Leopardus group. Unlike the other small cats within its range, the jaguarundi is not spotted; coat color is variable from black to gray to reddish (plate 12).

Range and Habitat

Currently the species occurs from the Rio Grande valley in Texas to northern Argentina. It is broadly tolerant of a variety of habitats and occurs in gallery forests, multistratal tropical evergreen forests, and dry deciduous forests (map 7.20).

Natural History

This is the species most frequently seen by fieldworkers, probably because the jaguarundi is often active in early morning and late afternoon. Unlike other cats, it has a circular pupil, which may reflect its more diurnal habits (Konecny n.d.. Little is known concerning the habits of this cat. It is considered to be an accomplished predator on small rodents, lagomorphs, and ground-nesting birds (Mondolfi 1986). It frequently hunts on the boundaries of gallery forests and thus may be found some distance from forest cover in savanna habitats. Two young may be born after a gestation period of approximately seventy-eight days (Hulley 1976). Two or more individuals are often seen together, undoubtedly a female and her dependent offspring.

Genus *Puma* Jardine, 1834
Puma (= *Felis*) *concolor* (Linnaeus, 1771)
Puma, Mountain Lion, León, León Americano

Description

Although I do not go as far as Leyhausen (1979), who separates concolor under the genus Profelis, I do believe it is sufficiently distinct to be included under Puma.

One of the largest cats in the Americas, the puma is second only to the jaguar in weight. Males may exceed 2.7 m in total length, with a tail length from 600 to 700 mm and weight from 60 to 100 kg. Females are generally smaller, 1.5 to 2.3 m in total length. Size varies widely over the geographic range of this species. The basic body color is cinnamon to

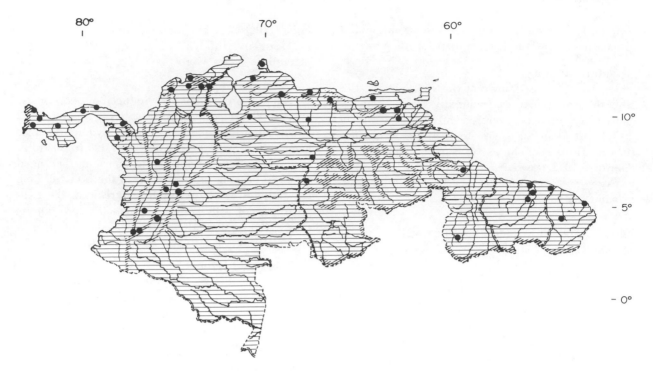

Map 7.20. Distribution of *Felis yagouaroundi*.

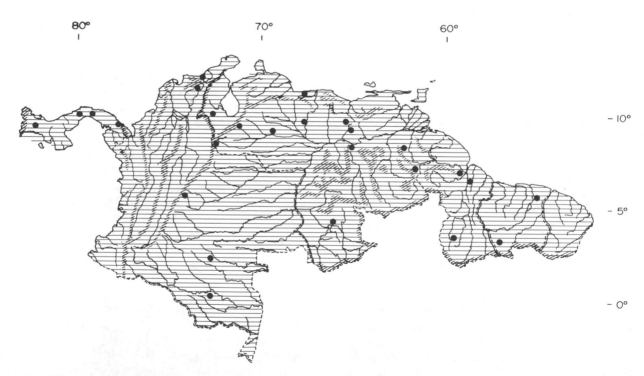

Map 7.21. Distribution of *Puma concolor*.

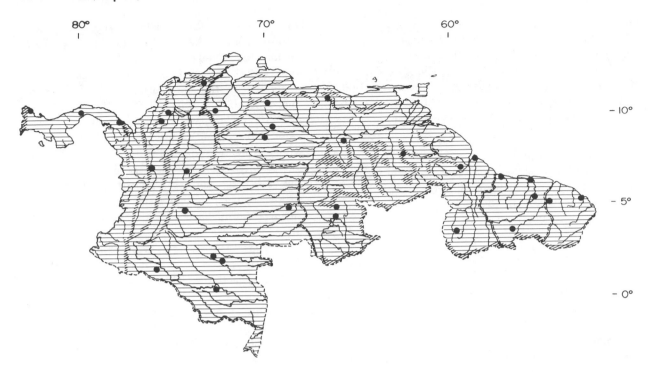

Map 7.22. Distribution of *Panthera onca*.

rufous brown; the underparts are white. The young exhibit a spotted pattern at birth that disappears with the molt to adult pelage at approximately six months of age.

Range and Habitat

The species ranges from British Columbia to Patagonia and thereby has the largest range of any cat in the Americas (map 7.21). Broadly tolerant of a variety of habitat types, the puma occurs from elevations exceeding 2,000 m down to sea level. It can occur in coniferous forests as well as in multistratal, low, and tropical rain forests.

Natural History

The puma has been well studied in North America, but little fieldwork has been done in the southern parts of its range (Robinette, Gashwiler, and Morris 1961; Currier 1983). One to five young are born after a gestation of ninety-three days in the northern parts of its range. There is some suggestion that litter size is reduced in more tropical habitats. The puma preys actively on medium to large game. Over much of its northern range it hunts deer of the genus *Odocoileus*.

In the northern parts of their range, the animals exhibit a territorial pattern, with like-sexed adults well spaced. Male ranges overlap those of one or more females. Over much of its southern range the puma co-occurs with the jaguar, *Panthera onca*. Generally, the jaguar is dominant over the puma, and the puma adjusts its movements to avoid contact

(Schaller and Crawshaw 1980). The puma tolerates a wider range of habitat types than the jaguar and replaces it as the dominant predator in southern Patagonia and at high elevations in the Andes.

Genus *Panthera* Oken, 1816
Larger Cats
Panthera onca (Linnaeus, 1758)
Jaguar, El Tigre, Tigre

Description
This is the largest cat in the Americas by average

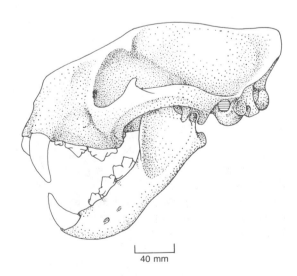

40 mm

Figure 7.6. Skull of *Panthera onca*.

weight: total length ranges from 1.7 to 2.4 m, tail 520 to 660 mm; and weight 60 to 120 kg (Venezuela). The jaws are powerful, and the skull is massive (fig. 7.6). Although melanistic individuals are known, the general color of the upperparts varies from buff to light gold, marked with a series of black rosettes that make it distinctive and unmistakable among cats in the Americas (plate 12).

Range and Habitat

Formerly the species occurred from New Mexico and the Gulf portion of Texas to northern Argentina. It prefers forested habitats at elevations below 1,200 m (map 7.22).

Natural History

The jaguar is similar in behavior to other large cats. Like-sexed adults tend to occupy exclusive areas for hunting; male ranges overlap those of one or more females and may exceed 25 km^2 (Schaller and Crawshaw 1980). There is no permanent social grouping beyond the female and her offspring. One to four young are born after a gestation period of 101 days, and the young may remain with the mother until eighteen months of age. A versatile hunter, the jaguar will kill and eat caiman, capybara, turtles, peccaries, and deer (Schaller and Vasconcelos 1978; Mondolfi and Hoogersteijn 1986). It can become a significant predator on livestock and therefore is often severely persecuted by man. It has been exterminated over many parts of its range and is considered a threatened species. It is often heard rather than seen, and its rasping vocalization (like someone rapidly sawing wood) is readily identifiable.

References

• References used in preparing distribution maps

Aagaard, E. M. J. 1982. Ecological distribution of mammals in the cloud forests and paramos of the Andes, Mérida, Venezuela. Ph.D. diss., Colorado State University.
• Beebe, W. 1919. The higher vertebrates of British Guiana with special reference to the fauna of the Bartica district, 7: List of Amphibia, Reptilia, and Mammalia. *Zoologica* 2:205–38.
Bekoff, M., ed. 1978. *Coyotes: Biology, behavior and management.* New York: Academic Press.
• Berta, A. 1982. *Cerdocyon thous. Mammal. Species* 186:1–4.
———. 1984. The Pleistocene bush dog *Speothos pacivorus* (Canidae) from the Lago Santo Caves, Brazil. *J. Mammal.* 65(4):549–59.
———. 1986. *Atelocynus microtis. Mammal. Species* 256:1–3.

Biben, M. 1982a. Object play and social treatment of prey in bush dogs and crab-eating foxes. *Behaviour* 79:201–11.
———. 1982b. Ontogeny of social behaviors related to feeding in the crab-eating fox and the bush dog. *J. Zool.* (London) 196:207–16.
———. 1983. Comparative ontogeny of social behaviour in three South American canids, the maned wolf, crab-eating fox and bush dog: Implications for sociality. *Anim. Behav.* 31:814–26.
• Bisbal, F. 1982. Nuevos registros de distribución para el zorro gris *Urocyon cinereoargenteus venezuelae* (Mammalia, Carnivora). *Acta Cient. Venez.* 33:255–57.
———. 1987. The carnivores of Venezuela: Their distribution and the ways they have been affected by human activities. M.S. thesis, University of Florida.
Bisbal, F., and J. Ojasti. 1980. Nicho trófico del zorro *Cerdocyon thous* (Mammalia, Carnivora). *Acta Biol. Venez.* 10:469–96.
Brady, C. A. 1978. Reproduction, growth, and parental care in crab-eating foxes, *Cerdocyon thous*, at the National Zoological Park, Washington. *Int. Zoo. Yearb.* 18:130–34.
———. 1979. Observations on the behavior and ecology of the crab-eating fox, *Cerdocyon thous.* In *Vertebrate ecology in the northern Neotropics*, ed. J. F. Eisenberg, 161–71. Washington, D.C.: Smithsonian Institution Press.
———. 1981. Vocal repertoires of the bush dog *Speothos venaticus*, crab-eating fox *Cerdocyon thous*, and maned wolf *Chrysocyon brachyurus. Anim. Behav.* 29:649–69.
Charles-Dominique, P., M. Atramentowicz, M. Charles-Dominique, H. Gerard, A. Hladik, C. M. Hladik, and M. F. Prevost. 1981. Les mammifères frugivores arboricoles noctunes d'une forêt guyanaise: Interrelations plantes-animaux. *Rev. Ecol. (Terre et Vie)* 35:341–435.
Clutton-Brock, J., G. B. Corbet, and M. Hills. 1976. A review of the family Canidae with a classification by numerical methods. *Bull. British Mus. (Nat. Hist.) Zool.* 29:117–99.
Coates-Estrada, R., and A. Estrada. 1986. *Manual de identificación de campo de los mamíferos de la Estación de Biología "Los Tuxtlas."* Mexico City: Universidad Nacional Autónoma de México.
Currier, M. J. 1983. *Felis concolor. Mammal. Species* 200:1–7.
Dathe, H. 1967. Bemerkungen zur Aufzucht von Brillenbären. *Zool. Gart.* (Leipzig) 34:105–33.
Defler, T. R. 1980. Notes on interactions between the tayra (*Eira barbara*) and the white fronted capuchin (*Cebus albifrons*). *J. Mammal.* 51:156.

Deutsch, L. 1983. An encounter between bush dog (*Speothos venaticus*) and paca (*Agouti paca*). *J. Mammal.* 64:532–33.

Dietz, J. M. 1985. *Chrysocyon brachyurus*. *Mammal. Species.* 234:1–4.

Drüwa, P. 1977. Beobachtungen zur Geburt und natürlichen Aufzucht von Waldhunden (*Speothos venaticus*) in der Gefangenschaft. *Zool. Gart.* (Leipzig), n.s., 47:109–37.

Duplaix, N. 1980. Observations on the ecology and behavior of the giant river otter *Pteronura brasiliensis* in Suriname. *Rev. Ecol. (Terre et Vie)* 34:496–620.

• Duplaix-Hall, N. 1977. Report of the Otter Specialist Group to the IUCN Survival Service Commission. Mimeographed.

Enders, R. K. 1935. Mammalian life histories from Barro Colorado Island, Panama. *Bull. Mus. Comp. Zool. (Harvard)* 78(4):385–502.

Ewer, R. F. 1973. *The carnivores.* New York: Cornell University Press.

Galef, B. G., Jr., R. A. Mittermeier, and R. C. Bailey. 1976. Predation by the tayra (*Eira barbara*). *J. Mammal.* 57:760–61.

Gardner, A. L. 1971. Notes on the little spotted cat, *Felis tigrina oncilla* in Costa Rica. *J. Mammal.* 52:461–65.

• Hall, E. R. 1951. *American weasels.* Publication 4. Lawrence: Museum of Natural History, University of Kansas.

• ———. 1981. *The mammals of North America.* 2d ed., 2 vols. New York: John Wiley.

• Handley, C. O., Jr. 1976. Mammals of the Smithsonian Venezuelan project. *Brigham Young Univ. Sci. Bull., Biol. Ser.* 20(5):1–90.

• Hershkovitz, P. 1958. A synopsis of the wild dogs of Colombia. *Nov. Colombianas, Mus. Hist. Nat. Univ. Cauca,* 3:157–61.

• ———. 1961. On the South American small-eared zorro *Atelocynus microtis* Sclater (Canidae). *Fieldiana: Zool.* 39:505–23.

Honacki, J. H., K. E. Kinman, and J. W. Koeppl, eds. 1982. *Mammal species of the world.* Lawrence, Kans.: Allen Press and Association of Systematics Collections.

Hulley, J. T. 1976. Maintenance and breeding of captive jaguarundis (*Felis yagouaroundi*) at Chester Zoo and Toronto. *Int. Zoo Yearb.* 16:120–22.

• Husson, A. M. 1978. *The mammals of Suriname.* Leiden: E. J. Brill.

Izor, J. R., and L. de la Torre. 1978. A new species of weasel (*Mustela*) from the highlands of Colombia with comments on the evolution and distribution of South American weasels. *J. Mammal.* 59:92–102.

Izor, J. R., and N. E. Peterson. 1985. Notes on South American weasels. *J. Mammal.* 66(4):788–89.

Janzen, D., and W. Hallwachs. 1982. The hooded skunk *Mephitis macroura* in lowland northwestern Costa Rica. *Brenesia* 19/20:549–52.

Kaufmann, J. H. 1962. Ecology and social behavior of the coati, *Nasua narica* on Barro Colorado Island, Panama. *Univ. Calif. Publ. Zool.* 60:95–222.

———. 1982. Raccoon and allies. In *Wild mammals of North America,* ed. J. A. Chapman and G. A. Feldhamer, 567–85. Baltimore: Johns Hopkins University Press.

Kaufmann, J. H., and A. Kaufmann. 1965. Observations on the behavior of tayras and grisons. *Z. Säugetierk.* 30:146–55.

• Kipp, H. 1965. Beitrag zur Kenntnis der Gattung *Conepatus* Molina, 1782. *Z. Säugetierk.* 30:193–256.

Kleiman, D. G. 1972. Social behavior of the maned wolf (*Chrysocyon brachyurus*) and the bush dog (*Speothos venaticus*): A study in contrast. *J. Mammal.* 53:791–806.

Kleiman, D. G., and J. F. Eisenberg. 1973. A comparison of canid and felid social systems from an evolutionary perspective. *Anim. Behav.* 21:637–59.

Konecny, M. J. n.d. Movement patterns and food habits of four sympatric carnivore species in Belize, Central America. In *Advances in Neotropical mammalogy,* ed. K. H. Redford, and J. F. Eisenberg. Leiden: E. J. Brill. In press.

• Krumbiegel, I. 1942. Die Säugetiere der Sudamerika-Expeditionen Prof. Dr. Kriegs. 17. Hyrare und Grisons (*Tayra* und *Grisons*). *Zool. Anzeiger* 139:81–108.

Langguth, A. 1975. Ecology and evolution in the South American canids. In *The wild canids,* ed. M. W. Fox, 192–206. New York: Van Nostrand Reinhold.

Leyhausen, P. 1963. Über sudamerikanische Pardelkatzen. *Z. Tierpsychol.* 20:627–40.

———. 1979. *Cat behavior.* New York: Garland STMPM Press.

Leyhausen, P., and M. Falkena. 1966. Breeding the Brazilian ocelot-cat *Leopardus tigrinus* in captivity. *Int. Zoo Yearb.* 6:176–82.

• Linares, O. J. 1968. El perro de monte, *Speothos venaticus* (Lund) en el norte de Venezuela (Canidae). *Mem. Soc. Cient. Nat. La Salle* 27:83–86.

———. 1981. Tres nuevos carnívoros prociónidos fósiles del Mioceno de Norte y Sudamérica. *Ameghiniana* 18:113–21.

Löhmer, R. 1976. Zur Verhaltensontogenese bei *Procyon cancrivorus. Z. Säugetierk.* 41:42–58.

Lord, R. D. 1961. Population study of the gray fox. *Amer. Midl. Nat.* 66:87–109.

Lotze, J. 1979. The raccoon on St. Catherine's Island, Georgia, 4: Comparisons of home ranges determined by live trapping and radiotracking. *Amer. Mus. Novitat.* 2664:1–25.

Ludlow, M. E. 1986. Home range, activity patterns and food habits of the ocelot (*Felis pardalis*) in Venezuela. M.S. thesis, University of Florida.

Melquist, W. E., and M. G. Hornocker. 1983. Ecology of river otters in west central Idaho. *Wildl. Monogr.* 83:1–60.

Mendez, E. 1981. The coyote in Panama. *Int. J. Study Anim. Problems* 2:252–55.

Mondolfi, E. 1970. Las nutrias o perros de aqua. *Defensa Nat.* 1(1):24–26, 47.

———. 1971. El oso frontino (*Tremarctos ornatus*). *Defensa Nat.* 1(2):31–35.

———. 1973. El mapurite un animal beneficioso. *Defensa Nat.* 2(6):37–41.

———. 1986. Notes on the biology and status of the small wild cats in Venezuela. In *Cats of the world: Biology, conservation and management*, ed. S. Douglas Miller and D. D. Everett, 125–46. Washington, D.C.: National Wildlife Federation.

Mondolfi, E., and R. Hoogersteijn. 1986. Notes on the biology and status of the jaguar in Venezuela. In *Cats of the world: Biology, conservation and management*, ed. S. Douglas Miller and D. D. Everett, 85–124. Washington, D.C.: National Wildlife Federation.

Montgomery, G. G., and Y. D. Lubin. 1978. Social structure and food habits of crab-eating fox (*Cerdocyon thous*) in Venezuelan llanos. *Acta Cient. Venez.* 29:382–83.

Peyton, B. 1980. Ecology, distribution and food habits of spectacled bears *Tremarctos ornatus* in Peru. *J. Mammal.* 6:639–52.

Poglayen-Neuwall, I. 1962. Beitrage zu einem Ethogramm des Wickelbären (*Potos flavus* Schreber). *Z. Säugetierk.* 27:1–44.

———. 1966. On the marking behavior of the kinkajou (*Potos flavus* Schreber). *Zoologica* 51:137–41.

———. 1973. Preliminary notes on the maintenance and behavior of the Central American cacomistle *Bassariscus sumichrasti*. *Int. Zoo Yearb.* 13:207–11.

———. 1975. Copulatory behavior, gestation, and parturition of the tayra *Eira barbara*. *Z. Säugetierk.* 40:176–89.

———. 1978. Breeding, rearing and notes on the behaviour of tayras *Eira barbara* in captivity. *Int. Zoo Yearb.* 18:134–40.

Poglayen-Neuwall, I., and I. Poglayen-Neuwall. 1965. Gefangenschaftsbeobachtungen an Makibären (*Bassaricyon gabbii* Allen, 1876). *Z. Säugetierk,* 30:321–66.

———. 1980. Gestation period and parturition of the ringtail *Bassariscus astutus*. *Z. Säugetierk.* 45:73–81.

Porton, D., D. G. Kleiman, and M. Rodden. 1987. Seasonality of bush dog reproduction and the influence of social factors on the estrous cycle. *J. Mammal.* 68:867–71.

Robinette, W. L., J. S. Gashwiler, and O. W. Morris. 1961. Notes on cougar productivity and life history. *J. Mammal.* 42:104–217.

Russell, J. K. 1981. Exclusion of adult male coatis from social groups: Protection from predation. *J. Mammal.* 62(1):206–8.

———. 1983. Altruism in coati bands: Nepotism or reciprocity? In *Social behavior of female vertebrates*, ed. S. K. Wasser, 263–90. New York: John Wiley.

Schaller, G. B. 1983. *Mammals and their biomass on a Brazilian ranch*. Arquivos de Zoologia 31.1 São Paulo: Museu de Zoologia da Universidade de São Paulo.

Schaller, G. B., and P. G. Crawshaw, Jr. 1980. Movement patterns of jaguar. *Biotropica* 12(3):161–68.

Schaller, G. B., and J. M. C. Vasconcelos. 1978. Jaguar predation on capybara. *Z. Säugetierk.* 43:296–301.

Stains, H. J. 1956. Distribution and taxonomy of the Canidae. In *The wild canids*, ed. M. W. Fox, 3–26. New York: Van Norstrand Reinhold.

Sunquist, M. E., and G. G. Montgomery. 1973. Arboreal copulation by coatimundi (*Nasua narica*). *Mammalia* 37:517–18.

Sunquist, M. E., F. Sunquist, and D. Daneke. n.d. Ecological separation in a Venezuelan llanos carnivore community. In *Advances in Neotropical mammalogy*, ed. K. H. Redford and J. F. Eisenberg. Leiden: E. J. Brill. In press.

• Tate, G. H. H. 1939. The mammals of the Guiana region. *Bull. Amer. Mus. Nat. Hist.* 76(5):151–229.

Trapp, G. R. 1978. Comparative behavioral ecology of the ringtail and gray fox in southwestern Utah. *Carnivore* 1:3–32.

Van Gelder, R. G. 1978. A review of canid classification. *Amer. Mus. Novitat.* 2646:1–10.

Vaughan, C. 1983. Coyote range expansion in Costa Rica and Panama. *Brenesia* 21:27–32.

Vaughn, R. 1974. Breeding the tayra, *Eira barbara* at Antelope Zoo, Lincoln. *Int. Zoo Yearb.* 14:120–22.

Vuilleumier, F. 1985. Fossil and recent avifaunas and the interamerican interchange. In *The great American biotic interchange*, ed. G. Stehli and S. David Webb, 387–424. New York and London: Plenum.

Walker, P. L., and J. G. H. Cant. 1977. A population survey of kinkajous (*Potos flavus*) in a seasonally dry tropical forest. *J. Mammal.* 58:100–102.

8

Order Pinnipedia
(Seals, Sea Lions, and Walruses)

Diagnosis

The body form is strongly modified for aquatic loco-motion. The hind feet and forefeet are in the form of flippers, with the digits completely enclosed in the integument. The tail is short, and the body is spindle shaped. The teeth are simplified in structure, often peglike or conical (fig. 8.1).

Distribution

Seals and their relatives are worldwide in the oceans close to continental shorelines or polar ice. A few species are landlocked in subarctic lakes. They reach their greatest species diversity in the temperate to Arctic and Antarctic latitudes and are generally absent from tropical waters except for the monk seals (King 1956).

History and Classification

The order Pinnipedia is sometimes considered a suborder of the Carnivora (Honacki, Kinman, and Koeppl 1982). The Pinnipedia are first detected in the middle Miocene, already extremely specialized for an aquatic existence. The present-day Pinnipedia may have had a dual origin and thus may be an artificial assemblage, since the ancestors of the true seals (Phocidae) may have derived from mustelid precursors while the eared seals (Otariidae) derived from an ancestral form that also gave rise to modern canids and ursids. In contrast, evidence presented by Berta (1986) and Sarich (1969) supports a monophyletic origin for the group. The existing pinnipeds are usually grouped into three families: the Otariidae, or eared seals, including the fur seals and sea lions; the Obodenidae, including the walruses; and the Phocidae, or "earless" seals. This last group is characterized by loss of the external ear (pinna) and

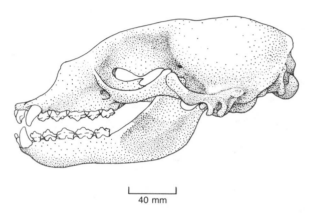

Figure 8.1. Skull of *Monachus tropicalus*.

by hind limbs incapable of rotating forward, thus limiting movement on land. The biology of the Pinnipedia is summarized by King (1964).

In the area covered by this volume only one recent species of phocid was found, in the Caribbean Sea. It ranged in the waters off the Antilles to the coasts of Nicaragua, Mexico, the southern part of Florida, and the Bahamas. It was never recorded from northern South America and is currently believed to be extinct, but for completeness I have included it (Allen 1942).

FAMILY PHOCIDAE
Genus *Monachus* Fleming, 1822
Tropical Seals, Foca
Monachus tropicalis (Gray, 1850)
Caribbean Monk Seal

Description

The dental formula was I 2/2, C 1/1, P 4/4, M 1/1 (Allen 1887). Males reached 2.5 m, and females were somewhat smaller. The dorsal surface of the body was a brownish gray with no spotting, and the venter tended to contrast, being white or yellowish. The vibrissae on the muzzle were short, and the nails of the digits were well developed.

Range and Habitat
The species fed in the waters off Florida and the Bahama Islands south in the Gulf of Mexico to Nicaragua and the Antilles. It generally rested and bore its young on small offshore islands.

Natural History
This species was briefly studied in the wild (Ward 1887). One may infer that its biology was similar to that of *M. schauinslandi*, the Hawaiian monk seal. Monk seals, when not foraging for fish at sea, typically hauled out on sheltered beaches. A single young was born in an advanced state and nursed by the mother on land for six weeks, taking to water to forage with the mother at about two months of age. Females probably produced a calf every other year. The species was exterminated by man in the early twentieth century. It was not recorded on the Caribbean coasts of Colombia and Venezuela (Rice 1973; Boulva 1979).

Note
Diaz, Hernandez-Camacho, and Cadena-G. (1986) list the Archipelago de San Andrés and the Cayos Serrana, Colombia, as former localities for *M. tropicalis*. They further list *Arctocephalus australis* as an occasional stray on the Pacific coast of Colombia.

References

Allen, G. M. 1942. Extinct and vanishing mammals of the Western Hemisphere. Publication 11. Washington, D.C.: American Committee for International Wild Life Protection.

Allen, J. A. 1887. The West Indian monk seal (*Monachus tropicalis*). *Bull. Amer. Mus. Nat. Hist.* 2:1–34.

Berta, A. 1986. Abstract no. 83, Annual Meeting of the American Society of Mammalogists, Madison, Wisconsin.

Boulva, J. 1979. Caribbean monk seal. In *Mammals in the seas*, 2:101–3. FAO Fisheries Series 5. Rome: Food and Agriculture Organization of the United Nations.

Diaz, A. C., J. Hernandez-Camacho, and A. Cadena-G. 1986. Lista actualizada de los mamíferos de Colombia. *Caldasia* 15:471–501.

Honacki, J. H., K. E. Kinman, and J. W. Koeppl, eds. 1982. *Mammal species of the world.* Lawrence, Kans.: Allen Press and Association of Systematics Collections.

King, J. E. 1956. The monk seals, genus *Monachus*. *Bull. British Mus. Nat. Hist.* (*Zool.*) 3:203–56.

———. 1964. *Seals of the world.* London: British Museum of Natural History.

Rice, D. W. 1973. The Caribbean monk seal. *IUCN Publ.*, n.s., 39:98–112.

Sarich, V. M. 1969. Pinniped phylogeny. *Syst. Zool.* 18:416–22.

Ward, H. L. 1887. Notes on the life history of *Monachus tropicalis*, the West Indian seal. *Amer. Nat.* 21:257–64.

9 Order Cetacea
(Whales, Dolphins, and Their Allies)

Diagnosis

Except for the Sirenia, the Cetacea exhibit the most complete adaptation for aquatic existence within the class Mammalia (see plate 13). The body is spindle shaped, hair is absent except for a few bristles in the vicinity of the lips of some species, and the skin surface is smooth. The forelimbs are modified into paddles; the digits cannot be detected externally. Hind limbs are absent, and the tail is flattened dorsoventrally and extended laterally to produce two pointed flukes. The nostrils open through either a single or a double aperture on the top of the head. The teeth are absent or extremely modified, exhibiting no cusp pattern in the adult. All the teeth are very similar in shape within the jaw of any species (homodont); it is impossible to distinguish molars, premolars, and canines. Some standard references on the order's biology and taxonomy are Slijper (1962), Norris (1966), Gaskin (1982), Minasian, Balcomb, and Foster (1984), and Hershkovitz (1966).

Distribution

The order is found worldwide in oceans and seas; some small species ascend rivers or are adapted for a permanent freshwater existence. Some cetaceans commonly occur in the Atlantic and Pacific off the coasts of Panama, Colombia, and the Guyanas. In the Caribbean they occur off the coasts of Venezuela and the Antilles. Cetaceans are less common in the Caribbean, and some are only occasionally recorded, when stranded. I shall discuss only the more common offshore cetaceans and concentrate on the life histories of these species and the river dolphins. A complete listing of stranded cetaceans for the southern Caribbean is included in the publications by van Bree (1975) and Caldwell et al. (1971). A key to the genera is given in table 9.2.

History and Classification

Cetaceans are classically divided into three suborders, the Archaeoceti, the Odontoceti, and the Mysticeti. Whether the last two lineages descended from a common ancestor is still open to some debate. It is enough to say that the Archaeoceti first appear in the fossil record in the middle Eocene and may have given rise to the extant families. The oldest odontocete fossils are from the late Eocene, and mysticete fossils first appear in the mid-Oligocene. Odontocetes have teeth and are thus often referred to as "toothed" whales. The mysticetes do not have teeth as adults, but there are modified epidermal derivatives hanging from the roof of the mouth in longitudinal plates called baleen (see fig. 9.1), which serve as a straining device. Mysticetes are specialized for feeding on various shrimplike crustaceans (krill), which they obtain by swimming through swarms of the creatures with mouths partially open, periodically expelling excess water and swallowing the retained crustaceans after scraping them off the baleen plates with the tongue. Baleen whales also eat fish (see species accounts).

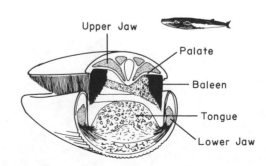

Figure 9.1. Schematic diagram of the baleen plates and their position in the mouth of *Balaenoptera* (modified from Slijper 1962).

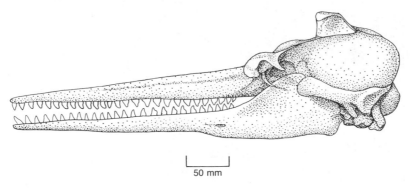

50 mm

Figure 9.2. Skull of *Inia geoffrensis*.

Natural History

The great whales and dolphins are entirely aquatic and never come to land unless accidentally stranded or washed ashore during storms. Usually only a single young is born, after an extended gestation, delivered at sea in a highly advanced state. Odontocetes are specialized for feeding on fish or pelagic squid. Mysticetes, as noted before, are specialized for feeding on planktonic crustaceans. Many of the great mysticete whales make enormous migrations to and away from the poles following plankton abundance. During their short growing season, the polar regions have high productivity and are favorite feeding grounds for mysticete whales during their summer movements. Although some whales are specialized for tropical waters, most cetacean taxa show their greatest species richness in the temperate and Arctic latitudes. Gaskin (1982) is a useful reference on the behavior and ecology of cetaceans. Eisenberg (1986) has reviewed the behavioral convergence demonstrable when terrestrial mammals are compared with cetaceans. Life history data are included in table 9.1. A key to the genera is included in table 9.2.

SUBORDER ODONTOCETI

Diagnosis
Baleen is absent; some simple teeth are present that are cone shaped or peglike in form. A single nasal opening at the top of the head is typical.

FAMILY PLATANISTIDAE
River Dolphins

Diagnosis
The teeth are numerous and cone shaped, and the skull is distinctive (fig. 9.2). These small dolphins have an extremely long, slender beak that is sharply marked off from the rather bulging forehead.

The dorsal fin is low. The eye tends to be very reduced in size, and in some genera it is absent.

Distribution
These freshwater dolphins have a very disjunct distribution, being found in the Ganges and Indus rivers of India, in the Yangtze Kiang of China, in the La Plata River and its major tributaries of Argentina and Uruguay, and in the Amazon and Orinoco river systems.

Genus *Inia* d'Orbigny, 1834
Inia geoffrensis (de Blainville, 1817)
Amazon River Dolphin, Bouto, Boto

Description
The long, slender beak is distinctive. The teeth are simple, single crowned, and number up to 132, with 20 to 30 on each side of the upper and lower jaws. The dorsal fin begins in the midback and extends as a ridge to the tail. Lengh ranges from 2 to 3 m and weight from 90 to 150 kg. The coloration is variable and related to age. Young dolphins have a grayish-black dorsum and a pink venter, while in older individuals the upperparts become pink (plate 13).

Range and Habitat
The species occurs in both the Amazon and the Orinoco river systems and is the common dolphin sighted in the upstream portions of these rivers (Pilleri and Pilleri 1982) (map 9.1).

Natural History
The Amazon river dolphin is specialized for feeding on fish. Much of the water it forages in is extremely turbid, and it employs echolocation to detect its prey (Layne and Caldwell 1964). Groups of up to half a dozen may be encountered. Group size varies in part because dolphin schools may break up and forage in small groups when the river water is high, and the dolphins follow their fish prey into the flooded forest. At low water dolphins occasionally

Table 9.1   Some Vital Statistics for the Cetaceans

Species	Size at Birth (m)	Size at Sexual Maturity (m)	Age at Sexual Maturity (yr)	Gestation (mo)	Interbirth Interval (yr)	Duration of Lactation (mo)	Size Dimorphism	Herd Size — Average	Herd Size — Maximum	Maximum Age (yr)
Inia geoffrensis	0.75	♂2.0; ♀1.7		10.5			♂ > ♀	1–5	20	
Delphinus delphis	0.86	1.67–1.8		10–11	1–2	>12	♂ > ♀		3,000	
Feresa attenuata								1	10	
Globicephala macrorhynchus	1.2		♂7–10; ♀13–17	12	3	>12	♂ > ♀	20–35	>100	>60
Grampus griseus	1.5	♂2.9; ♀2.7			3		♂ > ♀	5	>100	
Orcinus orca	2.4	♂6.7; ♀5.0		14.5	3	>12	♂ > ♀	2–3	25–30	>60
Peponocephala electra	1.8							100	300–500	
Pseudorca crassidens			16				♂ > ♀	1		
Sotalia fluviatilis	0.68	♂1.4		8.5			Little	100	6–7	
Stenella attenuata	0.8	♂2.0; ♀1.9	6–8	11	2	11	♂ > ♀	100	>1,000	
Stenella coeruleoalba	1		9–14	12	2–3	8	♂ > ♀	10–40		
Stenella longirostris	0.91	♂1.8; ♀1.7	8–12	10.6			♂ > ♀	5–15	>1,000	
Stenella plagiodon	0.91			11			♂ > ♀	20–50	100	
Tursiops truncatus	0.9–1.3	♂2.6; ♀2.4	♂11; ♀8	12	2–3	12–18	♂ > ♀	1–12	>100	>35
Kogia simus	1	♂3.0; ♀2.6		11	2		♂ > ♀	1–3	10	
Physeter macrocephalus	4		>9	16	3	24	♂ > ♀	1–3	40	70
Mesoplodon densirostris							Little	3–6		
Mesoplodon europaeus	2.2							1–3	10	
Ziphius cavirostris	2–3	♂5.5; ♀6.0					♀ > ♂	2–6		>35
Balaenoptera acutorostrata	3	♂7; ♀7.3	6	10.50	2	<6	♀ > ♂	1–6		>35
Balaenoptera borealis	4.5	♂13; ♀13.7	8	11	2	<6	♀ > ♂	1–2	50	50
Balaenoptera edeni	4.3	♂12; ♀13	8–13	11.50			♀ > ♂	1–10	100	70
Balaenoptera musculus	7.5	♂22.5; ♀24	23–30	11.25	2–3	7	♀ > ♂	1–2	12	40
Balaenoptera physalus	6.5	♂18.5; ♀19.8	10–13	11.50	2–3	6–7	♀ > ♂	1–10	100	>60
Megaptera novaeangliae	4.2	♂11.5; ♀12	14–30	11.50	2–3	7	♀ > ♂	1–4	12	>75

Sources: Minasian, Balcomb, and Foster (1984); Perrin, Brownell, and DeMaster (1984).
Note: All cetaceans typically bear a single young at each birth.

become trapped in lagoons and canos until the next high water (Trebbau 1975; Trebbau and van Bree 1974). Reproduction in the Amazon is seasonal and correlated with the fluctuation in water levels (Best and Silva 1984). Gestation is 10.5 months, with a two-year interval between calves (Brownell 1984). There appears to be a strong bond between the mother and the single young. Defler (1983) reported feeding associations between *Inia* and *Pteronura* in Colombia.

FAMILY ZIPHIIDAE
Beaked Whales

Diagnosis

Most species show a vastly reduced tooth number. The more primitive *Tasmacetus* has a tooth number of 19/27, but in most species it is reduced to 0/2 or 0/1. A rather long beak and a roundish head characterize this family. The beak is not sharply demarcated from the head, and the contours are smooth in *Ziphius* and *Mesoplodon*. The dorsal fin is set back about three-quarters of the total length.

Distribution
The family occurs worldwide and is oceanic.

Natural History
This family is poorly known. Many members are believed to feed upon pelagic squid. The standard reference is Moore (1968).

Genus *Mesoplodon* Gervais, 1850
Beaked Whale

Description
Length ranges from 3 to 7 m; the prominent dorsal fin is about 150 mm long, set three-quarters of the way back from the head. Color is variable throughout the genus but is usually black on the dorsum and lighter below. Sexes exhibit size dimorphism; males are larger and possess a single pair of large teeth in the anterior portion of the lower jaw. Males apparently fight among themselves, per-

Table 9.2   Key to the Adult Cetaceans of Northern South American Waters

1	Found mainly in fresh water	2
1'	Found in marine waters	3
2	No pronounced dorsal fin, but a ridge down the center of the back	*Inia*
2'	Dorsal fin present	*Sotalia*[a]
3	Large whales, greater than 14–30 m in length	4
3'	Whales less than 11 m in length	6
4	Dorsal fin absent; head enormous and blunt	*Physeter*
4'	Dorsal fin present; head pointed	5
5	Flippers long, one-third of body length	*Megaptera*
5'	*Balaenoptera*	
6	Medium-sized whales, 4–20 m in length	7
6'	Small whales and dolphins, less than 4 m long; dorsal fin present	13
7	Baleen plates present	*Balaenoptera acutorostrata*
7'	Baleen plates absent	8
8	Teeth present; one blowhole; two short throat grooves	9
8'	Teeth present; one blowhole; no throat grooves	10
9	Single pair of teeth (male only) at tip of lower jaw	*Ziphius*
9'	Single pair of teeth at arched point of lower jaw	*Mesoplodon densirostris*
10	Melon bulbous	*Globicephala*
10'	Not as above	11
11	Large dorsal fin, greater than 0.9 m; black-and-white coloration	*Orca*
11'	Dorsal fin less than 0.9 m	12
12	Body black	*Pseudorca*
12'	Body gray brown dorsally; white on venter	*Grampus*
13	Obvious beak	14
13'	Beak absent or inconspicuous	16
14	Dorsum gray	*Tursiops*
14'	Not as above	15
15	Dorsum black; venter white; white spot on dorsal fin	*Delphinus*
15'	Not as above	*Stenella*
16	Snout pointed	*Lagenorhynchus*
16'	Snout rounded or square	17
17	Snout square	*Kogia*
17'	Snout rounded	
	1 44–55 teeth per side	*Peponocephala*
	1' 20–26 teeth per side	*Feresa*

[a] Also found in coastal marine waters, but it is the smallest cetacean in South America (see text).

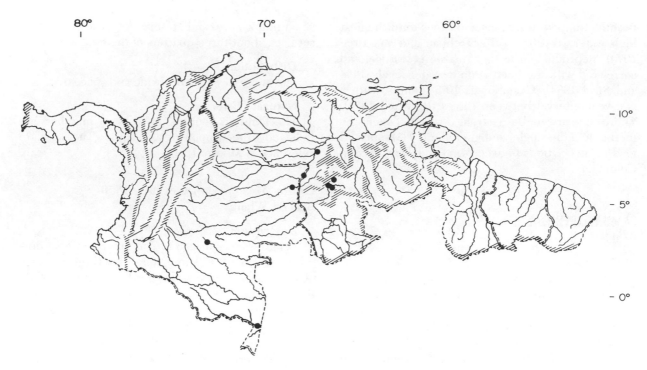

Map 9.1. Distribution of *Inia geoffrensis*.

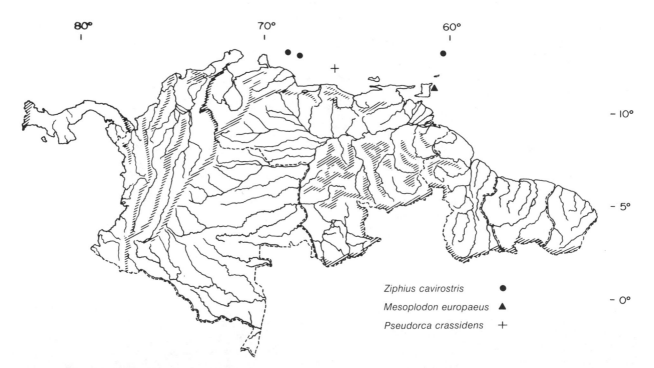

Ziphius cavirostris   ●

Mesoplodon europaeus   ▲

Pseudorca crassidens   +

Map 9.2. Distribution of three species of cetaceans.

haps over access to females, and their bites leave a characteristic scarring pattern (Moore 1966; Mead, Walker, and Houck 1982).

Mesoplodon europaeus (Gervais, 1855)
Gervais's Beaked Whale

Description
This whale is similar to *M. densirostris* in color and size, but the paired teeth of the males are at the anterior end of the lower jaw and are not on a conspicuous arch (fig. 9.3).

Range and Habitat
This pelagic species is known from the northwest Atlantic and is not usually found inshore. One has been recorded from Trinidad (map 9.2).

Mesoplodon densirostris (Blainville, 1817)
Dense-beaked Whale

Description
The single pair of lower teeth, prominent in the males, are situated to the rear of the lower jaw and sit on a conspicuous arch. The teeth do not erupt in females (fig. 9.3). Length averages 5 m. The dorsum is gray black, and the flanks are blue gray.

Range and Habitat
The species is distributed worldwide in tropical waters. This is the only member of the Ziphiidae likely to be encountered in the offshore waters covered by this volume.

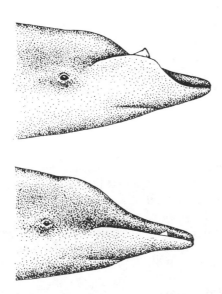

Figure 9.3. Heads of adult male *Mesoplodon* frequently encountered in Caribbean waters. Above: *Mesoplodon densirostris*; note the position of the tooth on the arched lower jaw. Below: *Mesoplodon europaeus*; note the smaller tooth and the lack of arch.

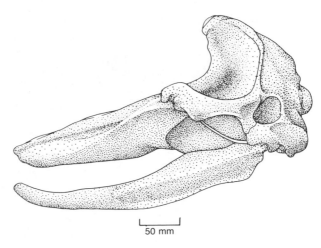

Figure 9.4. Skull of *Ziphius cavirostris* (female).

Genus *Ziphius* G. Cuvier, 1823
Ziphius cavirostris G. Cuvier, 1823
Goose-beaked Whale

Description
Males have only two functional teeth, at the tip of the lower jaw (fig. 9.4); females are toothless or have teeth that do not erupt. At maturity females measure from 6 to 7 m. Males are slightly smaller, reaching 6.7 m. The pectoral fin is situated far back on the body and measures about 500 mm. The anterior end of the body (face and throat) may be white, while the dorsum is black blending to gray on the venter. There is considerable variation in color.

Range and Habitat
This small whale usually ranges far out at sea. It is common in the tropics of all oceans and has been recorded from Curaçao and Bonaire (map 9.2).

Natural History
This species is migratory in the Atlantic and moves north in summer. The whales travel in schools of up to forty, but group size is highly variable. Deep divers, they are known to feed on squid.

FAMILY PHYSETERIDAE
Sperm Whales, Cachalotes

Diagnosis
The extant species of this family are included in two genera, *Physeter* and *Kogia*. They are so different that I will reserve description for the generic level. Neither has a beak, and the great size of *Physeter* immediately distinguishes it (plate 13). The undershot jaw of *Kogia* separates it from the beakless delphinids.

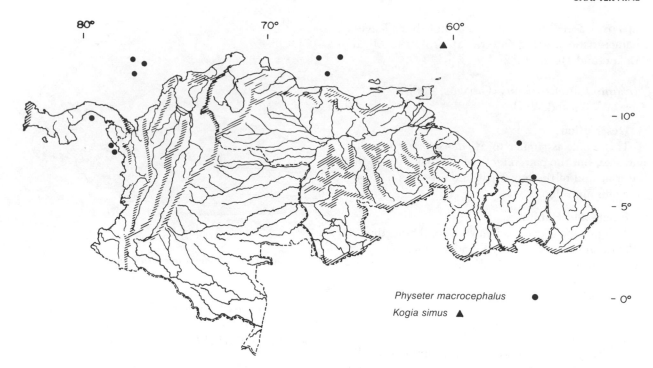

Map 9.3. Distribution of *Physeter macrocephalus* and *Kogia simus*.

Genus *Physeter* Linnaeus, 1758
Physeter macrocephalus Linnaeus, 1758

Description

The lower jaw is extremely slender and contains the only functional teeth, up to 60 in total number; the teeth of the upper jaw do not erupt through the gums. In this strongly dimorphic species, males may be nearly twice as large as females: adult males range from 15 to 18 m in length, and females range from 10 to 11 m. The foreflippers are about 2 m long. The most striking feature of these whales is the enormous head, nearly one-third of the total length. The head is blunt (plate 13), and the blowhole is near the tip of the rounded forehead. The rounded dorsal fin is set midway along the back. The color is gray to black, but males appear to become paler with age.

Range and Habitat

This species is distributed worldwide. Strongly migratory in its movements, it may be seasonally abundant in tropical waters (map 9.3).

Natural History

Sperm whales are adapted for feeding on pelagic squid. They may dive to great depths to feed on fish and nonpelagic cephalopods. The females form large schools that show reproductive synchrony. A single calf is born after a gestation period of sixteen months. Males are strongly polygynous and defend harems against competing adult males (Best 1974).

Genus *Kogia* Gray, 1846
Kogia simus (Owen, 1866)
Dwarf Sperm Whale

Description

Length does not exceed 2.7 m, and the skull is about 400 mm long. The snout is blunt as in the larger sperm whale, but the head is not nearly as large proportionately. There is no beak, and the lower jaw is set well back from the front of the head. The blowhole opens far back on the forehead. The small dorsal fin is set about two-thirds of the way back from the head (fig. 9.5) (Handley 1966; Nagorsen 1985). The dorsum is dark gray to black, and the venter is paler.

Range and Habitat

The species is found in all oceans but does not extend to the polar regions. It has been recorded near Saint Vincent (map 9.3).

Note: The pygmy sperm whale, *Kogia breviceps*, may also occur in Atlantic waters. It is larger than *K. simus*, and the dorsal fin is smaller than in that species (see fig. 9.5).

Natural History

This whale is believed to feed on pelagic cuttlefish and squid. The gestation period is approximately eleven months, and a single calf is born (Nagorsen 1985).

FAMILY DELPHINIDAE
Dolphins (Delfins) and Porpoises

Diagnosis
Composed of approximately eighteen genera and sixty-two species, this family includes most of the smaller, toothed whales. The beak is variable in its expression, being prominent in *Tursiops* and virtually absent in *Globicephala, Grampus,* and *Faresa* (plate 13). The dorsal fin is prominent in all species except the Asiatic genus *Neomeris.*

Distribution
Dolphins and porpoises are widely distributed in the oceans of the world. Some species, such as *Sotalia,* ascend freshwater rivers.

Natural History
These small cetaceans feed on fish and pelagic squid. They are the best studied of the Cetacea, since they can be kept in captivity and are easily trained. Research with captives allowed the experi-mental demonstration of their echolocating ability (Kellogg 1961).

Highly gregarious, these cetaceans are almost always found in schools. In addition to the echolocation pulses produced from the larynx, they use a wide variety of sounds for underwater communication (Caldwell and Caldwell 1977).

Genus *Delphinus* Linnaeus, 1758
Delphinus delphis Linnaeus, 1758
Common Dolphin, Delfin

Description
D. delphis may have up to 200 teeth (fig. 9.6). This small dolphin has a length of 1.5 to 2.5 m. The adult male is slightly larger than the female. The rostrum is very long (figs. 9.6, 9.7), and the dorsal fin is prominent, being 500 mm long and set almost at midbody. The dorsum is blackish brown, the flanks are gray, and the venter is white. Some populations exhibit a banding pattern on the side, alternating gray and cream.

Figure 9.5. Two species of *Kogia*. Above: *Kogia breviceps*. Below: *Kogia simus*.

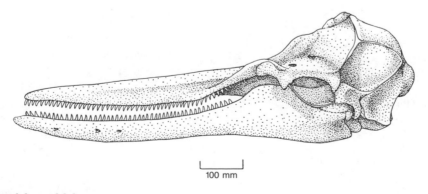

100 mm

Figure 9.6. Skull of *Delphinus delphis.*

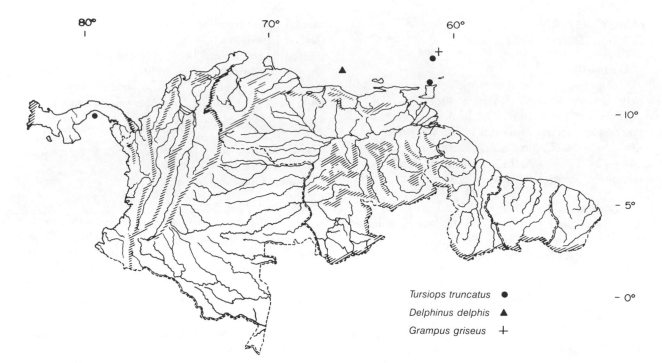

Map 9.4. Distribution of three species of cetaceans.

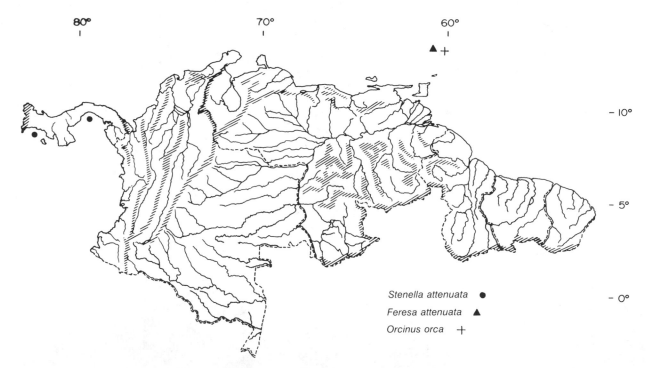

Map 9.5. Distribution of three species of cetaceans.

Figure 9.7. Head of *Delphinus delphis;* note the elongate rostrum.

Range and Habitat

The species is worldwide in distribution but does not enter the northern latitudes except seasonally. It has been found stranded at Islas Las Aves (map 9.4).

Natural History

Highly gregarious, this species may join other dolphin species to form a mixed school. A single young is born after a ten-month gestation period. These dolphins feed on fish and cephalopods.

Genus *Feresa* Gray, 1870
Feresa attenuata Gray, 1875
Pygmy Killer Whale

Description

Tooth number ranges from 20 to 26 per side, and this whale lacks a beak (fig. 9.8). It is small, averaging 2.2 to 2.4 m in length. The dorsal fin is situated at midbody and is prominent, averaging about 220 mm. The body is uniformly dark gray with white stripes around the mouth.

Range and Habitat

The species appears to be worldwide in distribution but is extremely rare (Caldwell and Caldwell 1971). It has been recorded from Saint Vincent (map 9.5).

Genus *Peponocephala* Nishiwaki and Norris, 1966
Peponocephala electra (Gray, 1846)
Melon-headed Whale

Description

The tooth number ranges from 44 to 50 per side (22–25 in each row in the upper and lower jaws). This tooth character distinguishes it from *Feresa*, to which it is similar since they both have a rounded head and lack a beak (fig. 9.8). The length may reach 2.8 m. The body is blue black on the dorsum and sides, and the venter is lighter. The lips may be white.

Range and Habitat

The species appears to occur worldwide but is rare in collections. It has been recorded from Saint Vincent.

Genus *Globicephala* Lesson, 1828
Globicephala macrorhynchus Gray, 1846
Pilot Whale, Ballena Pilota

Description

The tooth number may reach 44. This rather large delphinid ranges in length from 3.6 to 5.9 m. The dorsal fin is conspicuous, averaging 300 mm, and is set at midback. The front of the face is extremely blunt and rounded (plate 13). The color is black, but a white patch may be present on the chin.

Range and Habitat

This species is distributed through all tropical and subtropical waters (map 9.6).

Natural History

This gregarious species travels in schools and feeds on fish and squid. Some evidence indicates that the eastern Pacific species forms schools based on matrilineal descent (Kasuya and Marsh 1984). The species is believed to be polygamous. Gestation is approximately twelve months, and the young accompany the mother for several years. Interbirth intervals may exceed four years.

Genus *Grampus* Gray, 1828
Grampus griseus (G. Cuvier, 1812)
Risso's Dolphin

Description

Teeth are usually present only in the lower jaw and range from 3 to 7 pairs. The upper jaw may have a single pair of vestigial teeth. This species is large for a dolphin, averaging 4 m long, with a prominent

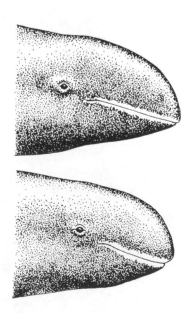

Figure 9.8. Heads of two dolphins. Above: *Peponocephala electra.* Below: *Feresa attenuata.* Note the longer mouth of *Peponocephala.*

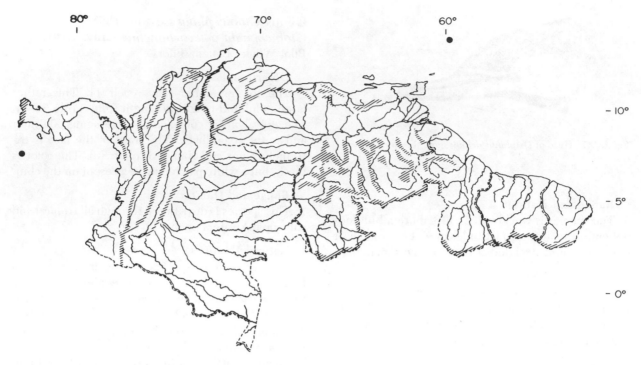

Map 9.6. Distribution of *Globicephala macrorhynchus*.

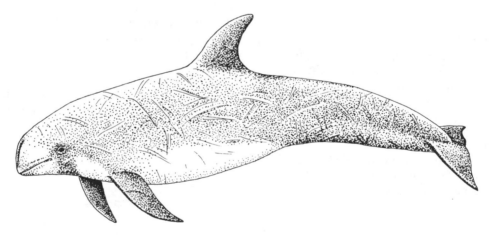

Figure 9.9. *Grampus griseus*.

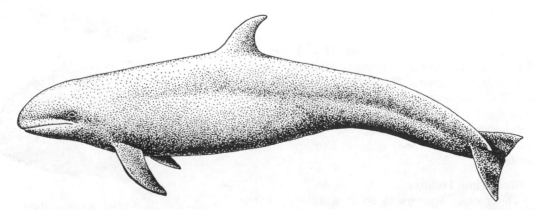

Figure 9.10. *Pseudorca crassidens*.

dorsal fin at midback averaging 500 mm. It lacks a beak and shares with *Feresa* and *Peponocephala* the characteristic blunt head (fig. 9.9). Adults are dark gray above and lighter below.

Range and Habitat

The species apparently occurs worldwide but is not common. It is known from strandings at Saint Vincent (map 9.4).

Genus *Orcinus* Fitzinger, 1860
Orcinus orca (Linnaeus, 1758)
Killer Whale, Orca

Description

This large delphinid is strongly size dimorphic: males may reach 9 m in length and females 6 m. A tall dorsal fin (male 1,800 mm, female 910 mm) set at midbody immediately distinguishes this species when at sea. The snout is bluntly rounded. The dorsum is jet black, contrasting sharply with the white venter, and the white may extend up on the flanks.

Range and Habitat

The species is distributed worldwide; it is oceanic but occurs near coastlines and seems to be more abundant toward the poles. It has been recorded from Saint Vincent (map 9.5).

Natural History

Killer whales exist in small, closed social groupings that have a very predictable movement pattern. They feed upon birds, seals, smaller cetaceans, and fish. Evidence indicates that these small groups are matriarchies, and the females within them are believed to be related. Schools in adjacent waters may have distinctive dialects, and underwater communication is important in regulating group movements and cohesion (Ford and Fisher 1983). The gestation period probably exceeds a year, and a single young is born.

Genus *Pseudorca* Reinhardt, 1862
Pseudorca crassidens (Owen, 1846)
False Killer Whale

Description

There are 16 to 22 teeth per side. In this large delphinid, males reach 6 m; females are about 1 m shorter. Like *Feresa*, *Peponocephala*, and *Grampus*, this species lacks a beak. Though it is sometimes confused with *Orcinus*, the dorsal fin is much shorter, less than 500 mm (fig. 9.10), and the coloration is quite different—uniform dark gray to black.

Range and Habitat

The species occurs worldwide and is pelagic. It has been recorded from Islas Las Aves and Saint Vincent (map 9.2).

Natural History

This whale feeds on squid and fish. Schools are known to be large, numbering up to three hundred.

Genus *Sotalia* Gray, 1866
River Dolphin

Description

In these small dolphins, head and body length ranges from 0.9 to 1.6 m. The conspicuous dorsal fin is set at midback and is approximately 127 mm high, immediately distinguishing *Sotalia* from *Inia* where they co-occur (plate 13). There is some controversy concerning the specific status of some members of this genus.

Distribution

River dolphins are found in the lower Amazon and Orinoco river systems and in coastal waters near the deltas of these rivers.

Sotalia fluviatilis (Gervais, 1853)
Tucuxi

Description

This is the smallest of the dolphins, averaging 1.9 m. The dorsum ranges from gray to blackish, and the venter tends to be lighter. *S. guianensis* is included in *S. fluviatilis*.

Range and Habitat

S. fluviatilis is found within the lower Amazon river system (Magnusson, Best, and da Silva (1980). *Sotalia guianensis* is usually the specific name given to the form found in the coastal waters of Venezuela and the Guyanas. The exact relationship of the two named forms remains to be determined (map 9.7).

Natural History

Species of the genus *Sotalia* travel in small groups of up to ten individuals. This coastal dolphin ascends the Orinoco and Amazon rivers and has been recorded from Lago de Maracaibo (Casinos, Bisbal, and Boher 1981). In the Amazon they may occur far upstream, but in the Orinoco they are nearer the delta (Layne 1958).

Genus *Stenella* Gray, 1866
Spotted Dolphin, Spinner Dolphin

Description

At least ten species of *Stenella* are listed in the literature. This is a vexing group taxonomically. Head and body length ranges from 1.2 to 3 m, and the dorsal fin is approximately 200 mm. Adults are usually gray to black above and paler below. Some species exhibit spotting on the distal third of the body, while others have longitudinal stripes on the sides of the head, shoulders, and back. There is a pronounced

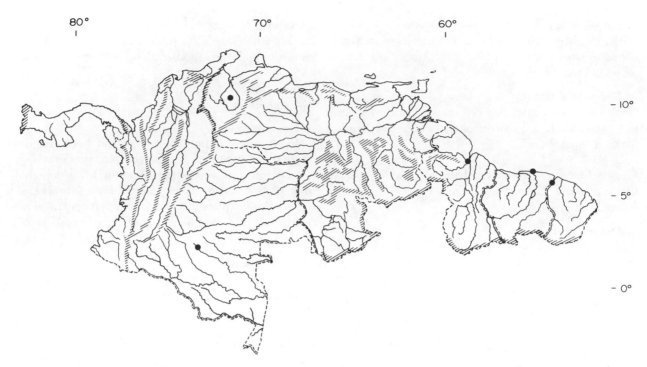

Map 9.7. Distribution of *Sotalia fluviatilis*.

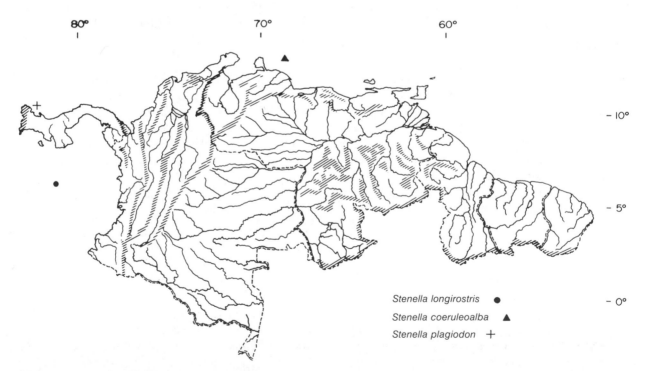

Stenella longirostris   ●
Stenella coeruleoalba   ▲
Stenella plagiodon   +

Map 9.8. Distribution of three species of *Stenella*.

beak, as in *Tursiops,* but the color pattern generally serves to distinguish *Stenella* (plate 13).

Distribution
The genus is found worldwide in tropical seas.

Natural History
This is a gregarious species, often occurring in large schools. The diet is presumed to consist almost entirely of fish.

Stenella attenuata (Gray, 1846)
Spotted Dolphin

Description
This description includes *S. graffmani.* The dorsum is dark gray, and the venter is lighter (Douglas, Schnell, and Hough 1984). A spotted pattern on the distal end is distinctive.

Range and Habitat
The species occurs in the tropical portions of the Pacific Ocean (map 9.5). *S. a. graffmani* is commonly recorded from the Pacific coast of Panama.

Stenella coeruleoalba (Meyen, 1833)
Striped Dolphin

Description
The flanks and sides exhibit white striping, contrasting with the blue-gray back.

Range and Habitat
This pelagic species has been recorded from the tropical Atlantic and the Mediterranean. It has been noted as a stranding on Curaçao (map 9.8).

Stenella longirostris (Gray, 1828)
Spinner Dolphin

Description
The basic pattern is a dark gray dorsum grading to a gray venter, but coloration varies greatly among the Pacific stocks. The white-bellied spinner dolphin exhibits striping on the flanks, which distinguishes it from the other species of *Stenella* found near the Pacific coast of Panama (plate 13).

Range and Habitat
The species is found in the tropical Pacific and has been recorded off the south coast of Panama (map 9.8).

Stenella plagiodon (Cope, 1866)
Atlantic Spotted Dolphin

Description
This dolphin is similar in appearance to *Stenella attenuata,* with gray spots on the flanks and sides, but it occurs in the tropical Atlantic.

Range and Habitat
The species occurs in the tropical portions of the Atlantic Ocean. It has been recorded from the east coast of Panama (map 9.8).

Genus *Tursiops* Gervais, 1855
Tursiops truncatus (Montagu, 1821)
Bottle-nosed Dolphin, Tursion

Description
Up to 52 teeth may be present. The length ranges from 1.75 to 3.6 m, the dorsal fin is at midbody and is modest in size. This graceful dolphin has a distinct beak and a rounded forehead. Dorsum varies from dark blue to gray (plate 13).

Range and Habitat
The species is found worldwide in coastal areas of the tropics and subtropics (map 9.4).

Natural History
This is one of the best studied of the small dolphins. It forms loosely bonded schools of adults that divide into subgroups, whose composition may be quite stable over two to three years. These units may be mothers and their maturing young. Gestation is approximately twelve months, and at birth the single calf is approximately 1 m long (Tavolga and Essapian 1957).

These dolphins are highly gregarious and have a complicated communication system including vocalization, touching, and visual display (Caldwell and Caldwell 1977). The use of special high-frequency sound pulses in echolocating underwater was clearly demonstrated in training experiments (Kellogg 1961). Injured dolphins are aided by other school members and may be supported at the surface (Caldwell and Caldwell 1966).

SUBORDER MYSTICETI
As indicated in the description of the order Cetacea, the mysticetes include the baleen whales. There are three families, the Eschrichtidae, the Balaenidae, and the Balaenopteridae. Only the last occurs within the range covered by this volume.

FAMILY BALAENOPTERIDAE
Rorquals, Ballenas con Aleta

Diagnosis
Teeth appear embryologically but are functionally replaced in the adult by epidermally derived plates called baleen, which hang from the roof of the mouth (fig. 9.1). These large whales range from 9 to 30 m in length. There is size dimorphism, and females gen-

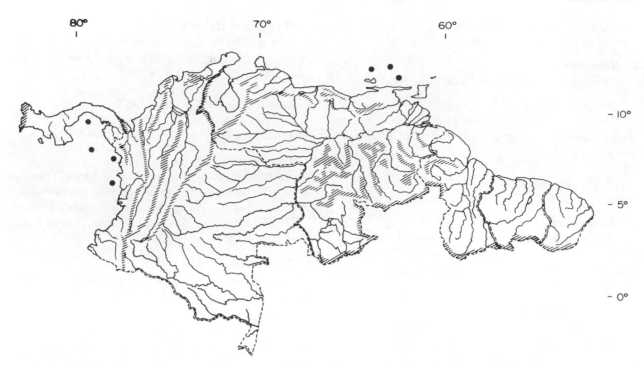

Map 9.9. Distribution of *Megaptera novaeangliae*.

erally exceed males by about 1 m. The body is elongated and streamlined, and in the region of the throat and chest there are numerous longitudinal furrows that allow the buccal cavity to distend during feeding. The skull is vastly modified to accommodate the baleen plates, and its structure is unmistakable (fig. 9.11).

Distribution
The family is found worldwide and is oceanic.

Genus *Megaptera* Gray, 1846
Megaptera novaeangliae (Borowski, 1781)
Humpbacked Whale

Description
Length ranges from 11.5 to 15 m. The pectoral fins are extremely long, approximately one-third the total length. The long flipper is diagnostic, and the dorsal fin is greatly reduced in size. The dorsum is black, and the throat and chest are white (plate 13).

Range and Habitat
The species is found in all oceans of the world but is extremely reduced in abundance (map 9.9).

Natural History
This migratory species feeds in the polar regions during summer and returns to warmer waters in the winter. It feeds extensively on planktonic crusta-

ceans but also may take small fish. A single young is born after a twelve-month gestation period. This whale has been the subject of numerous investigations, especially concerning the function of its songs. Males apparently pick traditional spots in shallow oceanic water, where they perform extended bouts of sound production or "singing." Songs of different stocks are dialectically distinct and vary from season to season. Subadult males apparently copy the pattern of a dominant male, so that a distinct song type characterizes the breeding grounds. It is believed that the singing serves to attract females or announce a male's presence (Payne and McVay 1971; Winn and Olla 1979; Guinee, Chu, and Dorsey 1983; Payne, Tyack, and Payne 1983).

Genus *Balaenoptera* Lacépède, 1804
Fin Whales and Relatives

Description
The body is long and streamlined, with the dorsal fin rather small and set far back. The female is larger than the male. The species of this genus are distinguishable by their size and form a graded series. The color pattern is dark gray-brown to black above and white below.

Distribution
These whales are oceanic and found worldwide. Gaskin (1982) is a useful summary of their biology.

Balaenoptera acutorostrata Lacépède, 1804
Minke Whale

Description
This is the smallest of the rorquals, averaging about 9.1 m in length. The light gray countershading may extend anteriorly to the sides (plate 13).

Range and Habitat
The species is oceanic and found worldwide. It has been recorded as a stranding in Suriname (map 9.10).

Natural History
In common with the other small rorquals, this species feeds extensively on fish such as herring. It often comes quite close to shore and frequently becomes stranded. The gestation period is twelve months, and a single calf is born. Interbirth intervals may exceed three years.

Balaenoptera borealis Lesson, 1828
Sei Whale

Description
One of the larger rorquals, this species can reach a length of 19 m and weigh over 45 tons. The dorsal fin averages about 500 mm in height. The dorsum is dark gray, extending to the back and sides, and the venter is a lighter gray.

Range and Habitat
This whale is oceanic and found worldwide. It has been recorded as a stranding from Suriname (map 9.11).

Natural History
The species is more prone to take plankton than to take fish. A single calf is born after a gestation period of twelve months. Calves remain with the mother for over a year.

Balaenoptera edeni Anderson, 1878
Bryde's Whale

Description
The maximum length is approximately 14 m, and weight ranges from 27 to 32 tons. The dorsal fin is about 470 mm high and set approximately two-thirds down the back (plate 13).

Range and Habitat
The species is found mainly in tropical waters. It is probably the most commonly encountered rorqual in the tropical latitudes (map 9.10).

Natural History
This whale is not as specialized for feeding on plankton and includes a great deal of fish in its diet.

Balaenoptera musculus (Linnaeus, 1758)
Blue Whale

Description
This whale is the largest living mammal. Total length ranges from 21 to 27 m. The dorsum is blue gray with white mottling, and the venter is lighter but can have a yellowish cast.

Range and Habitat
The blue whale was once worldwide and oceanic, but it has been vastly reduced in number. It has been recorded as a stranding on the north coast of Panama (map 9.11).

Natural History
The female gives birth to a single calf every three years following a twelve-month gestation period. Newborn calves measure approximately 7 m. The blue whale is nearly an obligate feeder on plankton and makes vast migratory movement to productive feeding grounds, generally near the poles in summer.

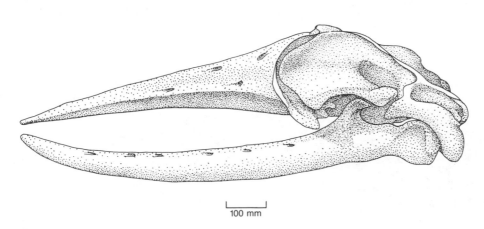

100 mm

Figure 9.11. Skull of *Balaenoptera acutorostrata*.

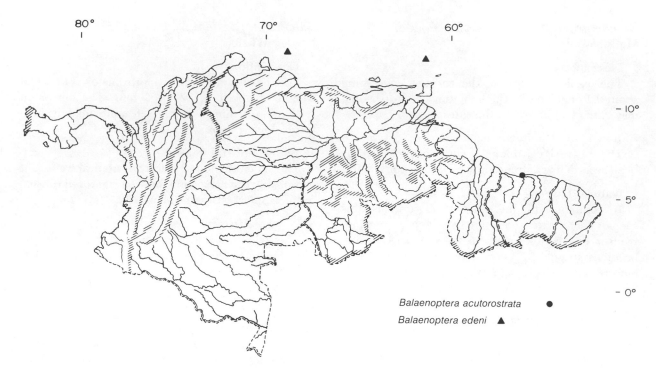

Map 9.10. Distribution of two species of *Balaenoptera*.

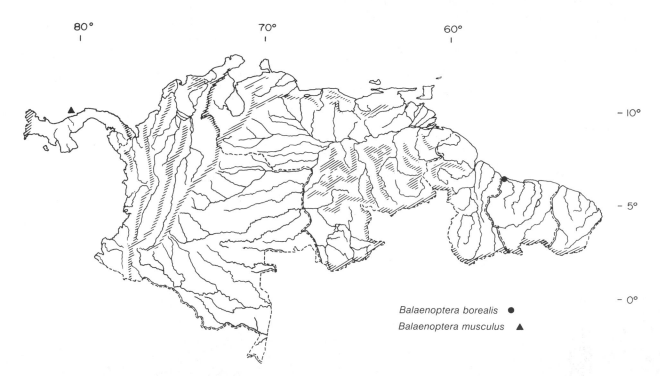

Map 9.11. Distribution of two species of *Balaenoptera*.

Balaenoptera physalus (Linnaeus, 1758)
Fin Whale

Description

This second-largest rorqual may exceed 21 m in length and weigh from 55 to 63 tons. The dorsal fin may be 500 mm long. Its back is dark gray, and it is white beneath. The right side of the lower jaw may be white, and generally just behind the head is a gray chevron pointing toward the tail, a distinctive marking.

Range and Habitat

The species is normally found in north temperate waters but is an occasionally straggler on the Pacific coast of northern South America.

Natural History

The fin whale is a plankton feeder but also eats fish and squid. Though it is generally solitary, small schools occasionally form. A single young is born after a twelve-month gestation period.

References

• References used in preparing distribution maps

Best, P. B. 1974. The biology of the sperm whale as it relates to stock management. In *The whale problem*, ed. W. E. Scheville, 257–93. Cambridge: Harvard University Press.

Best, R. C., and V. M. F. Silva. 1984. Preliminary analysis of reproductive parameters of the boutu, *Inia geoffrensis* and the tucuxi, *Sotalia fluviatilis* in the Amazon River system. In *Reproduction in whales, dolphins and porpoises*, ed. W. F. Perrin, R. L. Brownell, Jr., and D. P. DeMaster, 361–72. Special Issue 6. Cambridge: International Whaling Commission.

Brownell, R. L., Jr. 1984. Review of reproduction in platanistid dolphins. In *Reproduction in whales, dolphins and porpoises*, ed. W. F. Perrin, R. L. Brownell, Jr., and D. P. DeMaster, 149–61. Special Issue 6. Cambridge: International Whaling Commission.

Caldwell, D. K., and M. C. Caldwell. 1971. The pygmy killer whaler, *Feresa attenuata*, in the western Atlantic, with a summary of world records. *J. Mammal.* 52:206–8.

———. 1977. Cetaceans. In *How animals communicate*, ed. T. A. Sebeok, 794–801. Bloomington: Indiana University Press.

• Caldwell, D. K., M. C. Caldwell, W. F. Rathjen, and J. R. Sullivan. 1971. Cetaceans from the Lesser Antillean island of St. Vincent. *Fish. Bull.* 69(2): 303–12.

Caldwell, M. C., and D. K. Caldwell. 1966. Epimeletic (care giving) behavior in Cetacea. In *Whales, dolphins and porpoises*, ed. K. Norris, 755–89. Berkeley: University of California Press.

Casinos, A., F. Bisbal, and S. Boher. 1981. Sobre tres ejemplares de *Sotalia fluviatilis* del Lago de Maracaibo (Cetacea-Delphinidae). *Publ. Dept. Zool. Univ. Barcelona* 7:93–96.

Defler, T. R. 1983. Associations of the giant river otter (*Pteronura brasiliensis*) with fresh-water dolphins (*Inia geoffrensis*). *J. Mammal.* 64(4):692.

• Douglas, M. E., G. D. Schnell, and D. J. Hough. 1984. Differentiation between inshore and offshore spotted dolphins in the eastern tropical Pacific Ocean. *J. Mammal.* 65:375–87.

Eisenberg, J. F. 1986. Dolphin behavior and cognition: Evolutionary and ecological aspects. In *Dolphin cognition and behavior: A comparative approach*, ed. R. Buhr, R. Schusterman, J. Thomas, and F. Wood, 261–70. Hillsdale, N.J.: Lawrence Erlbaum Associates.

Ford, J. K. B., and D. Fisher. 1983. Group-specific dialects of killer whales (*Orcinus orca*) in British Columbia. In *Communication and behavior of whales*, ed. R. Payne, 129–62. Boulder, Colo.: Westview Press.

Gaskin, D. E. 1982. *The ecology of whales and dolphins*. London: Heinemann.

Guinee, L. N., K. Chu, and E. M. Dorsey, 1983. Changes over time in the songs of known individual humpback whales (*Megaptera novaeangliae*). In *Communication and behavior of whales*, ed. R. Payne, 59–80. Boulder, Colo.: Westview Press.

• Hall, E. R. 1981. *The mammals of North America*. 2d ed., 2 vols. New York: John Wiley.

Handley, C. O., Jr. 1966. A synopsis of the genus *Kogia*. In *Whales, dolphins and porpoises*, ed. K. S. Norris, 62–69. Berkeley: University of California Press.

Hershkovitz, P. 1966. Catalog of living whales. *Bull. U.S. Nat. Mus.* 246:1–259.

• Husson, A. M. 1978. *The mammals of Suriname*. Leiden: E. J. Brill.

Kasuya, T., and H. Marsh. 1984. Life history and reproduction of the short-finned pilot whale, *Globicephala macrorhynchus*. In *Reproduction in whales, dolphins and porpoises*, ed. W. F. Perrin, R. L. Brownell, Jr., and D. P. DeMaster, 259–310. Special Issue 6. Cambridge: International Whaling Commission.

Kellogg, W. N. 1961. *Porpoises and sonar*. Chicago: University of Chicago Press.

Layne, J. N. 1958. Observations on freshwater dolphins in the upper Amazon. *J. Mammal.* 39:1–22.

Layne, J. N., and D. K. Caldwell. 1964. Behavior of the Amazon dolphin *Inia geoffrensis* in captivity. *Zoologica* 49:81–108.

Magnusson, W. E., R. C. Best, and V. M. F. da Silva. 1980. Numbers and behaviour of Amazonian dolphins, *Inia geoffrensis* and *Sotalia fluviatilis fluviatilis*, in the Rio Solimões, Brasil. *Aquatic Mammals* 8(1):27–32.

Mead, J. G., W. A. Walker, and W. J. Houck. 1982. Biological observations on *Mesoplodon carlhubbsi* (Cetacea: Ziphiidae) *Smithsonian Contrib. Zool.* 344:1–23.

Minasian, S. M., K. Balcomb III, and L. Foster. 1984. *The world's whales.* Washington, D.C.: Smithsonian Institution Press.

Moore, J. C. 1966. Diagnoses and distributions of beaked whales of the genus *Mesoplodon* known from North American waters. In *Whales, dolphins and porpoises,* ed. K. S. Norris, 32–61. Berkeley: University of California Press.

———. 1968. Relationships among the living genera of beaked whales with classifications, diagnoses and keys. *Fieldiana: Zool.* 53(4):209–98.

Nagorsen, D. 1985. *Kogia simus. Mammal. Species* 239:1–6.

Norris, K. S., ed. 1966. *Whales, dolphins, and porpoises.* Berkeley: University of California Press.

Payne, K., P. Tyack, and R. Payne. 1983. Progressive changes in the songs of humpback whales (*Megaptera novaeangliae*): A detailed analysis of two seasons in Hawaii. In *Communication and behavior of whales,* ed. R. Payne, 9–57. Boulder, Colo.: Westview Press.

Payne, R. S., and S. McVay. 1971. Songs of humpbacked whales. *Science* 173:585–97.

Perrin, W. F., R. L. Brownell, and D. F. DeMaster. 1984. *Reproduction in whales, dolphins and porpoises.* Special Issue 6. Cambridge: International Whaling Commission.

• Pilleri, G., and O. Pilleri. 1982. *Zoologische expedition zum Orinoco und Brazo Casiquiare 1981.* Ostermundigen, Switzerland: Hirnanatomischen Institutes.

Slijper, E. J. 1962. *Whales.* London: Hutchinson.

Tavolga, M. C., and F. A. Essapian. 1957. The behavior of the bottlenosed dolphin (*Tursiops truncatus*): Mating, pregnancy, parturition and mother-infant behavior. *Zoologica* 42:11–31.

Trebbau, P. 1975. Measurements and some observations on the freshwater dolphin *Inia geoffrensis* in the Apure River. *Zool. Gart.,* n.s., 45:153–67.

• Trebbau, P., and P. J. H. van Bree. 1974. Notes concerning the freshwater dolphin *Inia geoffrensis* in Venezuela. *Z. Säugetierk.* 39:50–57.

• van Bree, P. J. H. 1975. Preliminary list of the cetaceans of the southern Caribbean. *Studies on the Fauna of Curaçao and Other Caribbean Islands* 68:79–87.

Winn, H. E., and B. L. Olla, eds. 1979. *Behavior of marine animals.* Vol. 3. *Cetaceans.* New York: Plenum Press.

Plates

Plate 1. Marsupialia: Didelphidae. Top to bottom: *Philander opossum; Metachirus nudicaudatus; Caluromys lanatus; Caluromysiops irrupta.*

Plate 2. Marsupialia: Didelphidae and Caenolestidae. Top row above left to right: *Marmosi alstoni*; *Caenolestes obscura*; below, *Marmosa robinsoni*. Second row: *Lutreolina crassicaudata*; *Chironectes minimus*. Third row: *Didelphis albiventris*.

Plate 3. Xenarthra: Myrmecophagidae, Bradypodidae, and Choloepidae. Top row: *Myrmecophaga tridactyla*. Second row: left to right, *Tamandua mexicana; Cyclopes didactylus; Tamandua tetradactyla*. Third row: left to right, *Bradypus tridactylus; Choloepus didactylus; Bradypus variegatus*.

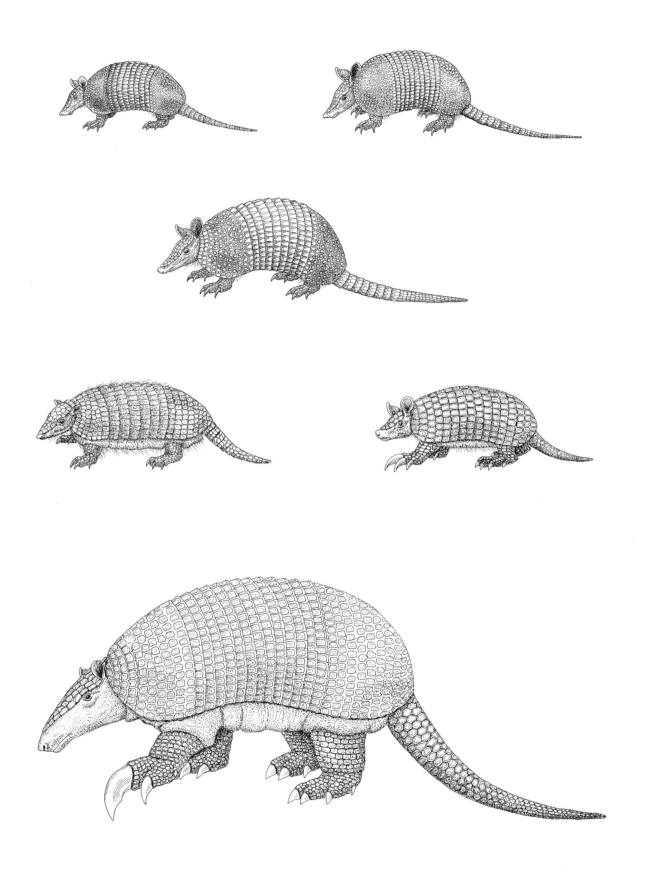

Plate 4. Xenarthra: Dasypodidae. Top row: *Dasypus sabanicola*; *Dasypus novemcinctus*. Second row: *Dasypus kappleri*. Third row: *Euphractus sexcinctus*; *Cabassous unicinctus*. Fourth row: *Priodontes giganteus*.

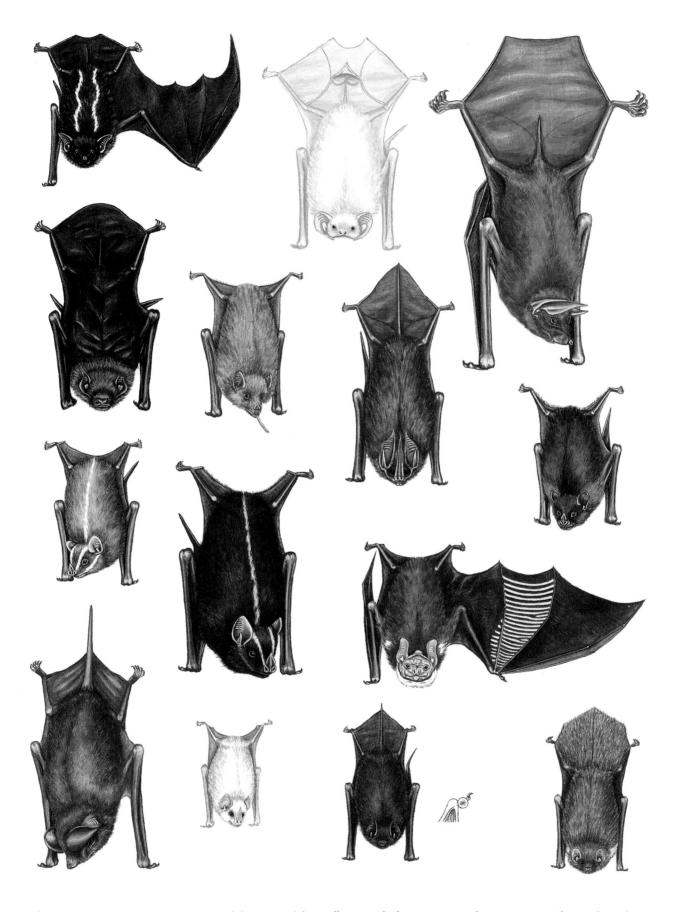

Plate 5. Chiroptera. Top row: *Saccopteryx bilineata*; *Diclidurus albus*; *Noctilio leporinus*. Second row: *Pteronotus davyi*; *Glossophaga soricina*; *Lonchorhina aurita*; *Sturnira lilium*. Third row: *Vampyrops vittatus*; *Vampyressa bidens*; *Centurio senex*. Fourth row: *Eumops glaucinus*; *Ectophylla alba*; *Thyroptera discifera*; *Lasiurus borealis*.

Plate 6. Primates: Callithricidae and Cebidae. Top row: *Saguinus midas; Cebuella pygmaea; Callimico goeldii.* Second row: *Saguinas oedipus; Saguinas geoffroyi; Saguinas leucopus.* Third row: *Saguinus fuscicollis; Saguinus nigricollis; Saguinus inustus.* Fourth row: *Aotus trivirgatus; Callicebus moloch; Callicebus torquatus.*

Plate 7. Primates: Cebidae. Top row: *Pithecia pithecia* (male); *Pithecia hirsuta*; *Chiropotes satanas*. Second row: *Cacajao melano-cephalus*. Third row: *Saimiri sciureus*; *Cebus capucinus*. Fourth row: *Cebus albifrons*. Fifth row: *Cebus nigrivittatus*; *Cebus apella* (subadult male).

Plate 8. Primates: Cebidae. Top row: *Alouatta seniculus*; *Alouatta palliata*. Second row: *Lagothrix lagothricha*; *Ateles geoffroyi*. Third row: *Ateles paniscus*; *Ateles belzebuth*.

Plate 9. Carnivora: Canidae and Ursidae. Top row: *Urocyon cinereoargenteus; Cerdocyon thous*. Second row: *Speothos venaticus; Atelocynus microtis*. Third row: *Tremarctos ornatus*.

Plate 10. Carnivora: Procyonidae. Top row: *Bassariscus sumichrasti*. Second row: *Nasuella olivacea*; *Nasua nasua*. Third row: *Procyon cancrivorus*; *Procyon lotor*. Fourth row: *Potos flavus*; *Bassaricyon gabbii*.

Plate 11. Carnivora: Mustelidae. Top row: *Galictis vittata*; *Mustela frenata*. Second row: *Conepatus semistriatus*; *Eira barbara*. Third row: *Lutra longicaudis*. Fourth row: *Pteronura brasiliensis*.

Plate 12. Carnivora: Felidae. Top row: *Felis pardalis*; *Felis wiedii*. Second row: *Felis tigrina*; *Felis yagouaroundi*. Third row: *Panthera onca*.

Plate 13. Cetacea. Top half: top to bottom, *Balaenoptera edeni*; *Balaenoptera acutorostrata*; *Megaptera novaeangliae*; *Physeter mac-rocephalus*. Bottom half: top, *Globicephala macrorhynchus*. Second row: *Stenella plagiodon*; *Inia geoffrensis*. Third row: *Tursiops truncatus*; *Sotalia fluviatilis*. (Bar top half = 2 m; bottom half = 1 m.)

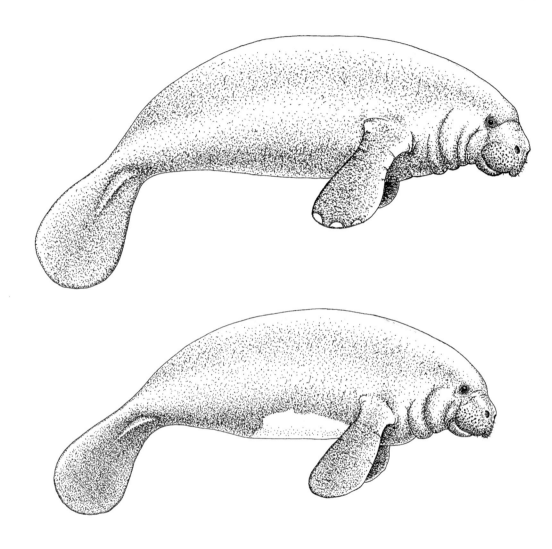

Plate 14. Sirenia. Top to bottom: *Trichechus manatus; Trichechus inunguis.*

Plate 15. Perissodactyla: Tapiridae. Top to bottom: *Tapirus bairdii*; *Tapirus terrestris*; *Tapirus pinchaque* with young.

Plate 16. Artiodactyla. Top row: *Mazama americana; Mazama gouazoubira*. Second row: *Mazama rufina; Pudu mephistophiles*. Third row: Young, female, and adult male *Odocoileus virginianus*. Fourth row: *Tayassu pecari; Tayassu tajucu*.

Plate 17. Rodentia: Sciuridae. Top row: *Sciurus pucheranii*; *Sciurillus pusillus*; *Microsciurus alfari*. Second row: *Sciurus granatensis*, two color phases; *Sciurus aestuans*. Third row: *Syntheosciurus brochus*; *Sciurus variegatoides*, two color phases. Fourth row: *Sciurus igniventris*.

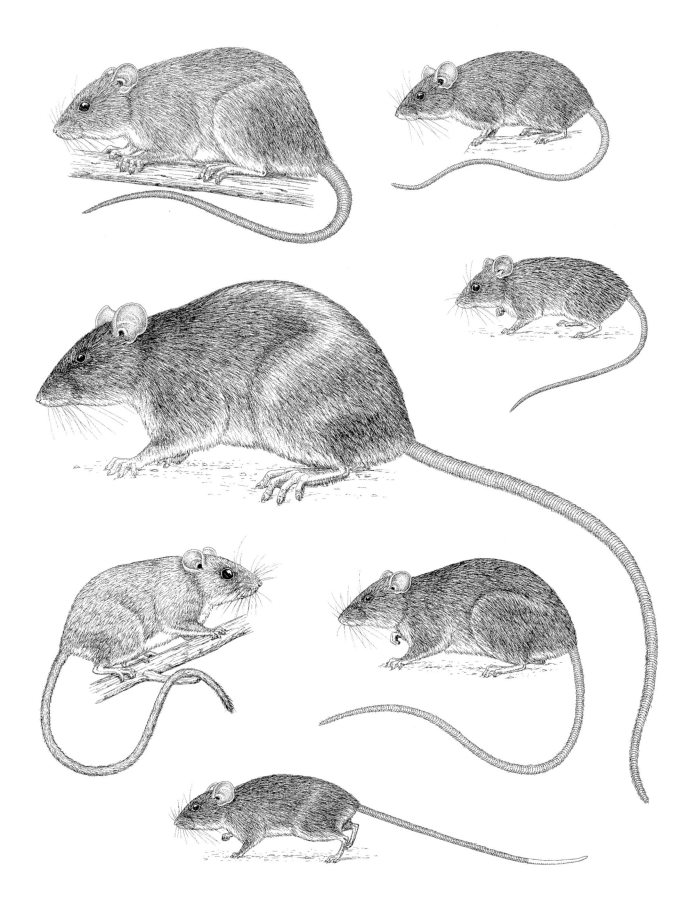

Plate 18. Rodentia: Cricetidae. Top row: *Oryzomys capito; Oryzomys minutus*. Second row: *Holochilus sciureus; Neacomys tenuipes*. Third row: *Rhipidomys mastacalis; Thomasomys lugens*. Fourth row: *Chilomys instans*.

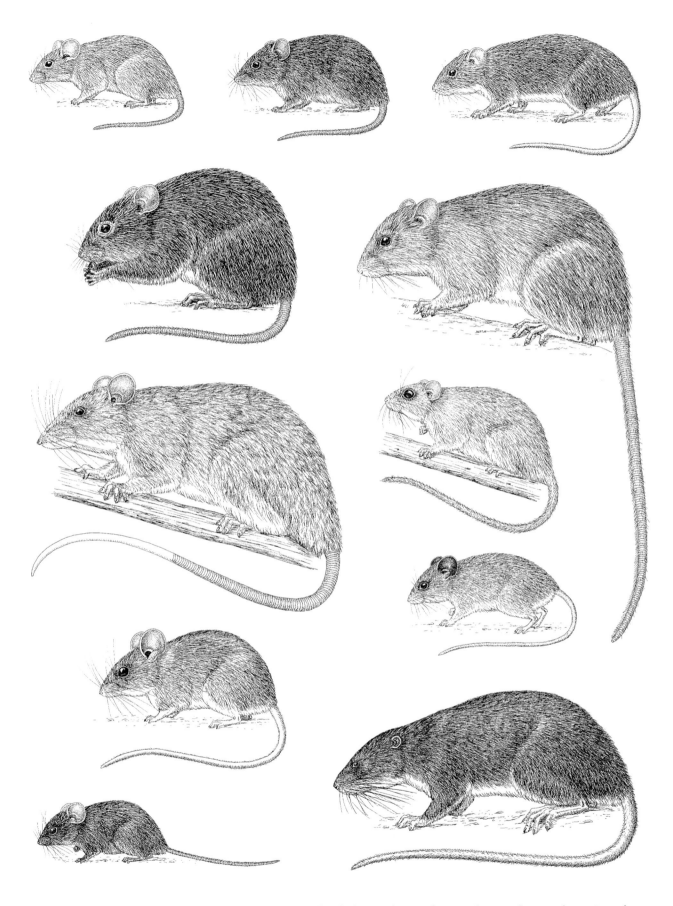

Plate 19. Rodentia: Cricetidae. Top row: *Calomys hummelincki; Akodon uruchi; Zygodontomys brevicauda.* Second row: *Sigmodon hispidus; Nectomys squamipes.* Third row: *Tylomys nudicaudus; Nyctomys sumichrasti.* Fourth row: *Peromyscus flavidus; Reithrodontomys* sp. Fifth row: *Scotinomys teguina; Ichthyomys hydrobates.*

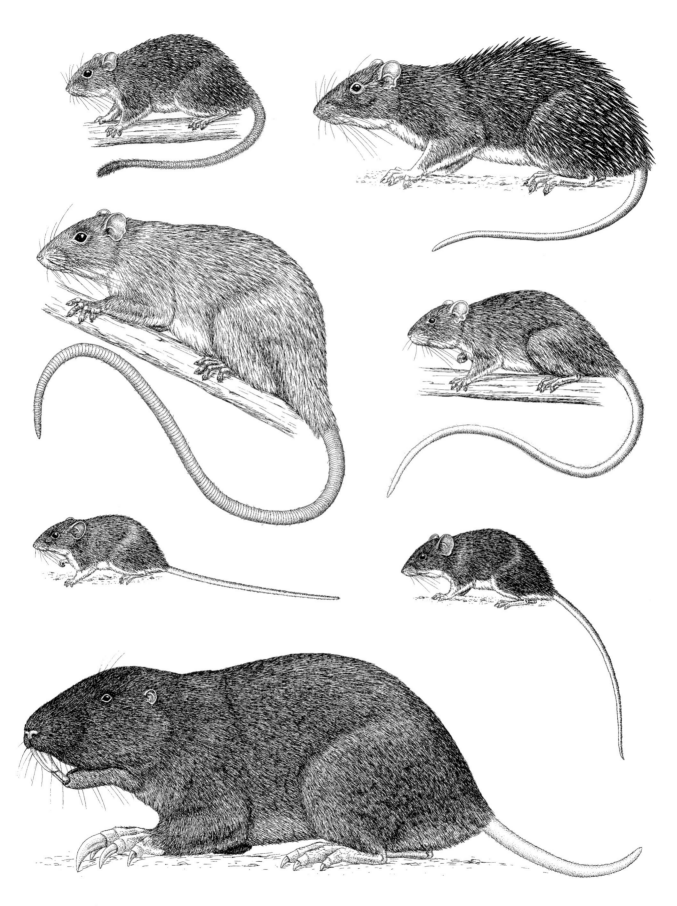

Plate 20. Rodentia: Hystricognatha, Heteromyidae and Geomyidae. Top row: *Mesomys hispidus*; *Hoplomys gymnurus*. Second row: *Dactylomys dactylinus*; *Thrinacodus albicauda*. Third row: *Liomys adspersus*; *Heteromys anomalus*. Fourth row: *Orthogeomys cavator*.

Plate 21. Rodentia: Hystricognatha. Top row: *Echimys semivillosus; Proechimys semispinosus; Diplomys labilis.* Second row: *Echino-procta rufescens; Coendou prehensilis; Cavia aperea.* Third row: *Dasyprocta punctata; Myoprocta acouchy.* Fourth row: *Agouti paca; Dinomys branickii.* Fifth row: *Hydrochaeris hydrochaeris.*

10   Order Sirenia (Sea Cows)

Diagnosis

These large, smooth-skinned aquatic mammals have no hind limbs. The tail has flukes and is whalelike in the Dugongidae but is paddlelike in the Trichechidae. The forelimbs are paddlelike. There is no dorsal fin. The head is large and rounded, with the nostrils on the upper surface of the snout. The rostrum is deflected ventrally (fig. 10.1), most pronounced in the Dugongidae. Rough, horny plates on the palate help to abrade the herbaceous vegetation it feeds on. The cecum is large for digestive fermentation.

Distribution

There are two extant families, the Trichechidae and the Dugongidae. The dugongs are found in the Indian Ocean and in the western Pacific from the Philippines to northern Australia. The Trichechidae occur in a disjunct distribution on the coast and rivers of West Africa and in coastal parts of the Ca-

ribbean as well as in the Amazon and Orinoco drainage systems (Bertram and Bertram 1973).

History and Classification

The sirenians, or sea cows, were much more diverse in the past. They are known from the Eocene of North Africa, Jamaica, and Florida. The group was widespread in the Miocene but is now extremely restricted in its distribution, occurring only in tropical waters. The two families, the Dugongidae and the Trichechidae, seem to have diverged in the Eocene and thus represent two ancient lineages (Domning 1982).

FAMILY TRICHECHIDAE

Manatees, Manatis

Diagnosis

Functional incisors and canines are absent, and the cheek teeth are numerous and variable in num-

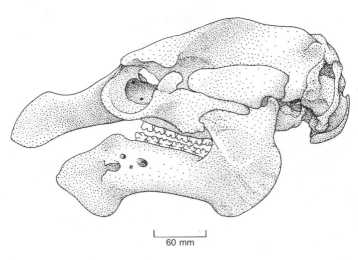

60 mm

Figure 10.1. Skull of *Trichechus manatus*.

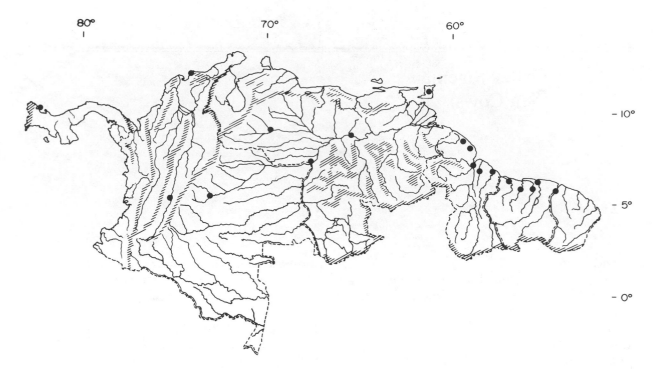

Map 10.1. Distribution of *Trichechus manatus*.

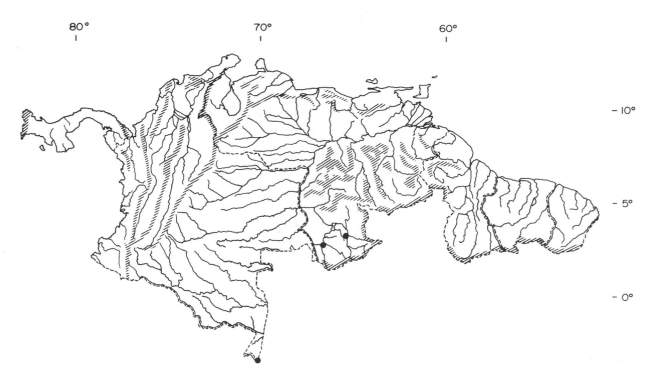

Map 10.2. Distribution of *Trichechus inunguis*.

312

ber, since they are replaced from the rear in a sequential fashion. The teeth are low crowned, with two crests. The manatee may reach 4.5 m in length and weigh over 600 kg. The tail is paddle shaped (plate 14). The upper lip is deeply divided, and the animal can move each side of the lips independently while feeding.

Distribution

The three species of this family show a disjunct distribution. Whereas *T. senegelensis* is found in the rivers of West Africa, the other two species are Neotropical. In South America the Caribbean manatee is confined to riverine areas and coastal zones around the Caribbean. It extends its range up the Orinoco to the first cataract (Mondolfi 1974). The Amazonian manatee is distributed within the Amazon and reaches the region covered by this book only in the extreme southern portions of Venezuela and Colombia (Domning 1982).

Genus *Trichechus* Linnaeus, 1758
Trichechus manatus Linnaeus, 1758
Caribbean Manatee

Description

This is the largest Neotropical manatee, reaching 4.5 m. Three nails are visible on each flipper (plate 14). The body is a dull gray.

Range and Habitat

The range extends from the southeastern United States, especially Florida, and disjunctly to southeastern Mexico. The species occurred in bays throughout the Greater Antilles and is distributed in coastal areas of eastern Central America and across northern South America (map 10.1). It has recently been exterminated over much of its range.

Natural History

Manatees are almost entirely herbivorous, feeding on aquatic vegetation, including mangroves, sea grass, and water hyacinths. They may browse on plants hanging over the water if they can reach them. They tend to move singly except for males courting females or females followed by dependent young. The most cohesive social unit is the female and her young. A single calf is born after a gestation period of thirteen months. It is dependent on the mother for a considerable time, so interbirth intervals may be three and a half years or more. Interbirth interval may also be a function of the habitat's carrying capacity. Age at sexual maturity apparently is not attained until five or six years of age. Members of a group maintain contact by underwater vocalizations (Hartman 1979; Moore 1956; Schevill and Watkins 1965; Bengtson and Fitzgerald 1985; Eisenberg 1981).

Trichechus inunguis (Natterer, 1883)
Amazonian Manatee

Description

This species is easily distinguished from the Caribbean manatee by the absence of nails on the flippers. It is slightly smaller and frequently has whitish patches on the venter (plate 14).

Range and Habitat

The species is confined to the Amazon river system and is allopatric with *T. manatus* (map 10.2).

Natural History

In feeding habits, social tendencies, and presumably other attributes of natural history, *T. inunguis* parallels *T. manatus*. It makes use of the "floating meadows" in the Amazon drainage, where *Paspalum repens* is a dominant aquatic plant. It also eats true water grasses such as *Panicum* and *Echinochloa* (Marmol 1976).

References

Bengtson, J. L., and S. M. Fitzgerald. 1985. Potential rule of vocalizations in West Indian manatees. *J. Mammal.* 66(4):816–18.

Bertram, G. C. L., and C. K. R. Bertram. 1973. The modern Sirenia: Their distribution and status. *Biol. J. Linn. Soc.* 5:297–338.

Domning, D. P. 1982. Evolution of manatees: A speculative history. *J. Paleontol.* 56:599–619.

Eisenberg, J. F. 1981. Moderator's remarks. In *West Indian manatee in Florida*, ed. R. L. Brownell and K. Ralls, 66. Tallahassee: Florida Department of Natural Resources.

Hartman, D. S. 1979. *Ecology and behavior of the manatee* Trichechus manatus *in Florida*. Special Publication 5. Shippensburg, Pa.: American Society of Mammalogists.

Marmol, B. A. E. 1976. Informe preliminar sobre las plantas que sirven de alimento al manati de la Amazonia (*Trichechus inunguis*). *Resum. I. Congr. Nac. Bot.* (Lima, Peru), 31–32.

Mondolfi, E. 1974. Taxonomy, distribution and status of the manatee in Venezuela. *Mem. Soc. Cienc. Nat. La Salle* 34(97):5–23.

Moore, J. C. 1956. Observations of manatees in aggregations. *Amer. Mus. Novitat.* 1811:124.

Reynolds, J. E., III. 1981. Behavior patterns in the West Indian manatee, with emphasis on feeding and diving. *Florida Scientist* 44(4):233–42.

Schevill, W. E., and W. A. Watkins. 1965. Underwater calls of *Trichechus*. *Nature* 205:373–74.

Order Perissodactyla
(Odd-toed Ungulates)

Diagnosis

These ungulates characteristically have an enlarged middle digit on both the forefeet and hind feet. The major weight-bearing axis is on the middle digit, in contradistinction to the Artiodactyla, where the major toes are reduced to two and the axis of weight passes between them. Among the Perissodactyla toe reduction reaches its most extreme form in the horse family (Equidae), where there is only one functional toe. The digestive system does not show extreme modification of the stomach, but the cecum, a blind pouch at the union of the small and large intestine, is enlarged and serves as a fermentation chamber. Tooth reduction is not pronounced, and the incisors are retained. Horns in extant forms, if they are developed at all, occur as epidermal derivatives and are situated in the midline of the nasal bones (e.g., Rhinocerotidae).

Distribution

The Recent distribution includes South America, Africa, Europe, and Asia. Recently extinct in North America (ca. 8,000 B.P.), the order was reintroduced by man in the form of the domestic horse and ass (Equidae).

History and Classification

There are three extant families, the Rhinocerotidae, the Equidae, and the Tapiridae. The last, the tapirs, is the only extant family in South America, although the horse has been reintroduced by Europeans. This order has had a long evolutionary history, first appearing in the Eocene. It radiated rapidly in Asia and North America to include twelve families by the end of the Eocene, but their diversity diminished until only four families existed in the late Miocene.

FAMILY TAPIRIDAE
Genus *Tapirus* Brunnich, 1772
Tapirs, Dantas

Diagnosis

The dental formula is I 3/3, C 1/1, P 4/3–4, M 3/3 (fig. 11.1). Head and body length averages 2 m; the short tail is less than 100 mm long. The nasal bones of the skull are short, and the animal bears a distinct proboscis formed from the nostrils and upper lip, which overhangs the lower lip. The forefeet bear four toes, although a vestige of the thumb can be detected upon dissection. The hind foot bears three toes. The adults of all South American species have uniformly brown dorsal pelage, but the young have a characteristic coat pattern with longitudinal yellowish stripes on a brown background (plate 15).

Distribution

Species of this family were once widely distributed in North America and Asia. Tapirs crossed into South America during the Pliocene; they persist today in South and Central America and in Southeast Asia. Their distribution in Colombia was revised by Hershkovitz (1954). They usually do not occur far from permanent water and are associated with a variety of tropical forested habitats, including dry deciduous forests and multistratal tropical evergreen forests. One species, *Tapirus pinchaque*, is adapted to higher elevations in the Andes. The other species occur predominantly in lowland and premontane forests.

Natural History

All living species are browsers and frugivores. Tapirs tend to confine much of their activity to the hours of darkness but may move around during the day, especially in areas where they are not hunted. They are excellent swimmers and generally have a wallowing area in their home range. They show

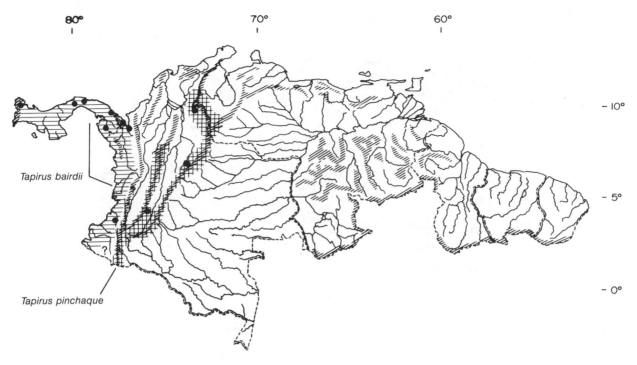

Tapirus bairdii

Tapirus pinchaque

Map 11.1. Distribution of two species of *Tapirus*.

great fidelity to trails and thus are vulnerable to human predation. Tapirs tend to defecate at special loci near water. Communal use of trails and defecation sites permits a loose communication system within a population. Densities can reach 0.8 animals per km^2 in good habitat (Eisenberg and Thorington 1973). A single young is born after a thirteen-month gestation period.

Tapirus bairdii (Gill, 1865)
Central American Tapir, Baird's Tapir

Description
The largest of the three South American species, this tapir may reach a weight of over 200 kg. It is distinguishable from the common tapir by the absence of a neck crest and by its uniformly short, dull brown coat (plate 15).

Range and Habitat
The species is distributed west of the western cordillera of the Andes from Ecuador to Veracruz, Mexico (Hershkovitz 1954). It occurs in a variety of habitats, including dry deciduous forest as well as multistratal tropical evergreen forest (map 11.1).

Natural History
This species has been vastly reduced by hunting. Enders (1935) described the natural history of the species from observations on Barro Colorado Island. Leopold (1959) provides useful information from his

researches in Mexico. Reproduction in captivity has been analyzed by Alvarez del Toro (1966).

Tapirs tend to move singly except for a female and her dependent young. They frequently use the same trails to and from wallowing and feeding sites, which makes them vulnerable to hunters. They are crepuscular, and where hunted they may become almost completely nocturnal.

A single young is born after a thirteen-month gestation period. For the first week after birth it re-

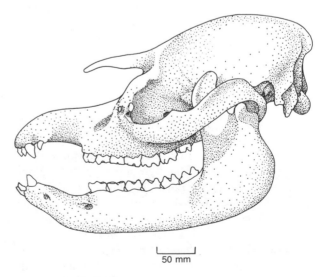

Figure 11.1. Skull of *Tapirus terrestris*.

50 mm

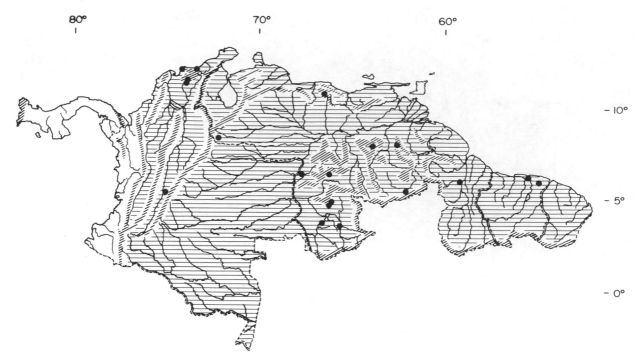

Map 11.2. Distribution of *Tapirus terrestris*.

mains in a secluded spot while the mother feeds, and she returns periodically to nurse it. Its striped coat provides concealing coloration when it is at rest. The young actively follows the mother at ten days of age. The long gestation and long interbirth interval of seventeen months results in a low recruitment rate compared with that of the white-tailed deer (*Odocoileus*). Where hunted, tapirs will take longer to recover than the deer.

Tapirs are selective feeders, eating a wide variety of leaves and fruits (Terwilliger 1978). Janzen has studied the feeding habits of captives and concludes that the tapir is extremely selective, taking a wide range of leaves but only small amounts of a given species in a single feeding bout. Although it may crush many seeds of fleshly fruits (e.g., *Entero-lobium*), the tapir is an important disperser of seeds, since some (e.g., *Raphia faedigera*) pass through its gut unharmed (Janzen 1981, 1982a,b).

Overall (1980) noted a remarkable association between a tapir and a male coati. Tapirs can become infested with ticks. In the case Overall described, the coati fed on the blood-gorged ticks by gleaning them from the tapir's body. The tapir, however, did not reciprocate. I have noted the same behavior when a tame tapir and peccary were associated at semiliberty in Panama. The peccary "cleaned" the tapir of ticks, but once again the tapir never reciprocated. Peccaries in groups are known to reciprocally groom each other, as do coatis.

Tapirus pinchaque (Roulin, 1829)
Mountain Tapir

Description
This is the smallest of the three South American species of tapir. It is distinguished by its long coat of brown hair and its white lips (plate 15).

Range and Habitat
The species occurs at moderate to high elevations in forested habitat, predominantly in the paramos of the central and eastern cordilleras of the Andes, extending from Colombia to Ecuador (Schauenberg 1969). It has been severely persecuted over much of its range and survives only in fragmented populations (map 11.1).

Natural History
Schauenberg (1969) has provided us with the most complete monograph on the behavior, ecology, and distribution of the mountain tapir. In its behavior it resembles *T. bairdii*. Its shaggy coat provides adequate insulation to permit foraging between 2,000 and 4,000 m. It may selectively browse in the paramos. As in the other species, there is little sexual dimorphism. Males may fight severely to gain mating rights with a sexually receptive female. Aside from the mating association or the mother-young unit, the animals move and feed alone.

Tapirus terrestris (Linnaeus, 1758)
Common Tapir

Description

Aside from its geographical distribution, this species is easily distinguished from the other three species by its short hair and by a muscular crest extending from the base of the cranium to the shoulders (plate 15).

Range and Habitat

The species is distributed east of the western cordillera of the Andes through tropical South America to northern Argentina (map 11.2). It occurs in a variety of habitats, from gallery forest to multistratal tropical evergreen forest at lower elevations. Though it can occur in seasonally very arid habitats of the chaco in Paraguay, over much of its range it is usually found near water.

Natural History

This tapir's behavior has been studied in captivity (Richter 1966). Its natural history appears to be similar to that of *T. bairdii*. Vocalizations include a variety of frequency-modulated whistles (Hunsaker and Hahn 1965). Individuals have predictable movement patterns and frequently range near water. Sexual maturity may be attained as early as two years of age; gestation is twelve to thirteen months (Mondolfi 1971; Mallinson 1969).

References

Alvarez del Toro, M. 1966. A note on the breeding of Baird's tapir at Tuxtla Gutierrez Zoo. *Int. Zoo Yearb.* 6:196–97.

Eisenberg, J. F., and R. W. Thorington, Jr. 1973. A preliminary analysis of a Neotropical mammal fauna. *Biotropica* 5:150–61.

Enders, R. K. 1935. Mammalian life histories from Barro Colorado Island, Panama. *Bull. Mus. Comp. Zool.* 78(4):385–502.

Hershkovitz, P. 1954. Mammals of northern Colombia, preliminary report no. 7: Tapirs (genus *Tapirus*), with a systematic review of American species. *Proc. U.S. Nat. Mus.* 103:465–96.

Hunsaker, D., and T. C. Hahn. 1965. Vocalizations of the South American tapir *Tapirus terrestris. Anim. Behav.* 13:69–75.

Janzen, D. 1981. Digestive seed predation by a Costa Rican Baird's tapir, *Tapirus bairdii. Biotropica* 13:59–63.

————. 1982a. Seeds in tapir dung in Santa Rosa National Park, Costa Rica. *Brenesia* 19/20:129–35.

————. 1982b. Wild plant acceptability to a captive Costa Rican Baird's tapir. *Brenesia* 19/20:99–128.

Leopold, A. S. 1959. *Wildlife in Mexico.* Berkeley: University of California Press.

Mallinson, J. J. C. 1969. Reproduction and development of Brazilian tapir *Tapirus terrestris. Dodo* 6:47–51.

Mondolfi, E. 1971. La danta o tapir. *Defensa Nat.* 1(4):13–19.

Overall, K. L. 1980. Coatis, tapirs and ticks: A case of mammalian interspecific grooming. *Biotropica* 12:158.

Richter, W. von. 1966. Untersuchung über angeborene Verhaltensweise den Shabrackentapirs und Flachland Tapirs. *Zool. Beitrage* (Nuremberg) 1(12):67–159.

Schauenberg, P. 1969. Contribution à l'étude du tapir pinchaque *Tapirus pinchaque* Roulin 1829. *Rev. Suisse Zool.* 76:211–56.

Terwilliger, V. J. 1978. Natural history of Baird's tapir on Barro Colorado Island, Panama Canal Zone. *Biotropica* 10:211–20.

Diagnosis

These mammals are especially adapted for feeding on fallen fruit and nuts or on grasses and leaves. The cheek teeth may be either high crowned or low crowned. In the latter case the species either is adapted for browsing or is omnivorous (e.g., Tayassuidae) rather than grazing and browsing (e.g., Cervidae, Bovidae). The premolars are usually simple in structure compared with the molars. Enlarged canines or tusks are present in the Suiformes, Tragulidae, and some species of Cervidae. The grazing and browsing forms exhibit a loss of the upper incisors. The main axis of weight passes between the third and fourth digits of the foot; the first, second, and fifth digits are reduced or lost in the more specialized forms. A corresponding reduction in metacarpals and metatarsals to some extent parallels the reduction in toe number. In males of many species of this order, bony excrescences develop on the forehead (e.g., Cervidae and Bovidae). Generally there are two types of hornlike structures, either antlers that are cast annually or true horns, which are an extended growth of the frontal bone covered with an epidermally derived cap.

Distribution

The recent natural distribution is worldwide except for the Australian area, Oceanic islands, and Antarctica. Domesticated and wild members of this order have been introduced into Australia and New Zealand.

History and Classification

The earliest artiodactyls appear in the early Eocene in both North America and Europe. This group rapidly differentiated into three specialized forms that are reflected in the subordinal classification of the Artiodactyls: the Suiformes, today represented by the hippopotamuses, peccaries, and swine; the Tylopoda, represented by the family Camelidae; and the suborder Ruminantia, which today includes the giraffes (Giraffidae), pronghorn antelopes (Antilocapridae), mouse deer (Tragulidae), musk deer (Moschidae), true deer (Cervidae), and bovines (Bovidae).

The artiodactyls had their origin in the northern continents, and since North America and Asia have been in contact off and on, faunal interchange was possible. Faunal interchange between the Northern Hemisphere and Southern Hemisphere increased as Africa became connected to Eurasia. The artiodactyls did not appear in South America until the late Pliocene, when the Panamanian land bridge became complete, thus linking North and South America. All three suborders are now represented in South America; they have arrived and differentiated only since the Pliocene. The three families in South America at present are the Tayassuidae, or peccaries; the Camelidae, including the vicuña and the llama; and the Cervidae, including some eleven species currently grouped into six genera. Members of the family Bovidae did not reach South America until Europeans transported them in the sixteenth century. The Cervidae underwent a rapid adaptive radiation in the Pleistocene of South America to fill some niches that are occupied by bovines on the more contiguous continental land masses (Eisenberg 1987; Eisenberg and McKay 1974).

FAMILY TAYASSUIDAE

Peccaries, Pecaris, Báquiros

Diagnosis

These piglike Artiodactyla have a dental formula of I 2/3, C 1/1, P 3/3, M 3/3. The canines are modified into tusks but are small; the upper canines are

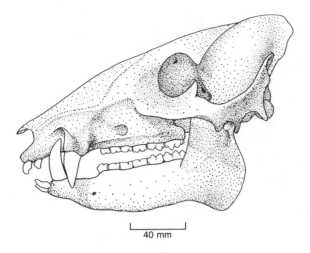

Figure 12.1. Skull of *Tayassu tajacu*.

directed downward and rub against the lower canines, thus maintaining a sharp cutting edge (fig.
12.1). The molars are low crowned. Head and body
length ranges from 700 to 1,000 mm, the tail averages 20 mm, and weight averages about 24 kg.
There are four visible toes on the forefeet and three
on the hind feet, but only two toes on each foot bear
the weight. Although the stomach is modified into
three chambers (Langer 1979), peccaries are not
ruminants.

Distribution

Species of this family are distributed from Texas,
in North America, to Northern Argentina. There are
currently three species, and two occur within the
area covered by this volume, grouped into the genus
Tayassu, including *T. tajacu* and *T. pecari.* The second genus, *Catagonus,* is restricted to Paraguay and
adjacent parts of Argentina. Although described as a
fossil, it was rediscovered as a living form by Wetzel
in 1974 (Wetzel 1977).

Natural History

All species have a similar natural history. They
are gregarious, living in bands of varying sizes, and
are adapted for feeding on palm nuts, fruits, roots,
and tubers as well as the soft parts of certain cacti.
There is a definite specialization for different foraging strategies between *T. tajacu* and *T. pecari* (Kiltie
1980; Kiltie and Terborgh 1983). One or two young
are born after a gestation period of approximately
116 days; they are precocial and can begin to follow
the group when a few days old. The young are
marked quite differently from the parents, having a
light buff coat that contrasts sharply with the darker
parental color. The natural history of the family has
been summarized by Mayer and Brandt (1982),
Sowls (1984), and Enders (1935).

Genus *Tayassu* G. Fischer, 1814
Peccary, Pecari, Baquiro

Description

This genus contains two Recent species. The description for the family adequately covers the general
characters. The two species are easily discriminated,
since *T. tajacu* has a whitish collar whereas *T. pecari*
has white on the upper lip portion and a much
darker coat (plate 16). *T. pecari* is the larger of the
two species.

Distribution

The distribution of the genus is from south Texas
to north-central Argentina. *T. tajacu* covers the
entire range, whereas *T. pecari* is disjunctly distributed from Honduras to Argentina.

Tayassu pecari (Link, 1795)
White-lipped Peccary, Pecari Labiado

Description

Total length averages 1,059 mm; T 43; HF 215;
E 78.5. Weight ranges from 25 to 34 kg. The color of
adults is variable from dark brown to black. The
lower jaw is white, hence the common name white-
lipped peccary.

Range and Habitat

The species has a discontinuous distribution from
southern Mexico to Argentina (map 12.1).

Natural History

This species differs markedly from the collared
peccary in that it tends to move and forage in large
groups. It is not uncommon for over one hundred to
be found in a single herd. Kiltie (1980) suggests that
such herding behavior allows them to harvest palm
nuts from species that show a "mast" fruiting but are
asynchronous over wide areas. Large herds have to
cover tremendous distances to seek out patches of
ripe palm fruits, which vary in concentration from a
few hectares of palms to areas where palms may be
the dominant plant form over several square kilometers (Kiltie and Terborgh 1983). The large herds
of *T. pecari* show little site fidelity, but their seasonal movements may be highly predictable. The
temporal musculature and powerful bite of the white-
lipped peccary permit it to harvest very hard palm
nuts and seeds that are not exploited by the sympatric collared peccary.

The white-lipped peccary herd is a potent antipredator system. These rugged animals with formid-
able tusks can drive off many potential predators.
When aroused, they erect the dorsal hairs so as to
present a large body image when standing broad-
side. The clacking of their teeth as they rapidly open

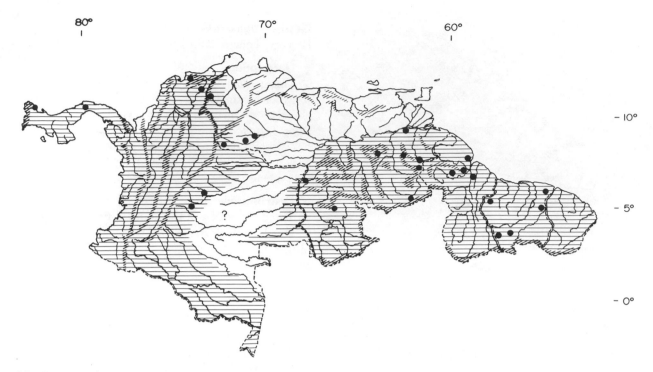

Map 12.1. Distribution of *Tayassu pecari*.

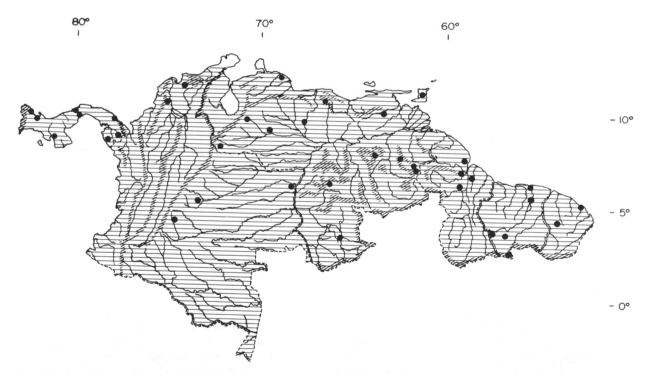

Map 12.2. Distribution of *Tayassu tajacu*.

and close their mouths sounds intimidating to the human observer.

Females mature at eighteen months; after a gestation period of 156 days two young are born. Where primary productivity is pulsed, breeding may be seasonal.

Tayassu tajacu (Linnaeus, 1758)
Collared Peccary, Pecari de Collar, Baquiro de Collar

Description
Average measurements are: TL 940; T 38; HF 196; E 82; Wt 19.3 kg. The basic color is dark brown, but the white-tipped hairs give a salt-and-pepper effect. A white collar extends from the jaw around the neck in the adult.

Range and Habitat
The species is found from the southwestern United States to northern Argentina (map 12.2). *T. tajacu* has an enormous range of habitat types, being found from the dry scrub areas of Arizona, New Mexico, Texas, and northern Mexico to the rain forests of the Amazonian portion of South America.

Natural History
The natural history of both *T. tajacu* and *T. pecari* has been summarized by Kiltie (1980, 1982) and Sowls (1984). *T. tajacu* co-occurs with *T. pecari* over an enormous area, but they have different modes of resource exploitation to reduce direct competition (Kiltie and Terborgh 1983). Kiltie (1981) notes that, given its jaw morphology, *T. tajacu* cannot develop the same leverage as does *T. pecari*, which is adapted for crushing palm nuts. In the northern parts of its range the collared peccary feeds on succulent roots, tubers, and the soft parts of cacti. In the more forested portions of its range it is an efficient exploiter of fallen fruits and palm nuts. *T. tajacu* typically travels in small herds of about eight animals (Byers and Beckhoff 1981), but seasonal coalescence of bands is not uncommon (Robinson and Eisenberg 1985; Oldenberg et al. 1985). In areas with high carrying capacity the core areas of the home ranges tend to be exclusive, and herd members are very cohesive in their movements and actions (Castellanos 1983). In the southwestern United States herds of eight are common.

In a typical greeting ceremony two animals approach each other and, facing in opposite directions, each rubs the lower portion of its jaw on a gland in the other's middorsum. Repeated rubbing tends to maintain odor homogeneity within a herd. The gestation period is about 150 days, and the average lit-

ter size is two. The animals' behavior patterns have been described in detail by Sowls (1974, 1984).

FAMILY CERVIDAE
Deer, Venados

Diagnosis
The dental formula is I 0/3, C 0/1, P 3/3, M 3/3 for deer found in South America. The upper incisors are lacking (figs. 12.2 and 12.3). Deer have four-chambered stomachs and are specialized for browsing. They are true ruminants, and after feeding they lie up in a sheltered area, eructate, and remasticate the rumen contents. The first, second, and fifth digits are very reduced. Males grow antlers that are cast and renewed annually (fig. 12.2). Adults show a uniform dorsal pelage color, but newborns are usually spotted (plate 16).

Distribution
The Recent distribution of the deer family includes North and South America, Europe, and Asia; they are present in Africa only north of the Sahara and are absent from Australia and New Zealand except where introduced by man.

Deer originated in Europe and Asia and spread to North America in the Pliocene. Their spread was rapid, and upon entering South America in the late Pliocene they underwent an adaptive radiation.

Natural History
Males of the species constituting the family Cervidae typically bear antlers that are usually shed annually and grown anew. The only exceptions are the musk deer (*Moschus*) and the Chinese water

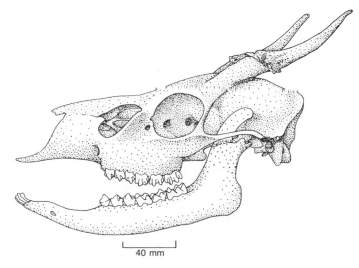

40 mm

Figure 12.2. Skull of *Mazama americana*, male.

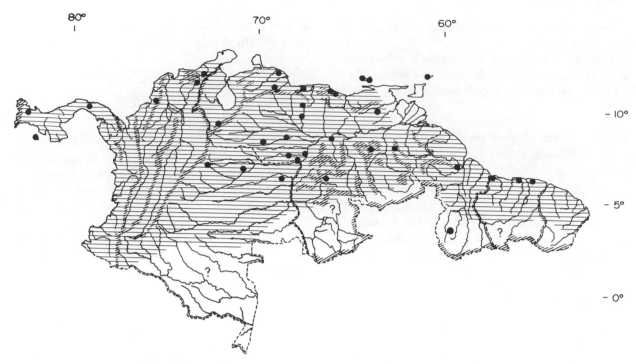

Map 12.3. Distribution of *Odocoileus virginianus*.

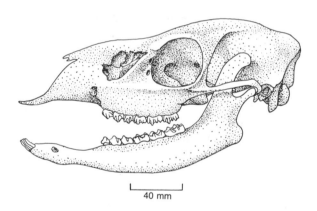

40 mm

Figure 12.3. Skull of *Mazama americana*, female.

deer (*Hydropotes*) of Asia, which are antlerless. Most species are adapted for browsing or mixed browsing and grazing.

Cervids evolved in the high latitudes of the Northern Hemisphere, and thus reproduction was highly synchronized so fawns would be born at a time of the year that would favor their survival. In the North, antler growth of the males, breeding, and the timing of birth tend to be highly seasonal. When deer adapt to the equatorial latitudes these events tend to be less sharply seasonal (Brock 1965). Most studies of deer living near the equator have reported the following: Males still cast their antlers annually, although in any given area the population may show weak synchrony (Brokx 1972a; Branan and Marchin-

ton 1987; Blouch 1987). Conception and birth of fawns, though not tightly synchronized, clearly are timed so that the birthdate favors survival. No sharp birth peak may be shown, because the factor determining survival is plant productivity, which in turn is tied to rainfall. In Suriname there are two peaks of rainfall; in the llanos of Colombia and Venezuela there is one rainy period. In both areas births, though scattered, still reflect a compromise that probably favors fawn survival (Brokx 1972a; Blouch 1987; Branan and Marchinton 1987).

The small deer (*Pudu*, *Mazama chunyi*, and *Mazama rufina*) all live in dense underbrush. Their small stature and reduced antler size could be an adaptation for efficient locomotion in dense vegetative cover. Herbivorous mammals from different taxa worldwide have converged in body form and antipredator behavior to occupy this ecological niche (Bourlière 1973; Dubost 1968; Eisenberg and McKay 1974).

Genus *Odocoileus* Rafinesque, 1832
New World Deer, Venado

Description
These medium-sized deer comprise two species, *O. virginianus* and *O. hemionus*. Only *O. virginianus* occurs within the range covered by this volume. The male's large size and forked antlers distinguish this species from the smaller *Pudu* and *Mazama* fig. 12.4).

Distribution

This genus is distributed from southern Canada to Central America, thence east of the Andes to Brazil. It extends west of the Andean cordillera to Peru.

Odocoileus virginianus (Zimmermann, 1780)
White-tailed Deer, Venado

Description

Height at the shoulder is about 950 mm; head and body length, 1,500 to 2,200 mm; tail, 100 to 250 mm. Weight ranges from 50 to 120 kg. Adult males bear forked antlers with a variable number of tines, immediately distinguishing them from adult males of *Mazama*, which bear a spike antler. The basic color is a light brown dorsum, seasonally turning to gray, contrasting with a white venter. The underside of the tail is white, and its dorsal portion either is all brown or has a bit of black at the tip. A black band generally extends from between the eyes to the snout. There may be varying degrees of white adjacent to the rhinarium (plate 16). The fawn is spotted at birth.

Range and Habitat

Odocoileus virginianus is distributed from southern Canada to Brazil generally north of the Amazon (map 12.3). It is broadly tolerant of a variety of habitat types, but the interface between savanna and forest appears to be especially favorable, since it allows the larger white-tailed deer to graze as well as browse.

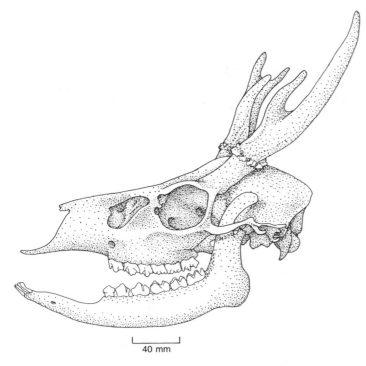

40 mm

Figure 12.4. Skull of *Odocoileus virginianus*, male.

Natural History

The white-tailed deer is one of the best-studied species of the Artiodactyla (Taylor 1956). Most research on this species has been carried out in the United States and Canada. The white-tailed deer was studied in Venezuela by Brokx (1972a,b, 1984), in Colombia by Blouch (1987), in Guyana by Brock (1965), and in Suriname by Branan and Marchinton (1987) and Branan, Werkhoven, and Marchinton (1985). In Venezuela the white-tailed deer is a mixed browser and grazer and is most abundant in the forest-edge habitats of the llanos. When palm fruits are shed in the llanos, mainly during the early wet season (May–June), the deer may feed heavily on *Copernicia* fruits (Brokx and Anderssen 1970). Mean group size in Venezuela is 2.7. The most common association is a mother and her nearly grown offspring or a courting male and female, but in favorable habitats larger aggregations may be observed grazing. In protected areas they may be active in the late afternoon and early morning. Generally, in the llanos the deer rest in a shaded area at midday. Densities can reach 2.2 per km^2 in Venezuela (Eisenberg, O'Connell, and August 1979).

In Venezuela, males shed their antlers annually, but the males in a population are not completely synchronized. The season of reproduction seems to be controlled by the pattern of rainfall. There is generally a birth peak, but tight reproductive synchrony of the population is not characteristic of the tropical latitudes. Although northern races of the white-tailed deer frequently twin, twins are exceedingly uncommon in the subspecies of whitetail occurring in northern South America. The spotted fawn is born after seven and a half months' gestation, and for the first ten days of its life it rests in a sheltered area and does not accompany its mother. The mother returns to feed it at least twice a day. At times of flooding in the lowland llanos of Venezuela and Colombia, large aggregations of whitetails may be trapped on islands of high ground, where they are extremely vulnerable to hunting.

Genus *Mazama* Rafinesque, 1817
Brocket, Corzuela

Description

These small deer have a shoulder height greater than 370 mm but less than 710 mm. There are four species; two are dwarfs (*M. rufina* and *M. chunyi*) and may possibly be confused with pudus. *M. chunyi* is restricted to southern Peru and thus is not within the range covered by this book (Hershkovitz 1959). It averages about 380 mm in shoulder height but does not co-occur with any known species of pudu. *M. rufina* averages about 450 mm at the shoulder. It is reddish brown and can be distin-

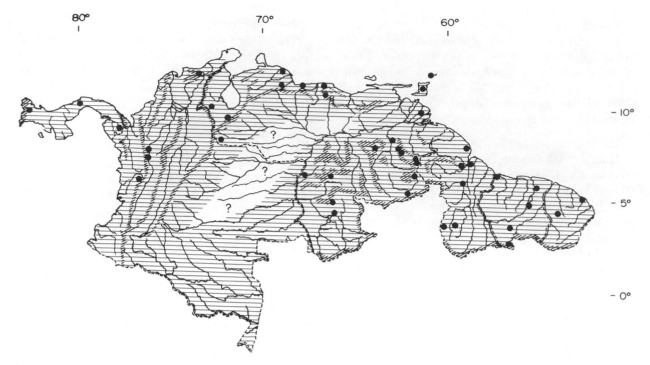

Map 12.4. Distribution of *Mazama americana*.

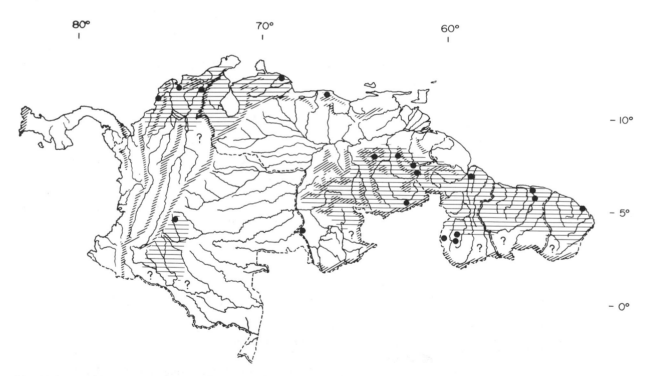

Map 12.5. Distribution of *Mazama gouazoubira*.

guished from *Pudu mephistophiles* by its coat color and its greater size. *M. americana* is reddish brown and the largest species, some 710 mm at the shoulder. It can co-occur with the second of the two larger species, *M. gouazoubira*. Over most of its range, *M. gouazoubira* is smaller than *M. americana;* approximately 610 mm would be an average in the northern part of South America (Hershkovitz 1982). Its coat tends to be gray to gray brown. Males of all species bear a single spike antler that is usually cast annually but may be retained into a second year. As far as can be determined from captive studies, fawns of all species of *Mazama* have spots at birth (Frädrich 1974, 1975).

Distribution

The genus *Mazama* is distributed from southern Mexico to central Argentina. It is adapted to a variety of habitats including montane forests, lowland rain forests, and tree savannas.

Natural History

These small deer are adapted to habitats with suitable cover. In savanna situations they can hide effectively in tall grass, but they are absent from shortgrass areas unless they have access to gallery forests. Forest-edge habitats seem to be excellent in providing shrubs for browsing and shelter. These browsers and frugivores occupy a wide variety of elevations and forest types. They do not form large aggregations and are typically seen as single animals or courting pairs. After a gestation period of seven months the newborn spotted fawn is concealed in a sheltered locus, and the female returns at intervals to nurse it. The fawn does not begin to follow the female until it is several weeks old (Thomas 1975; Frädrich 1974, 1975).

Mazama americana (Erxleben, 1777)
Red Brocket, Corzuela Roja

Description

The upper canines are reduced or absent; only 14 percent of individuals show upper canines in Suriname (Branan and Marchinton 1987). The largest species of the genus, *M. americana* attains 710 mm at the shoulder. Average measurements are TL 1,200; T 126; HF 674; E 99; Wt 29 kg. The venter tends to be white or cream and the dorsum is a deep reddish brown (plate 16). The hind legs are often very dark, almost black, below the heel to the hoof. Much variability is shown by subspecies over the total geographic range.

Range and Habitat

This species is distributed from southern Veracruz, Mexico, to northern Argentina (map 12.4). Habitat includes semideciduous tropical forests, multistratal tropical evergreen forests, and shrub savannas.

Natural History

Almost always seen as a solitary animal, the red brocket is secretive and appears to forage mainly in the evening, throughout the night, and in the early morning. Generally it is seen when flushed from the underbrush. Most data on its natural history derive from captive specimens (Thomas 1975), but Wagner (1960) presents some interesting observations in the wild from Chiapas, Mexico. Gardner (1971) reports a postpartum estrus in a captive female. Branan, Werkhoven, and Marchinton (1985) found that red brockets fed on over sixty plant species in Suriname. Fungi formed an important component of the diet during the wet season. Fruit was preferred when available, with leaves predominating at the end of the wet season when fruit is scarce.

Mazama gouazoubira (G. Fischer, 1814)
Brown or Gray Brocket

Description

The vernacular name brown brocket describes only part of the range of color variation seen in this species, from brown to graybrown and predominantly gray in the southern parts of the range. The major feature distinguishing it from *M. americana* is that it is not reddish brown and tends to be smaller; 610 mm shoulder height is roughly an average. Average measurements are TL 1,034; T 110; HF 27; E 108; Wt 16.3 kg. The black markings on the hind legs, characteristic of *M. americana*, are not usually seen in this species. The venter tends to be white (plate 16).

Range and Habitat

The species ranges disjunctly from Panama to central Argentina. It is broadly tolerant of a variety of habitat types and may be found from savanna to the fringes of multistratal tropical evergreen forests. *M. gouazoubira* can co-occur with *M. americana*, but microhabitat preferences appear to separate them (map 12.5).

Natural History

Little is known concerning the natural history of this species. It appears to prefer tallgrass savannas near gallery forests, browsing but also feeding on fallen fruit. Stallings (1984) found that its diet tracks the relative abundance of fruits in Paraguay; fruits of *Zyziphius* and *Caesalpina* predominate. It is generally seen alone or in pairs. The male bears spike antlers, shed on a twelve- to eighteen-month cycle. In certain habitats where it is not hunted, it may show crepuscular activity. Reproduction shows little seasonality.

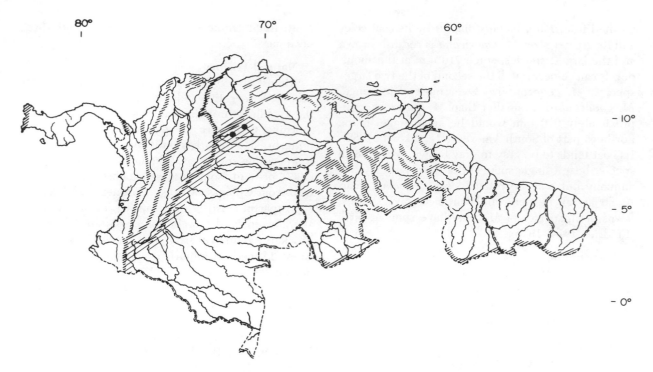

Map 12.6. Distribution of *Mazama rufina*.

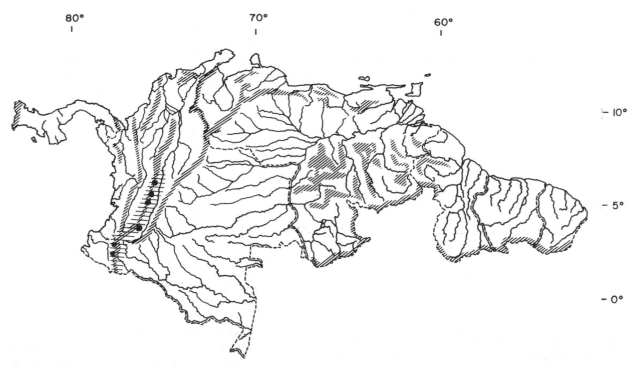

Map 12.7. Distribution of *Pudu mephistophiles*.

Mazama rufina (Bourcier and Pucheran, 1852)
Dwarf Red Brocket

Description
This is the smallest species of *Mazama* within the area covered by this volume. Average measurements are: TL 853; T 78; E 83; Wt 8.2 kg. It averages about 450 mm at the shoulder. The dorsum is a rich chestnut brown. Although the color is variable to some extent, its coat resembles that of *M. americana* rather than *M. gouazoubira* (plate 16).

Range and Habitat
In northern South America the species is confined to the eastern Andean cordillera of Colombia and Venezuela. A small red brocket also occurs in the Atlantic rain forest of Brazil and adjacent areas of Argentina and Paraguay. This form was named *Mazama nana* but is now referred to *M. rufina*. The northern form exists at moderate elevations and possibly overlaps with *Pudu mephistophiles* in the southern part of its Andean range. It is found almost exclusively in forested habitats (map 12.6).

Natural History
Little information is available concerning this solitary, secretive deer. It is assumed that its behavior and ecology are similar to those of *M. americana*. It prefers moist, forested habitats. In northeastern Argentina a single fawn is born between September and February (Crespo 1982).

Genus *Pudu* Gray, 1852
Pudu, Venado Conejo

Description
This genus contains two Recent species, *Pudu puda* and *P. mephistophiles*. Only the latter occurs within the range of this book. Pudus may be distinguished from all other South American deer because one of the carpal bones, the cuneiform, is fused with the navicular/cuboid. These are the smallest deer in South America, less than 380 mm high at the shoulder. The males bear a spike antler that is cast annually. The genus has been reviewed by Hershkovitz (1982).

Pudu mephistophiles (De Winton, 1896)
Northern Pudu

Description
This is the smaller of the two pudu species. The preorbital gland is small. The basic color of the back and shoulders is dark brown or black and the face is also black. The fawn is spotted at birth in *P. puda*, but apparently is unspotted in *P. mephistophiles* (plate 16).

Range and Habitat
In the area covered by this volume, this deer is confined to the Cordillera Central of Colombia. It is a high-elevation species; all specimens have been taken at between 1,700 and 4,000 m elevation (map 12.7).

Natural History
This small deer inhabits the forests of the high Andes. A single fawn is born after a gestation period of approximately seven months (Hick 1967). The standard reference for this genus is Hershkovitz (1982). Details on captive propagation are given in Frädrich (1974).

References

Blouch, R. A. 1987. Reproductive seasonality of the white-tailed deer on the Colombian llanos. In *Biology and management of the Cervidae*, ed. C. Wemmer, 339–43. Washington, D.C.: Smithsonian Institution Press.

Bourlière, F. 1973. The comparative ecology of rain forest mammals in Africa and tropical America. In *Tropical forest ecosystems in Africa and South America: A comparative review*, ed. B. J. Meggers, E. S. Ayensu, and W. D. Duckworth, 279–92. Washington, D.C.: Smithsonian Institution Press.

Branan, W. V., and R. L. Marchinton, 1987. Reproductive ecology of white-tailed and red brocket deer in Suriname. In *Biology and management of the Cervidae*, ed. C. Wemmer, 344–51. Washington, D.C.: Smithsonian Institution Press.

Branan, W. V., M. Werkhoven, and R. L. Marchinton. 1985. Food habits of brocket and white-tailed deer in Suriname. *J. Wildl. Manage.* 49:972–76.

Brock, S. E. 1965. The deer of British Guiana. *J. British Guiana Mus. Zoo Roy. Agric. Comm. Soc.* 40:18–24.

Brokx, P. A. J. 1972a. A study of the biology of Venezuelan white-tailed deer (*Odocoileus virginianus gymnotis* Wiegmann, 1833), with a hypothesis on the origin of South American cervids. Ph.D. diss., University of Waterloo, Ontario.

———. 1972b. Age determination of Venezuelan white-tailed deer. *J. Wildl. Manage.* 36(4): 1060–67.

———. 1984. South America. In *White-tailed deer, ecology and management*, ed. L. K. Hall and C. House, 525–46. Harrisburg, Pa.: Stackpole.

Brokx, P. A. J., and F. M. Anderssen. 1970. Analisis estomacales del venado caramerudo de los llanos venezolanos. *Bol. Soc. Venez. Cien. Nat.* 27: 330–53.

Byers, J. A., and M. Bekoff. 1981. Social, spacing and cooperative behavior of the collared peccary, *Tayassu tajacu*. *J. Mammal.* 62:767–85.

Castellanos, H. G. 1983. Aspectos de la organización social del baquiro de collar, *Tayassu tajacu* en el estado Guárico, Vez. *Acta Biol. Venez.* 11(4):127–43.

Crespo, J. A. 1982. Ecología de la comunidad de mamíferos del Parque Nacional Iguazu, Misiones. *Rev. Mus. Argent. Cien. Nat. "Bernardino Rivadavia" (Ecol.)* 3(2):45–162.

Dubost, G. 1968. Les niches écologiques des forêts tropicales sud-américaines et africaines, sources de convergences remarquables entre rongeurs et artiodactyles. *Terre et Vie* 1:3–28.

Eisenberg, J. F. 1987. Evolutionary history of the Cervidae with special reference to the South American radiation. In *Biology and management of the Cervidae*, ed. C. Wemmer, 60–64. Washington, D.C.: Smithsonian Institution Press.

Eisenberg, J. F., and G. M. McKay. 1974. Comparison of ungulate adaptations in the New World and Old World tropical forests with special reference to Ceylon and the rainforests of Central America. In *The behaviour of ungulates and its relation to management*, 2 vols., ed. V. Geist and F. Walther, 585–602. Publication n.s. 24. Morges, Switzerland: IUCN.

Eisenberg, J. F., M. A. O'Connell, and P. A. August. 1979. Density, productivity and distribution of mammals in two Venezuelan habitats. In *Vertebrate ecology in the northern Neotropics*, ed. J. F. Eisenberg, 187–207. Washington, D.C.: Smithsonian Institution Press.

Enders, R. K. 1935. Mammalian life histories from Barro Colorado Island, Panama. *Bull. Mus. Comp. Zool.* 78(4):385–502.

Frädrich, H. 1974. Notizen über seltener gehaltene Cerviden. Part 1. *Zool. Gart.*, n.s., 44(4): 189–200.

———. 1975. Notizen über seltener gehaltene Cerviden. Part 2. *Zool. Gart.*, n.s., 45(1):67–77.

Gardner, A. 1971. Postpartum estrus in a red brocket deer, *Mazama americana*, from Peru. *J. Mammal.* 52(3):623–24.

Hershkovitz, P. 1959. A new species of South American brocket, genus *Mazama* (Cervidae). *Proc. Biol. Soc. Washington* 72:45–54.

———. 1982. Neotropical deer (Cervidae). Part 1. Pudus, genus *Pudu* Gray. *Fieldiana: Zool.*, n.s., 11:1–86.

Hick, U. 1967. Geglückte Aufzucht eines Pudus (*Pudu pudu* Mol.). *Freunde Kölner Zoo.* 4:111–18.

Kiltie, R. A. 1980. Seed predation and group size in rain forest peccaries. Ph.D. diss., Princeton University.

———. 1981. Stomach contents of rain forest peccaries (*Tayassu tajacu* and *T. pecari*). *Biotropica* 13:234–36.

———. 1982. Bite force as a basis for niche differentiation between rain forest peccaries (*Tayassu tajacu* and *T. pecari*). *Biotropica* 14(3):188–95.

Kiltie, R. A., and J. Terborgh. 1983. Observations on the behavior of rain forest peccaries in Peru: Why do white-lipped peccaries form herds? *Z. Tierpsychol.* 62:241–55.

Langer, P. 1979. Adaptational significance of the forestomach of the collared peccary, *Dicotyles tajacu* (L. 1758) (Mammalia: Artiodactyla). *Mammalia* 43(2):235–45.

Mayer, J. J., and Brandt, P. N. 1982. Identity, distribution, and natural history of the peccaries, Tayassuidae. In *Mammalian biology in South America*, ed. M. A. Mares and H. H. Genoways, 433–55. Pymatuning Symposium in Ecology 6. Special Publication Series. Pittsburgh: Pymatuning Laboratory of Ecology, University of Pittsburgh.

Oldenberg, P. W., P. J. Ettestad, W. E. Grant, and E. Davis. 1985. Structure of collared peccary herds in south Texas: Spatial and temporal dispersion of herd members. *J. Mammal.* 66(4): 764–69.

Robinson, J., and J. F. Eisenberg. 1985. Group size in the collared peccary *Tayassu tajacu*. *J. Mammal.* 66:153–55.

Sowls, L. K. 1974. Social behaviour of the collared peccary *Dicotyles tajacu*. In *The behaviour of ungulates and its relation to management*, 2 vols., ed. V. Geist and F. Walther, 144–65. Publication n.s. 24. Morges, Switzerland: IUCN.

———. 1984. *The peccaries.* Tucson: University of Arizona Press.

Stallings, J. R. 1984. Notes on feeding habits of *Mazama gouazoubira* in the chaco boreal of Paraguay. *Biotropica* 16:155–57.

Taylor, W. P. 1956. *The deer of North America.* Harrisburg, Pa.: Stackpole Press.

Thomas, W. 1975. Observations on captive brockets, *Mazama americana* and *M. gouazoubira*. *Int. Zoo Yearb.* 15:77–78.

Wagner, H. O. 1960. Beitrag zu Biologie des mexikanischen Spiesshirsches *Mazama sartorii* (Saussure). *Z. Tierpsychol.* 17:358–63.

Wetzel, R. M. 1977. The chacoan peccary *Catagonus wagneri* (Rusconi). *Bull. Carnegie Mus. Nat. Hist.* 3:1–36.

13 Order Rodentia
(Rodents, Roedores)

Diagnosis

The dental formula is distinctive: a single pair of upper and lower incisors; no canines; premolars not exceeding two pair; molars not exceeding three per side of the upper and lower jaws. There is a distinct gap or diastema between the incisors and the cheek teeth (fig. 13.1). The incisors are ever growing and have enamel on the anterior surface. Premolar and molar patterns of cusps are diverse and indicate both phylogenetic relationship and adaptation for particular feeding strategies.

The articulation of the jaw with the skull is loose, permitting both lateral and back-to-front motion. The lower jaw can be moved forward into a gnawing phase, occluding the incisors, and retracted to permit occlusion of the molars and premolars during chewing.

Distribution

Rodents are distributed over the entire world except Antarctica. They were originally absent from the Oceanic islands and New Zealand but have been introduced by man. In South America the rodents derive from two separate colonizations. The hystricognath or caviomorph rodents are well represented from the early Oligocene to the present. The sigmodontine rodents (Muridae), Sciuridae, and Geomyoidea entered South America with a later invasion, probably in the Pliocene.

History and Classification

One of the oldest fossil rodent groups is known from the late Paleocene of North America. These forms have been assigned to the family Paramyidae, and modern squirrels derived from them. From a modest beginning rodents speciated widely during the Eocene in North America and Eurasia. Hystricog-

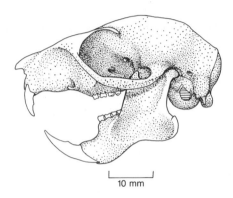

10 mm

Figure 13.1. Skull of *Sciurus aestuans*.

nath rodents first appear in the Oligocene of South America. This group shows certain affinities with African hystricognath rodents.

Attempts at classifying rodents into natural groupings at the level of subclasses have been fraught with difficulty. In the older literature such terms as Sciuromorpha, Myomorpha, and Hystricomorpha are frequently encountered (Simpson 1945). These terms refer to taxonomic groupings based on the way the masseter muscles are attached anteriorly on the rostrum. The hystricomorph rodents have an enlarged infraorbital foramen with a slip from the masseter passing through it to attach anteriorly to the rostrum. The lower jaw is often sharply flared posteriorly, referred to as the hystricognathous condition (fig. 13.8). The sciuromorph rodents show no enlarged infraorbital foramen, and the masseter attaches anteriorly on the rostrum and beneath the zygomatic arch (fig. 13.1). The typical myomorph rodent has a small slip of masseter muscle passing through a slightly enlarged infraorbital foramen, but most of the anterior attachment of the masseter is similar to that described for the Sciuromorpha (fig. 13.2).

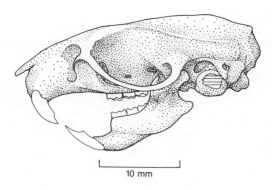

Figure 13.2. Skull of *Oryzomys couesi*.

The tripartite classification below the ordinal level has not proved entirely satisfactory because some rodent taxa exhibit intermediate characters or different mixtures of the major features of jaw form and masseter muscle attachment. For this reason there is a tendency to refer to the superfamily as the highest taxonomic category while recognizing that groups of superfamilies may be conveniently subsumed under yet higher subordinal groupings, but no nomenclatural consistency has yet been agreed upon (Simpson 1959; Wood 1955, 1974). A key to the suborders is given in table 13.1.

The New World hystricognath rodents include the superfamilies Erethizontoidea, Cavioidea, Chinchilloidea, and Octodontoidea (Woods 1982). The South American sciuromorph rodents include the superfamilies Sciuriodea and Geomyoidea. The native South American myomorph genera are grouped under the superfamily Muroidea and include a single subfamily, the Sigmodontinae (Reig 1980; Carleton and Musser 1984).

SUPERFAMILY SCIUROIDEA
FAMILY SCIURIDAE
Squirrels and Marmots, Ardillas

Systematics
The squirrel family is divisible into two subfamilies. The Petauristinae are characterized by a membrane between the forefoot and hind foot, so that extending the limbs and body presents a considerable surface area and lets them glide. This subfamily is not represented in South America. All South American sciurids belong to the subfamily Sciurinae. The interrelationships of the genera and subgenera in this subfamily were discussed by Moore (1959).

Diagnosis
The dental formula is I 1/1, C 0/0, P 1–2/1, M 3/3 (fig. 13.1). The cheek teeth usually show prominent cusps and ridges arranged in a triangular pattern in

the upper molars (see fig. 13.13). The infraorbital foramen is minute, and no muscle passes through it. There are five digits on the hind feet and four on the forefeet, each bearing a long claw. The tail is always haired and may present a bushy appearance (plate 17).

Distribution
Species of this family are distributed on all continents except Antarctica and Australia.

Natural History
The species of South American squirrels are all diurnal. They are strongly adapted for arboreal life yet may forage on the ground. They eat nuts, fruits, and fungi and typically cache food either in hollow trees or by burying it in scattered locations within the home range. The tail tends to be bushy, and color patterns are highly variable even within the same species.

Identification of Genera
Sciurillus is known as the Neotropical pygmy squirrel. With a head and body length less than 115 mm, it is distinctive and easily recognized. *Syntheosciurus* is a montane squirrel of intermediate size confined to western Panama; it could be superficially confused with *Microsciurus*, but its restricted range drops it from consideration over most of northern South America. The genus *Microsciurus* consists of four small species distributed in northern South America. Members of this genus can be confused with *Sciurus pucheranii*, but this species has nearly hairless ears whereas the ears of *Microsciurus* are well haired. The genus *Sciurus* in northern South America is divisible into three subgenera. The subgenus *Sciurus* barely extends into the area covered by this volume, but *Sciurus variegatoides* occurs in western Panama. The subgenus *Guerlinguetus* includes intermediate-sized reddish squirrels that often have an orange venter. The premolar formula is 1/1. The subgenus *Hadrosciurus* includes the giant squirrels, whose head and body measurements often exceed 300 mm. The premolar formula is 1/1. Species of the subgenus *Hadrosciurus* exhibit a variety of color polymorphisms even within one species (see species accounts).

Genus *Sciurus* Linnaeus, 1758
Tree Squirrel, Ardilla

Description
The dental formula is I 1/1, C 0/0, P 1–2/1, M 3/3. The dorsal color is highly variable; in the tropics generally red, dark brown, or orange predominates on the dorsum, although to the north in the temperate zone of North America many species exhibit varying forms of gray and gray brown (see plate 17).

Table 13.1   Key to the Suborders of Rodents and Families of the Sciuromorpha

1	Infraorbital canal not modified for transmitting any part of the medial masseter muscle (Sciuromorpha) 3
1'	Infraorbital canal enlarged for the purpose of transmitting medial masseter muscle 2
2	Infraorbital foramen greatly enlarged . . . . Hystricognatha
	(Caviomorpha)
2'	Infraorbital foramen moderately enlarged or minute . . . . Myomorpha
3	External fur-lined cheek pouches present 4
3'	External fur-lined cheek pouches absent . . . . Sciuridae
4	Tail equal to or longer than head and body; not strongly adapted for fossorial life . . . . Heteromyidae
4'	Tail usually one-third to one-half head and body length; strongly adapted for fossorial life . . . . Geomyidae

Distribution

The genus has speciated widely, and the fifty-five named species occur in Europe and Asia and from Canada south to northern Argentina.

Natural History

These active diurnal, scansorial rodents are striking components of the forest fauna in South America. Although strongly adapted for arboreal life, they come to the ground to forage for food, and many species scatterhoard fruits and nuts by burying them in discrete locations within the home range. Nests are constructed in hollow portions of trees, or a leaf nest may be made in the fork of a branch. The gestation period varies with size, but thirty-eight to forty-five days covers the range. In areas where there is a favorable food distribution two litters may be produced in a single year. Litter size ranges from two to five. The young generally do not open their eyes until they are approximately thirty days old; thus there is a rather long developmental period.

Subgenus *Guerlinguetus*
Sciurus (Guerlinguetus) aestuans Linnaeus, 1776

Description

The dental formula is I 1/1, C 0/0, P 1/1, M 3/3 (fig. 13.1). In this medium-sized squirrel, head and body length ranges from 163 to 180 mm, the tail from 171 to 200 mm, and the hind foot from 43 to 48 mm. Measurements of two specimens were TL 334, 363; T 187, 195; HF 35, 45; E 15, 24; Wt 164 g (Suriname; Brazil, CMNH). The dorsum is dark olive gray grizzled with yellow brown, and the tail is similar in color. There is often a pale yellow eye ring. The venter contrasts, being pale reddish brown, and is sharply demarcated from the dorsal pelage (plate 17).

Range and Habitat

This species occurs south of the Orinoco and in general replaces *S. granatensis* in the south. It extends from southern Venezuela through the Guyanas and thence south through Amazonas to northern Argentina (map 13.1). Its distribution in Colombia is poorly understood.

Natural History

The natural history of this species has not been well documented. It appears to occupy lowland forest and to have habits very similar to those of *S. granatensis*. In addition to feeding on fruits and nuts, it has been implicated in preying on the eggs and young of birds (Husson 1978).

Sciurus (Guerlinguetus) gilvigularis Wagner, 1842

Description

In this medium-sized squirrel, five Venezuelan specimens averaged TL 342; HB 166; T 173; HF 43. Measurements of one specimen from Brazil were TL 363; T 195; HF 35; E 15. It is similar in coloration to *S. aestuans* but paler; the dorsum is a grizzled, reddish buff, there is a buff eye ring, and the venter is reddish orange. The tail may exhibit a faint banding. Cabrera (1960) considered this a subspecies of *S. aestuans*.

Range and Habitat

This species is broadly distributed in northern Brazil and penetrates the range covered by this volume in southern Venezuela and parts of Guyana. A specimen referable to this species was described by Tate (1939) from Guyana (map 13.3).

Sciurus (Guerlinguetus) granatensis Humboldt, 1811
Neotropical Red Squirrel

Description

The dental formula is I 1/1, C 0/0, P 1/1, M 3/3. Total length ranges from 330 to 520 mm; T 140–280; HF 40–65; E 16–36; Wt 228–520 g (Nitikman 1985). The color of the dorsal pelage varies widely (plate 17). The upperparts tend to be rusty brown to almost black, though they may show some mixing of yellow hair. The venter tends to be light yellow to dark reddish brown, often with a white chest patch, but it usually contrasts with the dorsum. The range of size and color in northern South America has been analyzed by Hershkovitz (1947). There are thirty-three named subspecies (Nitikman 1985).

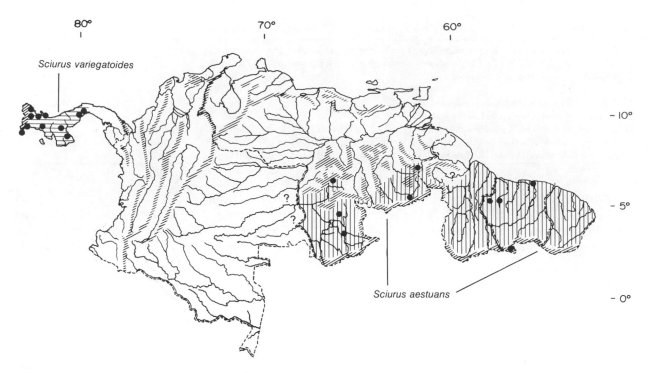

Map 13.1. Distribution of two species of *Sciurus*.

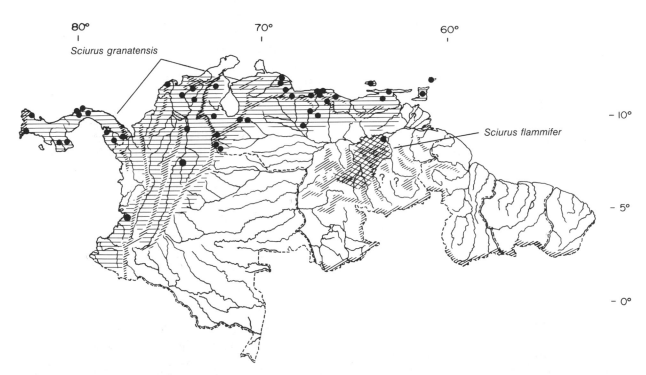

Map 13.2. Distribution of two species of *Sciurus*.

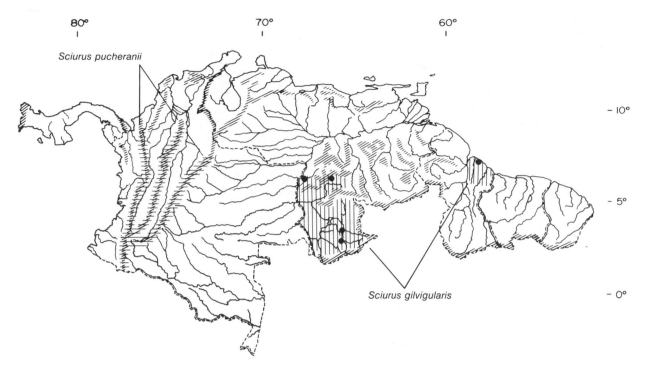

Map 13.3. Distribution of two species of *Sciurus* (ranges for *S. pucheranii* from Cabrera 1960 and Hernandez-Camacho 1960).

Range and Habitat

This species extends from western Costa Rica south through the Isthmus across northern Colombia, south to Ecuador and east into northern Venezuela (map 13.2).

Natural History

This is one of the best studied of the South American squirrels. In Panama this species forages both terrestrially and arboreally, frequently eating over twenty-one species of native plants. Most food items are large seeds and fruits, but young leaves, mushrooms, bark, and flowers are also taken. When it produces, the fruit of the palm *Scheelea* is used heavily. The squirrels cache food both in trees and in the ground (Heaney 1983).

Litter size varies from one to three, with a mode of two. In seasonal habitats such as Panama these squirrels begin breeding at the commencement of the dry season in late December. Births take place in February and early March, and lactation probably continues for nearly two months. Home range areas for adult females overlap only slightly and average about 0.64 ha. Home ranges of adult males overlap with each other and with those of adult females, averaging about 1.5 ha. In good habitat densities may be extremely high, up to four squirrels per ha. On the other hand, in habitats having lowered productivity, density may fall to 0.5 per hectare. At times of high fruit productivity squirrels may transgress on one another's home ranges and form tempo-

rary feeding aggregations (Heaney and Thorington 1978; Glanz et al. 1982; Nitikman 1985).

Sciurus (Guerlinguetus) pucheranii (Fitzinger, 1867)

Description

This small squirrel has a head and body length of approximately 140 mm, and the tail is approximately 120 mm. Three specimens from Colombia showed a range in measurements of TL 324–41; T 183–200; HF 38–43 (CMNH). One weight was recorded at 100 g (MVZ). The species may yet prove congeneric with *Microsciurus* (Moore 1959). The dorsal pelage is reddish brown, contrasting with the gray venter, and is very dark at the midline. The ears are only sparsely haired (plate 17).

Range and Habitat

This species is confined to the cordilleras of the Andes in Colombia and ranges from 2,000 to 3,000 m (map 13.3).

Subgenus *Hadrosciurus*
Sciurus (Hadrosciurus) flammifer Thomas, 1904

Description

The mean measurements for ten Venezuelan specimens were TL 580; HB 274; T 310; HF 66.5 (Allen 1915). In the type specimen the head and ears

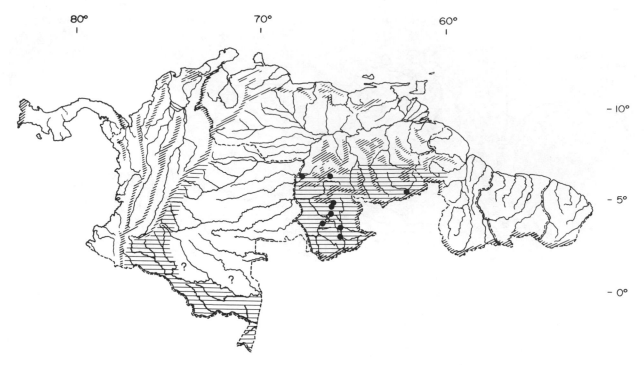

Map 13.4. Distribution of *Sciurus igniventris*.

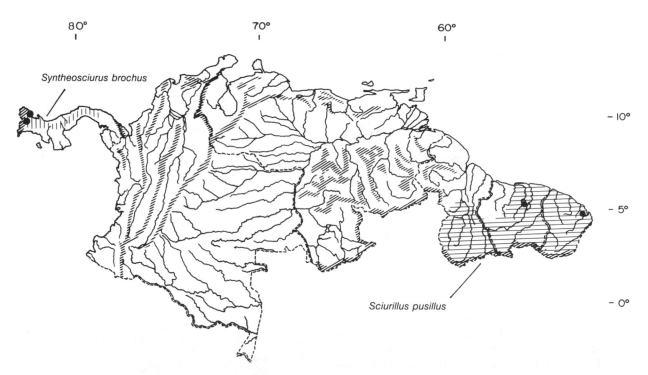

Map 13.5. Distribution of *Syntheosciurus brochus* and *Sciurillus pusillus*.

were red and the upperparts were grizzled yellow and black, darker on the rump. The chin was yellow to orange, with the venter white. The basal third of the tail was black and the rest orange. This species is color polymorphic, and all-black specimens are not uncommon (Tate 1939).

Range and Habitat

The species occurs south of the Orinoco in Venezuela and is known from a few localities in the state of Bolívar (map 13.2).

Sciurus (Hadrosciurus) igniventris Wagner, 1842

Description

In this extremely large squirrel, head and body length averages 305 mm, the tail is about 275 mm, and the hind foot averages 67 mm. The dorsum is grayish black, the head reddish brown, and the nose a darker brown. The shoulders are dark and contrast with the red forefeet and forearms. The upper surface of the hind feet is lighter than the thighs. The top of the head is not conspicuously dark. A dark band separates the dorsum from the venter, which is reddish brown. The tail is bushy and reddish brown to black at the base (plate 17).

Range and Habitat

This large squirrel is primarily distributed in northwest Brazil but extends to southeastern Colombia and southern Venezuela (map 13.4). It may co-occur with *S. aestuans*, *S. gilvigularis*, and *S. spadiceus*. Patton (1984) includes *S. igniventris* in the subgenus *Urosciurus*.

Sciurus (Hadrosciurus) spadiceus Olfers, 1818

Description

Average measurements of twelve individuals from Brazil were HB 235; T 240; HF 59. This giant squirrel has a variable color pattern, but it is usually dark if not melanistic. Shoulders and thighs are black or mixed reddish and black, hind feet and forefeet are black, and the top of the head is dark brown or black. The Dorsum may be almost black, but brown with a faint stippling of yellow is common. The venter is reddish brown. Patton (1984) includes this species in the subgenus *Urosciurus*.

Range and Habitat

The species frequents tropical lowland forests from southeastern Colombia through Peru and Bolivia to Brazil.

Comment

Patton (1984) supports Hershkovitz (1959) in separating *S. spadiceus* from *S. igniventris*. The species may co-occur in microsympatry, and the extreme difference in rostral length (*S. spadiceus* is much longer) suggests an ecological separation that remains to be confirmed.

Subgenus *Sciurus*
Sciurus (Sciurus) variegatoides Ogilby, 1839
Mexican Gray Squirrel

Description

The dental formula is I 1/1, C 0/0, P 2/1, M 3/3. Total length averages 510 to 560 mm, tail 240 to 305 mm, hind foot 60 to 70 mm, and weight 450 to 520 g. The color of the dorsum is highly variable, from blackish to grizzled yellow gray; some subspecies exhibit a tricolor pattern. The tail tends to be black above, washed with white. Often the white-tipped hairs gives the appearance of faint rings on the tail. The underparts vary from white to cinnamon buff (plate 17).

Range and Habitat

This tree squirrel is primarily distributed in Central America from Chiapas in Mexico south to extreme northwestern Panama. It is replaced in eastern Panama by *S. granatensis* (map 13.1).

Natural History

In western Panama this species can co-occur with *S. granatensis*, but in general *S. variegatoides* is larger and occupies drier and more open habitats. It builds a leaf nest in a tree, reminiscent of its North American congeners. It is associated with oak (*Quercus*) forests. Where it co-occurs with *S. granatensis* it shows a shift in diet, and 61% of its food may be drupes and berries whereas *S. granatensis* takes up to 73% palm nuts and legumes (Glanz 1984).

Genus *Sciurillus* Thomas, 1914
Sciurillus pusillus (Desmarest, 1817)
Pygmy Squirrel, Guerlingueto Pequeño

Description

The dental formula is I 1/1, C 0/0, P 2/1, M 3/3. This small squirrel has a head and body length ranging from 97 to 110 mm; the tail is 114 to 120 mm, and the hind foot is about 27 mm. It is distinguishable from all other squirrels in its range by its extremely small size. Within the range covered by this volume, the dorsum is dark gray grizzled with yellow, and the gray-brown venter does not markedly contrast. The tip of the nose is brown, and the ears are black tipped (plate 17).

Range and Habitat

This lowland squirrel is found in climax rain forests of Amazonian Peru and Brazil and extends to the Guyanas (map 13.5).

Natural History

Moore (1959) considered this species among the most morphologically conservative of the South American squirrels and felt it might be a relict from an earlier invasion into South America. This diminutive diurnal squirrel is poorly known. It appears to be associated with lowland Amazonian forest and gives birth to two young (Anthony and Tate 1935).

Genus *Syntheosciurus* Bangs, 1902
Syntheosciurus brochus Bangs, 1902
Groove-toothed Squirrel

Description

The dental formula is I 1/1, C 0/0, P 2/1, M 3/3. *Syntheosciurus brochus* is recognized by the longitudinal groove in each of its upper incisors, which Enders (1980) notes may be a variable character within a restricted area. Head and body length ranges from 150 to 170 mm, the tail from 120 to 150 mm, and the hind foot from 41 to 46 mm. The upperparts are reddish olive and the underparts orange rufous (Wells and Giacalone 1985) (plate 17).

Range and Habitat

The species *S. brochus* is known from the montane portions of Panama, occurring at elevations exceeding 2,000 m (map 13.5).

Natural History

Little is known concerning the habits of this rare squirrel. Enders (1980) offers evidence that the animals may live in small family groups. They are crepuscular and forage in low trees and even on the ground. The nest tree Enders describes was quite large (79 cm dbh). The habitat was very humid and cool at 2,135 m elevation in Panama. Giacalone, Wells, and Willis (1987) expand on Enders's account and note that diurnal activity is triphasic: 0800–1000, 1200–1500, 1600–1730. They record feeding on eight plant genera, including flowers, fruit, sap, and bark, and confirm foraging in small groups. An observed mating involved a preliminary pursuit of one female by six to eight males, a pattern typical of tree squirrels.

Genus *Microsciurus* J. A. Allen, 1895
Neotropical Dwarf Squirrel, Ardilla Menor

Description

The dental formula is I 1/1, C 0/0, P 2/1, M 3/3. Some seventeen species have been described in the literature. The head and body length averages from 120 to 160 mm, larger than *Sciurillus*. The fur varies with geographic location. The ears are well haired, contrasting with the sparsely haired ears of *Sciurus pucheranii*, which is similar in size to *Microsciurus*.

Distribution

Species of this genus occur from southern Nicaragua south through the Isthmus to the Amazonian portion of South America. Many are associated with elevations above 1,800 m.

Microsciurus alfari (J. A. Allen, 1895)

Description

Total length ranges from 228 to 240 mm, tail from 89 to 102 mm, and the hind foot from 35 to 39 mm (*N* = 3; CMNH). The upperparts are usually a dull olive brown to olive black; the dorsal surface of the tail varies from ochraceous tawny to cinnamon rufous and the venter from dull buff to gray (plate 17).

Range and Habitat

This species occurs from Costa Rica south through Panama into northern Colombia. The limits of its distribution in Colombia are poorly understood (map 13.6).

Microsciurus flaviventer (Gray, 1867)

Description

Total length ranges from 250 to 273 mm; the tail is 110 to 114 mm. Four specimens from Peru (MVZ) exhibited the following range of measurements: TL 260–75; T 128–36; HF 38–40; E 16–18; Wt 80–104 g. The dorsum is dark brown, contrasting with a gray to yellow-gray venter.

Range and Habitat

The species is found in the upper Amazonian tropical rain forests. Within the range of this volume, it occurs in southeastern Colombia.

Natural History

This squirrel's natural history is assumed to be similar to that of other species of the genus. One female had two embryos (MVZ).

Microsciurus mimulus (Thomas, 1898)

Description

This species is slightly larger than *M. alfari*. Total length ranges from 235 to 268 mm, the tail from 100 to 116 mm. Five specimens from Antioquia, Colombia, showed the following range: TL 250–63; T 105–14; HF 36–42 (CMNH). The upperparts are grizzled black and pale orange buff or, alternatively, buffy yellow. The tail is grizzled black on the dorsum, tawny below, and bordered with black.

Range and Habitat

M. mimulus generally occurs at higher elevations than *M. alfari* where their ranges overlap. It is found in Panama in montane regions, south to north-

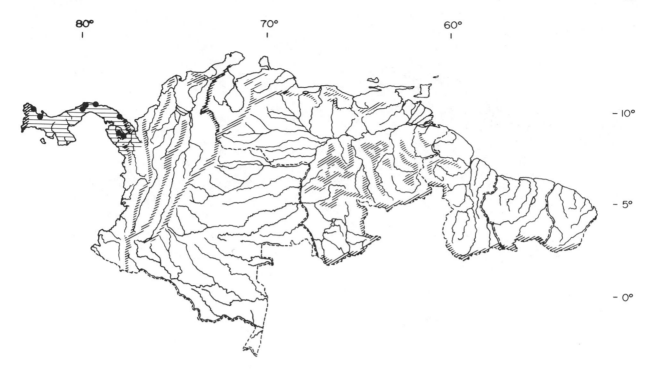

Map 13.6. Distribution of *Microsciurus alfari*.

ern Colombia. The limits of its distribution in Colombia are poorly understood. Apparently this small species is replaced by *S. pucheranii* in the high montane regions of Colombia (map 13.7).

Microsciurus santandarensis (Hernandez-Comacho, 1957)

Description

Microsciurus santandarensis was originally included as a subspecies of *S. pucheranii* (Borrero-H. and Hernandez-Camacho 1957). Hernandez-Camacho (1960) placed it in *Microsciurus*. See the comment under *S. pucheranii* and Moore (1959). Measurements average TL 272–308; T 136–52; HF 42–45; E 15–18 (Hernandez-Camacho 1960).

Range and Habitat

The species is known from the state of Santander, Colombia.

SUPERFAMILY GEOMYOIDEA
FAMILY GEOMYIDAE

Pocket Gophers, Tuzas

Diagnosis

The dental formula is I 1/1, C 0/0, P 1/1, M 3/3. The tail is shorter than the head and body. In accordance with the animals' adaptation for a burrowing life, the skull is extremely flattened and the claws are very long, with ears reduced in size, eyes

minute, tail short, and pelage soft and velvety. There is an external fur-lined cheek pocket on each side of the face, a character held in common with the family Heteromyidae.

Distribution

Species of this family are confined to the New World and extend from the plains of southern Canada south through Mexico to northern Colombia.

Natural History

As adults, pocket gophers spend most of their lives underground in a permanent tunnel system, which may be enlarged seasonally. They may move on the surface and travel overland to a new habitat, but they spend very little time aboveground except to extrude earth from their burrows or to gather grasses, seeds, and forbs near the burrow entrance. They are highly adapted for feeding on the roots of grasses and the tubers of various monocots. Their foraging burrows are generally very shallow, and their presence can be detected by the mounds of earth characteristically left as they excavate new sections for access to their feeding areas.

Genus *Orthogeomys* Merriam, 1895

Giant Pocket Gopher, Tuza

Description

Males range from 177 to 435 mm in total length, of which 70 to 140 mm is the tail. Females are smaller, with total length ranging from 170 to 390 mm

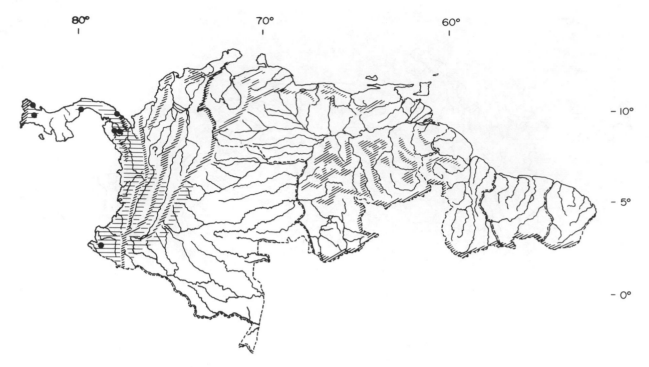

Map 13.7. Distribution of *Microsciurus mimulus*.

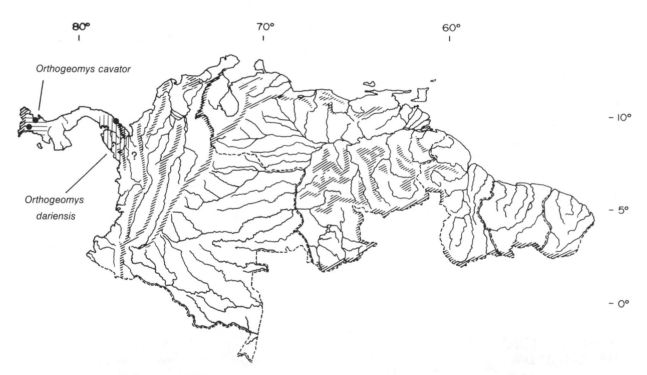

Map 13.8. Distribution of two species of *Orthogeomys*.

and tail length from 70 to 128 mm. The coarse dorsal pelage is usually some variant of agouti brown. Hairs are sparse on the dorsum, and some individuals appear partly naked (Hall 1981).

Distribution

The genus is distributed in southern Mexico south through Central America to northwestern Colombia. It has been reviewed by Russell (1968).

Orthogeomys (= Macrogeomys) cavator (Bangs, 1902)

Description

Total length for males ranges from 370 to 390 mm, of which 100 to 120 mm is the tail. In females total length ranges from 320 to 380 mm, with the tail 104 to 110 mm. The hind foot averages 52 mm for males and 50 mm for females. The dorsal fur is a very dark brown, almost black, and the pelage is coarse (Hall 1981) (plate 20).

Range and Habitat

This is a high-altitude species, occurring at up to 4,000 m. Populations are disjunct in southern Costa Rica and western Panama (map 13.8).

Orthogeomys (= Macrogeomys) dariensis (Goldman, 1912)

Description

Total length for males ranges from 358 to 401 mm, of which 122 to 135 mm is the tail. Females are somewhat smaller, from 348 to 388 mm in total length, with the tail 118 to 128 mm. The color is dull brown to blackish instead of the rich seal brown of *O. cavator*. The tail is relatively longer in *O dariensis* (Hall 1981).

Range and Habitat

This species has been described from the montane region at the border between Panama and Colombia. It surely extends into the extreme northwest of Colombia, but the limits of its range have not been precisely determined (map 13.8).

FAMILY HETEROMYIDAE

Diagnosis

The dental formula is I 1/1, C 0/0, P 1/1, M 3/3. In common with the pocket gophers, this family has externally opening fur-lined cheek pouches. A wide variety of morphological forms are displayed, ranging from the bipedal kangaroo rats to the quadrupedal spiny pocket mice. In no case are members of this family as specialized for a fossorial life as are the pocket gophers. The evolutionary history of the family was monographed by Wood (1935).

Distribution

Members of this family are distributed from the plains of Saskatchewan, Canada, to northern South America.

Natural History

There is an enduring trend toward an adaptation for arid habitats within this family that culminates in the specialized bipedal forms such as the kangaroo rats (*Dipodomys*) and the kangaroo mice (*Microdipodops*). Many of the pocket mice belonging to the genus *Perognathus* are similarly adapted for xeric habitats but are not bipedal. Two genera, *Liomys* and *Heteromys*, have adapted to moister subtropical to tropical habitats and occur within the range covered by this volume. They are referred to as spiny pocket mice or forest spiny pocket mice. They construct burrows but forage on the surface, gathering seeds, nuts, and fruits in their capacious cheek pouches and transporting them back to the burrow system for storage. *Liomys* may be one of the most important seed predators in the tropical dry deciduous forests of Central America (Janzen 1982). The behavior of the family was reviewed by Eisenberg (1963).

Genus *Liomys* Merriam, 1902

Description

Head and body length ranges from 97 to 135 mm and the tail from 97 to 156 mm. The dorsal pelage tends to be some variant of agouti brown, contrasting with the white venter. The fur of the dorsum tends to be harsh, with many of the hairs flattened. In some species this spinescent pattern is pronounced, hence the common name spiny pocket mouse. The tail is sparsely haired and usually bicolor, with its ventral portion corresponding to the ventral color of the body.

Distribution

The genus *Liomys* is distributed from Sonora in western Mexico and southeastern Texas on the Gulf coast southward in lowland Mexico to Central America, extending to southeastern Panama.

Liomys adspersus (Peters, 1874)

Measurements (males only)

	Mean	Minimum	Maximum	N
TL	265	248	285	18
T	136	123	148	18
HF	31	26	34	18

Location: Panama (Genoways 1973)

Description

Total length ranges from 222 to 285 mm, tail 107 to 148 mm, and hind foot 26 to 34 mm. The upper-

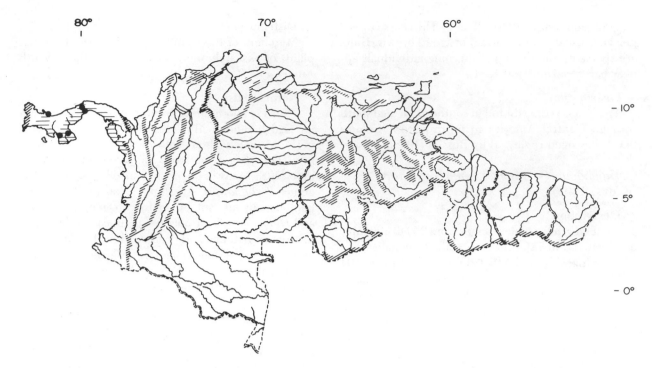

Map 13.9. Distribution of *Liomys adspersus*.

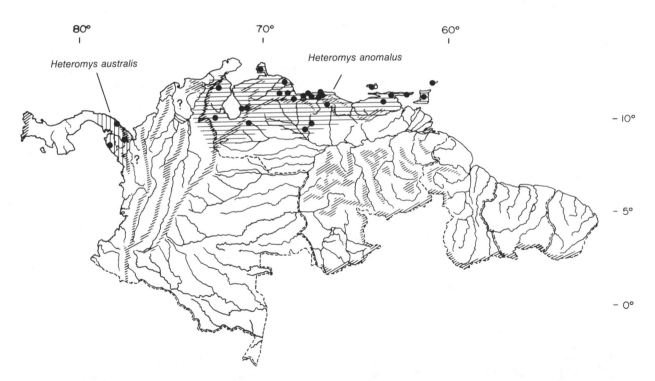

Map 13.10. Distribution of two species of *Heteromys*.

parts tend to be chocolate brown, contrasting with the white venter (plate 20).

Range and Habitat

This species is known from Panama but possibly extends to northwestern Colombia. It occurs at low elevations on the Pacific side of Panama and is uncommon in the Caribbean portion (map 13.9)

Natural History

In Central America *Liomys* may be very abundant in dry deciduous forests, reaching eleven per hectare in good habitat. In southern Panama the breeding season extends over a six-month period from February to July. Individuals become sexually mature at ninety days, and the average litter size is four (Fleming 1971). The closely related Costa Rican species *L. salvini* has been the subject of much research (Fleming 1983).

Genus *Heteromys* Desmarest, 1817
Forest Spiny Pocket Mouse

Description

Head and body length ranges from 125 to 160 mm and the tail from 130 to 200 mm. The tail tends to be longer than the head and body; in this *Heteromys* differs from *Liomys* (plate 20). The dorsal pelage is generally dark brown to almost black, contrasting sharply with the white venter. In some species flattened hairs give the fur a spiny texture.

Distribution

The animals occur from southern Veracruz, Mexico, south to northern South America at low elevations.

Natural History

Species of *Heteromys* are nocturnal. They construct elaborate burrow systems but forage on the surface, gathering seeds, fruits, and nuts in their capacious cheek pouches and transporting them to the burrow system for storage (Vandermeer 1979). Litter size averages about four, and a female may produce several litters each year depending on local availability of food, which is ultimately controlled by rainfall.

Heteromys anomalus (Thompson, 1815)

Description

Head and body length ranges from 125 to 160 mm, the tail from 130 to 200 mm. The dorsum is typically dark brown, and some of the longer hairs have whitish tips. The venter is snow white (plate 20).

Range and Habitat

This species occurs to the east of the central Andean cordillera, replacing *H. australis*. It occurs in the Maracaibo basin and across northern Venezuela, north of the Orinoco to the delta. It also occurs on the adjacent islands of Trinidad, Tobago, and Margarita (map 13.10). In northern Venezuela it is strongly associated with moist areas and multistratal tropical evergreen forest. In parts of the tropics that experience extended dry periods it exists in gallery forest in association with streams. It occurs at up to 2,200 m elevation (Handley 1976).

Natural History

This is one of the best-studied species of the genus. This nocturnal form is almost exclusively terrestrial but can climb in small shrubs. It digs an extensive burrow system in which it constructs a nest. Seeds, some fruit, and insects make up its diet. It can exist at densities of up to 2.2 individuals per hectare. Young mature to independence within fifty days of birth and may be caught near the maternal burrow. They subsequently disperse to establish new burrow systems (Rood 1963; O'Connell 1981; Eisenberg 1963).

Heteromys australis Thomas, 1901

Description

Total length ranges from 240 to 260 mm, tail length from 120 to 133 mm, and the hind foot from 32 to 33.5 mm. The dorsum is grizzled slate black, the underparts are usually white, and the tail is brown above and light below.

Range and Habitat

This species is distributed primarily to the west of the central Andean cordillera in Colombia. It extends from Ecuador through western Colombia to the extreme westernmost part of Panama (map 13.10). Throughout the remainder of Panama it is replaced by *H. desmarestianus*.

Heteromys desmarestianus Gray, 1868

Description

Total length ranges from 255 to 345 mm, of which 130 to 190 mm is the tail. The hind foot ranges from 31 to 42 mm. Males are heavier (83 g) than females (62 g). The dorsum is variable in color, ranging from dark gray to blackish, generally darker in the mid-dorsal region, and the underparts are white. In some subspecies the venter and dorsum show a buffy zone of demarcation.

Range and Habitat

This species occurs in moist habitats in the lowlands at up to 1,000 m elevation. It extends from Chiapas, Mexico, south through the Isthmus to the extreme northwest of Colombia (map 13.11)

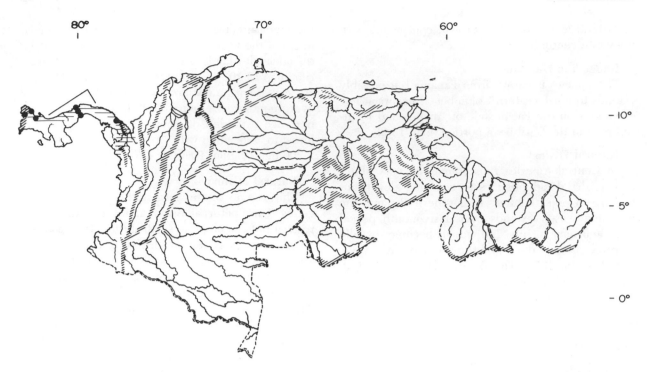

Map 13.11. Distribution of *Heteromys desmarestianus.*

Natural History

These mice are capable of breeding throughout the year but are seasonal breeders in habitats with strongly seasonal rainfall. In lowland tropical forests their densities range from ten to eighteen per hectare. They become sexually mature at eight months of age. Litter size averages 3.1 (Fleming 1974a). This species of *Heteromys* is more socially tolerant than *Liomys salvini.* The small home ranges of 0.08 to 0.20 ha overlap considerably with those of neighbors (Fleming 1974b).

SUPERFAMILY MUROIDEA
FAMILY MURIDAE
SUBFAMILY SIGMODONTINAE
(HESPEROMYINAE)
New World Rats and Mice

Systematics

The sigmodontine rodents are defined here as a subfamily of the family Muridae (Carleton and Musser 1984; Reig 1980). They are frequently referred to as the New World cricetines. For this account I consider the gerbils to be within their own subfamily, the Gerbillinae, and the voles to be in their own subfamily, the Arvicolinae. The New World Sigmodontinae have as their nearest relatives in the Old World the hamsters (Cricetinae) of Europe and Asia and the pouched rats (Cricetomyinae) of Africa. Cricetine rodents first appear in the Oligocene of Europe, and by the mid-Oligocene they are

represented in North America. Rodents that can be referred to the subfamily Sigmodontinae are identifiable in the Miocene of North America.

The classification of these rodents above the generic level, into tribes, has been plagued with difficulty. It seems fair to say that the original stock was adapted to forested environments, but with the increasing trends of drying during the Miocene and the creation of extensive grasslands, the original stocks began to adapt to more open grassland and xeric habitats (Hershkovitz 1962). How often this happened during the early divergence of the ancestral stock is unknown.

When the first sigmodontines entered South America is still a subject of some debate. Reig (1981) thinks it was as early as the Miocene. On the other hand, many other workers believe it was as late as the Pliocene when the land bridge was completed between North America and South America (Baskin 1978). Perhaps some of the early invasions by sigmodontines in South America occurred before the complete uplift of the Isthmus. If so, then from the Pliocene through the Pleistocene there may have been more than one passage of differently evolved stocks from South America to North America and from North America to South America, compounding the problem.

Earlier workers considered the neotomine-peromyscine group to be North American in origin and a clearly defined taxonomic unit. The remaining sigmodontines were thought to have evolved primarily in the Isthmus and the southern United States (Hooper and Musser 1964). Recently, the synonymy of the fossil *Bensonomys* of North America with the extant South American *Calomys* suggests that the major stocks of the South American cricetines may have differentiated in North America before a Pliocene transfer to South America (Baskin 1978).

In an effort to better understand the relationships among the sigmodontine rodents, numerous characters have been evaluated. Early workers such as Cabrera (1960), Moojen (1952), Hershkovitz (1962, 1966), and Pearson (1958) undertook important descriptions and attempts at classification. Teeth have proved useful for delineating some groups but have provided equivocal evidence of relationship for others. Studies have compared blood proteins, and extensive work has been done on chromosomal evolution, especially by Reig, Olivo, and Kiblinsky (1971), Pearson and Patton (1976), and Gardner and Patton (1976).

Hooper and Musser (1964) studied the morphology of the baculum and concluded that the neotomine-peromyscine group was distinguishable from the predominantly South American radiation by the structure of the penis. Implicit in their hypothesis was the assumption that the sequence of evolution went from a complex baculum to a simplified one. Recent workers have challenged this concept of the primitive morphological characters and conclude that the primitive character state was a simplified baculum (Carleton 1980).

The most recent attempt to analyze the problem of relationships was made by Carleton (1980), who first ascertained the primitive and derived conditions for each morphological character. He then performed a multivariate analysis for over seventy characters for the neotomine-peromyscine complex, with some comparisons of South American forms. He concluded that the neotomine-peromyscine group is not necessarily a natural unit, that the polarity of characters assumed by other workers is open to question, and that assumption about the derived state of the simplified baculum is probably erroneous. Furthermore, Carleton proposed that the neotomine-peromyscine group is divisible into at

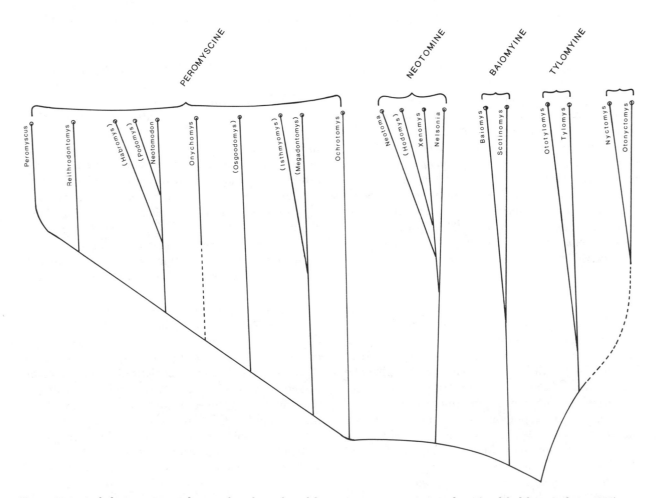

Figure 13.3. A phylogenetic tree indicating the relationship of the neotomine-peromyscine rodents (modified from Carleton 1980).

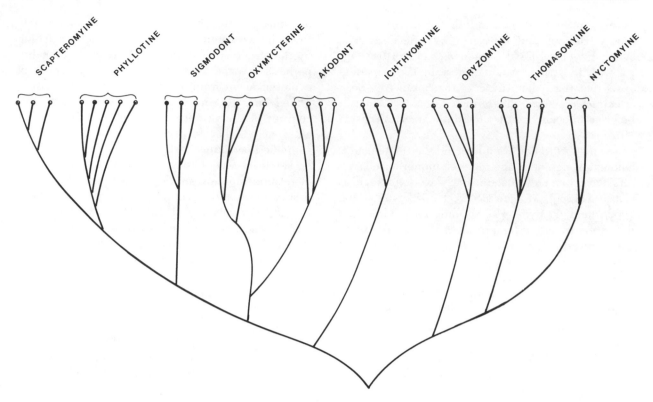

Figure 13.4. A phylogenetic tree indicating the relationship of the South American sigmodontine rodents. Derived taxa to the left.

least four units or supergeneric groupings. These groupings were not proposed as formal taxonomic units but were useful in an informal sense. These include the peromyscine rodents that are for the most part North American in distribution and include the deer mice (*Peromyscus*), harvest mice (*Reithrodontomys*), grasshopper mice (*Onychomys*), and golden mice (*Ochrotomys*). The neotomine group includes the wood rats (*Neotoma*) and their relatives, again a primarily North American group. The baiomyine group includes the pygmy mice (*Baiomys*) and brown mice (*Scotinomys*); and finally, the tylomyine group includes the climbing rats (*Tylomys*) of Central America (fig. 13.3).

Following Carleton in his groupings of genera, apparently the closest living relatives of the tylomyines are what may be referred to as the nyctomyines, or vesper mice. The nyctomyines are Central American forms adapted to tropical forests. They are excellent climbers with long tails. They form a phenetic link between the North American neotomine-peromyscines and the South American groupings. There is a close morphological resemblance between the nyctomyines and thomasomyines, which are predominantly long-tailed forms occupying forested habitats—also excellent climbers. In northern South America the genus *Thomasomys* is often confined to temperate-zone, high-altitude forests in the Andes,

whereas the nearly related *Rhipidomys* inhabits lower-altitude moist rain forests (see fig. 13.4).

The closest living relatives of the thomasomyines are the oryzomyines. Reig (1980) combined them into the same tribe. These are the rice rats and include some six genera. All are moderately long tailed; some are adapted for semiaquatic life, and others are predominantly terrestrial and scansorial.

There is considerable divergence in the remaining South American sigmodontines. The akodonts show a range of adaptations and include some five genera. They are predominantly terrestrial and show tendencies toward exploiting more open habitats, although some members of this group are still confined to somewhat moist, forested conditions. The key to the adaptive radiation of this group is terrestriality. The tribe most nearly related to the akodonts is the oxymycterines. The oxymycterines have carried terrestriality to further extremes, but many of them have become specialized for feeding on insects as well as seeds. This trend culminates in the burrowing, insect-feeding forms adapted for niches similar to those of the North American shrews and moles. This syndrome includes the genus *Blarinomys* of southeastern Brazil.

The tribe zygodontomyini contains the distinctive genus *Zygodontomys* and exhibits some superficial resemblance to the akodonts. Somewhat related to

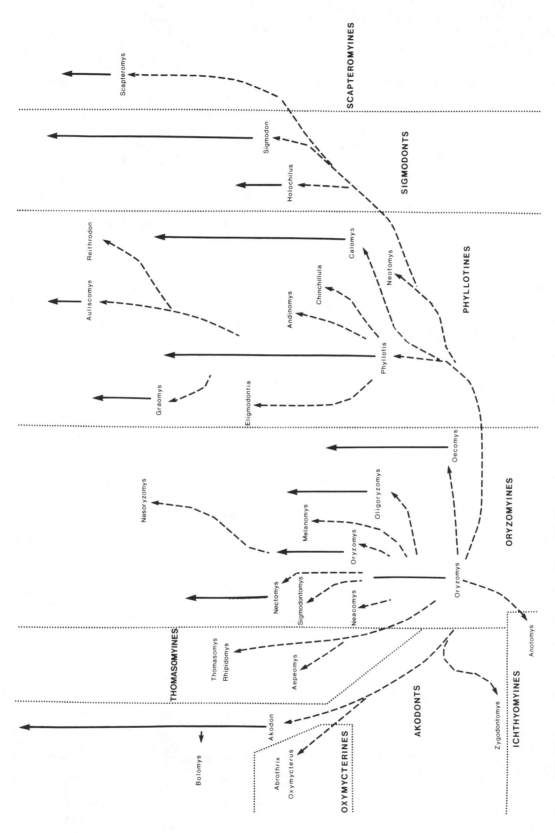

Figure 13.5. Cladogram illustrating the relationship of New World sigmodontine rodents (after Gardner and Patton 1976).

the oxymycterine-akodont groupings are the so-called fishing rats, or ichthyomyines. The genera comprising this set exhibit various adaptations to an aquatic life, and the most specialized members are adapted for feeding on crustaceans and fish.

The next three suprageneric groupings show some close relationships to one another, The sigmodont genera clearly are adapted for feeding on the green parts of plants, and some of them appear to occupy an almost volelike niche in South America. The phyllotines are a diverse group of genera that have colonized many open habitats from high altitude to dry, low-altitude habitats and open grasslands. They show an array of dietary specializations from feeding on seeds to feeding almost completely on herbaceous vegetation, displaying perhaps the greatest diversity of all the tribes of South American cricetid rodents. The last group, the scapteromyines, appear to be the most derived genera and consist of the pampas water rats and their relatives from southeastern Brazil, Argentina, Uruguay, and parts of Paraguay (see fig. 13.5).

This brief review of taxonomy should bring home the point that in South America the adaptive radiation of the sigmodontines has allowed the occupancy of many ecological niches. We can identify aquatic fish-eating, insect-eating, and crustacean-eating adaptations and an aquatic niche for larger rodents that are omnivorous or more herbivorous in their dietary requirements. There is an adaptive zone occupied by scansorial rodents that are able climbers and feed on fruits and seeds. We can identify many terrestrial niches that exhibit a trophic diversity from partially insectivorous forms to extreme insectivores that bur-

row. There are terrestrial forms that are granivores and, finally, terrestrial forms that are predominantly herbivores. It is clear that the adaptive radiation into different feeding niches in South America has occurred several times, and different lineages can have species in similar feeding niches but geographically separated. This is especially true of those species occupying the grassland niches in South America. There are clear cases of convergent evolution brought about through geographical separation when one compares sigmodonts (sensu stricto) with phyllotines (Eisenberg 1984).

In North America the grassland niches were predominantly filled by a subfamily of rodents related to the sigmodontines that entered North America late but did not reach South America. These North American rodents belong to the subfamily Arvicolinae (= Microtinae), exemplified by the lemmings and voles.

Clearly, the most conservative members of the various taxa, or those that have retained conservative traits, are to some extent scansorial and have rather long tails. Deviations from this early body form involved specialization for more terrestrial habits, with reduction in tail length, or specialization for aquatic habitats, with no necessary reduction in tail length but with modifications of the hind feet, including webbing and often a lateral flattening of the tail for better propulsion. All these rodents are of rather modest size; head and body length ranges from 44 to 280 mm. The only exceptions are the extinct *Megalomys* of the Lesser Antilles and the extinct *Megaoryzomys* of the Galápagos, whose head and body length exceeded 330 mm. Keys to

Table 13.2   Characters of the "Tribes" of Sigmodontine Rodents Found in Northern South America

"Tribes"	Predominant Distribution in Central and/or North America	Penis and Baculum Complex in Structure	Tail Visibly Longer Than Head and Body	Hind Feet Elongate	Pinnae Relatively Large	Molars Sigmodont	Molars "Pentalophodont"	Molars "Tetralophodont"	M_3 Half to Three-fourths Length of M_2
Peromyscine	+	−	+/−	+/−	+	−	−	+/−	+/−
Baiomyine	+	−	−	−	−	−	−	+	−
Tylomyine	+	−	+	−	+	−	−	+/−	−
Nyctomyine	+	−	+/−	+/−	−	−	+	−	+
Thomasomyine	−	+	+	+	+	−	+	−	+
Oryzomyine	+/−	+	+/−	+/−	+/−	−	+/−	−	+
Ichthyomyine	+/−	+	−	−	−	−	−	+	−
Akodontine	−	+	−	−	−	−	−	+	−
Sigmodontine	+/−	+	−	−	−	+	−	+	+
Phyllotine[a]	−	+	−	+/−	+	−	−	+	+
Zygodontomyine	+/−	+	−	−	−	−	−	+	+

[a] In the north only.
+/− = variable within taxa;
+   = character expressed;
−   = character not typical.

Table 13.3   Key to the Adult Oryzomyine Rodent Genera and Subgenera of Northern South America

1	Feet partially webbed; hairs project ventrally from the underside of the tail . . . . *Nectomys*
1′	Not as above 2
2	Dorsal pelage spinescent; weight less than 20 g . . . . *Neacomys* [a]
2′	Dorsal pelage not spinescent 3
3	Tail considerably less than head and body length; size small; head and body length less than 110 mm . . . . (*Melanomys*)
3′	Tail equal to or longer than head and body length 4
4	Cusps of molars present only in juveniles; flat crowns in adult . . . . (*Sigmodontomys*)
4′	Cusps evident in adult 5
5	Tail tip penicillate . . . . (*Oecomys*)
5′	Tail tip not penicillate 6
6	Hind foot longer than 25 mm . . . . (*Oryzomys*)
6′	Hind foot shorter than 25 mm . . . . (*Microryzomys* and *Oligoryzomys*)

[a] If *Scolomys* proves to be found in southern Colombia, this character will not distinguish between the two genera.

the sigmodontine rodents are given in tables 13.2, 13.3, and 13.5 (see also Eisenberg 1984).

Diagnosis

The dental formula I 1/1, C 0/0, P 0/0, M 3/3 distinguishes this group from the Sciuridae and Geomyoidea (see fig. 13.2). *Daptomys oyapocki* is exceptional, with two upper and lower molars per side. The cheek teeth are variable in cusp pattern and may be laminate, prismatic, or cuspidate. When cusps are present, they are arranged in two longitudinal rows in both upper and lower molars (see fig. 13.13). The infraorbital foramen is rather small; generally a strip of masseter muscle passes through to attach to the rostrum. The bulk of the masseter ataches on the zygomatic arch and the rostrum. There is no postorbital process on the frontal bone.

I follow Carleton and Musser (1984) in their classification and terminology.

Distribution

Sigmodontine rodents are found from northern Canada throughout North America to Patagonia in South America. This is a New World subfamily.

ORYZOMYINE RODENTS
Rice Rats

Content and Diagnosis

Within the western continents this assemblage includes the genera *Oryzomys*, *Nesoryzomys*, *Scolomys*, *Neacomys*, and *Nectomys;* insular forms include the extinct Antillean genus *Megalomys* and *Megaoryzomys* of the Galápagos. *Nesoryzomys* and *Scolomys* do not occur within the range covered by this volume. With the exception of *Megalomys* and *Megaoryzomys*, these rodents are small to medium sized, generally with a tail equal to or longer than the head and body. The molars tend to be pentalophodont. The lower third molar tends to be more than half the length of the lower second molar,

a character that distinguishes oryzomyines from the akodonts. The hind foot tends to be relatively long, more than 0.22% of the head and body length. The hallux usually bears a nail. Some forms have stiff hairs or spines in the dorsal pelage. They are adapted for semiaquatic, terrestrial, or semiarboreal habits. Detailed descriptions are included under the generic accounts.

Distribution

The continental assemblage of species is distributed from southern New Jersey and the Eastern Shore of the state of Maryland in the eastern United States south along the Atlantic coast and then across the United States Gulf coast through eastern Mexico and over all of South America to southern Argentina.

Genus *Oryzomys* Baird, 1858
Rice Rat, Ratón Arrocero

Description

Highly variable in size, these mouselike or ratlike rodents have a head and body length ranging from 93 to 197 mm, with tail length ranging from 100 to 235 mm (see plate 18). Weight averages from 40 to 150 g, but some small species weigh less than 20 g. The upperparts are gray brown to ochraceous, generally mixed with black, and the sides tend to be paler with underparts grayish white or buff. A key is given in table 13.3.

Distribution

The genus ranges from southern New Jersey in the eastern United States south across the Gulf Coast through Mexico to southern Argentina.

Natural History

Rice rats display great diversity in niche occupancy; some species prefer moist habitats and are semiaquatic, such as *O. alfari* and *O couesi*, while other species are adapted for montane habitats above 1,000 m elevation, such as *O. albigularis* and *O.*

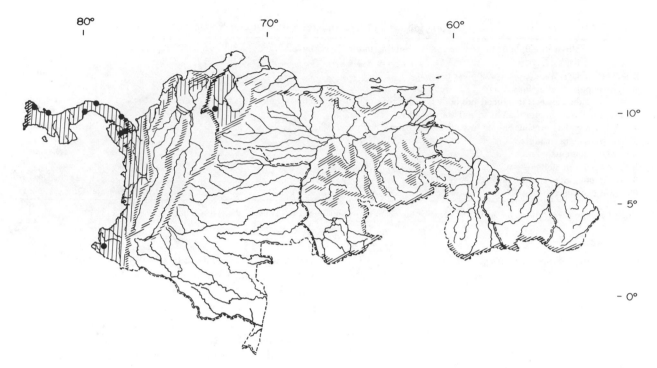

Map 13.12. Distribution of *Oryzomys alfari*.

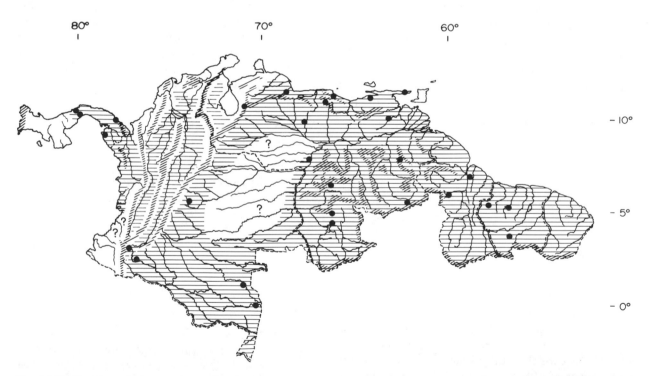

Map 13.13. Distribution of *Oryzomys bicolor*.

talamancae. Where several species co-occur the niche space is apparently divided according to the degree of arboreality and the size of the seeds and fruits taken. For example, the three species *O. fulvescens*, *O. concolor*, and *O. capito* can co-occur in Venezuela. *O. fulvescens* is small while *O. concolor* and *O. capito* are nearly the same size, but *O. capito* with its short tail tends to be strongly terrestrial, whereas *O. concolor* with its long tail is scansorial. In Venezuela the ubiquitous *O. capito* is replaced at higher elevations over much of its range by *O. albigularis*. The small *O. fulvescens* is replaced at higher elevations by *O. minutus*. Relative tail length is a good indicator of climbing ability, the long-tailed species tending to be scansorial and the shorter-tailed ones more terrestrial. For identification it is convenient to divide the genus into subgenera. A field key is given in table 13.3.

Subgenus *Sigmodontomys*
Oryzomys alfari (J. A. Allen, 1897)

Description
O. alfari has cusps on its molars when extremely young, but they rapidly wear down to give a flat-crowned appearance. The lack of cusps in the adult is diagnostic for the species but is shared with the closely related *Nectomys squamipes*. Head and body length ranges from 115 to 152 mm, while tail length ranges from 149 to 190 mm and the hind foot from 33.8 to 40 mm. Three specimens from Panama measured TL 300–312; T 161–80; HF 33–35; E 19. The dorsum is reddish brown, the venter varies from gray to buff, and the tail is not bicolor. Subspecies and their characters are reviewed by Hershkovitz (1944).

Range and Habitat
This species is found in the lowlands from eastern Honduras to Panama and thence to northern Ecuador (map 13.12).

Natural History
Oryzomys alfari is not nearly as specialized for an aquatic life as is *Nectomys squamipes*. There is no webbing between the toes. This is a lowland form confined to moist habitats, but its natural history is poorly known.

Subgenus *Oecomys*

Description
Cusps are noticeable on the molars even in the adult stage. Species of this subgenus tend to have a penicillate tail. Within the range covered by this volume the two species are scansorial and have tails longer than the head and body. The subgenus was reviewed by Hershkovitz (1960).

Oryzomys bicolor (Tomes, 1860)

Description
This species is smaller than *O. concolor*. Seven specimens from Suriname showed the following range in measurements: TL 172–223; T 89–111; HF 21–23; E. 13–15. The dorsum is cinnamon brown infused with black; the venter is white, and the tail is not noticeably bicolor. Juveniles have slightly darker pelage.

Range and Habitat
This species is distributed from Panama south to northwestern Bolivia, east to the Guyanas, and thence south to south-central Brazil. It tolerates both dry deciduous forest and multistratal tropical evergreen forest (map 13.13).

Natural History
This arboreally adapted species can exist in the Venezuelan llanos by utilizing gallery forest or palms in seasonally inundated areas. In suitable habitat adult density can range seasonally from 0.4 to 1.25 individuals per hectare. The litter size is 4.5, and the young become sexually mature at approximately three months of age (O'Connell 1981).

Oryzomys concolor (Wagner, 1845)

Measurements

	Mean	Minimum	Maximum	N
TL	296.5	242	332	11
T	163.2	123	190	11
HF	29.8	27	33	11
E	18.8	16	20	11

Location: Suriname (CMNH)

Description
For this account *O. trinitatus* is included within *O. concolor*. This species tends to be slightly larger than *O. bicolor* and has slightly different habitat preferences. The dorsal pelage is colored as in *O. bicolor* and gives the effect of a medium agouti brown, but the venter is gray to buff and the tail is not strongly bicolor.

Range and Habitat
The species is found in south-central Costa Rica, through Panama, across northern South America to the Guyanas, and south to central Brazil (map 13.14). To the west it is distributed broadly through Peru and Bolivia to northern Argentina. It can range to 2,230 m elevation in northern Venezuela and prefers multistratal tropical evergreen forest, although it tolerates man-made clearings.

Natural History
The common arboreal rice rat in lowland forest and premontane forest is less tolerant of xeric habi-

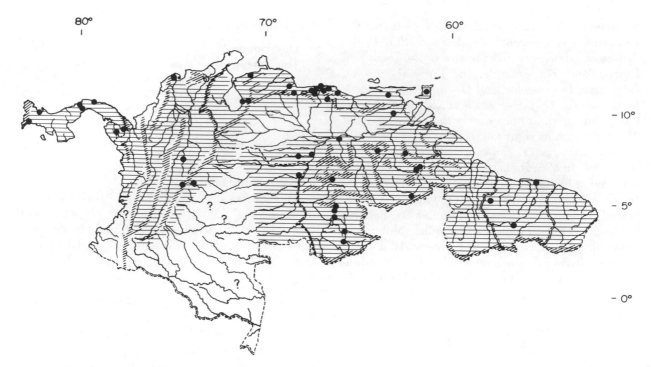

Map 13.14. Distribution of *Oryzomys concolor.*

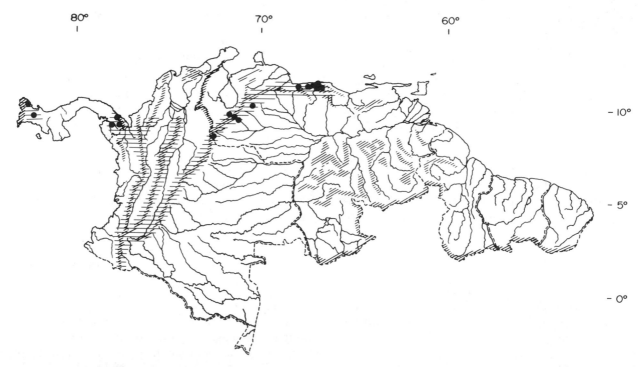

Map 13.15. Distribution of *Oryzomys albigularis.*

tats than is *O. bicolor.* Litter size ranges from two to four. Density ranges from 0.5 to 2.5 individuals per hectare. The species co-occurs over much of its range with *O. capito* (O'Connell 1981).

Subgenus *Oryzomys*

Description
Cusps are present on the molars. The tail is not penicillate and tends to be equal to or slightly shorter than the head and body length. The hind foot exceeds 25 mm, thus distinguishing this species assemblage from the subgenus *Oligoryzomys*, which includes most of the smaller species.

Oryzomys albigularis (Tomes, 1860)

Measurements

	Mean	Minimum	Maximum	N
TL	295.9	272	317	12
T	160.0	144	196	12
HF	31.8	27	35	12
E	22.5	20	28	12

Location: Colombia (CMNH)

Description
The dorsal pelage is dark tawny mixed with black, while the throat is usually white and the venter ranges from cream to gray. In the typical form the white is restricted to the throat, but it may extend to the entire venter (Gardner and Patton 1976).

Range and Habitat
This species is distributed from Costa Rica south through the Andean portion of Colombia to northwestern Bolivia (map 13.15). It ranges to the east in Venezuela in montane habitats, being adapted to cloud forest and multistratal evergreen premontane forest at elevations exceeding 1,000 m.

Natural History
Aagaard (1982) found this species up to 2,600 m elevation in the Venezuelan Andes. The rats were strongly associated with wet forests. They are primarily terrestrial and were sometimes active during the day.

Oryzomys alfaroi (J. A. Allen, 1891)

Measurements

	Mean	Minimum	Maximum	N
TL	218.6	191	238	8
T	109.4	102	120	8
HF	25.0	19	27	8
E	16.9	16	18	8
Wta	33.3	31	35	8

Location: Colombia (CMNH)

Description
This taxon as currently defined includes at least three species in Central America. Only *O. alfaroi* is South American in its distribution (Musser and Williams 1985). The dorsum is dark reddish brown to tawny mixed with black, which is pronounced in the middorsal region. The venter is grayish white.

Range and Habitat
As currently defined the group ranges from southern Tamaulipas, Mexico, south to Panama and in western Colombia to Ecuador (map 13.16).

Oryzomys bombycinus Goldman, 1912

Description
The total length ranges from 198 to 267 mm, the tail from 90 to 130 mm and the hind foot from 27 to 35 mm. One specimen from Costa Rica measured TL 265; T 125; HF 32; E 20. The dorsum is cinnamon brown, very dark in the dorsal midline, contrasting strongly with the grayish white venter. The tail is faintly bicolor. The vibrissae are extremely long and conspicuous, distinguishing it from most sympatric congeners.

Range and Habitat
The species is distributed from Nicaragua south to Panama and in western Colombia to northern Ecuador (map 13.17).

Oryzomys capito (Olfers, 1818)

Measurements

	Mean	Minimum	Maximum	N
TL	227.2	197	308	12
T	119.2	81	173	12
HF	29.7	26	42	12
E	19.0	14	20	12
Wta	45.2	28	58	12

Location: Suriname (CMNH)

Description
This account includes *O. laticeps*. The dorsum is dark buffy brown to yellowish brown, and a dark middorsal stripe is often evident. The tail is faintly bicolor; the venter is grayish white (plate 18). The juvenile pelage is very dark, almost smoky brown, with a contrasting gray venter.

Range and Habitat
The species occurs in Panama, Venezuela, Colombia, Guyana, French Guiana, and Suriname and extends south to Peru, Bolivia, and Brazil. In northern Venezuela it occurs below 1,500 m elevation. Broadly tolerant of man-made clearings, it prefers multistratal tropical evergreen forest (map 13.18).

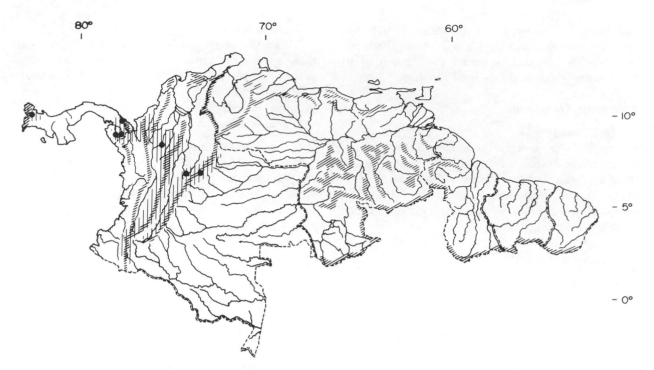

Map 13.16. Distribution of *Oryzomys alfaroi*.

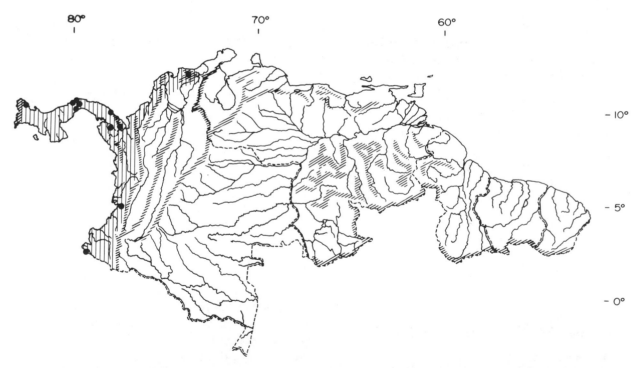

Map 13.17. Distribution of *Oryzomys bombycinus*.

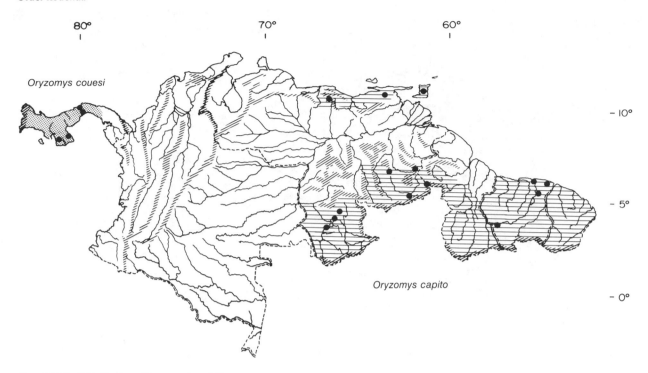

80° 70° 60°

Oryzomys couesi

– 10°

– 5°

Oryzomys capito

– 0°

Map 13.18. Distribution of two species of *Oryzomys*.

Natural History

This is the common terrestrial rice rat over much of the lowland northern Neotropics. It is ominivorous, feeding on fruits, seeds, adult insects, insect larvae, and fungi. Densities can vary between 0.5 and 5.0 individuals per hectare in Venezuela (O'Connell 1981). Guillotin (1982) found them to be predominantly nocturnal, and in French Guiana they are strongly frugivorous in second-growth forests. They do not burrow but build a leaf nest within a natural crevice. Home ranges ranged from 1.05 ha for males to 0.49 ha for females and show considerable overlap. Home range sizes were similar in the study by Everard and Tikasingh (1973) on Trinidad.

Oryzomys couesi (Alston, 1877)

Measurements

	Mean	Minimum	Maximum	N
TL	258.3	244	277	4
T	137.8	130	141	4
HF	30.5	28	32	4
E	15.8	15	16	4

Location: Mexico (CMNH)

Description

Total length can range from 225 to 332 mm, the tail from 108 to 182 mm, and the hind foot from 28 to 40 mm. The dorsum is grizzled gray brown, grading to reddish on the sides; the venter varies from grayish white to buff. The tail is bicolor.

Range and Habitat

The species is found from the Gulf coast of southern Texas in the east and Sonora, Mexico, in the west, south to central Panama (map 13.18). This is the common lowland rice rat, preferring moist habitats in Central America. Hershkovitz (1987) reports a range extension to 8°45′ N, 75°53′ W in Colombia.

Natural History

This rice rat was formerly considered conspecific with *O. palustris*, which has been studied in the southern United States. Marsh rice rats may tunnel in banks. They are excellent swimmers. A grass nest may be constructed off the ground in a stand of weeds or under a pile of leaves and branches. Litter size ranges from one to five in *O. palustris;* the gestation period is twenty-five days, and the young may be weaned as early as fourteen days of age (Negus, Gould, and Chipman 1961).

Oryzomys macconnelli Thomas, 1910

Measurements

	Mean	Minimum	Maximum	N
TL	286.3	262	320	8
T	151.8	131	174	8
HF	32.8	31	35	8
E	20.8	16	23	8
Wta	58.0	45	66	8

Location: Venezuela, Suriname (CMNH)

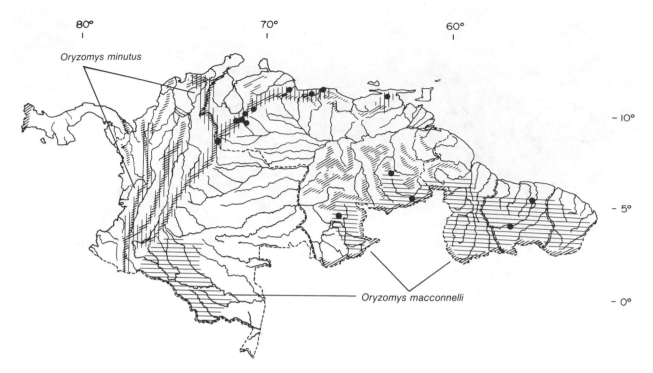

Map 13.19. Distribution of two species of *Oryzomys*.

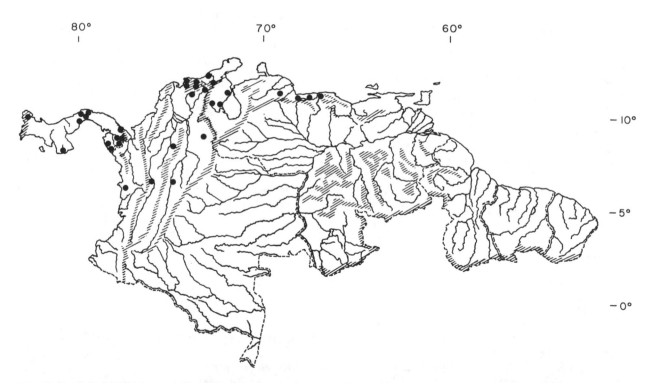

Map 13.20. Distribution of *Oryzomys talamancae* (from Musser and Williams 1985).

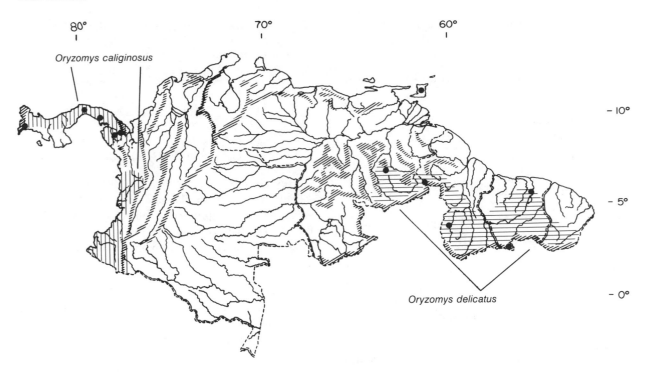

Map 13.21. Distribution of two species of *Oryzomys*.

Description

The dorsum is ochraceous tawny interspersed with black, and the sides are reddish. The venter is grayish white, the tail bicolor. The juvenile pelage is a dark brown with little or no reddish cast, and the venter is a dirty white. The long dorsal hair in the adult distinguishes this species from *O. capito* where they co-occur.

Range and Habitat

The species is distributed in Guyana, Suriname, southern Venezuela, and southern Colombia to eastern Ecuador and northern Brazil (map 13.19). It prefers moist forest.

Comment

O. macconnelli was considered a subspecies of *O. capito* (Hershkovitz 1960), but Husson (1978) considers it a valid species easily separable from *O. capito* on skull characters. One embryo count of two indicates a possible small litter size (MVZ).

Oryzomys talamancae J. A. Allen, 1899

Description

Head and body length averages 124.4 (101–36) mm in Panama; 135.2 (120–51) mm in Colombia. Tail length averages 124.7 (110–43) mm in Panama and 125.2 (114–40) mm in Colombia. The hind foot averages 29.2 mm in Panama and 29.5 mm in Colombia, and the ear averages 21.1 mm in Panama and 19.0 mm in Colombia (Musser and Williams 1985). The short dorsal pelage is russet brown mixed with blackish brown, "passing gradually into clear,

yellow-brown on the sides" (Allen as quoted in Musser and Williams 1985). The venter is grayish white and the tail is distinctly bicolor.

Range and Habitat

This rat is distributed from eastern Costa Rica through western Colombia south to Ecuador and east to northern Venezuela (map 13.20). It ranges from sea level to 1,500.

Comment

Musser and Williams (1985) note that *O. villosus* is invalid, since the holotype is a composite of an *O. talamancae* skin and an *O. albigularis* skull.

Oryzomys Species Having Extremely Restricted Distributions and Uncertain Affinities

Oryzomys gorgasi Hershkovitz, 1971

This species of uncertain affinities from northwestern Colombia was taken at 7°54′ N and 77° W (Hershkovitz 1971).

Oryzomys intectus Thomas, 1921

This species is known from a single locality in Antioquia in central Colombia at 2,700 m elevation. Its relationships are not precisely known (see Cabrera 1960).

Subgenus *Oligoryzomys*

Description

Cusps are present on the molars. The tail is not penicillate but is long. The body is rather small, with the hind foot less than 28 mm. Long hairs are present on the toes, protruding beyond the claws.

Oryzomys delicatus J. A. Allen and Chapman, 1897

Measurements

	Mean	Minimum	Maximum	N
TL	183.9	153	207	14
T	99.0	83	110	14
HF	22.1	20	24	14
E	14.1	12	15	14
Wta	14.6	7	23	14

Location: Suriname (CMNH)

Description

This rat is similar in appearance to *O. fulvescens* (see below). The dorsum is medium brown, the venter gray, and the tail faintly bicolor.

Range and Habitat

This tiny rice rat apparently replaces the Central American *O. fulvescens* in the Guyanas; their relationship is unclear. It is found on Trinidad and extends its distribution to north-central Brazil (map 13.21). It frequents open environments with shrubs.

Oryzomys fulvescens (Saussure, 1860)

Measurements

	Mean	Minimum	Maximum	N
TL	175.8	163	188	5
T	99.8	90	105	5
HF	21.2	21	22	5
E	14.2	12	19	5
Wta	9.75	9	11	5

Location: Mexico (CMNH)

Description

This species is distinctive by its small size. The upperparts are reddish brown with an admixture of black hairs. The underparts are white and the tail is not noticeably bicolor.

Range and Habitat

The species ranges from Tamaulipas in eastern Mexico and Nayarit in western Mexico south through Panama and northern Colombia to Venezuela (map 13.22). In Venezuela it is common in the lowlands below 1,500 m elevation and prefers multistratal tropical evergreen forest.

Natural History

This is the common low-elevation small rice rat over much of Central America and northern South America.

Oryzomys munchiquensis J. A. Allen, 1912

Measurements

	Mean	Minimum	Maximum	N
TL	198.4	185	217	5
T	112.8	100	130	5
HF	21.8	20	23	5
E	11.4	10	12	5

Location: Colombia (FMNH)

Description

This species is similar and closely related to *O. delicatus* and *O. flavescens*, but *O. flavescens* has a Brazilian distribution (see Myers and Carlton 1981). The dorsum is brown, the venter tan, and the tail faintly bicolor.

Range and Habitat

This rat is described from western Colombia in the basins of the Cauca and Patía rivers. It may range to 2,000 m.

Subgenus *Melanomys*
Oryzomys caliginosus (Tomes, 1860)

Description

Cusps are present on adult molars. The tail is not penicillate and is much shorter than the head and body. Total length ranges from 196 to 240 mm; T 85–105; HF 25–27.5; E 15–16; Wt 47–60 g. The dorsum is very dark brown, almost black; the underparts are paler, the tail is not bicolor, and the dorsal surface of the feet is black. One female was recorded with four embryos (MVZ).

Range and Habitat

The species is distributed from Honduras south through Panama west of the Andes to southern Ecuador (map 13.21).

Subgenus *Microryzomys*
Oryzomys minutus (Tomes, 1860)

Measurements

	Mean	Minimum	Maximum	N
TL	189.0	173	200	6
T	106.2	104	108	6
HF	20.6	20	22	6

Location: Venezuela (CMNH)

Description

Cusps are present on adult molars. The species is similar in size to *O. fulvescens*. The tail is not penicillate and is much longer than the head and body. The dorsum is dark brown, the venter tan, and the tail bicolor (plate 18).

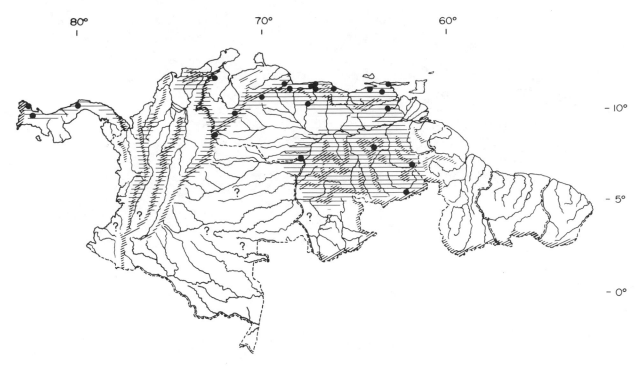

Map 13.22. Distribution of *Oryzomys fulvescens*.

Range and Habitat

The species ranges from west and central Colombia to northern Venezuela and south to Ecuador and Peru (map 13.19). This small *Oryzomys* is found between 2,000 and 3,500 m and replaces *O. fulvescens* at higher elevations. Although it climbs, it frequently forages on the ground and is broadly tolerant of both dry and moist sites, but it is strongly associated with cloud forest.

Natural History

Aagaard (1982) reports that this small rice rat is both terrestrial and arboreal. The species occupies a wide range of high-elevation microhabitats. Embryo counts vary from three to eight.

Genus *Neacomys* Thomas, 1900
Spiny Rice Rat

Description

The dental formula is I 1/1, C 0/0, P 0/0, M 3/3. In this small rice rat, head and body length ranges from 64 to 100 mm, and the tail is approximately the same length. The dorsal pelage is composed of hairs interspersed with bristles or spinelike hairs, easily distinguishing it from other small rice rats. The spines have alternating dark and light bands, giving a salt-and-pepper appearance. The underparts are white or cream, in contrast to the dorsum, which ranges from reddish to dark yellow brown. The tail is sparsely haired, brown above and lighter below.

Distribution

The genus occurs from Panama across northern South America to southwestern Brazil. In general it is confined to lower elevations.

Natural History

These small, terrestrial rice rats feed on seeds, insects, and small fruits and are nocturnal. Litter size ranges from two to four. This genus may show great fluctuations in abundance throughout the annual cycle (O'Connell 1981).

Neacomys guianae Thomas, 1905

Measurements

	Mean	Minimum	Maximum	N
TL	154.3	140	167	11
T	76.1	70	83	11
HF	20.2	19	22	11
E	14.3	12	15	11
Wta	14.2	12	18	11

Location: Suriname (CMNH)

Description

This is an extremely small species. The dorsum is reddish agouti brown, dark in the midline, and the sides may be orangish and the venter white. The tail is not noticeably bicolor.

Range and Habitat

The species is distributed in the Guyanas and adjacent parts of northern Brazil (map 13.23). It prefers dense, humid forests (Husson 1978).

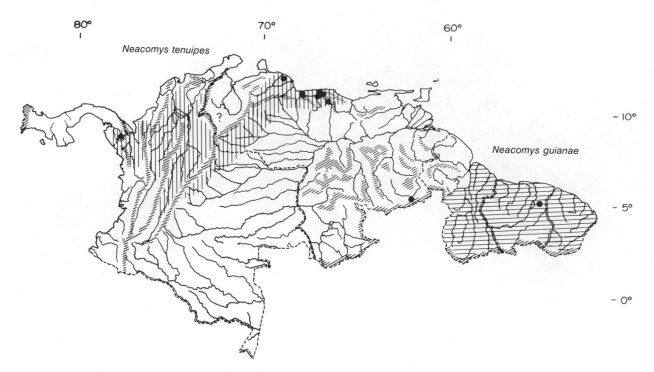

Map 13.23. Distribution of two species of *Neacomys*.

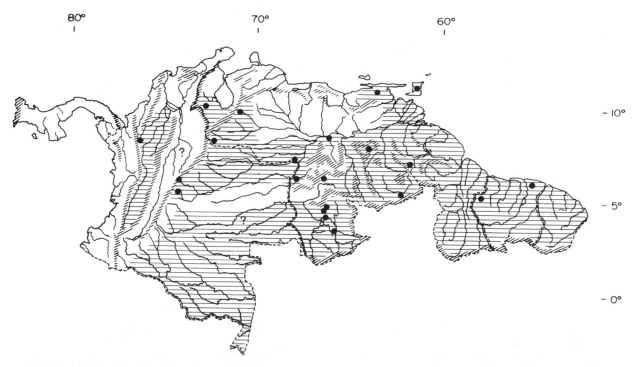

Map 13.24. Distribution of *Nectomys squamipes*.

Neacomys tenuipes Thomas, 1900

Description

Total length ranges from 158 to 163 mm, the tail from 83 to 87 mm, and the hind foot from 20.5 to 21 mm. Average weight is about 16 g. The dorsum is reddish brown mixed with black, the underparts are white to cream, and the tail is faintly bicolor.

Range and Habitat

The species occurs from eastern Panama occurs south to Colombia and eastern Ecuador and east to Venezuela (map 13.23). It is strongly associated with moist habitats and multistratal tropical evergreen forest to cloud forest and ranges up to 1,655 m elevation in northern Venezuela. It prefers dense cover of herbs, ferns, and shrubs (Handley 1976).

Genus *Nectomys* Peters, 1861
Nectomys parvipes Petter, 1979
Neotropical Water Rat, Rata Nadadora

Description

This small water rat was recently described from French Guiana 14°35′ N; 52°28′ W) and is known only from the type locality. It is similar in appearance to *N. squamipes* but much smaller: HB 135; T 152; HF 37; E 17 (Petter 1979).

Nectomys squamipes (Brants, 1827)

Measurements

	Mean	Minimum	Maximum	N
TL	416.0	390	445	5
T	206.8	187	210	5
HF	50.4	49	53	5
E	22.6	22	24	5
Wta	248.8	210	299	5

Location: Suriname (CMNH)

Description

Head and body length ranges from 160 to 255 mm. There are long, shiny guard hairs on the dorsum, and the underside of the tail has a series of stiff hairs that apparently increase the surface area when the tail is moved laterally while swimming. A distinct fringe of long hairs on the toes similarly increases the surface area of the hind foot, and the hind toes are partially webbed. The upperparts are dark agouti brown becoming reddish on the sides, and the underparts are tan to white, contrasting sharply with the dorsum. The tail is not bicolor. Variation has been analyzed by Hershkovitz (1944). The juvenile has a dark brown dorsum, and the pelage is very soft.

Range and Habitat

The species occurs across most of northern South America through Brazil to northeastern Argentina (map 13.24). In northern Venezuela most specimens were taken at below 500 m elevation in riverine habitats. It is broadly tolerant of agriculture. Although often associated with multistratal tropical evergreen forest it ranges into semixeric habitats, but it is confined to riverine situations.

Natural History

This semiaquatic rodent is never found far from streams or ponds. It is an excellent climber and swimmer (Ernest 1986; Ernest and Mares 1986). In captivity it feeds on plant matter, insects, tadpoles, and small fish. The animals are primarily nocturnal, spending the day in a grass nest constructed in a sheltered space such as under a log. In wet areas the nest is built on high ground. Linares (1969) describes a nest near a stream in a cave. Litter size ranges from two to seven, with five as an average. This species can show local increases in numbers and become an agricultural pest in areas of rice cultivation. Home ranges of 0.3 and 2.2 ha have been recorded in Brazil.

Comment

Reig (1984) notes that there are several chromosomal variations within the genus *Nectomys*, and at least three forms may prove to be valid species.

THOMASOMYINE RODENTS
Arboreal Cricetines or South American Climbing Rats

Diagnosis

This assemblage includes the genera *Aepeomys*, *Rhipidomys*, *Thomasomys*, and *Chilomys*. These small to medium-sized rodent are often distinguishable from the oryzomyines because the tail vastly exceeds the head and body in length. The ears are relatively large compared with those of the oryzomyine rodents, which they resemble closely.

Distribution

Species of thomasomyines are primarily South American, with one species, *Rhipidomys scandens*, entering the extreme eastern portion of Panama.

Genus *Aepeomys* Thomas, 1898

Description

The dental formula is I 1/1, C 0/0, P 0/0, M 3/3. These small mice have a head and body length of approximately 113 to 126 mm. The tail exceeds the head and body in length, averaging approximately 118 to 127 mm. The dorsum is gray brown to brown, and the venter is paler. The genus is considered distinct from *Thomasomys* by Gardner and Patton (1976).

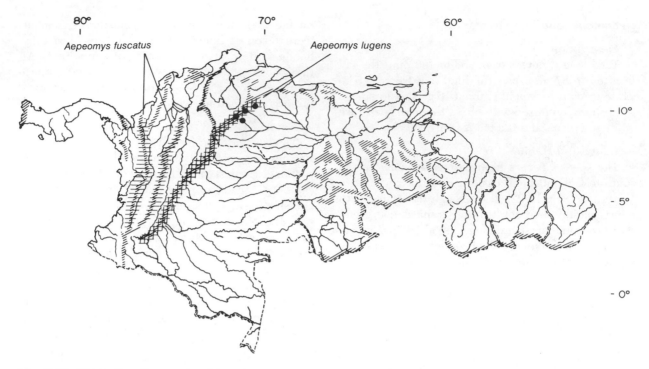

Map 13.25. Distribution of two species of *Aepeomys*.

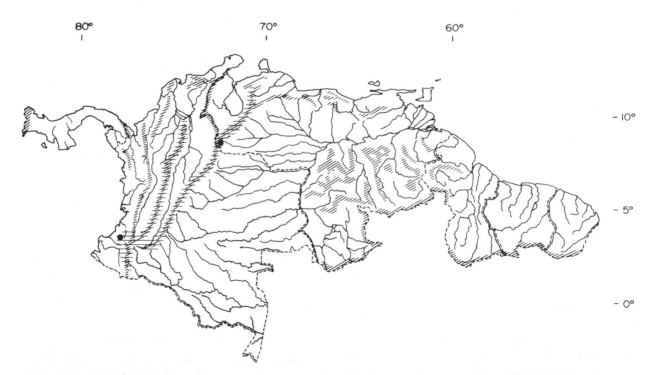

Map 13.26. Distribution of *Thomasomys aureus*.

Distribution
The genus is distributed in the Andean portions of Venezuela, Colombia, and Ecuador.

Aepeomys fuscatus J. A. Allen, 1912

Description
Head and body length ranges from 113 to 120 mm and the tail from 118 to 127 mm. The dorsum is gray brown, and the venter is paler. This small, long-tailed mouse is distinguishable from *A. lugens* by its distribution.

Range and Habitat
The species is found in the western and central cordilleras of the Andes in Colombia. It is replaced by *A. lugens* in the eastern cordillera (map 13.25).

Aepeomys lugens (Thomas, 1896)

Description
Total length ranges from 232 to 253 mm, tail length from 115 to 127 mm, and hind foot from 27 to 28 mm. Weight ranges from 32 to 42 g, and males are about 6% heavier than females. The dorsum is gray, grading to a lighter shade below on the venter.

Range and Habitat
The species is distributed in the eastern cordillera of the Andes from western Venezuela south to Andean Ecuador (map 13.25). It is strongly associated with moist habitats and cloud forest. In western Venezuela it was taken at between 1,900 and 3,560 m elevation.

Natural History
A. lugens breeds in the wet season and is almost entirely terrestrial (Aagaard 1982).

Genus *Thomasomys* Coues, 1884

Description
The dental formula is I 1/1, C 0/0, P 0/0, M 3/3. Head and body length ranges from 90 to 185 mm, and the tail is from 105 to 230 mm, greatly exceeding the head and body in length. The eye is very small; although *Thomasomys* seems arboreally adapted, given its long tail, it does not exhibit the large eye size so characteristic of the closely related genus *Rhipidomys*. The dorsal pelage is very dense and soft, varying from olivaceous gray to golden brown, and the venter is usually lighter. The tail tip is not noticeably penicillate (plate 18).

Distribution
The genus is distributed from the montane areas of Venezuela, Colombia, and Guyana south through Peru, Bolivia, and Brazil to northern Argentina. In northern South America it is common in the Andes at elevations up to 4,200 m.

Comment
Species names follow Cabrera (1960).

Thomasomys aureus (Tomes, 1860)

Measurements
	Mean	Minimum	Maximum	N
TL	365.5	346	395	8
T	208.8	190	234	8
HF	36.6	35	38	8
E	22.5	22	23	8

Location: Colombia (AMNH)

Description
Head and body length ranges from 142 to 166 mm. The dorsum is brown, but the bases of the dorsal hairs are black; the venter is pale orange. The tail is uniformly colored and sparsely haired.

Range and Habitat
The species ranges from the western and central cordilleras of the Andes in Colombia and the eastern cordillera of the Andes in Venezuela south to Ecuador and eastern Peru (map 13.26). In Venezuela specimens were taken at 2,400 m in association with cloud forest (Handley 1976), and it has been taken at up to 3,400 m elevation in Colombia.

Natural History
This species is known to feed on fruits. It is an excellent climber and builds nests in trees. Embryo counts of two and three have been recorded from Peru (MVZ).

Thomasomys bombycinus Anthony, 1925

Description
Head and body length ranges from 110 to 119 mm, the tail from 120 to 144 mm, and the hind foot from 27 to 30 mm (Anthony 1925). Three specimens from Colombia had the following average measurements: TL 244.7; T 133; HF 28.3; E 20. The dorsal pelage is dark brown, the venter is cinnamon buff. The tail is uniformly colored and sparsely haired. The vibrissae are extremely long.

Range and Habitat
The species is known from a restricted area, Paramillo, in the state of Antioquia in western Colombia at 2,750 m (map 13.27).

Thomasomys cinereiventer J. A. Allen, 1912

Measurements
	Mean	Minimum	Maximum	N
TL	282.0	274	290	8
T	153.1	146	162	8
HF	34.8	33	36	8

Location: Colombia (AMNH)

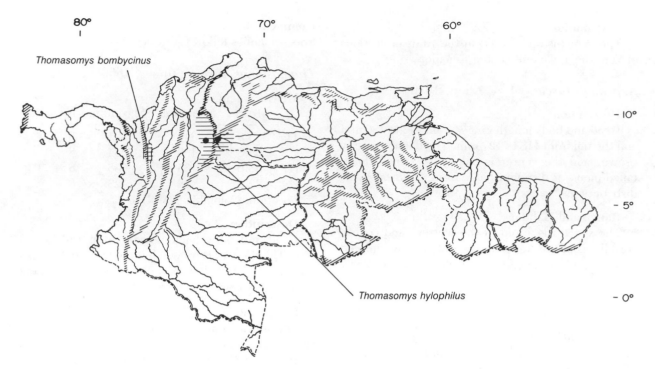

Map 13.27. Distribution of two species of *Thomasomys*.

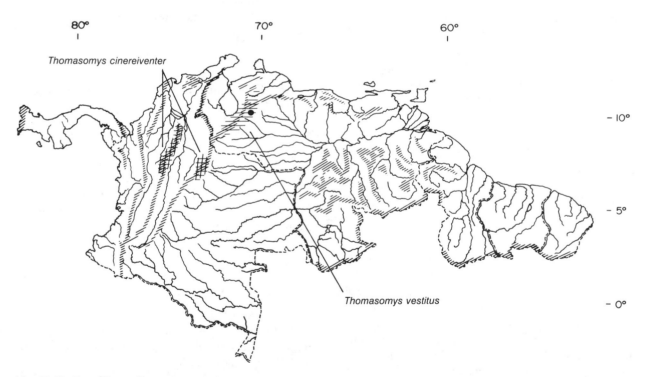

Map 13.28. Distribution of two species of *Thomasomys*.

Description

The dorsum is gray brown to dark brown and the venter is gray to yellow. The tail is uniformly colored and sparsely haired.

Range and Habitat

The species is distributed from central Colombia (Huila and Popayán, Cauca) to Ecuador. It is found between 2,000 and 3,500 m (map 13.28).

Thomasomys hylophilus Osgood, 1912

Description

Total length ranges from 248 to 275 mm, the tail from 135 to 152 mm, and the hind foot from 24 to 27 mm. The dorsum is a gray brown and the bases of the hairs are very dark; the venter is buffy gray. The sparsely haired tail is faintly bicolor but may have a white tip (Osgood 1912).

Range and Habitat

This species occurs in western Venezuela and northern Colombia. It is strongly associated with moist habitats and occurs in thick undergrowth of shrubs and tree ferns (map 13.27). In Venezuela it was taken at between 2,350 and 2,425 m elevation. It prefers cloud forest.

Thomasomys laniger (Thomas, 1895)

Measurements

	Mean	Minimum	Maximum	N
TL	225.3	194	250	9
HB	111.8	91	142	9
T	117.3	87	142	9
HF	23.6	23	25	9

Location: Colombia (AMNH)

Description

This species includes *T. monochromus*. This small species of *Thomasomys* is slightly smaller than *T. hylophilus*, which it resembles. Weight averages 31 to 34 g; adult males are slightly heavier than females (Aagaard 1982). The dorsum is dark brown, the venter has an orange cast, and the tail is uniformly colored.

Range and Habitat

This is a high-elevation form from the Andes of Colombia and Venezuela. It occurs in moist and dry forest (map 13.29).

Thomasomys vestitus (Thomas, 1898)

Description

A single male specimen measured TL 310, T 169, HF 33, E 21 and weighed 76.5 g (USNMNH). The dorsum is ochraceous brown and the venter a buff-washed gray. The tail is sparsely haired and slightly penicillate.

Range and Habitat

Specimens have been taken in western Venezuela in the Andes at elevations exceeding 1,630 m. The species is associated with moist habitats and cloud forest (map 13.28).

Genus *Rhipidomys* Tschudi, 1844
Climbing Rat, Rata Arborícola

Description

The dental formula is I 1/1, C 0/0, P 0/0, M 3/3. Head and body length ranges from 80 to 210 mm; the tail normally exceeds the head and body and in larger species can reach 270 mm. The fur is dense and soft; the tail is well haired with a terminal tuft. The hind feet are broad, suggesting arboreal adaptation, a character shared by *Thomasomys* and some species of *Oryzomys*. The dorsum ranges from gray to dark brown and the underparts usually contrast sharply, being white or gray.

Distribution

These climbing rats are broadly distributed in northern South America, extending southward to northern Argentina and east-central Brazil.

Natural History

This long-tailed, soft-furred rat is nocturnal in its habits and adapted for arboreal life. It nests in trees, and although it forages arboreally it will also come to the ground. It is typically a lowland form. Litter size is small, the mode being three. Young become sexually mature at about three months of age. In habitats showing marked seasonality in rainfall, reproduction may also be seasonal (O'Connell 1981).

Rhipidomys latimanus (Tomes, 1860)

Measurements

	Mean	Minimum	Maximum	N
TL	250.0	216	277	9
T	136.8	113	159	9
HF	23.3	20	25	9
E	18.3	17	20	9

Location: Colombia (FMNH)

Description

The dorsum is a reddish brown and the venter contrasts sharply, being snow white. The penicillate tail is not noticeably bicolor. This account follows Honacki, Kinman, and Koeppl (1982) and includes *R. fulviventer* and *R. venustus*.

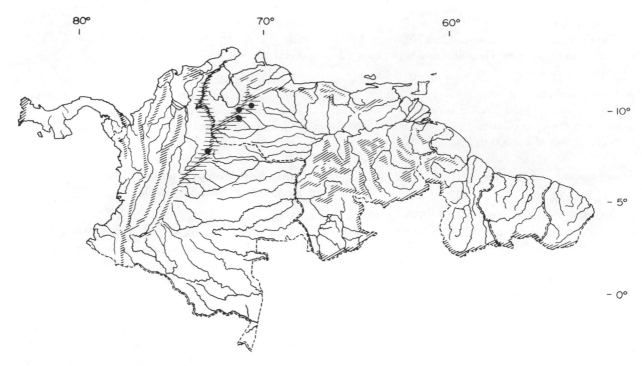

Map 13.29. Distribution of *Thomasomys laniger*.

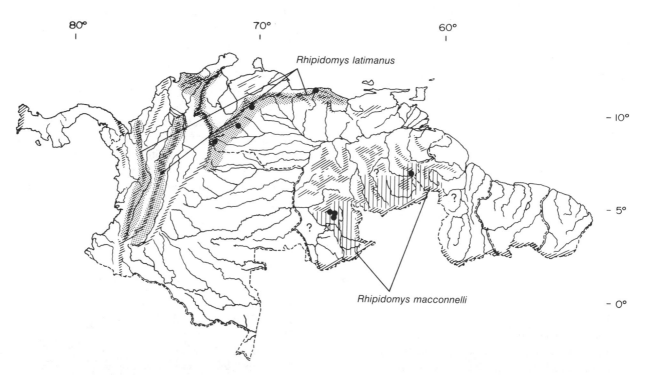

Rhipidomys latimanus

Rhipidomys macconnelli

Map 13.30. Distribution of two species of *Rhipidomys*.

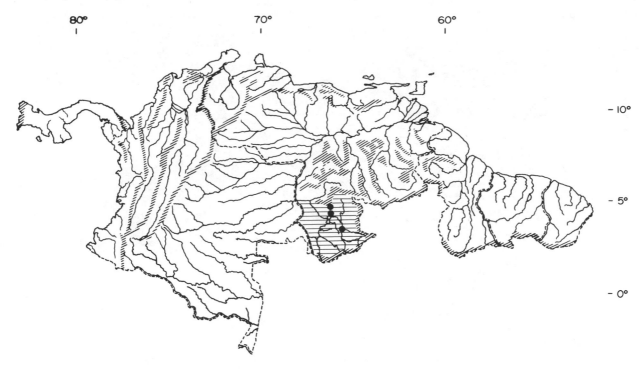

Map 13.31. Distribution of *Rhipidomys leucodactylus*.

Range and Habitat

This species is broadly distributed at high elevations from western Venezuela across Colombia and south to Ecuador. *R. latimanus venustus* occurs in northwestern Venezuela at moderate elevations (map 13.30). In Venezuela this species ranges from 1,160 to 2,422 m elevation. This rat is broadly tolerant of dry habitats but apparently prefers moist areas. It is found in both multistratal tropical evergreen forest and dry deciduous forest near rivers.

Natural History

The species is mainly arboreal. Litter size has been recorded as two (Aagaard 1982).

Rhipidomys leucodactylus (Tshudi, 1845)

Description

This is a large species. One specimen from Ecuador measured TL 410; T 213; HF 35. The dorsum is light agouti brown, sharply demarcated from a gray to buff venter. The tail is thinly haired to the penicillate tip and is not bicolor.

Range and Habitat

The species occurs from southern Venezuela to northwestern and eastern Ecuador and south through eastern Peru to northern Argentina (map 13.31). This is a lowland form; in Venezuela it occurs from 135 to 145 m elevation and prefers multistratal tropical evergreen forest (Handley 1976).

Rhipidomys macconnelli De Winton, 1900

Measurements

	Mean	Minimum	Maximum	N
TL	257.6	187	287	11
T	148.5	111	159	11
HF	26.7	25	28	11

Location: Venezuela (AMNH)

Description

In this small species of *Rhipidomys*, the dorsum is brown, contrasting sharply with the gray venter, the tail is bicolored, and the feet are reddish brown.

Range and Habitat

The species is found in southeastern Venezuela to Guyana in premontane forest. It ranges from 750 to 2,900 m elevation (map 13.30).

Natural History

Females from Mount Duida, Venezuela, contained one or two embryos (AMNH).

Rhipidomys mastacalis (Lund, 1840)

Measurements

	Mean	Minimum	Maximum	N
TL	230.5	140	330	9
T	160.5	122	182	9
HF	27.2	25	30	9
E	18.3	15	25	9

Location: Venezuela (AMNH)

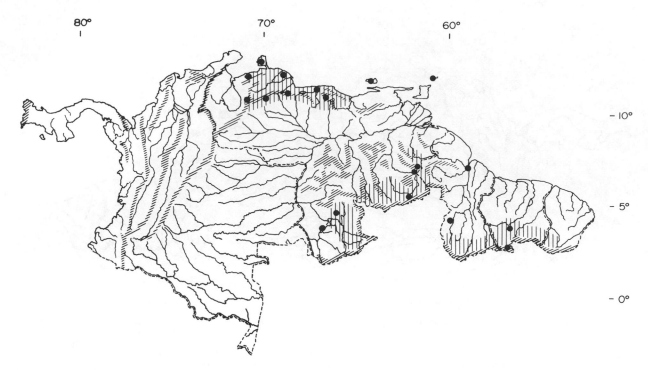

Map 13.32. Distribution of *Rhipidomys mastacalis*.

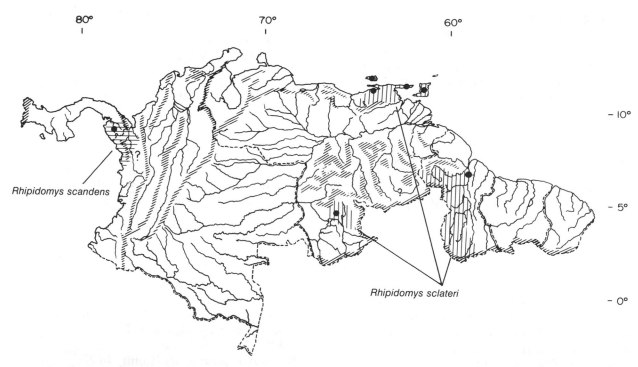

Rhipidomys scandens

Rhipidomys sclateri

Map 13.33. Distribution of two species of *Rhipidomys*.

366

Description

Adults weigh about 100 g. The dorsum is gray brown and the venter is pale gray to cream. The tail is not bicolor, and the penicillate tip is present but not always accentuated. This description includes *R. venezuelae*, *R. nitola*, and *R. tenicauda* (Honacki, Kinman, and Koeppl 1982; Cabrera 1960).

Range and Habitat

The subspecies *R. m. venezuelae* occurs in northern Venezuela to the Guyanas. The subspecies *R. m. tenicauda* occurs in southern Venezuela, the southern Guyanas, and to northeastern Brazil (map 13.32). This is a lowland species preferring moist habitats, but it can occur in dry deciduous tropical forest. In Venezuela it has been taken at elevations from 13 to 1,500 m.

Natural History

This nocturnal form is the common arboreal rat caught in the mountains of northern Venezuela. It feeds on fruits, seeds, leaves, fungi, and adult insects (O'Connell 1981). There are 3.8 young in an average litter, and adults are rather long-lived compared with semiarboreal rice rats such as *O. concolor*.

Rhipidomys scandens Goldman, 1913

Description

Total length averages about 330 mm, of which 132 mm is head and body length and 198 mm is the tail. The hind foot averages about 32 mm. The buffy brown dorsum contrasts with a white venter.

Range and Habitat

This species is described from the extreme eastern part of Panama at moderate elevations and evidently extends to an undetermined distance into western Colombia (map 13.33).

Rhipidomys sclateri (Thomas, 1887)

Description

This description includes *R. couesi*. In this extremely large climbing rat, total length ranges from 319 to 410 mm: T 161–209; HF 22–26; E 21–28. Its size distinguishes it from other species of *Rhipidomys*. The dorsum is light brown, the venter is cream, and the tail is not bicolor.

Range and Habitat

The species is distributed in southeastern Venezuela, through the Guyanas, south to Brazil (map 13.33). It is broadly tolerant of both moist and dry forests and occurs at elevations of 1 to 1,400 m.

Genus *Chilomys* Thomas, 1897
Chilomys instans (Thomas, 1895)
Colombian Forest Mouse

Measurements

	Mean	Minimum	Maximum	N
TL	218.6	204	228	9
T	123.4	119	126	9
HF	24.1	23	26	9
E	15.7	14	16	9

Location: Colombia (FMNH, AMNH)

Description

The dental formula is I 1/1, C 0/0, P 0/0, M 3/3. Head and body length ranges from 86 to 99 mm, and the tail from 105 to 130 mm. Weight ranges from 15 to 18 g. The dorsum is gray brown to very dark brown, and the venter usually matches it. There may be a buffy pectoral spot on the venter, and some specimens have a white stripe from the throat to the middle of the stomach (Osgood 1912). The tip of the tail is often white (plate 18).

Range and Habitat

The species is distributed in the eastern cordillera of the Andes of Venezuela and Colombia. It is strongly associated with cloud forest, and in Venezuela it occurs between 2,405 and 2,700 m elevation (map 13.34).

Natural History

Very little is known about this mouse other than it prefers moist cloud forest and co-occurs with the larger *Thomasomys hylophilus* (Aagaard 1982), which it resembles.

TYLOMYINE RODENTS
Central American Climbing Rats
Genus *Tylomys* Peters, 1866

Description

The tylomyine rodents include two genera, *Tylomys* and *Ototylomys*. *Ototylomys* includes four species distributed from southern Mexico to Costa Rica and does not occur within the range covered by this volume. *Tylomys* includes seven species. The dental formula is I 1/1, C 0/0, P 0/0, M 3/3. Head and body length ranges from 170 to 240 mm; the almost naked tail is longer than the head and body, ranging from 200 to 250 mm. The ears are large and naked (plate 19).

Distribution

Tylomys is primarily Central American in its distribution, ranging into northwestern South America.

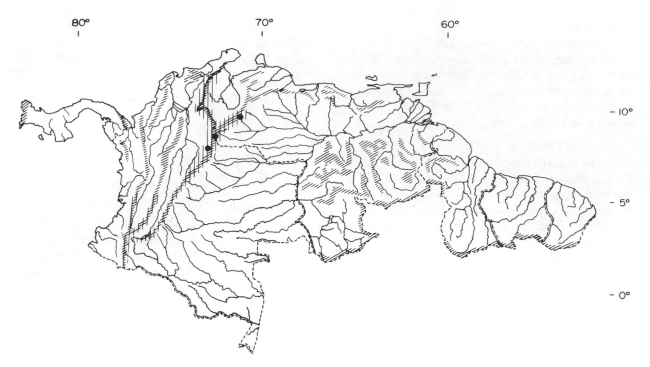

Map 13.34. Distribution of *Chilomys instans*.

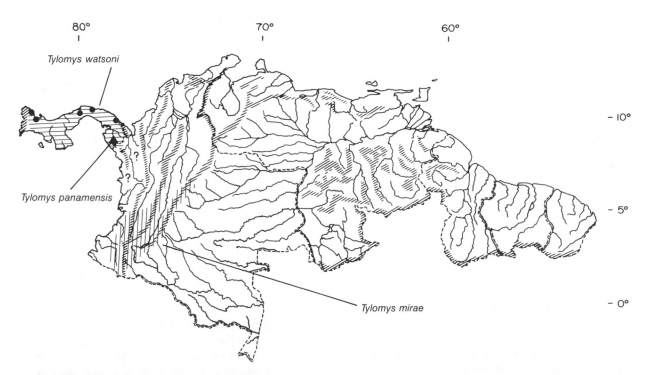

Map 13.35. Distribution of three species of *Tylomys*.

Tylomys panamensis (Gray, 1873)

Description
Head and body length averages 200 mm, and the tail is about 230 mm. These large climbing rats may reach 200 g. The dorsum is gray brown, but the longer hairs are black. The sides of head and body are paler, the venter is white, and the distal third of the tail is also white.

Range and Habitat
This species is known only from a restricted locality in extreme southeastern Panama (map 13.35).

Tylomys mirae Thomas, 1899

Measurements

	Mean	Minimum	Maximum	N
TL	379.8	310	477	8
T	189.5	140	242	8
HF	36.0	30	41	8
E	27.5	26	29	8

Location: Colombia (FMNH, AMNH)

Description
This species possibly includes *Tylomys fulviventer*. Total length 330 to 400 mm: HB 160–200; T 170–210; HF 30–40. The vibrissae are stout and long. The dorsum is gray brown, the venter buff to gray. There is generally a russet median stripe on the dorsum. The tail is black, with the terminal portion flesh colored; the ventral side of the tail tends to be gray (plate 19).

Range and Habitat
The species ranges from central Colombia south to northern Ecuador. If it is considered conspecific with *T. fulviventer*, then it extends to extreme eastern Panama (map 13.35).

Natural History
These nocturnal, arboreally adapted rodents appear to be primarily frugivorous. In captivity females of *T. nudicaudus* gave birth after a gestation period of thirty-two to thirty-three days to litters ranging from one to three, with a mean of 1.6 (Baker and Petersen 1965; Helm 1973, 1975).

Tylomys watsoni Thomas, 1899

Description
This large species measures TL 457–93; T 243–50; HF 35–38. The dorsum is cinnamon brown, contrasting with the whitish venter. The tail is brown proximally and white on the distal third.

Range and Habitat
The species is distributed through lowland Costa Rica and Panama and may extend into northern Colombia (map 13.35).

NYCTOMYINE RODENTS
Vesper Mice

Systematics
In his review of the neotomyine-peromyscine rodents, Carleton (1980) consider the vesper mouse one of the most conservative members of the Sigmodontinae. Hershkovitz (1962) considered vesper mice to be thomasomyine rodents. There are two genera, *Nyctomys* and *Otonyctomys*, the latter confined to the Yucatán peninsula. *Nyctomys* contains a single species, *N. sumichrasti*.

Genus *Nyctomys* Saussure, 1860
Nyctomys sumichrasti (Saussure, 1860)
Ratón Trepador

Description
The dental formula is I 1/1, C 0/0, P 0/0, M 3/3 (see fig. 13.13). The tail is much longer than the head and body: TL 208–86; T 110–56; HF 23–26. The tail is covered with short hair. The dorsum is buffy with a thin admixture of black hairs, and the venter is white. The eyes are quite large, and the hind feet are broad, showing adaptations for climbing (plate 19).

Range and Habitat
This species occurs from southern Mexico to Panama. It does not cross the Isthmus (map 13.36).

Natural History
This rodent is arboreal and nocturnal, although K. Langtimm (pers. comm.) reports diurnal activity. It constructs nests of twigs and leaves in trees. As far as is known, it is primarily a frugivore. K. Langtimm (pers. comm.) noted feeding on the fruits of *Cordia diversifolia*, *Hoffmania* sp., and *Psychotria gracilis*. Gestation is thirty-four days, and the modal litter size is two. Young open their eyes at approximately sixteen days of age, and the mother nurses them for up to forty days (Birkenholz and Wirtz 1965). In general *Tylomys* and *Nyctomys* replace *Rhipidomys* north of the Isthmus and occupy the nocturnal, arboreal, frugivorous niche for rat-sized rodents (Genoways and Jones 1972).

PEROMYSCINE RODENTS
White-footed Mice and Allies

Systematics
This assemblage includes the genera *Peromyscus* and allied subgenera, as well as *Reithrodontomys*, *Onychomys*, and *Ochrotomys*. These rodents are highly diversified and primarily distributed in North America. Seven species from three genera occur within the range covered by this volume.

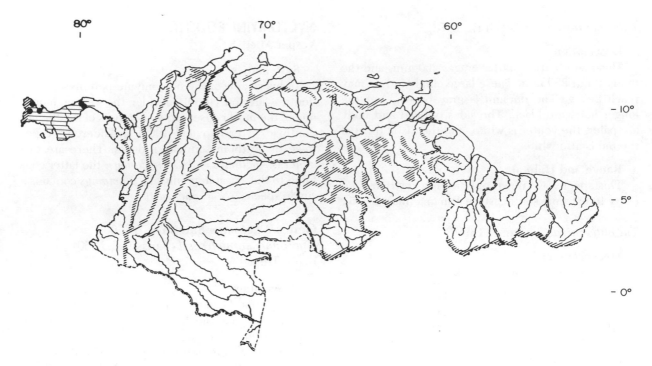

Map 13.36. Distribution of *Nyctomys sumichrasti.*

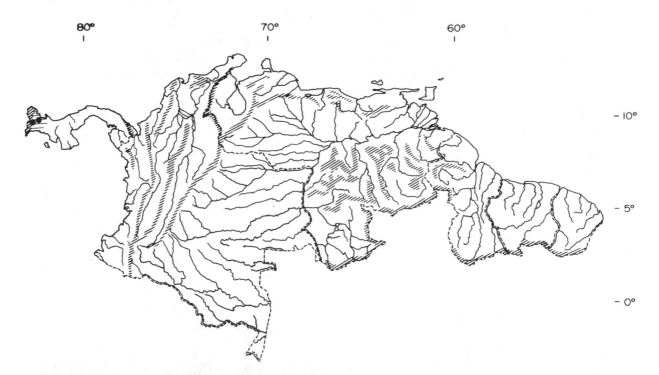

Map 13.37. Distribution of *Reithrodontomys creper.*

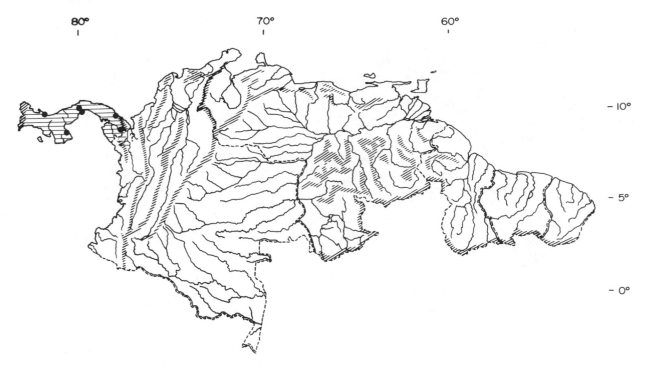

Map 13.38. Distribution of *Reithrodontomys darienensis*.

Genus *Reithrodontomys* Giglioli, 1874
Harvest Mouse, Grooved-toothed Mouse

Description
The dental formula is I 1/1, C 0/0, P 0/0, M 3/3. The upper incisors bear a longitudinal groove, which distinguishes these mice from all similar-sized forms within the range covered by this volume. Head and body length ranges from 50 to 145 mm, and the tail from 65 to 95 mm. In most species the pelage is brown above and white to cream below. The genus has been reviewed by Hershkovitz (1941) and Hooper (1952).

Distribution
The genus is distributed from southwestern Canada through Mexico to Ecuador west of the central cordillera of the Andes.

Natural History
Very little is known about the natural history of the southern species of the genus *Reithrodontomys*. The northern species have been well studied in North America. These are omnivorous rodents preferring edge habitats. Gestation in the northern species is twenty-three to twenty-four days; litter size averages about four. The young open their eyes at approximately eight days of age, and the female nurses them for approximately eighteen days, then they disperse. It is assumed that the southern species exhibit similar life-history traits, but the details need much work (see Jones and Genoways 1970).

Reithrodontomys creper Bangs, 1902

Measurements
	Mean	Minimum	Maximum	N
TL	219.5	208	237	6
T	127.3	118	140	6
HF	23.1	20	25	6
E	15.1	15	16	6

Location: Costa Rica (AMNH)

Description
This is the largest harvest mouse occurring in the region covered by this book. The dorsum is reddish brown intermingled with black hairs, and the venter is gray to cinnamon. The ears are black, and there is a black eye ring.

Range and Habitat
The species ranges from Costa Rica to western Panama at high elevations, reaching 3,000 m (map 13.37).

Reithrodontomys darienensis Pearson, 1939

Description
This is the smallest species in the area covered by this book. Head and body length ranges from 62 to 64 mm, tail 100 to 112 mm, hind foot 17 to 19 mm, ear 14 to 15 mm. The dorsum is brown interspersed with blackish hair, and the venter is white to cinnamon.

Range and Habitat
The species occurs in western Panama to extreme northwestern Colombia (map 13.38).

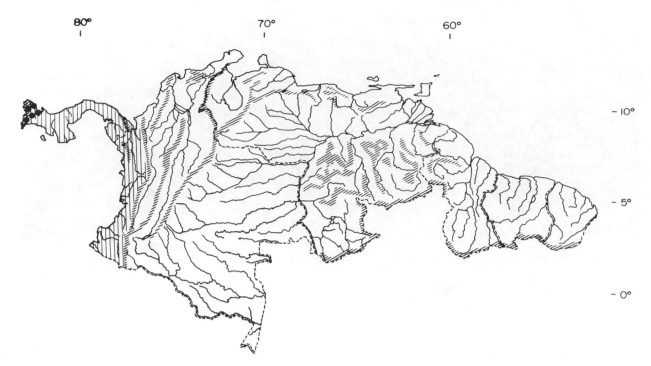

Map 13.39. Distribution of *Reithrodontomys mexicanus*.

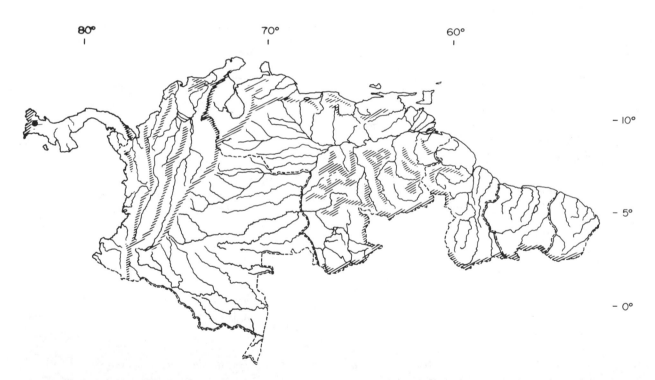

Map 13.40. Distribution of *Reithrodontomys sumichrasti*.

Reithrodontomys mexicanus (Saussure, 1860)

Description
Head and body length ranges from 68 to 77 mm, tail 92 to 126 mm, hind foot 17 to 21 mm, ear 14.5 to 18 mm. The dorsum is tawny to orange cinnamon and the venter is variable, ranging from white to cinnamon.

Range and Habitat
This species is distributed from Tamaulipas in eastern Mexico and Oaxaca in western Mexico south through the Isthmus, through western Colombia to Ecuador. It occurs at moderate elevations up to 2,000 m (map 13.39).

Natural History
In Central American this species breeds from June to August. Emrbyo counts range from three to five (Jones and Genoways 1970).

Reithrodontomys sumichrasti (Saussure, 1861)

Description
This mouse is similar in size to *R. mexicanus*. Head and body length ranges from 68 to 83 mm, tail 75 to 123 mm, hind foot 17 to 22 mm, ear 12 to 18 mm. The dorsum is very dark, almost black; the preponderance of blackish hairs obscures the brown hairs. The venter is dark gray.

Range and Habitat
The species is distributed through Central America, extending to western Panama (map 13.40).

Genus *Peromyscus* Gloger, 1841
White-footed Mouse

Description
Some fifty-five species of this genus have been described. Head and body length ranges from 80 to 179 mm, and tail length from 140 to 205 mm. Adults weigh from 15 to 50 g. The ears are large relative to the body size; tail length is highly variable depending on the species. The dorsal pelage is quite variable in color, from gray to dark brown, and the underparts are generally whitish. See King (1968) for a review of the biology of the genus.

Distribution
The genus occurs from Alaska and Labrador in North America south to the extreme northwest of Colombia.

Natural History
Members of the genus *Peromyscus* have gestation lengths between twenty-one and twenty-six days; litter size averages three or four. The young generally open their eyes by two weeks of age and are weaned at approximately one month. These granivorous rodents are nocturnal and scansorial in their foraging tendencies (see Eisenberg 1968 for a review of behavior patterns).

Peromyscus mexicanus (Saussure, 1860)

Description
For this description *P. nudipes* is included with *P. mexicanus* (Honacki, Kinman, and Koeppl 1982). The subspecies *P. m. nudipes* has a total length ranging from 240 to 280 mm: HB 120–40; T 120–40; HF 26–30; E 18.4–22. The upperparts are dark brown to blackish in the middorsal region, and the sides are a lighter brown. The underparts contrast, being cream to white.

Range and Habitat
The species is found from Veracruz in western Mexico and Chiapas in eastern Mexico south to extreme northwestern Panama (map 13.41).

Natural History
P. mexicanus is a rather large lowland form. It feeds on seeds and fruit. Litter sizes are two or three (Hall and Dalquest 1963).

Genus *Isthmomys* Hooper and Musser, 1964

Description
Isthmomys is considered a subgenus of *Peromyscus* by Hall (1981). The two species occurring within the range of this volume are both large. *I. flavidus* is paler and more yellow than *I. pirrensis*.

Distribution
As far as is known, the ranges of the two species are disjunct, and they do not occur in lowland central Panama.

Isthmomys flavidus (Bangs, 1902)

Description
In this very large peromyscine rodent, total length ranges from 320 to 375 mm: T 155–205; HF 31–33; E 20–24. The dorsum is dark yellow brown, and the underparts are yellowish white mixed with gray.

Range and Habitat
The species ranges from southeastern Costa Rica to the extreme northwest of Panama at elevations over 1,000 m (map 13.42).

Isthmomys pirrensis (Goldman, 1912)

Description
This mouse is similar in size to *I. flavidus*. Total length ranges from 342 to 376 mm; the tail is 185 to 204 mm and the hind foot is 36 to 36.5 mm. The dorsum is dark brown with cinnamon; the underparts are buffy white to gray.

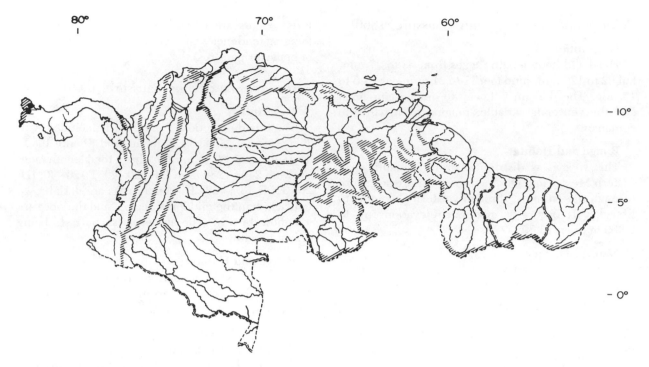

Map 13.41. Distribution of *Peromyscus mexicanus nudipes*.

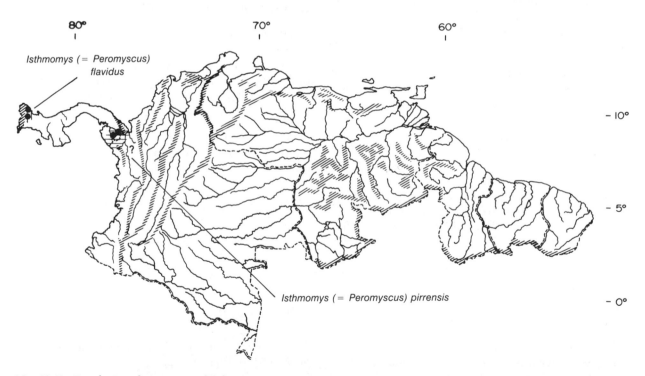

Map 13.42. Distribution of two species of *Isthmomys*.

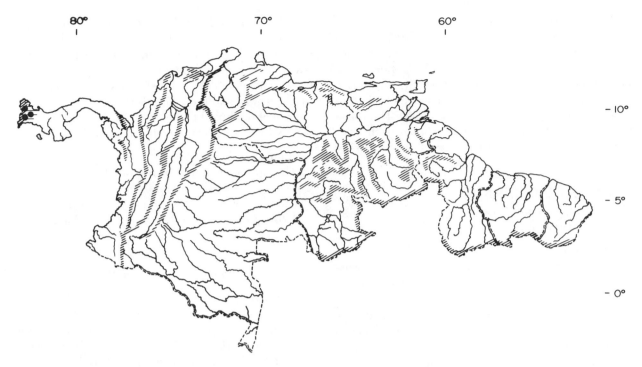

Map 13.43. Distribution of *Scotinomys teguina*.

Range and Habitat

This species is distributed from extreme eastern Panama extending an unknown distance into northwestern Colombia (map 13.42).

BAIOMYINE RODENTS
Dwarf or Pygmy Mice and Their Allies

Diagnosis

These small mice are referred to as pygmy mice (*Baiomys*) and brown mice (*Scotinomys*). They can be distinguished from the smaller species of *Reithrodontomys* by the lack of the grooved upper incisor and the relatively short tail.

Distribution

Baiomys is distributed from Arizona and Texas in the United States south to Nicaragua; *Scotinomys* inhabits Central America from southern Mexico to western Panama. Only two species occur within the range covered by this volume, *Scotinomys teguina* and *S. xerampelinus*. The genus was reviewed by Hooper (1972).

Genus *Scotinomys* Thomas, 1913
Scotinomys teguina (Alston, 1877)
Alston's Brown Mouse

Description

The dental formula is I 1/1, C 0/0, P 0/0, M 3/3. Total length ranges from 115 to 144 mm: T45–63; HF 15–19; E 12–18. The tail is distinctly shorter than the head and body. Average weight is 12 g. The short, harsh dorsal pelage is blackish brown; the venter is reddish brown to dark gray (plate 19).

Range and Habitat

This mouse is found from Chiapas, Mexico, south to extreme western Panama, generally at elevations exceeding 1,000 m (map 13.43).

Natural History

This small, terrestrially adapted mouse is insectivorous, though it also eats seeds. The gestation period is thirty days; mean litter size is three. The young open their eyes at approximately two weeks of age, and weaning occurs at three weeks. Sexual maturity occurs at thirty-four days of age. The males produce a high-pitched call lasting seven to ten seconds, which apparently serves a territorial spacing function (Hooper and Carleton 1976).

Scotinomys xerampelinus (Bangs, 1902)

Description

Total length ranges from 136 to 164 mm: T 65–80; HF 17–20; E 12–17; Wt 15 g. This mouse is distinguishable from *S. teguina* by the slightly longer tail. The dorsum is dark yellow brown, and the underparts are buffy brown.

Range and Habitat

This species is restricted to montane areas in east-central Costa Rica and extreme western Panama. It has been taken at elevations up to 3,000 m (map 13.44).

Natural History

S. xerampelinus occurs at higher elevations than does *S. teguina* and is not as tolerant of high temperatures as its congener (Hill and Hooper 1971). Gestation averages 33.1 days, and the litter size ranges from two to four. The female tolerates the male in the nest throughout rearing. The male produces a very high-pitched call similar to that described for *S. teguina* (Hooper and Carleton 1976).

ICHTHYOMYINE RODENTS
Fish-eating Rats and Mice
Ratones, Pescadores

Diagnosis

These rodents are adapted for an aquatic existence. They include five genera, *Ichthyomys*, *Anotomys*, *Daptomys*, *Rheomys*, and *Neusticomys*. The last is known from Ecuador and is outside the range covered by this volume. These genera show a reduction in the external ear; the toes either are partially webbed or show a fringe of hairs, thus increasing the surface area of the hind foot. The fur tends to be velvety, and the vibrissae on the snout are stout and rather long. All these morphological features correlate with a semiaquatic existence. The dorsal pelage is usually dark brown to blackish, contrasting with the gray to white venter. Voss (1983, 1988) has reviewed the group and concludes they are monophyletic.

Distribution

These species occur from moderate to high elevations in the vicinity of freshwater streams in Panama, Colombia, Venezuela, the Guyanas, Ecuador, and Peru.

Natural History

As the common name suggests, some species, especially of the genus *Ichthyomys*, may feed upon small freshwater fishes in addition to crustaceans. Other species, such as *Rheomys*, may eat small crustaceans and snails. Within the generic assemblage different degrees of adaptation for a semiaquatic habitat are displayed. *Ichthyomys* seems highly specialized with respect to its partially webbed feet; *Anotomys* shows specializations with the loss of the pinnae; *Neusticomys* shows the least specialization for semiaquatic life. These rodents may not represent a natural grouping.

Genus *Anotomys* Thomas, 1906
Anotomys trichotis (Thomas, 1897)

Description

The dental formula is I 1/1, C 0/0, P 0/0, M 3/3. Head and body length averages 130 mm, the tail

125–50 mm, and the hind foot 30–32 mm. This group is immediately distinguishable from the other genera by the extreme reduction of the external ear, which is almost absent. Generally the opening to the ear is surrounded by whitish hairs. The dorsum is blackish gray, and the venter is pale gray. The dorsal surface of the broad feet is gray, and the ventral surface is black (*Chibchanomys trichotis* [Voss 1988]).

Range and Habitat

This species is found in the montane portions of Colombia and northern Venezuela south to northern Ecuador. In northern Venezuela it occurs in streams at 2,400 m elevation in association with cloud forest (map 13.44).

Genus *Daptomys* Anthony, 1929

Description

The dental formula is I 1/1, C 0/0, P 0/0, M 2/2 or 3/3. Head and body length averages 100 to 131 mm and the tail is 105 too 111 mm, usually shorter than the head and body. There are long guard hairs with a dense underfur. The upperparts are blackish brown, and the venter is paler. The characters of the genus are reviewed by Musser and Gardner (1974).

Distribution

The genus is known from the northern and southern montane region of Venezuela and lowland Peru, and from French Guiana (Musser and Gardner 1974).

Daptomys venezuelae Anthony, 1929

Measurements

	Mean	Minimum	Maximum	N
HB	114.3	109	131	3
T	108.3	105	111	3
HF	27.3	25	29	3
E	10.3	10	11	3

Location: Venezuela (Musser and Gardner 1974)

Description

The dental formula is I 1/1, C 0/0, P 0/0, M 3/3. The dorsum is blackish brown, and the venter is gray.

Range and Habitat

This species ranges from 750 to 1,400 m in southern Venezuela in association with montane streams (map 13.44). It is rare throughout its range.

Daptomys oyapocki Dubost and Petter, 1978

Description

The dental formula is I 1/1, C 0/0, P 0/0, M 2/2, which is unique. This mouse is similar to *D. venezuelae* in color.

Range and Habitat

The species is known only from the type locality French Guiana on the banks of the Oyapock River, 2°10′ N, 53°11′ W (Dubost and Petter 1978).

Genus *Ichthyomys* Thomas, 1893
Rata Pescadora

Description

The dental formula is I 1/1, C 0/0, P 0/0, M 3/3. The upper incisors are V-shaped, apparently an adaptation for seizing aquatic prey. Head and body length ranges from 145 to 210 mm and the tail from 145 to 190 mm. The eyes are reduced in size, as are the pinnae. The vibrissae are long and stout (plate 19). The toes of the hind feet are partially webbed, and there is a distinct fringe of stiff, short hairs on the outer sides of the hind feet. The pelage is thick, with the upperparts dark olive brown to blackish and the underparts white. The bicolor tail is covered with fine hairs.

Distribution

The genus is found from Peru north to Ecuador and montane Colombia and Venezuela.

Ichthyomys pittieri Handley and Mondolfi, 1963

Description
See the account for the genus.

Range and Habitat

The species is known from extremely restricted localities in the north coastal range of northern Venezuela (Handley and Mondolfi 1963) (map 13.45).

Natural History

Although some species of *Ichthyomys* eat fish, research in Venezuela indicates that *I. pittieri* feeds on aquatic insects and the freshwater crabs of the family Pseudothelphusidae (Voss, Silva, and Valdes 1982).

Ichthyomys hydrobates (Winge, 1891)

Description

One male from FMNH measured HB 182; T 150; HF 35; E 9. The dorsum is very dark agouti brown, the tail is bicolor, and the venter is gray. The hind feet are very broad, and the fringe of hair on the toes is conspicuous.

Range and Habitat

The species is broadly distributed in montane areas where suitable habitat is available in the foothills of the Andes in Colombia (map 13.45).

Genus *Rheomys* Thomas, 1906

Description

The dental formula is I 1/1, C 0/0, P 0/0, M 3/3. The upper incisors have a shallow longitudinal groove. Head and body length ranges from 105 to 188 mm, the tail from 95 to 150 mm. The external ear is very reduced. The hind foot is partially webbed, and the toes are fringed with bristles. The dorsum is dark brown mixed with black, with the venter paler.

Distribution

The genus is found from southern Mexico south through Central America to western Panama.

Rheomys hartmanni Enders, 1939

Description

Total length ranges from 223 to 230 mm, the tail from 105 to 115 mm, and the hind foot from 27 to 29 mm. The dorsum is cinnamon brown mixed with black, and the underparts are silvery gray.

Range and Habitat

The species is found in the vicinity of freshwater streams in southern Costa Rica to the extreme southwest of Panama (map 13.45).

Natural History

This mouse is confined to freshwater stream edges, where it feeds extensively on aquatic insects and their larvae (Hooper 1968).

Rheomys raptor Goldman, 1912

Measurements

	Mean	Minimum	Maximum	N
TL	203	196	211	4
T	97.5	93	106	4
HF	24.8	23	26	4
E	11.0	9	12	4

Location: Colombia (FMNH)

Description

The dorsum is cinnamon mixed with black, especially dark on the rump. The underparts are gray brown and the tail is not bicolor. A fringe of hair is noticeable on the toes.

Range and Habitat

The species is found in the vicinity of freshwater streams in the extreme southeast of Panama. It extends to Colombia (map 13.44).

Rheomys underwoodi Thomas, 1906

Description

In this large species, total length ranges from 280 to 292 mm: T 148–56; HF 37; E 5–7. The dorsum is brown, darkened by black-tipped hairs. The tail is distinctly bicolor, brown above and paler below, and the underparts are grayish (Hall 1981).

Range and Habitat

This mouse is found in the vicinity of freshwater

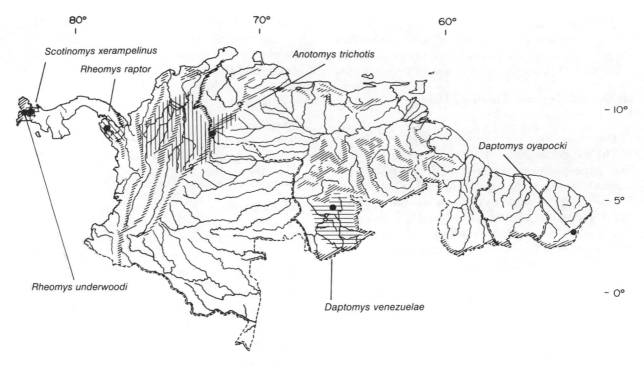

Map 13.44. Distribution of two species of *Daptomys*, two species of *Rheomys*, *Anotomys trichotis*, and *Scotinomys xerampelinus*.

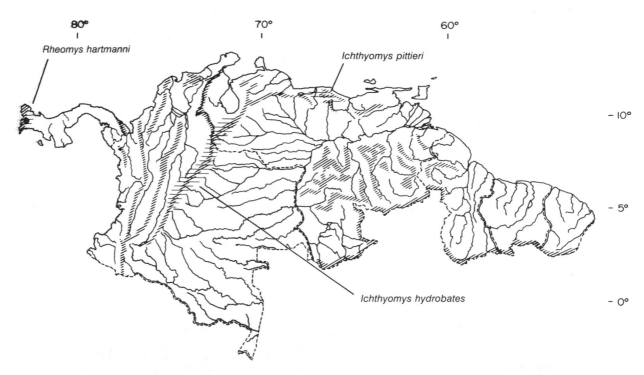

Map 13.45. Distribution of *Rheomys hartmanni* and two species of *Ichthyomus* (based in part on Handley and Mondolfi 1963).

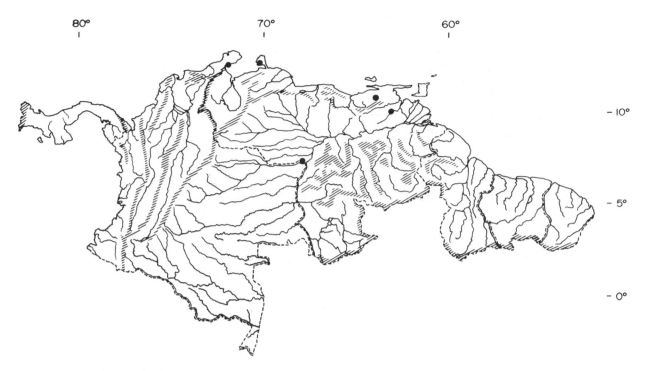

Map 13.46. Distribution of *Calomys hummelincki.*

streams from central Costa Rica to extreme west-central Panama (map 13.44).

Natural History
Like *R. hartmanni*, this species appears to be insectivorous (Hooper 1968). An evaluation of anatomy and behavior was prepared by Starrett and Fisler (1970).

Comment
It is clear that the fish-eating rats and mice are specialized for feeding niches normally occupied by water shrews (Soricidae) and their relatives in other parts of the world. The absence of Insectivora in South America, with the exception of *Cryptotis*, has left the small aquatic insectivore niche vacant. Populations of ichthyomyines appear to be fragmented in their distribution. That is to say, the current distribution of "ichthyomyines" probably reflects isolation owing to long-term climatic changes in the northern Neotropics. The systematic position of some of these species is poorly understood.

PHYLLOTINE RODENTS
Leaf-eared Mice and Their Allies

Systematics
The phyllotine rodents enjoy their greatest diversity farther to the south in Peru, Bolivia, Chile, and Argentina, extending to parts of Brazil and Paraguay. Typical phyllotine rodents have moderate to large pinnae. The molars are tetralophodont, and the

lower third molar is more than half the length of the lower second molar. In the northern Neotropics only one genus occurs, *Calomys*. The northern distribution of *Calomys* suggests that the current range is extremely fragmented and that they once enjoyed a broader distribution when climatic conditions may have been quite different. The phyllotines have been reviewed by Hershkovitz (1962), Pearson (1958), and Pearson and Patton (1976).

Genus *Calomys* Waterhouse, 1837
Vesper Mouse

Description
Head and body length ranges from 60 to 125 mm; the tail is generally less than half the head and body length. These small mice superficially resemble *Mus* (plate 19; fig. 13.6).

Distribution
Representatives of the gneus are distributed from northern South America south to Argentina.

Calomys hummelincki (Husson, 1960)

Description
The species was originally named *Baiomys hummelincki* (see Husson 1960), but Hershkovitz (1962) considered it a junior synonym of *C. laucha*. It is small and reminiscent of a house mouse. Head and body length ranges from 60 to 79 mm; the tail is 31 to 54 mm, the hind foot 13 to 16 mm, and the ear 10 to 13 mm. The skull is drawn in figure 13.6 and the

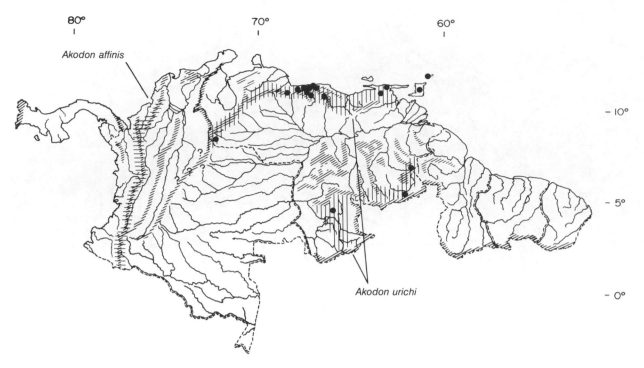

Map 13.47. Distribution of two species of *Akodon*.

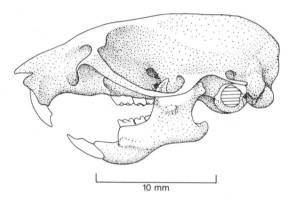

Figure 13.6. Skull of *Calomys laucha*.

dentition in figure 13.14. The dorsum is brown; the venter is paler (grayish white) but never sharply defined.

Range and Habitat

The species has been described from the islands of Aruba and Curaçao and is known from Venezuela, north of the Orinoco in scattered localities, including the llanos and the more semiarid regions around the mouth of Lake Maracaibo, where it extends to Colombia (map 13.46). *C. laucha* is closely related but is distributed in southeastern Brazil, Uruguay, northern Argentina, and Paraguay.

Natural History

This primarily herbivorous and nocturnal mouse is found in sandy grassland habitats. The gestation is

believed to be less than twenty-five days, and five young have been reported as the typical litter. The species *C. hummelincki* was first described in the Dutch West Indies from the islands of Curaçao and Aruba. It is unknown if the species in northern Venezuela was accidentally transported to the islands by man, nor is there any agreement on how it became distributed in the Dutch West Indies (see Hershkovitz 1962). That *Calomys* has a scattered distribution in the semiarid portions of Colombia and Venezuela suggests this is a natural, relictual distribution (Petter and Band 1981).

AKODONT RODENTS

Diagnosis

This group of rodents consists of three genera: *Akodon*, *Bolomys*, and *Microxus*. The molars have reduced cusps and exhibit a simplified infolding (see figs. 13.7 and 13.13). The lower third molar is very small, and this separates most akodonts from the phyllotines. Head and body length ranges from 75 to 140 mm and tail length from 50 to 100 mm. In the akodonts, the tail is usually shorter than the head and body length, and the ears and eyes are of modest size. The general body form suggests a terrestrial adaptation, and several genera are specialized for burrowing. The dorsal pelage varies from light brown to almost black. Cytogenetics of the akodonts are reviewed by Bianchi et al. (1971).

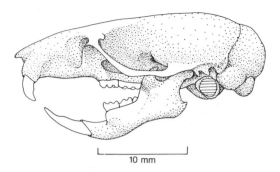

Figure 13.7. Skull of *Akodon urichi*.

Distribution
Akodont rodents are entirely South American and are distributed over the whole continent.

Genus *Akodon* Meyen, 1833
Laucha de Campo

Description
Head and body length ranges from 90 to 140 mm and the tail from 55 to 100 mm. The short tail and neck suggests the body form of the temperate zone voles (plate 19). The dorsal pelage is dark brown; the venter is grayish.

Distribution
The genus is found over most of South America to Tierra del Fuego.

Akodon affinis (J. A. Allen, 1912)

Description
In the type, an adult male, head and body measured 90 mm, tail 70 mm, and hind foot 20 mm (Allen 1912). The dorsum is dusky brown, and the ears are gray brown.

Range and Habitat
This mouse is found in the western cordillera of the Colombian Andes (map 13.47).

Akodon urichi J. A. Allen and Chapman, 1897

Description
Total length averages about 190 mm, of which the tail is 70 mm. A female from Venezuela measured TL 188; T 65; HF 24; E 15. Weight averages 39 g for males; females are lighter, averaging 29 g. The dorsum is dark brown, the venter gray to reddish brown. Note that this species may include three related forms (Reig, Olivo, and Kiblisky 1971).

Range and Habitat
The species is distributed in northern Venezuela and adjacent portions of Colombia. It recurs in southern Venezuela and from thence to central Brazil (map 13.47). It prefers moist habitats and is found

from 24 to 2,232 m elevation in northern Venezuela. Though it is predominantly associated with multi-stratal tropical evergreen forest, it tolerates second-growth forest and man-made clearings. In the Andes of Venezuela it can occur up to 3,020 m.

Natural History
This species of *Akodon* is active both diurnally and nocturnally and appears to be completely terrestrial. Its diet includes fruits, seeds, grass stems, fungi, and insects (O'Connell 1981). It appears to breed during the wet season.

Genus *Microxus* Thomas, 1909
Microxus bogotensis (Thomas, 1895)

Measurements

	Mean	Minimum	Maximum	N
TL	148.5	136	163	8
T	62.1	56	67	8
HF	19.4	19	20	8
E	14	13	15	8

Location: Colombia (FMNH)

Description
The eyes are exceedingly small. The fur is long and soft, with the dorsal color varying from yellowish brown to almost black and the venter only a little paler.
Note: Barros and Reig (1979) report chromosomal polymorphisms within this species

Range and Habitat
The species is distributed in the cordilleras of the Andes of Venezuela and Colombia. It is associated with cloud forest, and in northern Venezuela it occurs from 2,360 too 3,850 m elevation. It prefers grassy areas or the forest edge (Aagaard 1981) (map 13.48).

Podoxymys roraimae Anthony, 1929

Description
Head and body length averages 101 mm, the tail 94 mm, and the hind foot 23 mm. The claws are quite long, suggesting a semifossorial adaptation. The pelage is long and soft. The dorsum is gray and black, darker on the rump, and the underparts are slightly paler. The feet are brown. Anthony (1929) noted the strong relationship with the akodont rodents.

Range and Habitat
Known from Mount Roraima, Guyana (not mapped).

ZYGODONTOMYINE RODENTS
Reig (1980) removes *Zygodontomys* from the akodont rodents and lists it as *incertae sedis*. Clearly,

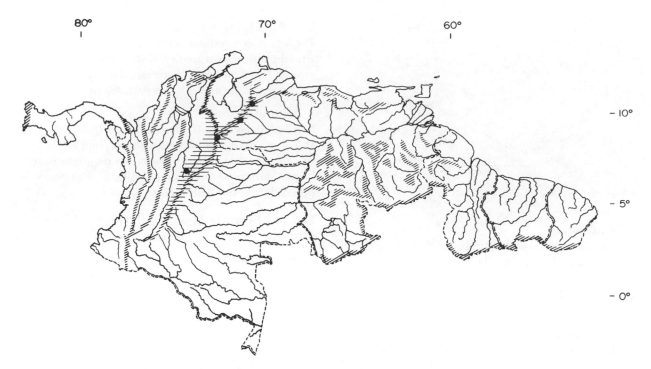

Map 13.48. Distribution of *Microxus bogotensis*.

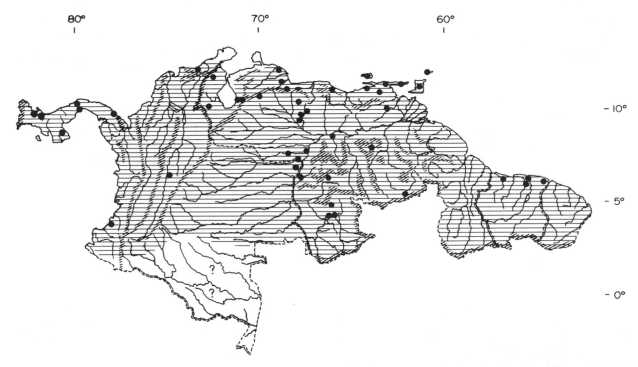

Map 13.49. Distribution of *Zygodontomys brevicauda*.

the affinities of this genus are not well understood. In their general morphology they follow the characters of the akodonts, but when composite characters are considered their affinities remain in doubt.

Genus *Zygodontomys* J. A. Allen, 1897
Cane Mouse, Piahuna

Description
Head and body length ranges from 95 to 155 mm; the tail is from 35 to 130 mm, markedly shorter than the head and body. The dorsum is generally yellowish brown interspersed with black-tipped hairs, and the venter is gray to buff (plate 19).

Distribution
The species of *Zygodontomys* occur from the southwest of Costa Rica through Panama to Peru and east-central Brazil.

Zygodontomys brevicauda (J. A. Allen and Chapman, 1893)

Measurements

	Mean	Minimum	Maximum	N
TL	235.5	225	244	4
T	95.8	93	100	4
HF	27.0	26	28	4
E	14.5	14	20	4
Wta	65.2	54	70	4

Location: Suriname (CMNH)

Description
The upperparts are gray to agouti brown, the underparts white to gray. The tail is scantily haired and tends to be bicolor, whitish below and gray brown above. The pelage of juveniles tends to be gray.

Range and Habitat
The species is distributed from southeastern Costa Rica through Panama and across most of northern South America, usually below 500 m elevation. It is broadly tolerant of a variety of habitat types, including croplands, pasture, and clearings in multistratal evergreen forest. It is the dominant rodent in the llanos of Colombia and Venezuela (map 13.49).

Natural History
This species is terrestrial and nocturnal. It opportunistically builds nests of grass at the base of trees in earthen cracks or in burrows. It eats a variety of foods including insects, seeds, and fruits. In the field the average litter size is 4.1 and the young attain sexual maturity before two months of age. Reproduction can occur throughout the year, but the timing is frequently controlled by rainfall (O'Connell 1981). In the laboratory, Aguilera (1985) found that gesta-

tion averaged twenty-five days. Litter size for primiparous females averaged 3.3, while the value for multiparous females was 5.0. The sex ratio at birth was 1:1, and in captivity females attain first estrus at 25.6 days of age and males reach sexual maturity at 42.3 days. At high densities home ranges of individuals can show considerable overlap (Vivas and Roca 1983).

Comment
The species name *Z. microtinus* is often applied to the populations north of the Orinoco in Venezuela (Aguilera 1985). The species *Z. brevicauda* may be a composite of four species (see Honacki, Kinman, and Koeppl 1982). *Zygodontomys reigi*, described by Tranier (1976) from Cayenne, French Guiana, may be conspecific with *Z. brevicauda*. *Z. borreroi* (Hernandez-Camacho 1957) has a restricted distribution in Santander, Colombia (see commentary in Honacki, Kinman, and Koeppl 1982).

SIGMODONT RODENTS

Diagnosis
The molars exhibit a pronounced folding that produces an S-shape (see fig. 13.13). These rodent are highly diversified in their adaptations. There are two general types: the marsh rats are adapted for aquatic life, and the cotton rats are more terrestrial.

Genus *Holochilus* Brandt, 1835
Holochilus sciureus Wagner, 1842
Marsh Rat, Rata Acuática

Measurements

	Mean	Minimum	Maximum	N
TL	333.5	300	360	7
T	162.1	148	174	7
HF	37.5	18	42	7
E	19.0	12	34	7
Wta	163.5	144	177	7

Location: Suriname (CMNH)

Description
Head and body length can range from 130 to 220 mm; the tail is nearly equal to the head and body. The toes are partially webbed, and the tail has a fringe of hair on the ventral side. The dorsum is buffy or tawny, usually mixed with black. The sides are paler, and the underparts vary from white to orange (plate 18). The genus *Holochilus* was revised by Hershkovitz (1955), but Gardner and Patton (1976) believe his *H. brasiliensis* is a composite. The northern form is usually referred to as *H. sciureus*, while the form south of the Amazon is called *H. brasiliensis*.

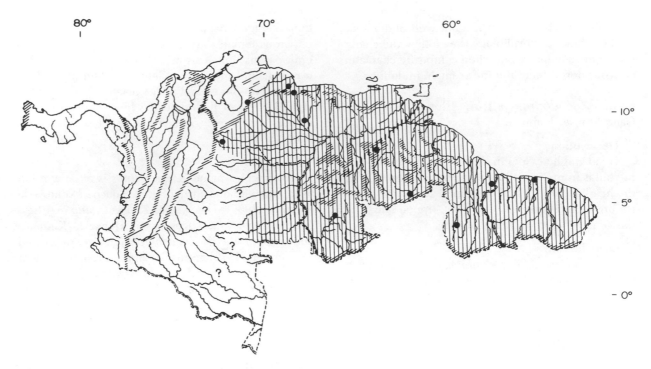

Map 13.50. Distribution of *Holochilus sciureus*.

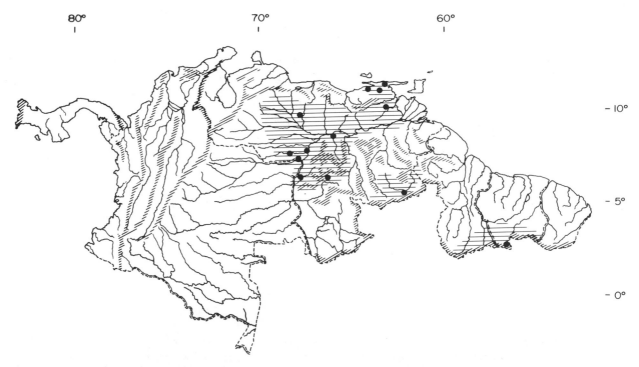

Map 13.51. Distribution of *Sigmodon alstoni*.

Range and Habitat

The group is distributed east of the Andes across northern South America and south to Brazil and Uruguay (map 13.50).

Natural History

These rats are semiaquatic. They may construct a nest from pieces of reeds and plant stems, creating a very large mound; nests have been reported with a diameter of 400 mm. In the sugarcane fields of Guyana nests may be constructed about 1 m off the ground. (Twigg 1962). Marsh rats feed on plant stems and seeds and may include snails in their diet. Litter size varies from four to six. This species can reach high densities (713 per hectare) in rice fields and can be a serious agricultural pest (Cartaya and Aguilera 1985).

Comment

Holochilus sciureus Wagner, 1842, was included in *H. brasiliensis* by Hershkovitz (1955). Considerable confusion surrounds the taxonomy of this genus, and at least six species have been named.

Genus *Sigmodon* Say and Ord, 1825
Cotton Rat, Rata Afelpada

Description

Head and body length ranges from 125 to 200 mm, tail from 75 to 125 mm, and weight from 70 to 200 g. The dorsum is grayish brown to blackish brown, and the underparts are gray to buff (plate 19).

Distribution

The genus occurs from the southeastern United States through Mexico and Central America to northern South America.

Sigmodon (Sigmomys) alstoni (Thomas, 1881)

Description

The measurements of a female from Suriname were TL 252; T 107; HF 30, E 21; Wt 74 g. The species is similar to *S. hispidus* (see below), though smaller, but it is characterized by a longitudinal groove in the upper incisors. For this reason it has frequently been placed in a separate genus, *Sigmomys*. The dorsum is agouti brown and the venter gray. The eye is ringed with pale yellow hairs.

Range and Habitat

The species is found in northern and eastern Venezuela east to the Guyanas and south to northern Brazil (map 13.51). In Venezuela this species is associated with low-elevation grasslands. Although it occurs in moist habitats, it is very tolerant of seasonally arid habitats and in general replaces *S. hispidus* in these areas.

Natural History

This species feeds primarily on forbs and green plants as well as seeds. It typically has a litter size of five and may show seasonal fluctuations in numbers depending on rainfall and the abundance of food. It appears to occupy a volelike niche (O'Connell 1982).

Sigmodon hispidus Say and Ord, 1825

Description

Total length ranges from 224 to 365 mm: T 81–166; HF 28–41; E 16–24. The dorsum is brown mixed with blackish hairs, giving a salt-and-pepper effect; the venter is pale gray. The tail is sparsely haired and tends to be a single color, usually some shade of brown.

Range and Habitat

The species is distributed from the southeastern United States through Mexico and across most of northern South America. In Venezuela it was found in moist habitats and at elevations between 200 and 1,580 m (map 13.52).

Natural History

Sigmodon hispidus has been well studied in the northern parts of its range. The gestation period there is twenty-seven days, and up to six young are born. The young are born furred, and their eyes open within twenty-four hours. They begin to disperse from the nest at less than ten days of age, and females commence breeding at about six weeks old. Nest construction has been described by Dawson (1973). The nest, built from grass, provides protection from light rainfall and also buffers the external temperature by approximately 5°C.

FAMILY MURIDAE
SUBFAMILY MURINAE
Old World Rats and Mice

Diagnosis

The dental formula is I 1/1, C 0/0, P 0/0, M 3/3, and typically the molar cusps are aligned in three longitudinal rows (fig. 13.14), thus differing from the varied sigmodontine forms. These rodents are superficially very similar to cricetine rodents in body form and proportions.

Distribution

These rodents are native to the Old World, but species of the subfamily Murinae have been widely introduced by man around the world. Three species occur near human settlements in the area covered by this volume: *Rattus rattus*, *Rattus norvegicus*, and *Mus musculus*.

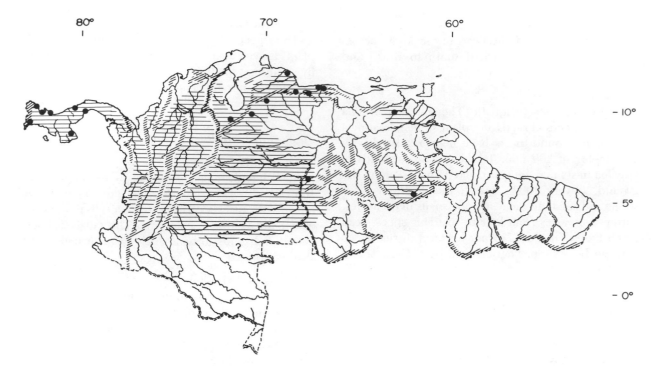

Map 13.52. Distribution of *Sigmodon hispidus*.

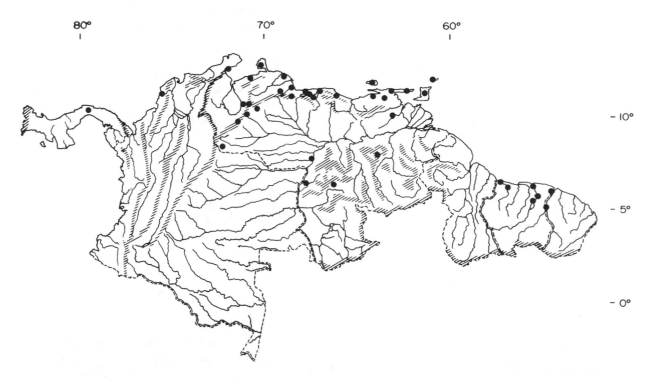

Map 13.54. Distribution of *Rattus norvegicus*.

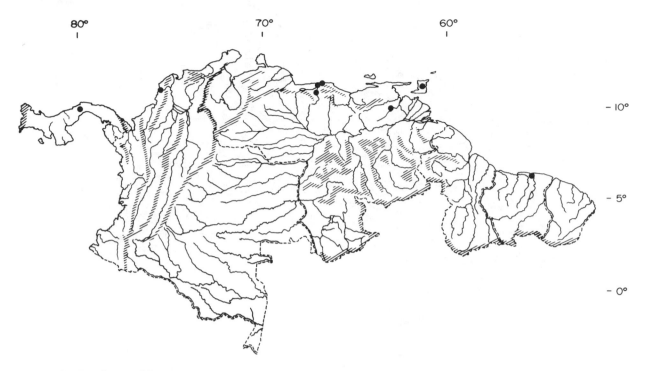

Map 13.53. Distribution of *Rattus rattus*.

Genus *Rattus* Fischer, 1803
Rattus rattus (Linnaeus, 1758)
Black Rat

Description
Total length ranges from 327 to 430 mm: the tail is 160 to 220 mm, the hind foot 35.5 mm. A large male may weigh 200 g. This species may be distinguished from *R. norvegicus* by its longer tail. The dorsum is grayish black to brown, and the venter may be gray or almost black.

Range and Habitat
As a species introduced by man, this rat tends to be associated with human dwellings. It is highly arboreal and may extend into forest regions far from points of original introduction. Generally it occurs near coastal settlements (map 13.53).

Natural History
This rat is highly adaptable and an excellent climber. It is omnivorous but readily eats fruits and nuts when occupying natural forests. Three to five young are produced after a twenty-three day gestation period. The behavior of black rats has been described by Ewer (1971).

Rattus norvegicus (Berkenhout, 1769)
Brown Rat

Description
Total length ranges from 320 to 480 mm; the tail is 153 to 218 mm, the hind foot 37 to 44 mm. The tail is shorter than the head and body, and this rat is more robust than *R. rattus*. The dorsal pelage is brown interspersed with black hairs, and the underparts are pale gray.

Range and Habitat
This terrestrial species is associated with large urban centers. It does not penetrate into undisturbed habitats (map 13.54).

Natural History
This species is much more terrestrial than *R. rattus*. It tunnels actively, and its presence is usually first detected when its burrows are noticed. Up to seven young are born after twenty-three days' gestation. The brown rat is an omnivore and is often a serious pest when it exploits stored cereal grains. The ecology and behavior of the brown rat have been summarized by Calhoun (1962).

Genus *Mus* Linnaeus, 1766
Mus musculus Linnaeus, 1766
House Mouse

Description
Total length ranges from 148 to 205 mm; the tail is 69 to 85 mm; the hind foot 16 to 20 mm. The dorsum is gray brown to brown; the venter is lighter and often buffy.

Range and Habitat
This species is closely associated with human dwellings and farms. It has been widely introduced in coastal cities (map 13.55).

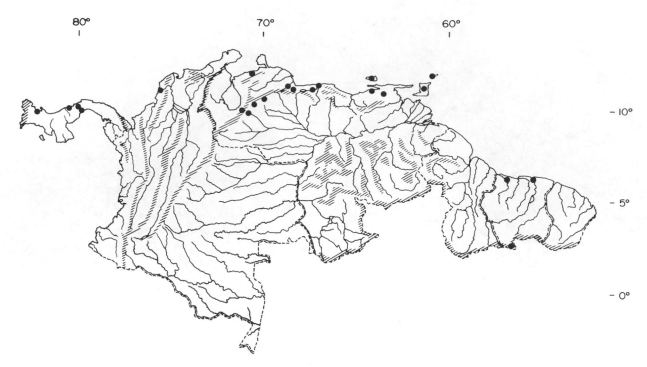

Map 13.55. Distribution of *Mus musculus*.

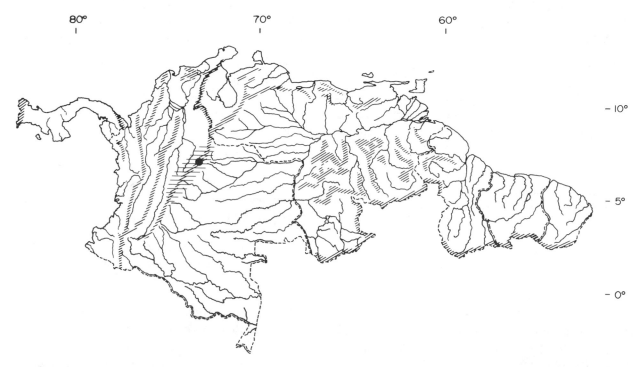

Map 13.56. Distribution of *Echinoprocta rufescens*.

Natural History

This mouse is omnivorous and mainly terrestrial. It is almost always associated with man as a commensal. It climbs well but does not nest arboreally. Five to seven young are born after twenty to twenty-one days' gestation. The species' natural history in Australia has been summarized by Newsome (1971) and that in England by Chitty and Southern (1954).

HYSTRICOGNATH RODENTS

Systematics

As indicated above under Rodentia: History and Classification, the earliest rodents on the continent of South America are hystricognathous. The classification employed here follows Woods (1982), who groups the South American hystricognath rodents into four superfamilies. Three superfamilies occur within the range covered by this volume: the Erethizontoidea, the Cavioidea, and the Octodontoidea (Woods 1982).

Natural History

The hystricognath rodents exhibit a number of unique reproductive characters. Litter size tends to be small (one to three), and gestation is long. The young are precocial. The biology of hystricognath rodents is admirably summarized in the volume edited by Rowlands and Weir (1974). Reproduction has been summarized by Weir (1974) and by Kleiman, Eisenberg, and Maliniak (1979). Behavior patterns were reviewed by Kleiman (1974), Eisenberg (1974), and Eisenberg and Kleiman (1977), and distribution and ecology by Mares and Ojeda (1982).

SUPERFAMILY ERETHIZONTOIDEA
FAMILY ERETHIZONTIDAE

Porcupines, Puercos Espinosos

Diagnosis

The dental formula is I 1/1, C 0/0, P 1/1, M 3/3. The genera of this family are adapted for an arboreal life; the thumb is replaced by a broad movable pad, and the sole of the hind foot is wide. The dorsal pelage is modified so that many hairs have become spinelike, and the tips of the spines have minute barbs. The spines are easily detached and embed themselves in potential predators.

Distribution

Species are distributed from northern Canada and Alaska south to northern Argentina. The northern species belong to the genus *Erethizon*. The species covered in the range by this volume are in the genera *Echinoprocta*, *Sphiggurus*, and *Coendou*.

Genus *Echinoprocta* Gray, 1865
Echinoprocta rufescens (Gray, 1865)
Short-tailed Porcupine

Description

Three specimens of this small porcupine from Colombia measured HB 309–70; T 105–50; HF 60–67; E 17 (FMNH, MVZ). The dorsal pelage ranges from medium brown to almost black and is very spiny. The base of the spined is yellow, but the distal 20 mm are pigmented. The venter is pale brown and spinescent. There is usually some white on the face (plate 21).

Range and Habitat

The species is known from the montane areas of the eastern cordillera of the Andes at elevations of 800 to 2,000 m (map 13.56).

Genus *Sphiggurus* F. Cuvier, 1825
Long-haired Prehensile-tailed Porcupine

Description

The dental formula I 1/1, C 0/0, P 1/1, M 3/3. The dorsum is covered with long hairs that cover and conceal the spines in the midportion, distinguishing *Sphiggurus* from *Coendou*. The long hairs are white tipped, the middle part being blackish brown while the basal third is again paler. The spines on the dorsum are rather short, about 25 mm long, and are yellowish with brown tips. The tail is fully prehensile and hairless.

Distribution

If one considers *C. mexicanus* to be part of *Sphiggurus*, then the genus has a distribution from San Luis Potosí, Mexico, south through the Isthmus across much of northern South America and into Amazonian Brazil.

Natural History

Little work has been done with this genus. It is similar to *Coendou* in its behavior.

Comment

Sphiggurus has been proposed as a generic category for two northern species formerly considered within the genus *Coendou* (Husson 1978). They include *S. insidiosus* and *S. vestitus*, *C. mexicanus* exhibits the long dorsal hairs that are diagnostic for *Sphiggurus*. There is no reason not to include *C. mexicanus* within the proposed new genus.

Sphiggurus insidiosus (Lichtenstein, 1818)

Description

Head and body length ranges from 363 to 380 mm; the tail is 370 to 375 mm, the hind foot 62 to 75 mm. The long black tail is prehensile and sparsely haired

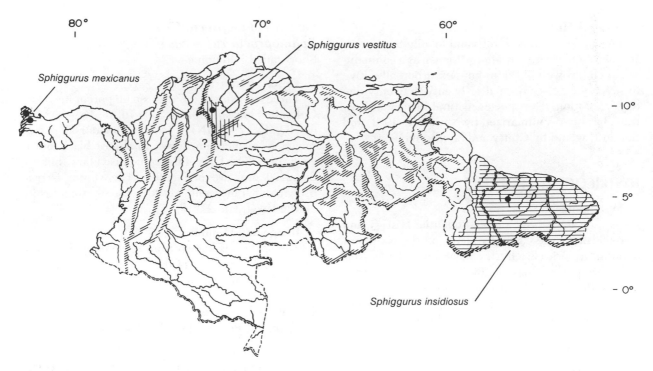

Map 13.57. Distribution of three species of *Sphiggurus*.

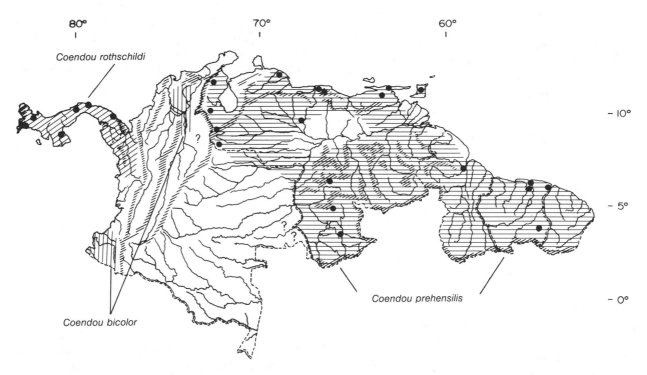

Map 13.58. Distribution of three species of *Coendou*.

390

to the tip. The bases of the dorsal hairs are black, and the tips are yellow. The venter is sparsely covered with a mixture of black and yellow hairs. The spines are hidden by the long dorsal hairs.

Range and Habitat
The species ranges from Amazonian Brazil north to Suriname and perhaps Guyana (map 13.57).

Natural History
Husson (1978) reports these porcupines feed on fruit, ant pupae, cultivated vegetables, and roots.

Sphiggurus vestitus (Thomas, 1899)

Description
This description includes *C. pruinosus*. The spines are covered by dorsal hairs. This species is similar in appearance and size to *S. insidiosus*, which it replaces to the west.

Range and Habitat
The species occurs from extreme southwestern Venezuela south of Lake Maracaibo across northern Colombia. It is strongly associated with multistratal tropical evergreen forest (map 13.57).

Sphiggurus mexicanus (Kerr, 1792)

Description
This porcupine is similar in size to *S. insidiosus*. The dorsum is blackish to dark brown and the underparts are grayish. The spines are hidden by the dorsal hairs.

Range and Habitat
The species is distributed from San Luis Potosí, Mexico, south to western Panama. Hall (1981) named this species *Coendou mexicanus* (map 13.57).

Genus *Coendou* Lacépède, 1799
Prehensile-tailed Porcupine, Puerco Espin

Description
The dental formula is I 1/1, C 0/0, P 1/1, M 3/3. Head and body length ranges from 300 to 600 mm, and the tail is 330 to 450 mm. Adults may weigh over 4 kg. The naked tail is fully prehensile. The dorsal pelage is distinctive; the long dorsal hairs do not cover the prominent spines. (See fig. 13.8.)

Distribution
The genus *Coendou* as defined in this volume is distributed from eastern Panama south to Bolivia and east across Brazil to the Guyanas.

Coendou bicolor (Tschudi, 1844)

Description
Head and body length averages about 543 mm,

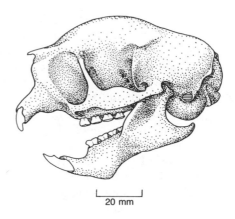

Figure 13.8. Skull of *Coendou prehensilis*.

and the tail is 481 mm. The prominent spines are pale yellow at the base and black tipped, and those in middorsum are very dark. The tail is fully prehensile; spines extend on the dorsal side for most of its length, while the distal 50 mm is haired. This species is darker than *C. prehensilis*. The venter may be light to dark brown.

Range and Habitat
The species is distributed from northern Colombia to western Ecuador and thence to Amazonian Peru and Bolivia and adjacent parts of Brazil (map 13.58)

Coendou prehensilis (Linnaeus, 1758)

Description
This species is similar in size and appearance to *C. bicolor*, but the middorsum is not as dark. Adults attain a weight of approximately 5 kg (plate 21). Infants are born with long reddish dorsal hairs that obscure the spines.

Range and Habitat
This porcupine ranges from eastern Venezuela across the Guyanas, southward to central eastern Brazil and thence to Bolivia (map 13.58). In northern Venezuela it can occur from sea level to 1,524 m but generally is found below 500 m elevation. Although tolerant of a range of vegetation types, it prefers multistratal tropical evergreen forest. It may occur in dry deciduous forests in the vicinity of streams.

Natural History
The best-studied species of the genus *Coendou*, *C. prehensilis* is nocturnal and rests in trees during the day, often not sheltering in cavities but remaining inconspicuous high in a tree crown. It is solitary in its habits. The home range may be quite large depending on the availability of suitable feeding

trees. In northern Venezuela in the llanos habitat Montgomery and Lubin (1978) found that animals could range over an area of 15 to 20 ha. Nine vocalization types are produced, including a long moan employed as a contact call between isolated individuals (Roberts, Brand, and Maliniak 1985).

The animals feed nocturnally or in the late afternoon on blossoms, fruits, seeds, and nuts. They include some young leaves in their diet (Charles-Dominique et al. 1981). The single young, born after an extended gestation period of 195–210 days, may weigh up to 390 g. The spines are soft at birth, and the young possess long orange guard hairs, distinctly different from those of the adult. The young is left in a sheltered place and nursed at least once a day until seventy days of age, though it first eats solids at fourteen days (Roberts, Brand, and Maliniak 1985).

Coendou rothschildi Thomas, 1902

Description

C. rothschildi is similar to *C. bicolor* and darker than *C. prehensilis*. Hall (1981) considers it a subspecies of *C. bicolor*.

Range and Habitat

The species is known from eastern Panama and probably extends to Colombia (map 13.58).

FAMILY DINOMYIDAE
Genus *Dinomys* Peters, 1873
Dinomys branickii Peters, 1873
Branick's Giant Rat, Pacarana, Guagua Loba

Systematics

This family of rodents shows great species diversity in the Miocene, but only a single living representative survives. It is considered to be within the superfamily Erethizontoidea in accordance with Grand and Eisenberg (1982).

Diagnosis

The dental formula is I 1/1, C 0/0, P 1/1, M 3/3. The teeth are high crowned. Head and body length ranges from 730 to 790 mm, the tail averages 200 mm, and adult weight may exceed 15 kg. The head is massive, and the limbs and ears are short. *D. branickii* has long, powerful claws and a short, stout tail that is haired and not prehensile. The dorsal pelage is brown to black, and on each side are a series of white spots arranged in lines (plate 21).

Distribution

The single living species of this family, *D. branickii* is currently found in isolated localities in Amazonian Brazil, Peru, Ecuador, Venezuela, and Colombia (map 13.59) at up to 1,000 m elevation.

Natural History

These large, slow-moving rodents have some climbing ability, especially prominent in young animals. Their behavior in captivity has been summarized by Meritt (1984). They tend to be active at night. Their diet includes fruits, some leaves, and herbaceous vegetation. They shelter in caves or holes at the bases of trees. They produce a wide variety of vocalizations, and males seeking mates produce a complicated, intricate series of calls (Eisenberg 1974). Gestation lies between 222 and 280 days; generally only two young are born. This species is hunted for food and may be on the verge of extinction (Collins and Eisenberg 1972).

SUPERFAMILY CAVIOIDEA
FAMILY CAVIIDAE

Diagnosis

The dental formula is I 1/1, C 0/0, P 1/1, M 3/3. The cheek teeth are ever growing and have two enamel prisms with sharp folds and angular projections. The lower mandible exhibits the least hystricognathous condition of the extant hystricognath rodents. The lateral process of the supraoccipital is lacking. The body is stout and the tail extremely reduced (fig. 13.9).

Distribution

Disjunct distributions characterize the northern distribution, but the genus occurs over most of South America except the easternmost parts of Brazil and the extremely arid portions of southern Peru and Chile.

Genus *Cavia* Pallas, 1766
Cavia aperea Erxleben, 1777
Cavy, Guinea Pig, Conejo de las Indias

Description

For this description *C. porcellus* is included in *C. aperea*. Head and body length ranges from 225 to

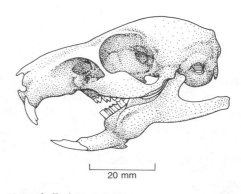

Figure 13.9. Skull of *Cavia aperea*.

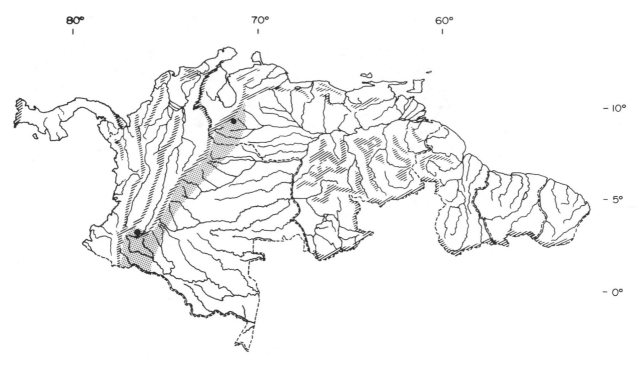

Map 13.59. Distribution of *Dinomys branickii*.

355 mm; the tail is virtually absent. Adult weight ranges from approximately 400 to 700 g. Three specimens from Suriname showed the following range in measurements: TL 160–265; T 7–8; HF 36–45; E 22. There are only three toes on the hind feet, while the forefeet have four. All digits bear sharp claws. The dorsal pelage is agouti brown, and the venter tends to contrast strongly, being lighter (plate 21).

Range and Habitat

The species occurs in isolated colonies in northern South America (see Ojasti 1964). It is more broadly distributed to the south in Peru, Chile, and Argentina. The exact origin of the northern colonies is obscure, but some may have been transported to their present positions by man (map 13.60).

Natural History

Cavies either construct resting places or elaborate on burrows dug by other forms. Generally they are more active at night, but sometimes activity periods extend to early morning and late afternoon. Cavies feed on herbaceous vegetation, and where they are numerous, small trails may be formed in the grasslands (Rood 1972).

The young are born after sixty-one days' gestation; the litter size averages three. At birth the young have hair, and their eyes are open. They nurse for about twenty-one days but begin to take solid food almost immediately after birth. Sexual maturity in females occurs at approximately seventy days of age (Weir 1974).

The wild cavy was domesticated by Amerindians some three thousand years ago. It was widely transplanted from its original point of domestication, which was believed to be Peru (Wing 1986).

FAMILY HYDROCHAERIDAE
Genus *Hydrochaeris* Brunnich, 1772
Hydrochaeris hydrochaeris (Linnaeus, 1766)
Capybara, Chigüiro

Diagnosis

The dental formula is I 1/1, C 0/0, P 1/1, M 3/3. The molars are vastly modified, consisting of transverse lamellae joined with cementum, and the third

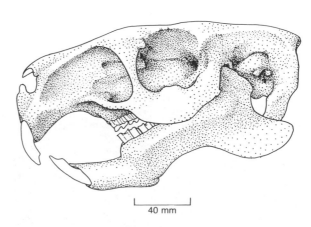

Figure 13.10. Skull of *Hydrochaeris hydrochaeris*.

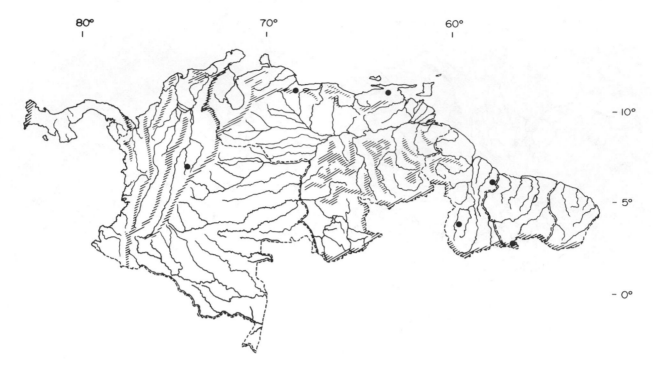

Map 13.60. Distribution of *Cavia aperea*.

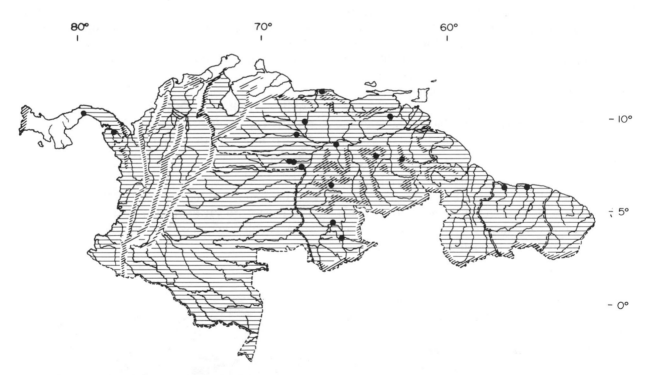

Map 13.61. Distribution of *Hydrochaeris hydrochaeris*.

molar is longer than the other three, so that the teeth are distinctive (fig. 13.10). This is the largest living rodent and cannot be confused with any other form. Head and body length ranges from 1 to 1.3 m, height at the shoulder averages 0.5 m, and weight may reach 50 kg. With its massive skull and vestigial tail, the capybara gives the appearance of an enormous guinea pig. There are four toes on the forefeet and three on the hind feet, and the digits are semi-webbed. In the sexually mature male a bare raised glandular area on top of the snout is conspicuous. The upperparts are reddish brown and the underparts are lighter, usually some shade of yellow brown (plate 21).

Distribution

The species ranges from Panama to the east of the Andes south to Uruguay and northern Argentina. The capybara is associated with moist habitats and is never found far from lagoons, rivers, or ponds (map 13.61).

Natural History

The capybara may be nocturnal where it is severely persecuted, but in areas relatively free from human predation it frequently extends its activity into the early morning and late afternoon. The animals live in groups of up to twenty. They graze on herbaceous vegetation in low, marshy areas (Ojasti 1973).

On perceiving danger, they generally give a warning cough and flee to water. They are excellent swimmers. Reproduction may in part be controlled by rainfall, and they exhibit some reproductive seasonality where rainfall is also sharply seasonal. Up to seven young are born after a 150-day gestation period, fully furred with eyes open. Several females with like-aged young may band together. In waters infested with caimans, the young may cluster in a tight group with adult females swimming fore and aft.

During the reproductive season males are extremely hostile toward one another. A single male will attempt to mate with as many females as he can defend. Social organization is quite flexible, and social groupings are strongly influenced by the extent and availability of free water. Crowding and breakdown of formal social structure may occur when water is limited (Schaller and Crawshaw 1981; MacDonald 1981). The population biology of the species has been intensively studied by Ojasti (1973, 1978).

FAMILY AGOUTIDAE
Genus *Agouti* (= *Cuniculus*) Lacépède, 1799

Diagnosis

The dental formula is I 1/1, C 0/0, P 1/1, M 3/3. Head and body length averages from 600 to 795 mm,

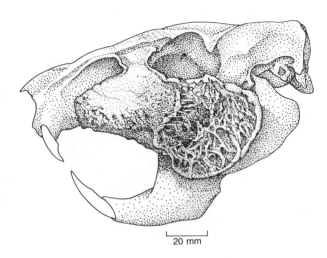

Figure 13.11. Skull of *Agouti paca*.

and the vestigial tail barely exceeds 20 mm; the hind foot averages 188 mm and the ear 45 mm. Weight can reach 10 kg. The adult male is about 15% larger than the adult female. The zygomatic arch is massive and distinctively sculpted (fig. 13.11); a small slit on its underside is visible externally, opening to a blind cavity on each side of the head. There are four toes on the forefeet and four on the hind feet. The dorsal pelage is brown to almost black, with four lines of white dots on each side of the body, and the venter tends to be white to buff (plate 21).

Distribution

Members of the genus are distributed from southern Mexico to northern Argentina in suitable lowland habitats.

Agouti paca (Linnaeus, 1766)
Paca, Lapa, Tepeizcuinte

Description
See the account for the genus.

Range and Habitat

The species ranges from southern Mexico to northern Argentina in lowland habitats (map 13.62). It is usually found in moist habitats but may extend into seasonally arid areas near permanent streams or watercourses.

Natural History

This large rodent tends to be nocturnal in its habits. It feeds upon fruits, nuts, and seeds, and at some times of the year it browses on herbaceous vegetation. It constructs its own burrows or modifies armadillo burrows for shelter, plugging the entrance with leaves during the day. The gestation period is 115 days, and the female usually has a single young or on rare occasions twins (Kleiman, Eisenberg, and Maliniak 1979; Smythe 1970).

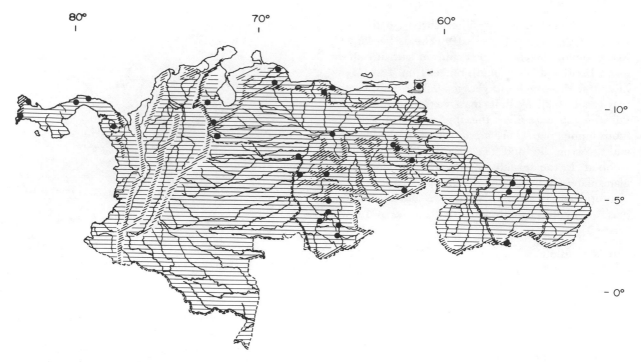

Map 13.62. Distribution of *Agouti paca.*

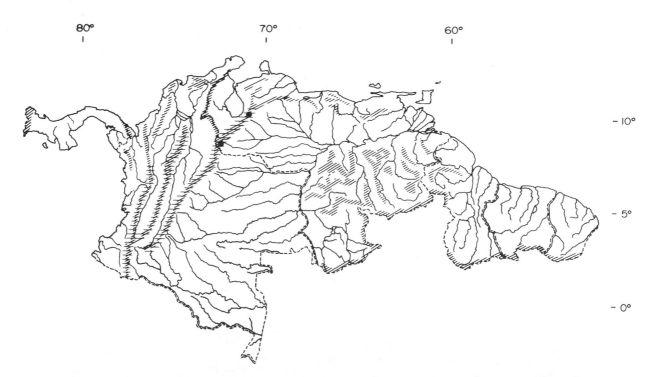

Map 13.63. Distribution of *Agouti taczanowskii.*

Pacas tend to move singly, though a mother may be accompanied by her young. During the early phase of the newborn's development it usually shelters in a burrow, and the mother returns to nurse. Although born fully furred with the eyes open, the young is quite vulnerable to predation until about one month of age (Lander 1974; Matamoros de Rodriguez 1980).

An adult female occupies an exclusive home range except for the male and subadult offspring. The male defends a territory against other adult males, using a growling roar in fighting. This animal is higly prized as game, but it has a low reproductive rate, in marked contrast to the capybara (Smythe, Glanz, and Leigh 1982; Marcus 1983; Collett 1981).

Agouti taczanowskii (Stolzmann, 1865)
Mountain Paca

Description
This species is similar in size and color to other members of the genus, but the mountain paca has a dense undercoat.

Range and Habitat
This paca is found in the high cloud forest of the Andean cordilleras of Venezuela, Colombia, and south to Ecuador (map 13.63). Aagaard (1982) lists its maximum abundance at between 2,000 and 3,050 m.

FAMILY DASYPROCTIDAE

Diagnosis
The dental formula is I 1/1, C 0/0, P 1/1, M 3/3 (fig. 13.12). Molars are high crowned and semirooted and have five crests, a character shared with the paca. These large cursorial rodents are similar to the pacas but lack the spotted pattern and are more slender in build (plate 21).

Distribution
The family ranges from Veracruz in Mexico south to northern Argentina and Uruguay in appropriate moist habitats.

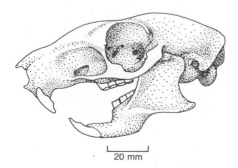

20 mm

Figure 13.12. Skull of *Dasyprocta punctata*.

Genus *Dasyprocta* Illiger, 1811
Agouti, Agutí, Picure

Description
Head and body length ranges from 415 to 620 mm; the tail is 10 to 25 mm, the hind foot 125 to 140 mm, and the ear about 35 mm. The weight may reach 4 kg. The forefoot has four functional digits and a vestigial thumb, and the hind foot has three digits. The upperparts vary from reddish brown to almost black, while the underparts are white, yellow, or gray. In the dark brown form the rump may be reddish (plate 21).

Distribution
The genus ranges from Veracruz, Mexico, south to Uruguay and northern Argentina at low elevations in both dry deciduous and evergreen forests.

Natural History
The best-studied species of *Dasyprocta* is *D. punctata* (Smythe 1970, 1978). The agouti is diurnal, but where it is extremely persecuted by man it may extend its activity to the hours of darkness. Though adults rarely use burrows, the young are generally born near a crevice or some burrow dug by another animal, to which they retreat for shelter except when the female calls them for nursing. Gestation is about 120 days, and litter size is one or two. The young, like all hystricognath rodents, are born furred and with their eyes open (Weir 1974; Smyth 1978).

The agouti is a significant harvester of seeds and fruits. Within its home range it accumulates stores for times of fruit scarcity, caching food in small pits that it excavates and then covers. Adult females defend parts of their home ranges when food is scarce. Young may be tolerated within the parent's home range. An adult male defends as large an area as he can against other adult males, thereby ensuring the paternity of the young in his range. At high densities the ranges of a male and a female may be coincident, and thus they give the appearance of living in pairs (Smythe 1978). During times of fruit scarcity the subadults may lose weight. On Barro Colorado Island they appear to be food limited (Smythe, Glanz, and Leigh 1982).

When startled, agoutis produce a barklike warning call to alert family members within the home range to a potential predator, and the long hairs of the rump are erected, increasing the apparent size of the animal. On encountering boas (*Constrictor constrictor*) they sit at a distance and drum with a hind foot, attracting other family members, who join them in foot drumming until the snake moves off (Kleiman 1974; Eisenberg 1974; Smythe 1978).

Comment
Many species are assigned to the genus *Dasy-*

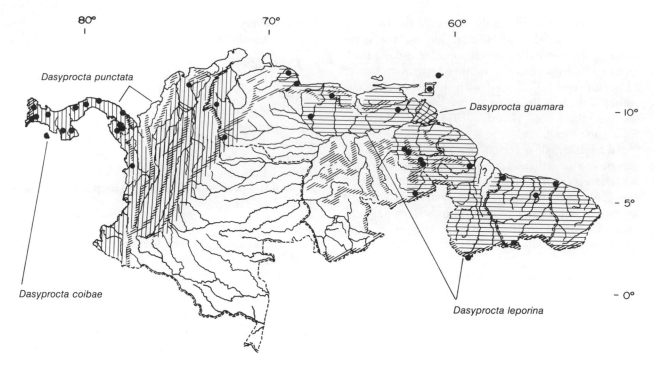

Map 13.64. Distribution of four species of *Dasyprocta*.

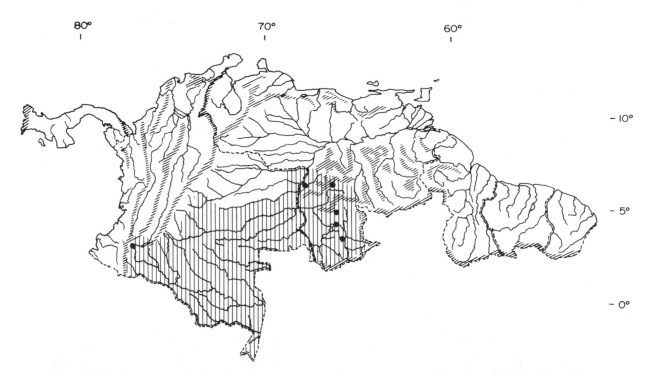

Map 13.65. Distribution of *Dasyprocta fuliginosa*.

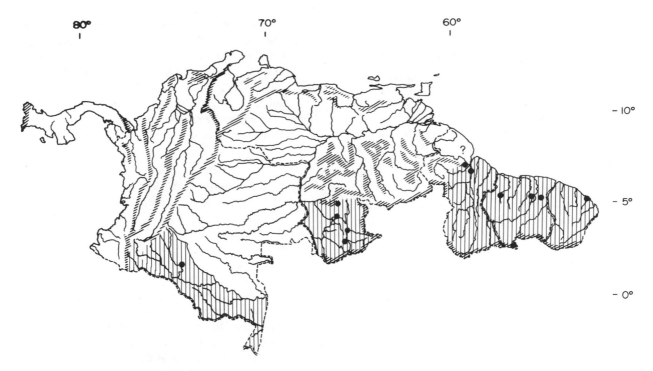

Map 13.66. Distribution of *Myoprocta acouchy*.

procta. In general, none are sympatric. Since the natural history of all forms is similar, in the species accounts I will merely list them and give their distribution.

Dasyprocta coibae Thomas, 1902
This speices is endemic to Coiba Island, Panama. It is smaller than the mainland form (map 13.64).

Dasyprocta fuliginosa Wagler, 1842
The species is distributed in southern Venezuela and adjacent portions of Colombia, southward to west-central Brazil (map 13.65).

Dasyprocta guamara Ojasti, 1972
This distinctive form is known only from the delta of the Orinoco (Ojasti 1972) (map 13.64).

Dasyprocta leporina (Linnaeus, 1758)
In this account, *D. agouti* and *D. cristata* are included within *D. leporina* (Husson 1978). This species is distributed in northern Venezuela, east to the Guyanas, and south into northeastern Brazil (map 13.64).

Dasyprocta punctata Gray 1842
This description includes *D. variegata*. The species is distributed to the west of the easternmost Andean cordillera from southern Mexico through Panama, south to Ecuador. Its status in the Maracaibo basin is unclear, since Handley (1976) reports both *D. fuliginosa* and *D. punctata* from this region (map 13.64).

Comment
Further study should clarify the taxonomic status of the species. At present there is considerable confusion and synonymy in the literature (see Honacki, Kinman, and Koeppl 1982).

Genus *Myoprocta* Thomas, 1903
Acouchi, Cutias de Rabo

Description
Members of this genus are smaller than the agoutis. Head and body length ranges from 250 to 380 mm. The tail is relatively longer than in *Dasyprocta*, some 45 to 70 mm; the hind foot averages 100 mm and the ear 28 mm. Weight ranges from 880 to 1,200 g. The dorsal pelage varies geographically from olive green to reddish brown (plate 21). The venter generally contrasts with the dorsum, being some shade of orange to cream, and the tip of the tail may bear a few white hairs, enhancing its conspicuousness.

Distribution
The genus is found in lowland evergreen forest in the Amazonian portion of Colombia, Ecuador, Peru, Brazil, Venezuela, and the Guyanas.

Comment
The green color phase is often referred to as *M. pratti* and the red phase as *M. acouchy*.

Myoprocta acouchy (Erxleben, 1777)

Description

M. pratti and *M. exilis* are included in *M. acouchy* (Cabrera 1960). The description is the same as given for the genus.

Range and Habitat

The range is given under genus. This lowland species is strongly tied to multistratal tropical evergreen forest (map 13.66).

Natural History

Similar in its habits to the agouti, the acouchi may be considered a small version of its larger cousin. In common with the agouti, the acouchi scatter hoards its food and is a significant consumer of fruits and nuts in the tropical rain forest. Entirely terrestrial, like the agouti it is cursorial and depends on its keen eyesight and hearing for avoiding predators (Kleiman 1974).

Gestation is ninety-nine days, and litter size averages two. The young are precocial and will shelter in a burrow or natural crevice until they are several weeks old. The mother returns to their hiding place to nurse them (Kleiman 1970, 1972).

The courtship between the male and female is highly ritualized and shares many elements with that described for *Dasyprocta* and *Agouti*. When aroused by a slow predator such as a snake, the acouchi will face the predator and drum with its hind foot, which may warn other acouchis in the vicinity. This same behavior is shown by *Dasyprocta* (Kleiman 1971, 1974).

SUPERFAMILY OCTODONTOIDEA
FAMILY ECHIMYIDAE

Spiny Rats, Ratas Espinosas, Casiraguas

Diagnosis

The dental formula is I 1/1, C 0/0, P 1/1, M 3/3. The decidous premolars are apparently retained throughout life. The tail may be longer than the head and body. As the common name implies, many species of these rat-sized hystricognath rodents, though not all, have spiny hairs interspersed in their dorsal pelage. The dorsum is highly variable in color but is usually some shade of brown to almost black; the venter often contrasts sharply, being white or yellowish. A key to the genera is given in table 13.4.

Distribution

Echimyid rodents are distributed from Honduras south through the Isthmus over much of South America to northern Argentina.

Natural History

The best-studied echimyid rodents are members of the genus *Proechimys*. These terrestrial forms use hollow logs, natural crevices, or burrows they construct to cache food, bear their young, and seek refuge. Other spiny rat genera are strongly arboreal, including *Dactylomys*, *Echimys*, *Diplomys*, *Isothrix*, *Mesomys*, and *Thrinacodus*. The arboreal genera have not been studied in as great detail as have the terrestrial forms. *Proechimys* may be one of the most abundant rodents in a lowland tropical rain forest.

Genus *Proechimys* J. A. Allen, 1899
Spiny Rat, Macangu, Casiragua, Rata Espinosa

Taxonomy

This genus has speciated widely, and some speciation events are evidently quite recent. Many of the species are superficially very similar in appearance, and their distinctiveness can be determined only by careful measurement of adult specimens or by determining the karyotype. All the species primarily inhabit lowland multistratal tropical evergreen forest, although some forms may range up to 1,000 m elevation.

The evolution of species within many South American mammals has apparently been accelerated by periodic isolations resulting from alternating continentwide wet and dry cycles deriving from variations in the extent of polar ice caps (see chap. 1). A combination of geographic isolation and genetic drift can result in populations that are incapable of breeding and producing viable offspring when they subsequently come in contact. Such allopatric speciation events have been well documented (Mayr 1963). As a result of enormous efforts by O. Reig and his colleagues, the genus *Proechimys* in Venezuela has been studied from the standpoint of karyotypic variations. Six such variants were identified in Venezuela alone (Reig and Useche 1976; Reig et al. 1980). Whether these represent true species remains to be determined.

Gardner and Emmons (1984), building on Patton and Gardner (1972) and Martin (1970), have provided us with a general review of this vexing genus and have also evaluated morphological characters, including the structure of the tympanic bulla. The septa dividing the bulla into compartments range from a few simple ones to many, apparently varying with how fossorial the species is. Emmons (1982) confirms the contention of Hershkovitz (1948) that different species of *Proechimys* apparently can coexist in the southern parts of South America, but such sympatry is difficult to document for most of the area covered by this volume (Gardner and Emmons 1984). Guillotin and Ponge (1984) have developed evidence that two species exhibit sympatry in French Guiana.

Table 13.4   Key to the Adult Echimyid Rodents of Northern South America

1 Dorsal pelage spinescent	2
1' Dorsal pelage soft, with no spines	5
2 Little hair on dorsum; spines stiff and prominent	3
2' Hair on dorsum predominates; some hairs flattened and stiff but differs strongly from 2 . . . . *Proechimys* complex	
3 Tail haired and penicillate at tip; tail almost equal to or greater than head and body length . . . . *Mesomys*	
3' Tail sparsely haired or naked; tail noticeably shorter than head and body length	4
4 Tail sparsely haired and usually penicillate . . . . *Echimys (Makalata)*	
4' Tail almost naked, not penicillate . . . . *Hoplomys*	
5 Tail equal to or less than head and body length	7
5' Tail greatly exceeds head and body length	6
6 Head and body length less than 280 mm . . . . *Thrinacodus*	
6' Head and body length greater than 300 mm . . . . *Dactylomys*	
7 Tail well haired . . . . *Isothrix*	
7' Tail sparsely haired . . . . *Diplomys*	

The basic contention of Gardner and Emmons (1984) is that the genus *Proechimys* falls into four groups of species. The species *P. semispinosus* is the most geographically defined, extending from Nicaragua through the Isthmus and then west of the Andes to Ecuador (Gardner 1983). However, isolated populations in central Colombia show affinities with the *semispinosus* group. Gardner and Emmons further define a group of karyotypic variants in northeastern Colombia across northern Venezuela, which they refer to as the *guairae* group. This group, studied by Reig et al. (1980), is divisible into at least six variants based on karyotype alone. As with the *semispinosus* group, outlying populations are found south of the Orinoco in south-central Venezuela. The third group is referred to as the *brevicauda* complex. It is primarily distributed in the Amazon basin and the Guyanas but has a few pockets of apparent relatives in northeastern Colombia. Again this suggests range expansion and contraction paralleling global climatic changes (Haffer 1974). The fourth group of species is predictably isolated in southeastern Brazil, a site of extensive endemism. This "*Trinomys*" group is outside the range of this volume.

Patton (1987) recognizes nine species groups of the subgenus *Proechimys*. Although he agrees with Gardner and Emmons to separate the southeastern Brazilian spiny rats as a separate genus, *Trinomys*, and concurs with Gardner (1983) on the identity of *P. semispinosus*, he believes that the *brevicauda* group of Gardner and Emmons is an artificial assemblage. Patton presents evidence that three groups are monotypic: *decumanus*, *canicollis*, and *simonsi*. The other six groups are polytypic, but "the number of species in each remains unclear." Patton uses baculum, molars, and the palatal foramina as characters to define his groupings.

It should be noted that the karyotypic variants of *Proechimys* cannot easily be characterized by external features. Indeed, in the south (e.g., Peru) two or three species may coexist that are quite different in

their habits but not easily distinguished by an "in-the-hand" inspection. Since this volume is only a general guide to what the best experts have defined as true species, I have decided to follow Honacki, Kinman, and Koeppl (1982) with respect to the species names, with modifications as noted by Gardner and Emmons (1984) and Patton (1987).

Description

Head and body length ranges from 165 to 290 mm; tail 121 to 242 mm; hind foot 35 to 57 mm; and ear 17 to 29 mm. The dorsum is generally some shade of brown, contrasting sharply with a white or cream-colored venter. The tail is very sparsely haired. The dorsal pelage has spinescent hairs, but this is not nearly as pronounced as in some of the other genera (plate 21).

Distribution

The genus is distributed in moist evergreen forest from Honduras south to Peru and central Brazil and east to the Guyanas.

Natural History

Proechimys guyannensis is nocturnal and lives in burrows and hollow logs. Usually only a single adult inhabits a burrow system (Guillotin 1982). In *P. semispinosus* the gestation is sixty-four to sixty-five days, and the average litter size is 2.8, with a mode of three (Maliniak and Eisenberg 1971).

Proechimys emerges from its burrow at night to forage for seeds and fruits on the forest floor, and fungi are widely used as food. It caches food in its burrow system. Emmons (1982) confirmed that two species live in sympatry in Peru, but usually one species is slightly smaller.

Brevicauda Group
Proechimys amphichoricus Moojen, 1948
The species is distributed from extreme southern Venezuela to adjacent Brazil (map 13.67).
Chromosome number: $2n = 26$, $FN = 44$.

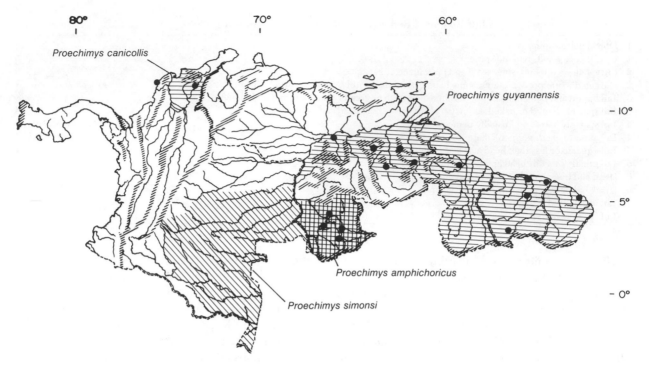

Map 13.67. Distribution of *Proechimys* "*brevicauda*" group.

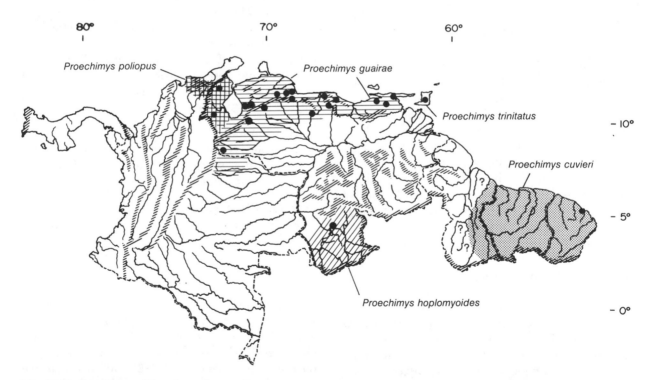

Map 13.68. Distribution of *Proechimys* "*guairae*" group.

Proechimys cuvieri Petter, 1978

This rat is distributed from the Guyanas south to Brazil. Guillotin and Ponge (1984) claim that *P. cuvieri* is distinct karyotypically and can be discriminated from *P. guyannensis* only by a multivariate analysis of several measurements or by an examination of the shape of the incisiform foramina in the palate. It co-occurs with *P. guyannensis* (map 13.68). Chromosome number: $2n = 28$.

Proechimys guyannensis (E. Geoffroy, 1803)

This form is distributed from south-central Venezuela to the Guyanas. Reig, Trainer, and Barros (1979) demonstrated clear karyotypic separation from *P. cuvieri* (map 13.67) Chromosome number: $2n = 40$, FN = 54.

Natural History

Guillotin (1982) studied *P. guyannensis* in French Guiana. Home ranges average 0.84 for males and 0.31 ha for females. The species is highly frugivorous but eats insects in the dry season. The animals are strictly nocturnal and use a complex burrow system.

Proechimys simonsi (= *hendeei*) Thomas, 1926

The species is distributed from northeastern Peru to southern Colombia. Chromosome number: $2n = 32$, FN = 52.

Proechimys canicollis (J. A. Allen, 1899)

This form is found in north-central Colombia (map 13.67). Chromosome number: $2n = 24$, FN = 44.

Guairae Group

Proechimys guairae Thomas, 1901

This species is found in the mountains of north-central Venezuela. *P. ochraceus* and *P. urichi* are included in *P. guairae* (map 13.68). Chromosome number: A rassenkreis with variable chromosome numbers, $2n = 44–50$, FN = 72–76.

Natural History

O'Connell (1981) studied *P. guairae* in northern Venezuela, where it was primarily frugivorous but also ate seeds and fungi. Population densities ranged from 2.1 to 5.0 animals per hectare throughout a twenty-two-month period.

Proechimys hoplomyoides (Tate, 1939)

This form is found in southeastern Venezuela, extending to adjacent portions of Guyana and Brazil (map 13.68).

Proechimys poliopus Osgood, 1914

This species occurs in extreme northwestern Venezuela to the west of Lake Maracaibo (map 13.68). Chromosome number: $2n = 42$, FN = 76.

Proechimys trinitatus (J. A. Allen and Chapman, 1893)

This form is distributed on the island of Trinidad (map 13.68). Chromosome number: $2n = 62$, FN = 80.

Natural History

Everard and Tikasingh (1973) found home ranges of 1.4 ha for males and 0.15 ha for females on Trinidad.

Semispinosus Group

Proechimys semispinosus (Tomes, 1860)

This species ranges from Honduras through Panama to the west of the western cordillera of the Andes, southward to Peru (map 13.69). Chromosome number: $2n = 30$, FN = 50–54.

Natural History

Fleming (1971) studied *P. semispinosus* in Panama. Median densities ranged from 3.0 to 8.5 per hectare. They breed throughout the year and can have three to four litters a year with a mean litter size of about 2.7. The highest densities (9.7/ha) were recorded by Gliwicz (1973) on small offshore islands in Lake Gatun. Maliniak and Eisenberg (1971) studied this species in captivity, where it was nocturnal and cached food in its nest box. Adults are intolerant of one another, but a male and a female could be housed together in a large cage. During courtship a variety of vocalizations are employed. Males utter a whimper when approaching the female, who responds with a "twitter." Both animals may continue calling during copulation. In captivity gestation was from sixty-three to sixty-five days, and litter sizes ranged from one to five, with an average of 2.8. The young are born fully haired with their eyes open. Adult proportions are reached at eighty-five to ninety days of age. Earliest age of conception by a captive-born female was one hundred days.

Proechimys oconnelli Allen, 1913

The species is found in a restricted area of east-central Colombia along the Meta River and its tributaries (map 13.69). Chromosome number: $2n = 32$, FN = 52.

Comment

Two to four species of *Proechimys* can co-occur within the same habitat in Ecuador and Peru (Patton

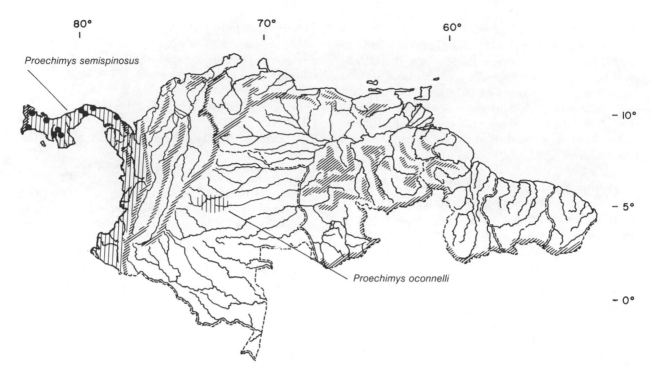

Map 13.69. Distribution of *Proechimys "semispinosus"* group.

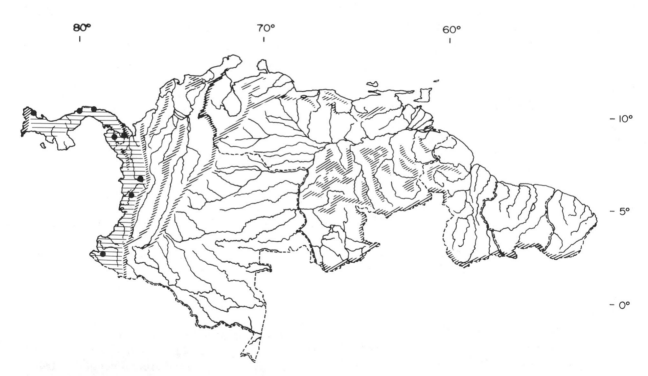

Map 13.70. Distribution of *Hoplomys gymnurus*.

and Gardner 1972; Emmons 1982). The exact way the niches are segregated is difficult to determine at this point, but there must surely be some sharp division in dietary preferences to allow for the coexistence of such nearly equal-sized sympatric species.

Arrangement of Northern *Proechimys* Species according to Patton (1987)

1. The *guyannensis* group
   Proechimys guyannensis (E. Geoffroy, 1803)
   P. warreni Thomas, 1905
2. The *simonsi* group
   P. hendeei Thomas, 1926
3. The *cuvieri* group
   P. cuvieri Petter, 1978
4. The *semispinosus* group
   P. semispinosus (Tomes, 1860)
   P. oconnelli Allen, 1913
5. The *canicollis* group
   P. canicollis (J. A. Allen, 1899)
6. The *goeldii* group
   P. quadruplicatus Hershkovitz, 1948
   P. amphichoricus Moojen, 1948
7. The *trinitatus* group
   P. guairae Thomas, 1901
   P. hoplomyoides (Tate, 1939)
   P. poliopus Osgood, 1914
   P. trinitatus (J. A. Allen and Chapman, 1893)
   P. urichi (J. A. Allen, 1899)
8. The *longicaudatus* group
   P. brevicauda (Gunther, 1877)

Genus *Hoplomys* J. A. Allen, 1908
Hoplomys gymnurus (Thomas, 1897)
Thick-spined or Armored Rat

Description
Measurements of the type, an adult male, were HB 380; T 170; HF 53; female head and body length is about 320 mm. The sparsely haired tail is generally shorter than the head and body. The upperparts are dark brown and the underparts are whitish. The spinescent hairs on the dorsum are extremely long, up to 33 mm (plate 20).

Range and Habitat
The species ranges from east-central Honduras south through Panama, west of the central Andean cordillera to northwestern Ecuador (map 13.70). It prefers lowland multistratal tropical evergreen forest habitats.

Natural History
This nocturnal, terrestrial form constructs elaborate burrows (Buchanan and Howell 1965). Armored rats feed on seeds, fruits, and nuts and are known to cache food in their burrows. Like other burrowing

rodents, they tend to use a discrete chamber of the burrow system for defecation. Litter size averages 2.1 in Panama; the young are precocial (Tesh 1970b). In captivity, males and females may tolerate each other, but males are highly intolerant of other adult males. A complicated series of vocalizations are employed during social encounters, but their precise function has not been determined.

Genus *Mesomys* Wagner, 1845

Description
Head and body length ranges from 150 to 200 mm and the tail from 120 to 220 mm. The dorsum is usually some shade of brown. The spiny hairs on the dorsum are banded, giving a salt-and-pepper effect; when stroked, these spines become much more obvious than in *Proechimys*. The underparts are buff to white. The tail is sparsely haired, is penicillate at the tip, and is not bicolor (plate 20).

Distribution
Members of the genus are chiefly found in the Amazonian portion of Brazil, Colombia, Peru, Ecuador, the Guyanas, and southern Venezuela (map 13.71). They are associated with multistratal tropical evergreen forest at up to 1,950 m elevation.

Natural History
These spiny rats are nocturnal and strongly arboreal. *M. hispidus* has a relatively short gastrointestinal tract, suggesting that leaves do not make up a significant proportion of its diet (Emmons 1981). Females bear a single young that is born furred and with its eyes open.

Mesomys stimulax Thomas, 1911

Description
This form matches the description given for the genus but is slightly smaller than *M. hispidus*. Measurements of one specimen were HB 158; T 122; HF 29.

Range and Habitat
The species occurs in multistratal tropical evergreen forest in northeastern Brazil, extending to Suriname (map 13.71).

Mesomys hispidus (Desmarest, 1817)

Measurements

	Mean	Minimum	Maximum	N
TL	353.7	343	375	4
T	164.5	149	173	4
HF	32.8	30	34	4
E	13.3	13	14	4

Location: Colombia (FMNH)

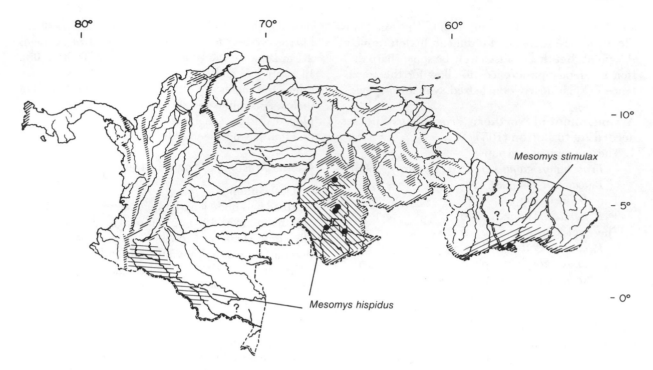

Map 13.71. Distribution of two species of *Mesomys*.

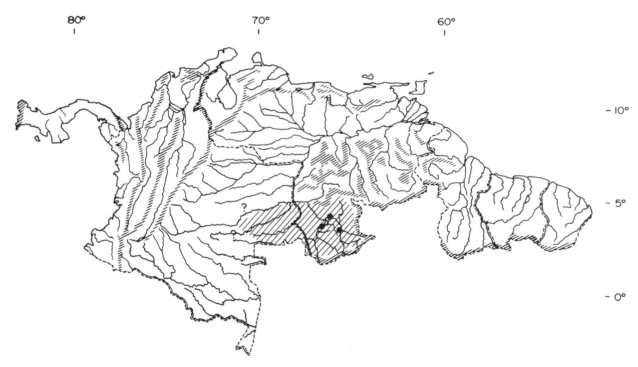

Map 13.72. Distribution of *Isothrix bistriatus*.

Description

This species is larger than *M. stimulax*. See also the description for the genus.

M. ferrugeneus is included in *M. hispidus* (Honacki, Kinman, and Koeppl 1982).

Range and Habitat

The species is found in multistratal tropical forest from the lowlands to the foothills of the Andes, in northern Brazil, Peru, Ecuador, Colombia, and Venezuela. It prefers moist habitats and is strongly arboreal (map 13.71).

Genus *Isothrix* Wagner, 1845

Description

Head and body length ranges from 220 to 290 mm; the tail is approximately 250 mm. The color pattern is highly variable for the genus, and one species, *I pictus*, is blotched dark brown and white. In general the dorsum is dark brown broken by a buff patch across the shoulders. The fur is soft and without spines. The tail is very well haired, almost squirrellike.

Distribution

The genus is confined to the Amazonian portions of Venezuela, Colombia, Peru, and Brazil.

Isothrix bistriatus Wagner, 1845
Toro, Rata de Doble Estría

Measurements

	Mean	Minimum	Maximum	N
TL	505.5	478	550	4
T	242	232	250	4
HF	47	45	50	4
E	16.5	16	17	4

Location: Peru, Brazil (FMNH)

Description

Head and body length ranges from 220 to 251 mm. This rat is almost squirrellike in appearance. The soft dorsal pelage is brown, somewhat lighter on the sides, the venter is orange, and the forehead is yellow to cream, framed by two brown stripes. The tail is well haired, pale brown at the base grading to black at the tip.

Range and Habitat

The species occurs in southwestern to north-central Brazil, Amazonian Peru, southern Venezuela, and southern Columbia (map 13.72). It is found in moist habitats in multistratal tropical evergreen forests.

Natural History

This soft-furred form is nocturnal and strongly arboreal. It nests in tree cavities. The tail may be semiprehensile in some of the proposed species (Patton and Emmons 1985). The female generally bears only a single young, which is born furred and with its eyes open.

Genus *Diplomys* Thomas, 1916
Arboreal Soft-furred Spiny Rat, Ratón Marenero

Description

Head and body length ranges from 250 to 480 mm and the tail from 200 to 280 mm. The fur is very soft, with no spines; the tail is haired and tends to be tufted at the tip. The dorsal pelage is brown mixed with black hairs and the venter is lighter, ranging from buffy to white (plate 21).

Distribution

The genus occurs from Panama through Colombia to northern Ecuador, west of the eastern cordillera of the Andes.

Natural History

These highly arboreal rodents nest in tree cavities. Their presence often goes undetected, but by thumping on the side of a tree one can occasionally flush a specimen from its hole. They are nocturnal and feed on fruit, seeds, and nuts (Tesh 1970a). In Panama *D. labilis* has a litter size of one or two.

Diplomys caniceps (Gunther, 1877)

Measurements

	Mean	Minimum	Maximum	N
TL	390	379	401	3
T	189	162	208	3
HF	41.6	36	47	3
E	15.3	15	16	3

Location: Colombia (FMNH)

Description

This rat is similar in appearance to *D. labilis*, but the dorsum is not as reddish. The venter is tan. The tail is sparsely haired to the tip and is not noticeably bicolor.

Range and Habitat

The species is found in the central portion of Colombia between the western and central cordilleras of the Andes, south to Ecuador (map 13.73).

Diplomys labilis (Bangs, 1901)

Description

This account includes *D. darlingi*. The type has a total length of 540 mm: T 200; HF 47; E 17. Two specimens from Panama showed the following mean measurements: TL 554; T 220; HF 47.5; E 15 (FMNH). The upperparts are reddish brown, darker at the middorsal region. There is a small yellowish-

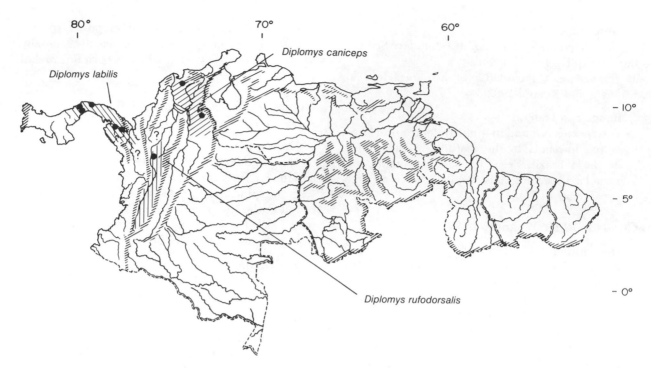

Map 13.73. Distribution of three species of *Diplomys*.

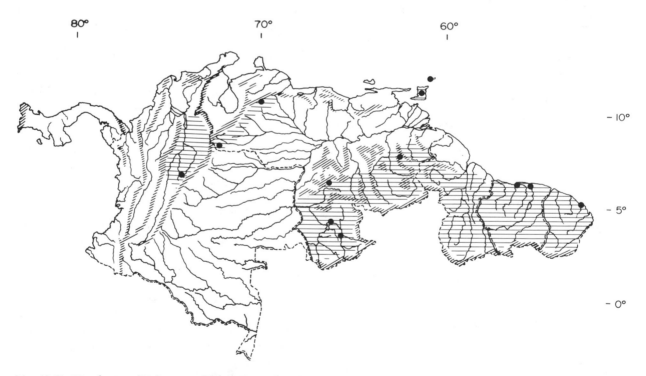

Map 13.74. Distribution of *Echimys armatus*.

white patch at the base of the mystacial vibrissae, above the eye and behind the ear. The underparts vary from buff to orange. The distal portion of the tail may have white hairs.

Range and Habitat

The species is found from central Panama to the east and then south into Colombia to the west of the western cordillera of the Andes (map 13.73).

Diplomys rufodorsalis (J. A. Allen, 1899)

Description

See the description for the genus.

Range and Habitat

The species is known from the extreme northeast of Colombia (map 13.73).

Genus *Echimys* G. Cuvier, 1809

Arboreal Spiny Rat, Rata Espinosa

Taxonomy

For this discussion *Makalata* is considered a junior synonym of *Echimys*. Husson (1978) erected *Makalata* for *Echimys armatus* in recognition of some differences separating this species from *E. chrysurus*. Since *E. semivillosus* is clearly related to *E. armatus*. I will retain *Echimys* for all three pending a revision.

Description

Head and body length ranges from 170 to 350 mm and tail length from 150 to 300 mm. The dorsum is usually brown to gray brown, but dorsal coloration varies geographically. The underparts are usually buff to white.

Distribution

The genus is found over most of northern South America south to Brazil.

Natural History

These nocturnal, arboreal rats nest in hollow trees. Their presence can sometimes be detected at night by distinctive chattering vocalizations. The female appears to bear only one or two young. Small family groups can use the same nesting hole in a tree.

Echimys (= Makalata) armatus (I. Geoffroy, 1830)

Description

Two specimens measured HB 238, 264; T 195, 220; HF 35, 44; E 18, 19 (FMNH). This species may attain a weight of 300 g. The dorsal pelage is furry but strongly admixed with spines. The tail is furred for about 30 mm at the base and the distal portion is sparsely haired. The dorsum is dark brown and the venter gray brown.

Range and Habitat

The species occurs in northern Ecuador, Peru, Colombia, Venezuela, the Guyanas, and Brazil. Distributions in northern Colombia were documented by Hershkovitz (1948). It is a lowland form preferring multistratal tropical evergreen forest. In northern Venezuela most specimens were taken at below 350 m elevation (map 13.74).

Natural History

Unlike *Echimys semivillosus*, this rat is strongly tied to moist habitats. Charles-Dominique et al. (1981) found it to be a significant predator on immature seeds, particularly *Virola* and *Inga*. The species is strictly nocturnal and nests in hollow trees. The litter size is one or two, and the young are precocial.

Echimys chrysurus (Zimmermann, 1780)

Description

Head and body length ranges from 230 to 290 mm and the tail from 310 to 335 mm; the hind foot is 51 mm and the ear 20 mm. Weight may reach over 500 g. The dorsal pelage includes very broad, heavy spines interspersed with hairs; the tip of the spine tends to be brown while the shaft is gray. The dorsum is dark rufous brown to gray brown, darkest in the midline. The dorsal surface of the head is usually much darker than the back, almost blackish, and in some specimens there is a distinctive white to yellow median stripe running from the nose over the face to the neck. Specimens with white facial stripes usually have white tail tips.

Range and Habitat

The species occurs in multistratal tropical evergreen forest from the Guyanas south to northeastern Brazil (map 13.75).

Natural History

This is a strongly arboreal, nocturnal form. The female bears two young that are born fully furred and with their eyes open (Husson 1978).

Echimys semivillosus (I. Geoffroy, 1838)

Description

This species is approximately the same size as *E. chrysurus*. A female from Colombia measured TL 444; T 228; HF 36; E 19. The spines are extremely numerous on the dorsum, interspersed with fine, short hairs. The dorsum is usually gray brown, contrasting with a cream venter (plate 21).

Range and Habitat

The species ranges from northern Colombia across northern Venezuela. It prefers dry deciduous tropical forest and replaces *E. armatus* in the drier areas (map 13.75).

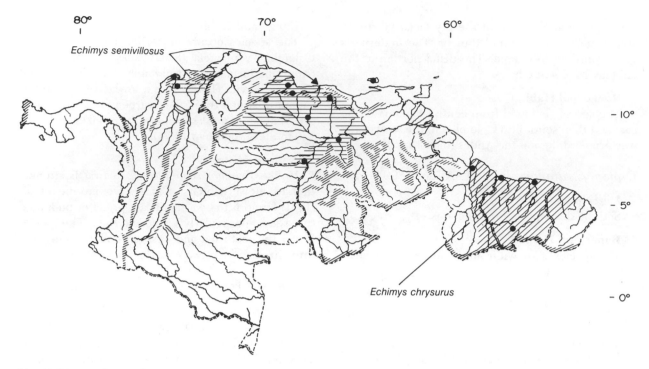

Map 13.75. Distribution of two species of *Echimys*.

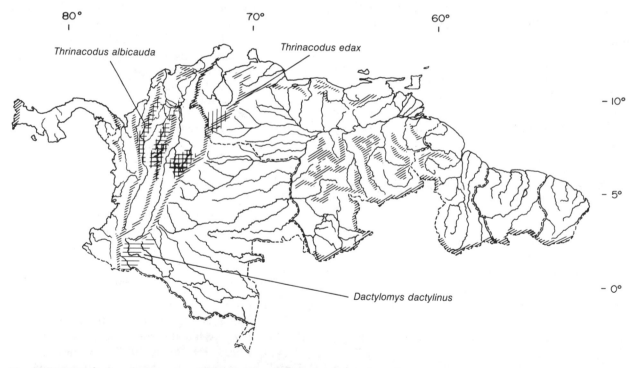

Map 13.76. Distribution of two species of *Thrinacodus* and *Dactylomys dactylinus* (based in part on Cabrera 1960).

Natural History

In the dry deciduous tropical forest bordering the streams of the llanos of Venezuela, this species can occur at densities of one individual per four hectares. Though strongly arboreal and frugivorous, it will also forage on the forest floor. It nests in tree cavities (August 1981, 1984).

Genus *Thrinacodus* Gunther, 1879
Andean Soft-furred Spiny Rat

Description

Head and body length ranges from 180 to 240 mm. The tail is longer than the head and body—a diagnostic feature of this genus. These rats are smaller than *Dactylomys*. The fur is soft, without spines or bristles. The upperparts are usually some shade of reddish brown and the underparts white to yellow. The tail is sparsely haired, faintly bicolor, and paler at the tip (plate 20).

Distribution

The genus is known from Colombia and Venezuela at higher elevations. Specimens are generally taken at above 2,000 m.

Thrinacodus albicauda Gunther, 1879

Measurements

	Mean	Minimum	Maximum	N
TL	554.0	538	586	3
T	326.6	314	351	3
HF	48.6	47	51	3
E	23.3	22	24	3

Location: Colombia (FMNH)

Description

See the description for the genus. The tail is distinctly bicolor, brown above and light below and often paler at the tip.

Range and Habitat

The species is found in the northwestern and central portions of Colombia to the west of the central cordillera of the Andes. It occurs at up to 3,000 m elevation and is often associated with moist, dense bamboo thickets (map 13.76).

Thrinacodus edax Thomas, 1916

Description

See the description for the genus. The tail is bicolor, and there is a tendency for the distal third to be entirely white.

Range and Habitat

The species occurs at high elevations in the western Venezuelan Andes and adjacent portions of northeastern Colombia (map 13.76).

Natural History

This nocturnal, arboreal form produces a whistle-like call at night.

Genus *Dactylomys* I. Geoffroy, 1838
Dactylomys dactylinus (Desmarest, 1817)
Bamboo Rat, Coro-coro

Description

Two specimens measured TL 710, 720; T 410, 415; HF 50, 55; E 17, 20. Its large size and very long

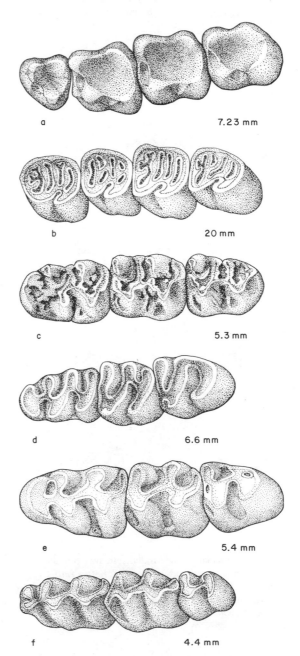

a   7.23 mm

b   20 mm

c   5.3 mm

d   6.6 mm

e   5.4 mm

f   4.4 mm

Figure 13.13. Some molar teeth of rodents; figures in millimeters indicate actual length of tooth row: (*a*) *Sciurus aestuans*; (*b*) *Dasyprocta*; (*c*) *Nyctomys sumichrasti*; (*d*) *Sigmodon hispidus*; (*e*) *Zygodontomys brevicauda*; (*f*) *Akodon* sp. Dorsal tooth row: anterior, left; posterior, right.

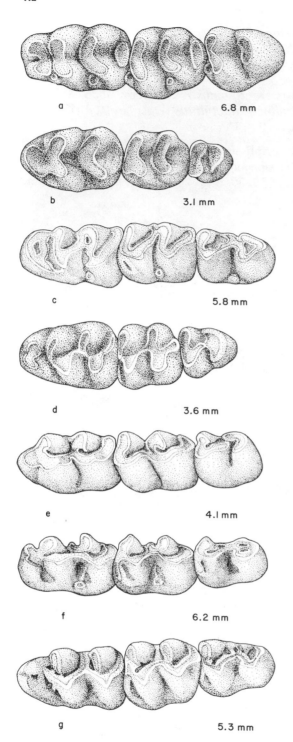

Figure 13.14. Some molar teeth of rodents; figures in millimeters indicate actual length of tooth row: (*a*) *Rattus rattus;* (*b*) *Mus musculus;* (*c*) *Oryzomys capito;* (*d*) *Calomys laucha;* (*e*) *Scotinomys xerampelinus;* (*f*) *Isthmomys (Peromyscus) flavidus;* (*g*) *Rhipidomys leucodactylus.* Dorsal tooth row: anterior, left; posterior, right.

tail distinguish this species from all other spiny rats occurring within its range. The tail is furred for about 30 mm at the base; the dorsal pelage is soft and lacks spines. Males have a prominent sternal gland. The upperparts vary from olive gray to agouti brown with a reddish tinge; the midline may be dark agouti brown, and the underparts are cream to white (plate 20).

Range and Habitat

This rat occurs from southern Colombia to northwestern Brazil, Amazonian Peru, and Ecuador. This lowland form is strongly associated with multistratal tropical evergreen forest (map 13.76).

Natural History

This species is nocturnal and strongly arboreal. At night individuals emit a characteristic pulsed call, "boop . . . boop . . . boop," which may be the only indication of their presence without trapping (Laval 1976). Emmons (1981), working in Ecuador and Peru, established that these rodents are highly folivorous. Gut modifications suggest they can ferment masticated leaves in the cecum. They are arboreal foragers in bamboo clumps. Home ranges vary from 0.23 to 0.43 ha, and they appear to forage in family groups.

References

• References used in preparing distribution maps

• Aagaard, E. M. 1982. Ecological distribution of mammals in the cloud forests and paramos of the Andes, Mérida, Venezuela. Ph.D. diss., Colorado State University.

Aguilera, M. 1985. Growth and reproduction in *Zygodontomys microtinus* (Rodentia-Cricetidae) from Venezuela in a laboratory colony. *Mammalia* 49:58–67.

Allen, J. A. 1912. Mammals from western Colombia. *Amer. Mus. Nat. Hist.* 31:71–95.

———. 1915. Review of the South American Sciuridae. *Bull. Amer. Mus. Nat. Hist.* 34(art. 8): 147–309.

• Anthony, H. E. 1921. Mammals collected by William Beebe at the British Guiana Tropical Research Station. *Zoologica* 3(13):265–85.

• ———. 1925. New species and subspecies of *Thomasomys. Amer. Mus. Novitat.* 178:1–4.

———. 1929. Two new genera of rodents from South America. *Amer. Mus. Novitat.* 383:1–6.

Anthony, H. E., and G. H. H. Tate. 1935. Notes on South American Mammalia—No. 1: *Sciurillus. Amer. Mus. Novitat.* 780:1–13.

Table 13.5   Key to the Genera of Myomorph Rodents in Panama and Extreme Northwestern Colombia

1	External form much modified for aquatic life; hind toes fringed with stiff hairs; underfur dense; webbing present between some toes; braincase much flattened . . . . *Rheomys*
1'	External form not modified for aquatic life; hind toes not fringed with stiff hairs; braincase not markedly flattened 2
2	Upper cheek teeth specialized; cusps not apparent at any time; folds of upper molars pronounced . . . . *Sigmodon*
2'	Upper cheek teeth not markedly specialized; cusps usually apparent; molars usually not flat crowned 3
3	Upper incisors conspicuously grooved . . . . *Reithrodontomys*
3'	Upper incisors not conspicuously grooved 4
4	Cheek teeth complex with subsidiary ridges normally present in the main outer folds of M1 and M2; teeth cuspidate (secondary ridges tend to disappear in some species of *Peromyscus*) 5
4'	Cheek teeth not especially complex; subsidiary ridges in main outer folds of M1 and M2 not traceable 10
5	Cusps of upper molars alternate; anterointernal cusps reduced or absent: M3 notably reduced . . . . *Peromyscus* and *Isthmomys*
5'	Cusps of upper molars opposite or nearly so; anterointernal cusps not obliterated; M3 normal, not reduced 6
6	Pelage bristly or spiny . . . . *Neacomys*
6'	Pelage not bristly or spiny 7
7	Anterointernal cusps of M1 much reduced . . . . *Nyctomys*
7'	Anterointernal cusps of M2 but little reduced 8
8	Palate not reaching posterior part of tooth row: lacking lateral pits in posterior part of palate; tail naked 9
8'	Palate extending posteriorly beyond tooth rows; palate having well-marked lateral pits in posterior part of palate . . . . *Oryzomys*
9	Tail naked; M3 similar to M2; subsidiary ridges of upper molars reduced . . . . *Tylomys*
9'	Tail not naked; M3 more reduced than M2; subsidiary ridges of upper molars not reduced or very little reduced . . . . *Rhipidomys*
10	Some part of backward prolongation of outer folds in upper molars isolated as deep, conspicuous pits between main cusps and all but greatly worn teeth . . . . *Scotinomys*
10'	No part of backward prolongation of outer folds in upper molars isolated as deep, conspicuous pits between main cusps. Cusps of upper molars not high; molars not unusually narrow; M3 not greatly reduced; cusps of molars opposite or nearly so . . *Zygodontomys*

Source: Adapted from Hall (1981).

August, P. V. 1981. Population and community ecology of small mammals in northern Venezuela. Ph.D. diss., Boston University.

———. 1984. Population ecology of small mammals in the llanos of Venezuela. In *Contributions in mammalogy in honor of Robert L. Packard*, 71–104. Special Publications of the Museum 22. Lubbock: Texas Tech University Press.

Baker, R. H., and M. K. Petersen. 1965. Notes on a climbing rat, *Tylomys* from Oaxaca, Mexico. *J. Mammal.* 46:694–95.

Barros, M. A., and O. A. Reig. 1979. Doble polimorfismo robertsoniano en *Microxus bogotensis* (Rodentia, Cricetidae) del Páramo de Mucubaji (Mérida, Venezuela). *Acta Cient. Venez.* 30(suppl. 1): 96.

Baskin, J. A. 1978. *Bensonomys, Calomys* and the origin of the phyllotine group of Neotropical cricetines. *J. Mammal.* 59:125–35.

Bianchi, N. O., O. A. Reig, C. J. Molina, and G. N. Dulout. 1971. Cytogenetics of the South American akodont rodents (Cricetidae), part 1. *Evolution* 25:724–36.

Birkenholz, B. E., and W. O. Wirtz II. 1965. Laboratory observations on the vesper rat. *J. Mammal.* 46:181–89.

• Boher B., S., J. Naveuva S., and L. Escobar M. 1988. First record of *Dinomys branickii* for Venezuela. *J. Mammal.* 69:433.

Borrero-H., J. L., and J. Hernandez-Camacho. 1957. Informe preliminar sobre aves y mamíferos de Santander, Colombia. *Ann. Soc. Biol. Bogotá* 7(5):197–230.

Buchanan, O. M., and T. R. Howell. 1965. Observations on the natural history of the thick spined rat, *Hoplomys gymnura* in Nicaragua. *Ann. Mag. Nat. Hist.*, ser. 13, 8:549–59.

• Cabrera, A. 1960. Catálogo de los mamíferos de América del Sur. Vol. 2. Sirenia-Perissodactyla-Artiodactyla-Lagomorpha-Rodentia-Cetacea. *Rev. Mus. Argent. Cieno. Nat. "Bernardino Rivadavia," Zool.* 4(1–2):309–732.

Calhoun, J. B. 1962. *The ecology and behavior of the Norway rat.* Washington, D.C.: National Institutes of Health.

Carleton, M. D. 1980. Phylogenetic relationships in neotomine-peromyscine rodents (Muroidea) and a reappraisal of the dichotomy with New World Cricetinae. *Misc. Publ. Mus. Zool. Univ. Michigan* 157:1–146.

Carleton, M. D., and G. G. Musser 1984. Muroid rodents. In *Orders and families of Recent mammals of the world*, ed. S. Anderson and J. Knox Jones, Jr., 289–379. New York: John Wiley.

Cartaya, E., and M. Aguilera. 1985. Estudio de la comunidad de roedores plaga en un cultivo de arroz. *Acta Cient. Venez.* 36:250–57.

Charles-Dominique, P., M. Atramentowicz, M. Charles-Dominique, H. Gerard, A. Hladik, C. M. Hladik, and M. F. Prevost. 1981. Les mammifères frugivores arboricoles nocturnes d'une forêt guyanaise: Interrelations plantes-animaux. *Rev. Ecol. (Terre et Vie)* 35:342–435.

Chitty, D., and H. N. Southern, ed. 1954. *Control of rats and mice.* 3 vols. Oxford: Clarendon Press.

Collett, S. F. 1981. Population characteristics of *Agouti paca* (Rodentia) in Colombia. *Publ. Mus. Michigan State Univ., Biol. Serv.* 5:487–601.

Collins, L. R., and J. F. Eisenberg. 1972. Notes on the behavior and breeding of pacaranas. *Int. Zoo Yearb.* 12:108–14.

Dawson, G. A. 1973. The functional significance of nest building by a Neotropical rodent (*Sigmodon hispidus*). *Amer. Midl. Nat.* 89:503–9.

Dubost, G., and F. Petter, 1978. Une espèce nouvelle de "rat-pecheur" de Guyane française: *Daptomys oyapocki. Mammalia* 42:435–39.

Eisenberg, J. F. 1963. The behavior of heteromyid rodents. *Univ. Calif. Publ. Zool.* 69:1–100.

———. 1968. Behavior patterns. In *The biology of Peromyscus* (*Rodentia*), ed. J. A. King, 451–95. Special Publication 2. Shippensburg, Pa.: American Society of Mammalogists.

———. 1974. The function and motivational basis of hystricomorph vocalizations. In *The biology of hystricomorph rodents*, ed. I. W. Rowlands and B. J. Weir, 211–44. London: Academic Press.

———. 1984. New World rats and mice. In *The encyclopaedia of mammals*, ed. D. Macdonald, 640–49. New York: Facts on File.

Eisenberg, J. F., and D. G. Kleiman. 1977. Communication in lagomorphs and rodents. In *How animals communicate*, ed. T. Sebeok, 634–54. Bloomington: Indiana University Press.

Ellermann, J. 1940. *The families and genera of living rodents.* 2 vols. London: British Museum of Natural History

Emmons, L. 1981. Morphological, ecological, and behavioral adaptations for arboreal browsing in *Dactylomys dactylinus* (Rodentia, Echimyidae). *J. Mammal.* 62:183–89.

———. 1982. Ecology of *Proechimys* in southeastern Peru. *Trop. Ecol.* 23:280–90.

Enders, R. K. 1980. Observations on *Syntheosciurus* taxonomy and behavior. *J. Mammal.* 61:725–27.

Ernest, K. A. 1986. *Nectomys squamipes. Mammal. Species* 265:1–5.

Ernest, K. A., and M. A. Mares. 1986. Ecology of *Nectomys squamipes*, the Neotropical water rat in central Brazil. *J. Zool.* (London), ser. A, 210:599–612.

Everard, C. O. R., and E. S. Tikasingh. 1973. Ecology of the rodents *Proechimys guyannensis trinitatis* and *Oryzomys capito velutinus* on Trinidad. *J. Mammal.* 54:875–86.

Ewer, R. F. 1971. The biology and behavior of a

free-living population of black rats (*Rattus rattus*). *Anim. Behav. Monogr.* 4(3):127–74.

Fleming, T. H. 1971. Population ecology of three species of Neotropical rodents. *Misc. Publ. Mus. Zool. Univ. Michigan* 143:1–77.

———. 1974a. The population ecology of two species of Costa Rican heteromyid rodents. *Ecology* 55:493–510.

———. 1974b. Social organization in two species of Costa Rican heteromyid rodents. *J. Mammal.* 55:543–61.

———. 1983. *Heteromys desmarestianus.* In *Costa Rican natural history*, ed. D. H. Janzen, 474–75. Chicago: University of Chicago Press.

Gardner, A. L. 1983. *Proechimys semispinosus* (Rodentia: Echimyidae): Distribution, type locality, and taxonomic history. *Proc. Biol. Soc. Washington* 96:134–44.

Gardner, A. L., and L. Emmons. 1984. Species groups in *Proechimys* (Rodentia, Echimyidae) as indicated by karyology and bullar morphology. *J. Mammal.* 65:10–25.

Gardner, A. L., and J. L. Patton. 1976. Karyotypic variation in oryzomyine rodents with comments on chromosomal evolution in Neotropical cricetine complex. *Occas. Pap. Mus. Zool. Louisiana State Univ.* 49:1–48.

• Genoways, H. H. 1973. *Systematics and evolutionary relationships of spiny pocket mice, genus* Liomys. Special Publications of the Museum 5. Lubbock: Texas Tech University Press.

Genoways, H. H., and J. Knox Jones, Jr. 1972. Variation and ecology in a local population of the vesper mouse *Nyctomys sumichrasti. Occas. Papers Mus. Texas Tech Univ.* 3:1–22.

Giacalone, J., N. Wells, and G. Willis. 1987. Observations on *Syntheosciurus brochus* in Volcán Pous National Park, Costa Rica. *J. Mammal.* 68:145–46.

Glanz, W. E. 1984. Food and habitat use by two sympatric *Sciurus* species in central Panama. *J. Mammal.* 65:342–46.

Glanz, W. E., R. W. Thorington, Jr., J. Giacalone-Madden, and L. R. Heaney. 1982. Seasonal food use and demographic trends in *Sciurus granatensis.* In *The ecology of the tropical forest*, ed. E. Leigh, Jr., A. Stanley Rand, and D. M. Windsor, 239–52. Washington, D.C.: Smithsonian Institution Press.

Gliwicz, J. 1973. A short-term characteristic of a population of *Proechimys semispinosus*, a rodent species of the tropical rain forest. *Bull. Acad. Polonaise Sci. (Biol. Ser.)* 21:413–18.

• Goodwin, G. G., and A. M. Greenhall. 1961. A review of the bats of Trinidad and Tobago. *Bull. Amer. Mus. Nat. Hist.* 122(3):187–302.

Grand, T. H., and J. F. Eisenberg. 1982. On the affinities of the Dinomyidae. *Säugetierk. Mitt.* 30:151–57.

Guillotin, M. 1982. Rythmes d'activité et régimes alimentaires de *Proechimys cuvieri* et d'*Oryzomys capito velutinus* (Rodentia) en forêt guyanaise. *Rev. Ecol. (Terre et Vie)* 36:337–71.

Guilloton, M., and J. F. Ponge. 1984. Identification de deux espèces de rongeurs de Guyane française *Proechimys cuvieri* and *Proechimys guyannensis.* *Mammalia* 48:287–91.

• Gyldenstolpe, N. 1932. A manual of Neotropical sigmodont rodents. *Kungl. Svensk. Vetenskapsakad. Handl.* 11:1–164.

Haffer, J. 1974. *Avian speciation in tropical South America.* Publication 14. Cambridge, Mass.: Nuttall Ornithological Club.

• Hall, E. R. 1981. *The mammals of North America.* 2d ed., 2 vols. New York: Academic Press.

Hall, E. R., and W. W. Dalquest. 1963. The mammals of Veracruz. *Univ. Kansas Publ. Mus. Nat. Hist.* 14(14):165–362.

• Handley, C. O., Jr. 1976. Mammals of the Smithsonian Venezuela project. *Brigham Young Univ. Sci. Bull., Biol. Ser.* 20(5):1–90.

Handley, C. O., Jr., and E. Mondolfi, 1963. A new species of fish-eating rat *Ichthyomys* from Venezuela (Rodentia: Cricetidae). *Acta Theol. Venez.* 3:417–19.

Heaney, L. R. 1983. *Sciurus granatensis.* In *Costa Rican natural history,* ed. D. H. Janzen, 489–90. Chicago: University of Chicago Press.

Heaney, L. R., and R. W. Thorington, Jr. 1978. Ecology of Neotropical red-tailed squirrels, *Sciurus granatensis,* in the Panama Canal Zone. *J. Mammal.* 59:846–51.

Helm, J. D. 1973. Reproductive biology of *Tylomys* and *Ototylomys.* Ph.D. diss., Michigan State University, East Lansing, Michigan.

———. 1975. Reproductive biology of *Ototylomys* (Cricetidae). *J. Mammal.,* 56:575–90.

Hernandez-Camacho, J. 1960. Primitiae mastozoologicae Colombianae 1: Status taxonomical de *Sciurus pucheranii santanderensis. Caldasia* 8:359–68.

Hershkovitz, P. 1941. The South American harvest mice of the genus *Reithrodontomys. Occas. Pap. Mus. Zool. Univ. Michigan* 441:1–7.

———. 1944. A systematic review of the Neotropical water rats of the genus *Nectomys* (Cricetinae). *Misc. Publ. Mus. Zool. Univ. Michigan* 48:1–101.

• ———. 1947. Mammals of northern Colombia, preliminary report no. 1: Squirrels (Sciuridae). *Proc. U.S. Nat. Mus.* 97:1–46.

• ———. 1948. Mammals of northern Colombia, preliminary report no. 2: Spiny rats (Echimyidae),

with supplemental notes on related forms. *Proc. U.S. Nat. Mus.* 97:125–40.

• ———. 1955. South American marsh rats, genus *Holochilus,* with a summary of sigmodont rodents. *Fieldiana: Zool.* 37:639–73.

• ———. 1959. Nomenclature of the Neotropical mammals described by Olfers, 1818. *J. Mammal* 40:337–53.

• ———. 1960. Mammals of northern Colombia, preliminary report no. 8: Arboreal rice rats, a systematic revision of the subgenus *Oecomys* genus *Oryzomys. Proc. U.S. Nat. Mus.* 110:513–68.

• ———. 1962. Evolution of Neotropical cricetine rodents (Muridae) with special reference to the phyllotine group. *Fieldiana: Zool.* 46:1–524.

• ———. 1966. South American swamp and fossorial rats of the scapteromyine group (Cricetinae, Muridae) with comments on the glans penis in murid taxonomy. *Z. Säugetierk.* 31:81–149.

———. 1971. A new rice rat of the *Oryzomys palustris* group from northwestern Colombia with remarks on distribution. *J. Mammal.* 52:700–709.

———. 1987. First South American record of Coues' marsh rat *Oryzomys couesi. J. Mammal.* 68:152–53.

Hill, R. W., and E. T. Hooper. 1971. Temperature regulation in mice of the genus *Scotinomys. J. Mammal.* 52:806–16.

Honacki, J. H., K. E. Kinman, and J. W. Koeppl, eds. 1982. *Mammal species of the world.* Lawrence, Kans.: Allen Press and Association of Systematics Collections.

Hooper, E. T. 1952. A systematic review of the harvest mice (genus *Reithrodontomys*) of Latin America. *Misc. Publ. Mus. Zool. Univ. Michigan* 77:1–255.

———. 1968. Habitats and food of amphibious mice of the genus *Rheomys. J. Mammal.* 49:550–53.

———. 1972. A synopsis of the rodent genus *Scotinomys. Occas. Pap. Mus. Zool. Univ. Michigan* 665:1–32.

Hooper, E.T., and M. D. Carleton, 1976. Reproduction, growth and development in two contiguously allopatric rodent species, genus *Scotinomys. Misc. Publ. Mus. Zool. Univ. Michigan* 151:1–52.

Hooper, E. T., and G. G. Musser. 1964. The glans penis in Neotropical cricetines, with comments on the classification of murid rodents. *Misc. Publ. Mus. Zool. Univ. Michigan* 123:1–57.

Husson, A. M. 1960. *De zoogdieren van de Nederlandse Antillen.* Vol. 2 of *Fauna Nederlands Antillen.* The Hague: Grurenhage and Willemstad.

• ———. 1978. *The mammals of Suriname.* Leiden: E. J. Brill.

Janzen, D. H. 1982. Attraction of *Liomys* mice to

horse dung and the extinction of this response. *Anim. Behav.* 30:483–89.

Jones, J. Knox, Jr., and H. H. Genoways. 1970. Harvest mice (genus *Reithrodontomys*) of Nicaragua. *Occas. Pap. Western Foundation Vert. Zool.* 2:1–16.

King, J. 1968. *The biology of* Peromyscus. Special Publication 1. Shippensburg, Pa.: American Society of Mammalogists.

Kleiman, D. G. 1970. Reproduction in the female green acouchi, *Myoprocta pratti. J. Reprod. Fertil.* 23:55–65.

———. 1971. The courtship and copulatory behavior of the green acouchi, *Myoprocta pratti. Z. Tierpsychol.* 29:259–78.

———. 1972. Maternal behavior of the green acouchi, *Myoprocta pratti*, the South American caviomorph rodent. *Behaviour* 43:48–84.

———. 1974. Patterns of behavior in hystricomorph rodents. In *The biology of hystricomorph rodents*, ed. I. W. Rolands and B. Weir, 171–209. London: Academic Press.

Kleiman, D. G., J. F. Eisenberg, and E. Maliniak. 1979. Reproductive parameters and productivity of caviomorph rodents. In *Vertebrate ecology in the northern Neotropics*, ed. J. F. Eisenberg, 173–83. Washington, D.C.: Smithsonian Institution Press.

Lander, E. 1974. Observaciones preliminares sobre Lapas *Agouti paca* (Linne, 1766) (Rodentia, Agoutidae) en Venezuela. Thesis, Universidad Central de Venezuela, Facultad de Agronomía, Instituto de Zoologica Agricola, Maracay.

Laval, R. K. 1976. Voice and habitat of *Dactylomys dactylinus* (Rodentia: Echimyidae) in Ecuador. *J. Mammal.* 57:402–4.

• Lemke, T. O., A. Cadena, R. H. Pine, and J. Hernandez-Camacho. 1982. Notes on opossums, bats, and rodents new to the fauna of Colombia. *Mammalia* 46(2):225–34.

Linares, O. J. 1969. Notas acerca de la captura de una rata acuática (*Nectomys squamipes*) en la cueva del agua. *Bol. Soc. Venez. Espeleol.* 2:31–34.

Macdonald, D. W. 1981. Dwindling resources and the social behaviour of capybaras (*Hydrochoerus hydrochaeris*) (Mammalia) *J. Zool.* (London) 194:371–91.

Maliniak, E., and J. F. Eisenberg. 1971. The breeding of *Proechimys semispinosus* in captivity. *Intl. Zoo Yearb.* 11:93–98.

Marcus, M. J. 1983. Sexual dimorphism, cheek pouch function and behavior in paca (*Agouti paca*, Rodentia). Abstract 14, 63d annual meeting, American Society of Mammalogists, University of Florida, Gainesville.

Mares, M., and R. A. Ojeda. 1982. Patterns of diversity and adaptation in South American hystricognath rodents. In *Mammalian biology in South America*, ed. M. A. Mares and H. H. Genoways, 393–432. Pymatuning Symposia in Ecology 6. Special Publication Series. Pittsburgh: Pymatuning Laboratory of Ecology, University of Pittsburgh.

Martin, R. E. 1970. Cranial and bacular variation in populations of spiny rats of the genus *Proechimys* (Rodentia: Echimyidae) from South America. *Smithsonian Contrib. Zool.* 35:1–19.

Matamoros de Rodriguez, Y. 1980. Investigaciones preliminares sobre la reproducción, comportamiento, alimentación y manejo del Tepezcuinte (*Cuniculus paca*) in cautiverio. In *Zoología neotropical: Actas del VIII Congreso Latinoamericano de Zoología, Mérida*, ed. P. Salinas, 961–94. N.p. [Venezuela].

Mayr, E. 1963. *Animal species and evolution.* Cambridge: Belknap Press of Harvard University Press.

Merritt, D. 1984. The pacarana, *Dinomys branickii*. In *One medicine*, ed. O Ryder and M. L. Byrd, 154–61. Berlin: Springer-Verlag.

Montgomery, G. G., and Y. Lubin, 1978. Movements of *Coendou prehensilis* in the Venezuelan llanos. *J. Mammal.* 59:887–88.

Moojen, J. 1952. Os reodores do Brasil. *Bibl. Cient. Brasil.* Ser. A-2 (Rio de Janeiro), 1–214.

Moore, J. C. 1959. Relationships among living squirrels of the Sciurinae. *Bull. Amer. Mus. Nat. Hist.* 118(4):159–206.

Musser, G. G., and A. L. Gardner. 1974. A new species of ichthyomyine, *Daptomys* from Peru. *Amer. Mus. Novitat.* 2537:1–23.

Musser, G. G., and M. M. Williams. 1985. Systematic studies of orizomyine rodents (Muridae): Definitions of *Oryzomys villosus* and *Oryzomys talamancae. Amer. Mus. Novitat.* 2810:1–22.

Myers, P., and M. D. Carleton. 1981. The species of *Oryzomys* (*Oligoryzomys*) in Paraguay and the identity of Azare's "rat sixième ou rat tarse noir." *Misc. Publ. Mus. Zool. Univ. Michigan* 161:1–41.

Negus, N. C., E. Gould, and R. K. Chipman. 1961. Ecology of the rice rat, *Oryzomys palustris* (Harlan), on Breton Island, Gulf of Mexico, with a critique of the social stress theory. *Tulane Studies Zool.* 8:95–123.

Newsome, A. E. 1971. The ecology of house mice in cereal haystacks. *J. Anim. Ecol.* 40:116.

Nitikman, L. Z. 1985. *Sciurus granatensis. Mammal. Species* 246:1–8.

O'Connell, M. A. 1981. Population ecology of small mammals from northern Venezuela. Ph.D. diss., Texas Tech University.

————. 1982. Population biology of North and South American grassland rodents: A comparative review. In *Mammalian biology in South America*, ed. M. A. Mares and H. H. Genoways, 167–86. Pymatuning Symposia in Ecology 6, Special Publication Series. Pittsburgh: Pymatuning Laboratory of Ecology, University of Pittsburgh.

Ojasti, J. 1964. Notas sobra el genero *Cavia* (Rodentia: Caviidae) en Venezuela con descripción de una nueva subespecies. *Acta Biol. Venez.* 4(3): 145–55.

————. 1972. Revisión preliminar de los picures o agustís de Venezuela. *Mem. Soc. Cient. Nat. La Salle* 32(93): 150–204.

————. 1973. *Estudio biológico del chigüire o capibara*. Caracas: Fundo Nacional de Investigaciones Agropecuarias.

————. 1978. The relation between population and reproduction of the capybara (*Hydrochoerus hydrochaeris*). Ph.D. diss., University of Georgia, Athens.

————. 1983. Ungulates and large rodents of South America. In *Ecosystems of the world*, vol. 13, *Tropical savannas*, ed. F. Bourliere, 427–39. Amsterdam: Elsevier.

Osgood, W. H. 1912. Mammals from western Venezuela and eastern Colombia. *Field Mus. Nat. Hist. Zool. Ser.*, publ. 155, vol. 10, no. 5: 32–66.

————. 1914. Four new mammals from Venezuela. *Field. Mus. Nat. Hist. Zool. Ser.*, publ. 175, vol. 10, no. 11: 135–41.

Patton, J. L. 1984. Systematic status of the large squirrels (subgenus *Urosciurus*) of the western Amazon basin. *Stud. Neotrop. Fauna Environ.* 19(2): 53–72.

————. 1987. Species groups of spiny rats, genus *Proechimys* (Rodentia: Echimyidae). In *Studies in Neotropical mammalogy: Essays in honor of Philip Hershkovitz*, ed. B. D. Patterson and R. M. Timm, 305–45. Fieldiana-Zoology, n.s., 39. Chicago: Field Museum of Natural History.

Patton, J. L., and L. Emmons. 1985. A review of the genus *Isothrix* (Rodentia, Echimyidae). *Amer. Mus. Novitat.* 2817: 1–14.

Patton, J. L., and A. L. Gardner. 1972. Notes on the systematics of *Proechimys* (Rodentia: Echimyidae), with emphasis on Peruvian forms. *Occas. Pap. Mus. Zool. Louisiana State Univ.* 44: 1–30.

Pearson, O. P. 1958. A taxonomic revision of the rodent genus *Phyllotis*. *Univ. Calif. Publ. Zool.* 56(4): 391–496.

Pearson, O. P., and J. L. Patton. 1976. Relationships among South American phyllotine rodents based on chromosome analysis. *J. Mammal.* 57: 339–50.

Petter, F. 1979. Un nouvelle espèce de rat d'eau de Guyane française *Nectomys parvipes*. *Mammalia* 43: 507–10.

Petter, F., and F. J. Band. 1981. *Calomys laucha* en Colombie (Rongeurs, Cricetinae, Phyllotini). *Mammalia* 45: 513–14.

Reig, O. A. 1980. A new fossil genus of South American cricetid rodents allied to *Wiedomys*, with an assessment of the Sigmodontinae. *J. Zool.* (London) 192: 257–81.

————. 1981. *Teoría del origen y desarrollo de la fauna de mamíferos de América del Sur*. Monografie Naturae. Mar del Plata, Argentina: Museo Municipal de Ciencias Naturales Lorenzo Scaglia.

————. 1984. Significado de los métodos citogenéticos para a distinción y la interpretación de los especies, con especial referrencia a los mamíferos. In *Actas de la III Reunion Ibero-Americana de Conservación y Zoología de Vertebrados*, 19–44. Buenos Aires: Museo Argentino de Ciencias Naturales "Bernardo Rivadavia."

Reig, O. A., M. Aguilera, M. A. Barros, and M. Useche. 1980. Chromosomal speciation in a rassenkreis of Venezuelan spiny rats (genus *Proechimys*, Rodentia, Echimyidae). *Genetica* 52/53: 291–312.

Reig, O. A., N. Olivo, and P. Kiblisky. 1971. The idiogram of the Venezuelan vole mouse, *Akodon urichi venezuelensis* Allen (Rodentia, Cricetidae). *Cytogenetics* 10: 99–114.

Reig, O. A., M. Trainer, and M. A. Barros. 1979. Sur l'identification chromosomique de *Proechimys guyannensis* et de *Proechimys cuvieri*. *Mammalia* 43: 501–5.

Reig, O. A., and M. Useche. 1976. Diversidad cariotípica y sistemática en poblaciones venezolanas de *Proechimys* (Rodentia, Echimyidae), con datos adicionales sobre poblaciones de Perú y Colombia. *Acta Cient. Venez.* 27: 132–40.

Roberts, M., S. Brand, and E. Maliniak. 1985. The biology of captive prehensile tailed porcupines, *Coendou prehensilis*. *J. Mammal.* 66: 476–82.

Rood, J. 1963. Observations on the behavior of the spiny rat, *Heteromys melanoleucus* in Venezuela. *Mammalia* 27: 186–92.

————. 1972. Ecological and behavioral comparisons in three genera of Argentine cavies. *Anim. Behav. Monogr.* 5: 1–83.

Rowlands, I. W., and B. J. Weir, eds. 1974. *The biology of hystricomorph rodents*. London: Academic Press.

Russell, R. J. 1968. Evolution and classification of the pocket gophers of the subfamily Geomyinae. *Univ. Kansas Publ. Mus. Nat. Hist.* 16: 473–579.

Schaller, G. B., and P. G. Crawshaw, Jr. 1981. Social organization in a capybara population. *Säugetierk. Mitt.* 29: 3–16.

Simpson, G. G. 1945. The principles of classification and a classification of the mammals. *Bull. Amer. Mus. Nat. Hist.* 85:1–350.

———. 1959. The nature and origin of supraspecific taxa. *Cold Spring Harbor Symp. Quant. Biol.* 24:255–72.

Smythe, N. 1970. Ecology and behavior of the agouti (*Dasyprocta punctata*) and related species on Barro Colorado Island, Panama. Ph.D. diss., University of Maryland.

———. 1978. The natural history of the Central American agouti (*Dasyprocta punctata*). *Smithsonian Contrib. Zool.* 257:1–52.

Smythe, N., W. E. Glanz, and E. Leigh, Jr. 1982. Population structure in some terrestrial frugivores. In *The ecology of a tropical forest*, ed. E. Leigh, Jr., A. Stanley Rand, and D. M. Windsor, 227–38. Washington, D.C.: Smithsonian Institution Press.

Starrett, A., and G. F. Fisler. 1970. Aquatic adaptations of the watermouse, *Rheomys underwoodi*. *Los Angeles Co. Mus. Contrib. Sci.* 182:1–14.

• Sturm, H., A. Abouchaar, R. de Bernal, and C. de Hoyos. 1970. Distribución de animales en las capas bajas de un bosque humedo tropical de la región Carare-Opon (Santander, Colombia). *Caldasia* 10(50):529–78.

• Tate, G. H. H. 1939. The mammals of the Guiana region. *Bull. Amer. Mus. Nat. Hist.* 76:151–229.

Tesh, R. B., 1970a. Notes on the reproduction, growth and development of echimyid rodents in Panama. *J. Mammal.* 51:199–202.

———. 1970b. Observations on the natural history of *Diplomys darlingi*. *J. Mammal.* 51:197–99.

Twigg, G. I. 1962. Notes on *Holochilus sciureus* in British Guiana. *J. Mammal.* 43:369–74.

Vandermeer, J. H. 1979. Hoarding behavior of captive *Heteromys desmarestianus* (Rodentia) on the fruits of *Welfia georgii*, a rainforest dominant palm in Costa Rica. *Brenesia* 16:107–16.

Vivas, A., and R. Roca. 1983. Uso del espacio por *Zygodontomys microtinus* (Rodentia, Cricetidae) en el estado Guárico. *Acta Cient. Venez.* 34(suppl. 1): 146.

Voss, R. S. 1983. Comparative morphology and systematics of ichthyomyine rodents (Muroidea). Ph.D. diss., University of Michigan.

———. 1988. Systematics and ecology of ichthyomyine rodents (Muroidea): Patterns of morphological evolution in a small adaptive radiation. *Bull. Mus. Nat. Hist.* 188(2): 259–493.

Voss, R. S., and A. V. Lindzey. 1981. Comparative gross morphology of male accessory glands among Neotropical Muridae (Mammalia: Rodentia) with comments on systematic implications. *Misc. Publ. Mus. Zool. Univ. Michigan* 159:1–41.

Voss, R. S., J. L. Silva, and J. A. Valdes. 1982. Feeding behavior and diets of Neotropical water rats, genus *Ichtyomys* Thomas, 1893. *Z. Säugetierk.* 47:364–69.

Weir, B. J. 1974. Reproductive characteristics of hystricomorph rodents. In *The biology of hystricomorph rodents,* ed. W. Rowlands and B. J. Weir, 265–99. London: Academic Press.

Wells, N. M., and J. Giacalone. 1985. *Syntheosciurus brochus. Mammal. Species* 249:1–3.

Wing, E. S. 1986. The domestication of animals in the high Andes. In *High altitude tropical biology,* ed. M. Monasterio and F. Vuilleamier, chap. 15, Oxford: Oxford University Press.

Wood, A. E. 1935. Evolution and relationship of the heteromyid rodents with new forms from the Tertiary of western North America. *Ann. Carnegie Mus.* 24:73–262.

———. 1955. A revised classification of the rodents. *J. Mammal.* 36:165–87.

———. 1974. The evolution of the Old World and New World hystricomorphs. In *The biology of hystricomorph rodents,* ed. W. Rowlands and B. J. Weir, 21–54. London: Academic Press.

Woods, C. 1982. The history and classification of South American hystricognath rodents: Reflections on the far away and long ago. In *Mammalian biology in South America,* ed. M. A. Mares and H. H. Genoways, 377–92. Pymatuning Symposia in Ecology 6. Special Publication Series. Pittsburgh: Pymatuning Laboratory of Ecology, University of Pittsburgh.

14 Order Lagomorpha

Diagnosis

Lagomorphs are small, digitigrade mammals with the tail either extremely short and well furred or virtually absent. The upper lip is cleft but is mobile and can be compressed in the midline. The dental formula for the family is variable: I 2/1, C 0/0, P 3/2, M 2–3/3. The first upper incisor has a longitudinal groove down the midline. The second pair of upper incisors is immediately behind the first, and this character is diagnostic (fig. 14.1).

Distribution

Aside from introduction by man, members of the Lagomorpha were found on all major continents and islands except Australia, New Zealand, and Antarctica. They were introduced by Europeans in Australia and New Zealand, where they have multiplied to become a serious pest.

History and Classification

There are two Recent families, the Ochotonidae and the Leporidae; both originated in the Oligocene. The oldest fossils believed to belong to this order trace

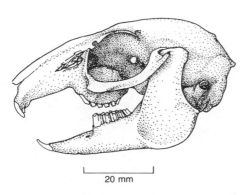

|⊢——— 20 mm ———⊣|

Figure 14.1. Skull of *Sylvilagus brasiliensis*.

from the Paleocene. Lagomorphs radiated and diversified in Eurasia and North America; they did not enter South America until the completion of the Panamanian land bridge at the end of the Pliocene. European hares and rabbits have been widely introduced in southern South America.

FAMILY LEPORIDAE
Rabbits and Hares

Diagnosis
The dental formula is I 2/1, C 0/0, P 3/2, M 3/3. The ears are rather long. The hind limbs are longer than the forelimbs and typify an adaptation for quadrupedal, ricochetal locomotion. There is a strong reduction of the first digit on the forefeet and hind feet. The tail is short but well haired, with a fluffy appearance.

Distribution
See the account for the order.

Genus *Sylvilagus* Gray, 1867
Cottontail Rabbit, Conejo

Description
These rather small rabbits have a head and body length ranging from 250 to 450 mm; the tail is 25 to 60 mm. Weight ranges from 400 to 2,300 g. There are thirteen species. The South American forms were reviewed by Tate (1933).

Distribution
The genus is distributed from southern Canada to Argentina, occupying a wide range of habitats from semiarid, brushy areas to forests. Two species occur in the region covered by this volume, *Sylvilagus brasiliensis* and *S. floridanus*. Their distributions in northern South America were reviewed by Hershkovitz (1950).

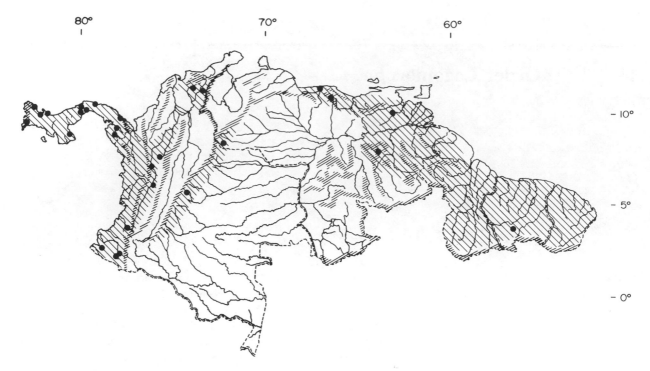

Map 14.1. Distribution of *Sylvilagus brasiliensis*.

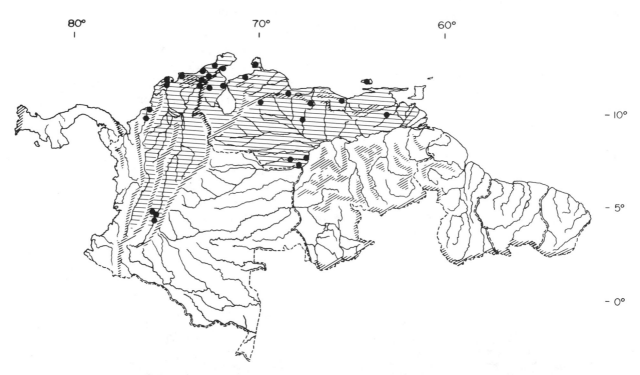

Map 14.2. Distribution of *Sylvilagus floridanus*.

Sylvilagus brasiliensis (Linnaeus, 1758)
Brazilian Cottontail

Description
Head and body length ranges from 360 to 400 mm, tail 20 to 30 mm, hind foot 77 to 86 mm, and ear 39 to 46 mm. Adult weight averages 785 g for males and 924 g for females. This small cottontail has a rather dark brown dorsum and often has white spots over the eyes. The underside of the tail is buff.

Range and Habitat
The Brazilian cottontail occurs from southern Veracruz, Mexico, south through Panama, Colombia, Venezuela, and Suriname to Brazil and northern Argentina (map 14.1). It can occur at up to 2,500 m (Hershkovitz 1950; Hoogmoed 1983).

Natural History
The ecology of this species in the paramos of Venezuela was addressed by Durant (1983, 1984), who studied the high-altitude race *S. b. meridensis*. He noted that this subspecies has a long gestation of 44.9 days compared with the 27-day gestation of *S. floridanus*. The litter size of *S. b. meridensis* is small, 3.9, and the reproductive cycle has an interval length of 270 days. The recruitment of this subspecies is extremely low. The Brazilian cottontail, throughout its habitat, is a crepuscular browser and may show seasonal decline in abundance.

Sylvilagus floridanus (J. A. Allen, 1890)

Description
Head and body length ranges from 335 to 397 mm, tail length 39 to 65 mm, hind foot 87 to 104 mm, and ear, from the notch, 49 to 68 mm. The dorsal pelage is gray brown and the venter white. The nape may be reddish in some races. The ventral surface of the tail is white (fig. 14.2).

Figure 14.2. *Sylvilagus floridanus.*

Range and Habitat
The cottontail rabbit is distributed from southern Manitoba, Canada, south to Costa Rica. Although it has not been taken in Panama to date, it appears again in the llanos of Colombia and Venezuela (map 14.2). In the South American part of its range it prefers scrub habitat and forest-edge habitat in the vicinity of grasslands. It is replaced in moist tropical forests and at high elevations by *S. brasiliensis*. There appears to be habitat segregation between the two species in northern South America.

Natural History
Cottontail rabbits are completely terrestrial and browse on the leaves of shrubs or graze on forbs and grasses. They do not dig a burrow but have a resting site, usually in a sheltered area. The rabbit is active in the early morning and late afternoon. There is usually a great deal of activity and feeding at night.

This is not a sociable species. Animals are usually seen alone or in small family units; they are widely spaced when feeding. During the breeding season a male may actively court a female by following her and springing over her.

Three young are born after a gestation period of twenty-six to thirty days. Before their birth the female digs a small despression in the ground and lines it with fur plucked from her breast. The young are placed in the nest and covered with the fur. The young are born with their eyes closed and with sparse fur. Fur grows rapidly, and the eyes open within four days. The female does not remain near the young but returns to nurse them at least once every twenty-four hours for about two weeks, then they are able to disperse from the nest site and forage on their own. In the northern parts of the cottontail's range the litter size may be larger, with a maximum of seven young recorded.

Ojeda-Castillo and Keith (1982) studied this species in Venezuela at two sites: the arid Falcón peninsula and the llanos of the state of Guárico. Sexual maturity is attained at 2.5 months of age, and the animals can potentially breed in any month. Litter size is small compared with northern populations, 2.64 and 2.38 for Falcón and Guárico, respectively. Seasonality of birth is strongly influenced by rainfall patterns.

References

Durant, P. 1983. Estudios ecológicos del conejo silvestre *Sylvilagus brasiliensis meridensis* (Lagomorpha: Leporidae) en los páramos de los Andes venezolanos. *Caribbean J. Sci.* 19(1–2):21–29.

———. 1984. Estudios ecológicos y necesidad de protección del conejo del páramo. *Rev. Ecol. Conserv. Ornit. Latinoamer.* 1:35–37.

Hershkovitz, P. 1950. Mammals of northern Colombia, preliminary report no. 6: Rabbits (Leporidae), with notes on the classification and distribution of the South American forms. *Proc. U.S. Nat. Mus.* 100:327–75.

Hoogmoed, M. S. 1983. The occurrence of *Sylvilagus brasiliensis* in Surinam. *Lutra* 26(1):34–45.

Ojeda-Castillo, M., and L. B. Keith, 1982. Sex and age composition and breeding biology of cottontail rabbit populations in Venezuela. *Biotropica* 14(1):99–107.

Tate, G. H. H. 1933. Taxonomic history of the Neotropical hares of the genus *Sylvilagus*. *Amer. Mus. Novitat.* 661:1–10.

15 Speciation and Faunal Affinities of Mammals in the Northern Neotropics

In chapter 1 I discussed the effect of temperate-zone glaciation events on the probable waxing and waning of forest in tropical South America. In brief, current theory holds that during glacial maxima there is increased aridity within the tropics. Thus xeric-adapted vegetation increases in area and multistratal tropical evergreen forests constrict to refugial areas, usually highlands or premontane forest zones. The net result is an isolation of segments of what were once contiguous populations. The remaining stocks are split into small surviving demes that can undergo genetic differentiation in separation from the parent stock. With changing climates allowing for more mesic conditions, there will be an increase in multistratal tropical evergreen forest, and populations adapted to those conditions can again expand. Populations that were once cut off from one another may now comingle (Prance 1982).

Riverine gallery forest is one means of dispersal by which mesic-forest-adapted forms can pass through drier areas. Temperate-zone forms can use mountain chains as dispersal corridors. Barriers include mountains for tropical forms and broad rivers for forms that do not readily swim. Other barriers may be more subtle, including areas of extremely low plant productivity. For example, the Guyana shield can act as a filter barrier (Eisenberg and Redford 1979). Given the very peculiar dissected topographic relief and the extraordinarily low fertility of soils over wide regions in the Guyana highlands, occupancy by certain mammalian populations may be severely retarded. This results in the paradoxical situation where the Guyana shield area may be a refugium for some species that have become adapted over the long term to the peculiar vegetational climax (e.g., *Callicebus torquatus*) while at the same time serving as a filter barrier to dispersal from the south into Venezuela and adjacent portions of the Guyanas (Eisenberg and Redford 1979).

Evidence for Speciation Events in the Northern Neotropics

Especially useful to our analysis of evolutionary trends has been the science of cytogenetics (Reig 1984). Two species may be nearly indistinguishable from external features yet differ profoundly in the structure of their chromosomes. Chromosome numbers are usually given in two forms of notation. The $2n$ number refers to an actual count of individual chromosomes in an organismic cell. The FN number, or fundamental number, refers to the number of "arms." That is to say, if one finds the centromere at a midpoint on the chromosome, then there are two "arms," but if the centromere is at one end the chromosome is "single armed." One refers to Robertsonian rearrangements in terms of changes in the $2n$ number but not the FN number. Robertsonian changes derive from centric fusions or fissions and whole-arm translocations. Non-Robertsonian chromosomal rearrangements involve changes in the FN number but not the $2n$ number. These include paracentric inversions, unequal translocations, and whole-arm heterochromatin additions or deletions. Robertsonian rearrangements in different populations need not necessarily involve incompatibility with respect to the production of viable offspring when the two populations are brought into contact. Non-Robertsonian rearrangements, however, may have profound consequences on reproductive isolation, and hybridization may be impossible (Nevo 1978).

If we turn to the bats or the cricetine rodents, we note that both increases and decreases in $2n$ and FN have occurred. The general trend through evolutionary time, however, is to decrease both $2n$ and FN numbers. Thus some workers conclude that the primitive stocks, or conservative stocks, have higher modal values whether one refers to the $2n$ or the FN number (Bickham and Baker 1979).

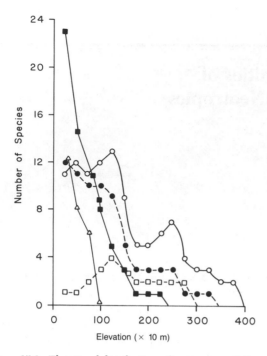

Figure 15.1. Elevational distributions of some mammals in Venezuela: *solid square*, Phyllostominae; *open square*, Sturnirinae; *triangle*, Emballonurinae; *open circle*, sigmodontine rodents; *solid circle*, Marsupalia. Note that all taxa show maximum species richness at lower elevations but that rodents have evolved a group of species adapted for higher elevations. For most bats, species richness declines precipitously above 1,000 m elevation, but the Sturnirinae seem to achieve greatest species richness in premontane forested areas. Data analyzed from Handley (1976).

A consequence of reduction in the $2n$ number is a tightening of the linkage groups and therefore a reduction of recombination potential and the stabilizing of a phenotype. We can refer to this as the process of evolutionary canalization. One way to estimate this process is to develop a ratio of FN to $2n$ numbers. A high ratio indicates a great degree of retaining two-armed chromosomes, and one suspects a reduced speciation potential because Robertsonian mechanisms are not favored. A low FN/$2n$ ratio, however, implies a greater speciation potential by reducing the possibility of hybridization between stocks isolated and at some subsequent time coming into contact (Bickham and Baker 1979).

Studies of chromosome numbers in bats (Baker et al. 1982) and rodents (Gardner and Patton 1976; Reig and Useche 1976) have allowed us some insight into the speciation process and have confirmed the notion that refugia were imporatnt not only in conserving populations but in potentiating speciation during the Pleistocene glaciation sequence. I envision that today we are witnessing a very dynamic situation in the Neotropics. As we progress century by century into a more and more mesic phase, savanna areas are in transition toward multistratal

tropical evergreen forest. Given this dynamism, it is no wonder that the ranges of some species are difficult to define.

Figures 15.1 and 15.2 indicate major barriers to dispersal of terrestrial mammals. The Andean chain is a barrier to tropical forms but a conduit for temperate-adapted forms. the Orinoco and Essequibo rivers appear to be barriers to the dispersal of terrestrial mammals. As indicated in chapter 1, the Guyana highlands complex, because of the reduced fertility of soils, can itself serve as a filter barrier to movement from south to north. Figure 15.3 suggests the probable dispersal routes for lowland tropical forms and temperate or highland forms. As one can see, there has been northern expansion from Brazilian refugial areas into the Guyanas, southern Venezuela, and southern Colombia. It is possible to imagine successive waves of movement into the biogeographically defined areas, since one can conceive of Amazonian rain forest derived forms expanding and contracting their ranges. One can actually postulate both early Amazonian derivatives in the north and late Amazonian derivatives advancing northward. Figure 15.4 portrays such a hypothetical distribution. The llanos, depauperate and having few endemics, is constantly being recolonized through

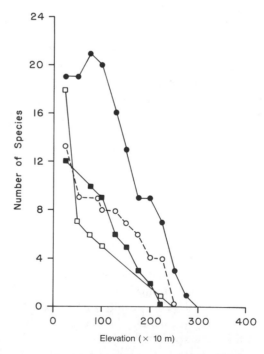

Figure 15.2. Elevational distributions of some bats in northern Venezuela: *solid circle*, Stenoderminae; *open circle*, Vespertilionidae; *solid square*, Glossophaginae; *open square*, Molossidae. Note that the Stenoderminae can reach maximum species richness in premontane forests. The Molossidae decline rapidly in species richness above 500 m. Data analyzed from Handley (1976).

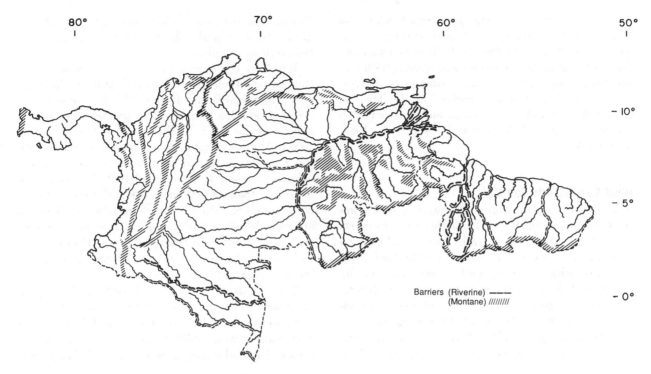

Figure 15.3. Major barriers to mammalian dispersal as indicated by contemporary distributions. The eastern cordillera of the Andes has been an important barrier to faunal interchange by nonvolant mammals. The Orinoco River also has acted as an important barrier, and apparently the Essequibo River and its tributaries in Guyana have been important in retarding emigration. The distribution of many forms in southern Colombia seems to be confined to the south of the Río Japurá.

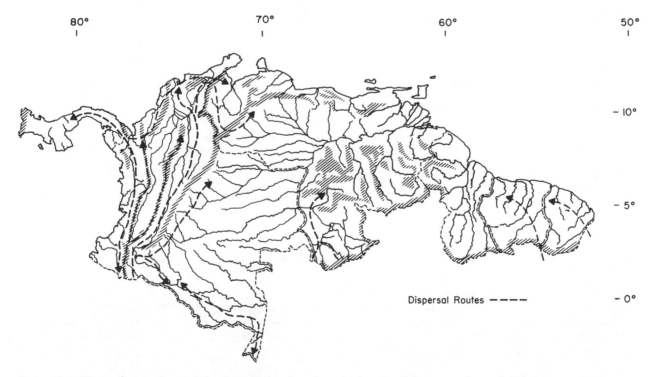

Figure 15.4. Dispersal routes for tropical and temperate-adapted species from South America to the north. Montane corridors have clearly been important for those species adapted to higher elevations. The premontane forest refugia along the eastern flank of the Andes have been important reservoirs for future dispersal that may still be in progress. Dispersal from the west Brazilian Amazonian region to Colombia may still be going on, as are dispersal events from northeastern Brazil into the Guyanas.

gallery forest conduits. The premontane forests of Ecuador and Peru serve as important refugial centers for emigration to the north. One can imagine distributional patterns waxing and waning with late Amazonian derivatives, early Amazonian derivatives, and those refugial pockets in the Guyana highlands continuing to expand and contract with variations in global climatic stability (Prance 1982; Reig 1984).

What Can Islands Tell Us?

Islands can often be useful indicators of speciation rates, especially offshore islands that were formerly connected with the mainland. Species richness on islands is a function of the size and of the land mass, its topographic diversity, and its proximity to the mainland. Off the Pacific coast of Panama are two famous islands, Isla Coiba and the archipelago associated with Isla de las Perlas. Coiba is about 520 km², and the archipelago of las Perlas is approximately 400 km². These land masses were once part of the Panamanian mainland, but following the Wisconsin glacial collapse and subsequent rise in sea level, they remain islands.

Bats can colonize over water with some ease; thus, given their poor dispersal ability, the non-volant mammal fauna of these islands is of some interest. Coiba has six nonvolant species of mammals and las Perlas has five (Hall and Kelson 1959; Hall 1981). All the species show distinct affinities with the Panamanian mainland and are at best subspecifically distinct. It is assumed that they were isolated less than 12,000 years B.P. Of interest to the zoogeographer, when the offshore islands of Panama are considered, is that almost all nonvolant species are either South American endemics (Didelphidae, Cebidae, hystricognath rodents) or early immigrants (Miocene) to South America (*Potos*).

The islands off the coast of Venezuela have great intrinsic interest for zoogeographers since there are many more islands of diverse size, and they may reflect endemism deriving from past xeric and mesic cycles. Trinidad is large and has probably been sepa-

Table 15.1   Mammals of Trinidad

MARSUPIALIA	Heteromyidae	*Peropteryx macrotis*	*Vampyrops helleri*
Didelphidae	*Heteromys anomalus*	*Diclidurus albus*	*Vampyrodes caraccioli*
Didelphis marsupialis	Echimyidae	Noctilionidae	*Chiroderma villosum*
Marmosa mitis	*Echimys armatus*	*Noctilio leporinus*	*Chiroderma trinitatum*
Marmosa fuscata	*Proechimys trinitatus*	Mormoopidae	*Artibeus jamaicensis*
Caluromys philander	Erethizontidae	*Pteronotus parnelli*	*Artibeus lituratus*
Chironectes minimus	*Coendu prehensilis*	*Pteronotus personatus*	*Artibeus cinereus*
PRIMATES	Dasyproctidae	*Pteronotus davyi*	*Enchisthenes hartii*
Cebidae	*Dasyprocta agouti*	*Mormoops megalophylla*	*Ametrida centurio*
Alouatta seniculus	Agoutidae	Phyllostomidae	*Centurio senex*
Cebus olivaceus	*Agouti (Cuniculus) paca*	*Micronycteris megalotis*	*Desmodus rotundus*
EDENTATA	CARNIVORA	*Micronycteris minuta*	*Diaemus youngi*
Myrmecophagidae	Procyonidae	*Micronycteris hirsuta*	Natalidae
Tamandua tetradactyla	*Procyon cancrivorus*	*Micronycteris brachyotis*	*Natalus tumidirostris*
Cyclopes didactylus	Mustelidae	*Micronycteris nicefori*	Furipteridae
Dasypodidae	*Lutra longicaudis*	*Micronycteris sylvestris*	*Furipterus horrens*
Dasypus novemcinctus	*Eira barbara*	*Lonchorhina aurita*	Thyropteridae
RODENTIA	Viverridae	*Tonatia bidens*	*Thyroptera tricolor*
Sciuridae	*Herpestes auropunctatus*	*Tonatia minuta =*	Vespertilionidae
Sciurus granatensis	(Intr.)	*brasiliense* (?)	*Myotis keaysi*
Muridae	Felidae	*Mimon crenulatum*	*Myotis riparius*
Oryzomys concolor	*Felis pardalis*	*Phyllostomus discolor*	*Myotis nigricans*
Oryzomys capito (=	SIRENIA	*Phyllostomus hastatus*	*Lasiurus borealis*
Oryzomys laticeps)	Trichechidae	*Trachops cirrhosus*	*Lasiurus ega*
Oryzomys delicatus	*Trichechus manatus*	*Vampyrum spectrum*	*Rhogeessa tumida*
Akodon urichi	ARTIODACTYLA	*Glossophaga longirostris*	*Eptesicus brasiliensis*
Zygodontomys	Tayassuidae	*Glossophaga soricina*	Molossidae
brevicauda	*Tayassu tajacu*	*Anoura geoffroyi*	*Molossops greenhalli*
Rhipidomys couesi	Cervidae	*Choeroniscus*	*Tadarida europs*
Nectomys squamipes	*Mazama americana*	*intermedius*	*Promops centralis*
Rattus rattus (Intr.)[a]	CHIROPTERA	*Carollia perspicillata*	*Promops nasutus*
Rattus norvegicus (Intr.)	Emballonuridae	*Sturnira lilium*	*Molossus ater*
Mus musculus (Intr.)	*Rhynchonycteris naso*	*Sturnira tildae*	*Molossus major*
	Saccopteryx bilineata	*Uroderma bilobatum*	*Molossus trinitatus*
	Saccopterys leptura		*Eumops auripendulus*

Sources: Alkins (1979) and Goodwin and Greenhall (1961), as amended by Gary Morgan. Also Carter et al. (1981).
[a](Intr.) = introduced by Europeans.

Table 15.2   Mammals of Tobago

MARSUPIALIA	Molossidae
Didelphidae	*Molossus molossus*
Didelphis marsupialis	EDENTATA
Marmosa robinsoni	Dasypodidae
Marmosa murina	*Dasypus novemcinctus*
CHIROPTERA	RODENTIA
Emballonuridae	Sciuridae
Saccopteryx leptura	*Sciurus granatensis*
Peropteryx macrotis	Heteromyidae
Mormoopidae	*Heteromys anomalus*
Pteronotus parnellii	Muridae
Noctilionidae	*Akodon urichi*
Noctilio leporinus	*Nectomys squamipes*
Phyllostomidae	*Oryzomys concolor*
Micronycteris megalotis	*Rhipidomys venezuelae*
Phyllostomus hastatus	*Zygodontomys brevicauda*
Glossophaga longirostris	Echimyidae
Carollia perspicillata	*Echimys armatus*
Vampyrodes caraccioli	Dasyproctidae
Chiroderma villosum	*Dasyprocta agouti*
Sturnira lilium	CARNIVORA
Artibeus cinereus	Procyonidae
Artibeus jamaicensis	*Procyon cancrivorus*
Artibeus lituratus	ARTIODACTYLA
Centurio senex	Tayassuidae
Natalidae	*Tayassu tajacu*
Natalus tumidirostris	Cervidae
Vespertilionidae	*Mazama americana*
Myotis nigricans	
Eptesicus brasiliensis	
Lasiurus borealis	

Source: Goodwin and Greenhall (1961), as amended by Gary Morgan.

Table 15.3   Mammals of Isla Margarita

CHIROPTERA	PRIMATES
Emballonuridae	Cebidae
Peropteryx trinitatus	*Cebus apella*
Saccopteryx leptura	EDENTATA
Noctilionidae	Dasypodidae
Noctilio leporinus	*Dasypus novemcinctus*
Mormoopidae	LAGOMORPHA
Pteronotus parnellii	Leporidae
Mormoops megalophylla	*Sylvilagus floridanus*
Phyllostomidae	RODENTIA
Micronycteris megalotis	Sciuridae
Phyllostomus discolor	*Sciurus granatensis*
Glossophaga longirostris	Heteromyidae
Glossophaga soricina	*Heteromys anomalus*
Leptonycteris curasoae	Muridae
Carollia perspicillata	*Rhipidomys couesi*
Artibeus jamaicensis	*Oryzomys* sp.
Artibeus lituratus	*Zygodontomys brevicauda*
Desmodus rotundus	Echimyidae
Diaemus youngi	*Echimys semivillosus*
Vespertilionidae	CARNIVORA
Rhogeessa minutilla	Mustelidae
Molossidae	*Conepatus semistriatus*
Molossus molossus	Felidae
MARSUPIALIA	*Felis pardalis*
Didelphidae	ARTIODACTYLA
Marmosa robinsoni	Cervidae
Caluromys philander	*Odocoileus virginianus*

Source: Musso Q. (1962); Bisbal (1983); and Smith and Genoways (1974), as amended by Gary Morgan.

rate from the mainland for less than 15,000 years. Twenty-seven families of mammals occur on the island and include ninety-three species (not counting four species established and introduced by Europeans). Sixty-one species of the ninety-three (66%) are members of the order Chiroptera. Since bats make up on the average 52% of a Neotropical assemblage (Eisenberg 1981), one may speculate that the bats have found it simpler to colonize or recolonize over water than is the case with nonvolant forms. The Trinidad mammal fauna is clearly derived from Venezuela, and while many species are subspecifically distinct, with the possible exception of *Proechimys trinitatus* and *Rhipidomys couesi* (*R. sclateri couesi*), all forms are conspecific with mainland forms (table 15.1).

Tobago was connected with Trinidad some 10,000 to 15,000 years ago. It is much smaller and currently supports thirty-six species of mammals. Of the thirty-six, bats constitute 60%, a figure remarkably similar to that for Trinidad. With the possible exception of *Rhipidomys nitela* (*R. mastacalis nitela*), all species are derived from mainland forms (table 15.2).

Isla Margarita is rather xeric and supports less tropical evergreen forest than does Trinidad. It cur-

Table 15.4   Mammals of Grenada

CHIROPTERA	Vespertilionidae
Emballonuridae	*Myotis nigricans*
Peropteryx macrotis	Molossidae
Noctilionidae	*Molossus molossus*
Noctilio leporinus	MARSUPIALIA
Mormoopidae	Didelphidae
Pteronotus davyi	*Marmosa robinsoni*
Phyllostomidae	*Didelphis marsupialis* (Intr.)[a]
Micronycteris megalotis	RODENTIA
Glossophaga longirostris	Dasyproctidae
Carollia perspicillata	*Dasyprocta agouti* (Intr.)
Anoura geoffroyi	EDENTATA
Artibeus cinereus	Dasypodidae
Artibeus jamaicensis	*Dasypus novemcinctus*
Artibeus lituratus	

Source: Hall (1981), as amended by Gary Morgan.
[a](Intr.) = probably introduced by man.

Table 15.5   Mammals of Bonaire

CHIROPTERA	Vespertilionidae
Mormoopidae	*Myotis nigricans*
Mormoops megalophylla	Molossidae
Phyllostomidae	*Molossus pygmaeus*
Glossophaga longirostris	RODENTIA
Leptonycteris nivalis	Muridae
(*Ametrida minor*) *A.*	*Thomasomys* sp. (Foss.)[a]
centurio	*Rattus rattus* (Intr.)[b]
	Mus musculus (Intr.)

Source: Husson (1960), as amended by Gary Morgan.
[a](Foss.) = fossil form.
[b](Intr.) = introduced by Europeans.

rently supports only thirty-one species, of which 55%, or seventeen, are bats. All species are derived from the mainland. *Cebus apella* is far from its mainland range, suggesting that the species was originally transported to the island by Amerindians (table 15.3).

Grenada is not within the range covered by this book, but its fauna is clearly derived from South America. This volcanic island was never connected with the mainland. Sixteen species of mammals occur there, and predictably the volant bats make up almost 80% of the total (twelve). Although *Marmosa robinsoni* may have rafted there, it is entirely possible that the agouti, armadillo, and opossum were transported as game animals by the Amerindians (table 15.4).

The Netherlands Antilles present some interesting distributions (tables 15.5–15.7). These islands also support considerable xeric vegetation, and this contributes to a lower species richness. Aruba and Curaçao are closer to the mainland, and the latter is of reasonable size. Leaving aside the fossil forms and the Amerindian introductions, Aruba has six mammals, of which five are bats; Curaçao has twelve species, of which eight are bats. Both islands have *Calomys hummelincki*, which also occurs in the xeric mainland areas around the Gulf of Venezuela. Bonaire has six species of bats.

Isolation of nonvolant mammals on islands off the coast of northern South America since the last rise in sea level demonstrates several interesting trends. Little specific distinctiveness has evolved, but subspecific differences can be noted. Some larger island forms have become extinct. Bats are able to colonize over water with greater ease than nonvolant forms. Finally, Curaçao and Aruba have preserved one xeric-adapted rodent with a very restricted distribution in the north (e.g., *Calomys*).

Table 15.6   Mammals of Curaçao

CHIROPTERA	*Megalomys curazensis* (Foss.)[a]
Noctilionidae	
Noctilio leporinus	*Rattus rattus* (Intr.)[b]
Mormoopidae	*Rattus norvegicus* (Intr.)
Mormoops megalophylla	*Mus musculus* (Intr.)
Phyllostomidae	Hydrochaeridae
Glossophaga longirostris	*Hydrochaeris hydrochaeris* (Foss.)
Leptonycteris nivalis	
Artibeus jamaicensis	Dasyproctidae
Natalidae	*Dasyprocta* sp. (Ind.)[c]
Natalus tumidirostris	LAGOMORPHA
Vespertilionidae	Leporidae
Myotis nigricans	*Sylvilagus floridanus*
RODENTIA	ARTIODACTYLA
Muridae	Cervidae
Calomys hummelincki	*Odocoileus virginiana*
Oryzomys sp. (Foss.)[a]	

Source: Husson (1960), as amended by Gary Morgan.
[a](Foss.) = fossil form.
[b](Intr.) = introduced by Europeans.
[c](Ind.) = probably transported to the islands by early Amerindians.

Table 15.7   Mammals of Aruba

CHIROPTERA	*Mus musculus* (Intr.)
Emballonuridae	Dasyproctidae
Peropteryx macrotis	*Dasyprocta* sp. (Ind.)[b]
Mormoopidae	CARNIVORA
Mormoops megalophylla	Canidae
Phyllostomidae	*Dusicyon* cf. *thous* (Ind.)
Glossophaga longirostris	Felidae
Leptonycteris nivalis	*Felis* cf. *tigrina* (Ind.)
RODENTIA	LAGOMORPHA
Muridae	Leporidae
Calomys hummelincki	*Sylvilagus floridanus*
Rattus rattus (Intr.)[a]	

Source: Husson (1960), as amended by Gary Morgan.
[a](Intr.) = introduced by Europeans.
[b](Ind.) = probably transported to the islands by early Amerindians.

What Can Mountaintops Tell Us?

Isolated mountain ranges with strong altitudinal gradients permit temperate-zone conditions to exist in the tropics. During glacial maxima the cooler climates characteristic of mountaintops can spread to lower elevations, thus offering an appropriate habitat for species of mammals so adapted. When climatic changes occur during a warming trend, the montane-adapted species may be more and more restricted elevationally. In addition, lower mountain ranges can permit more mesic conditions to persist during a drying phase as a result of the adiabatic cooling effect as air masses move over mountain ranges, changing temperature and thereby often releasing moisture. Clearly, the high Andes with their great elevation have allowed temperate conditions to persist for considerable periods in the northern Neotropics. The north coast range of Venezuela has in a similar manner allowed mesic conditions to persist, but because of its much lower elevation it has not always maintained temperate conditions.

Even between 10° and 5° north latitude, elevation profoundly influences the species richness of bats and rodents. Consider figures 15.1 and 15.2, which demonstrate species richness over elevational gradients in the north coast range and the Andes of Venezuela (Handley 1976). It will be appreciated that bat species richness declines greatly at elevations exceeding 1,000 m. Rodents decline at a lesser rate, only to show an increase again as one reaches the higher elevations where the Andean-adapted forms occur.

In the state of Bolívar and in parts of Territorio Amazonas in southern Venezuela there exist high

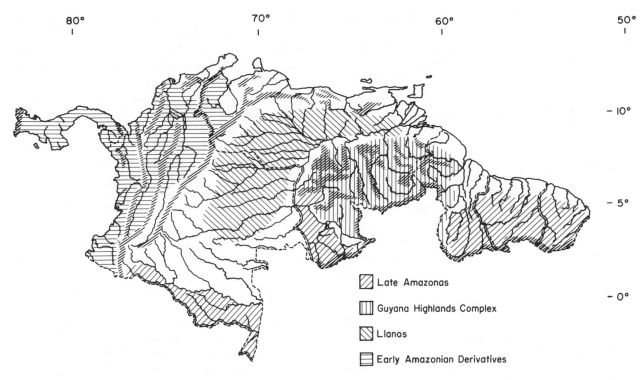

80°   70°   60°   50°

Figure 15.5. Hypothetical faunal divisions of northern South America according to the "age" of the major faunal components. Early Amazonian derivatives apparently passed in extreme western South America into Panama and into the Maracaibo basin by skirting around through the Falcón peninsula in more mesic times. By surviving in refugial areas in Panama and Costa Rica they have been able to reinvade this western zone. The Guyana highlands complex reflects the heterogeneity of the terrain, and some endemics are to be found, especially those at higher altitudes. Late Amazonian reentrants have probably occurred in the past 12,000 years, primarily in the Guyana region and in southeastern Colombia. The llanos region fauna is derived from adjoining areas (see fig. 15.4).

pinnacles left from the erosion of the Guyana shield. These are referred to as tepuis. One may well ask, On such isolated table mountains, what sorts of mammals occur, and what are their formal affinities? Tate (1939) noted that on those tepuis he had studied, certain mammalian species showed Andean affinities. Gardner (n.d.) has confirmed this phenomenon and extended the species list. The suggestion is that during times of maximal glaciation and cooling, certain forms adapted to higher elevations could then disperse through the lowlands and colonize the tepuis. With subsequent warming trends one would again find an isolation on mountaintops and the animals reoccupying higher altitudes. The sporadic distribution of Andean-related forms among the tepuis suggests that extinction events have occurred, probably in proportion to the size of the tepui itself, thus conforming in part to some of the basic rules of island biogeography. Clearly, the study of small mammals isolated on mountains such as the tepuis, which in fact are inselbergs, offers promise for the future understanding not only of climatic cycles but of speciation events and the rapidity with which they can occur (Gardner, n.d.).

Although biogeography is often a neglected topic in the curricula of universities, its pivotal impor-

tance in understanding the speciation process is now being rediscovered. Maps of species distributions are the starting point for formulating hypotheses. Much more collecting will be necessary before distributional limits for many species can be established with confidence. Seven nations have contributed significantly to our understanding of the mammalian fauna of the northern Neotropics. I hope that the international spirit will prevail in the future for the benefit of all.

References

Alkins, M. E. 1979. *The mammals of Trinidad.* Occasional Paper 2. Saint Augustine, Trinidad: Department of Zoology, University of the West Indies.

Baker, R. J., M. W. Haiduk, L. W. Robbins, A. Cadena, and B. F. Koop. 1982. Chromosomal studies of South American bats and their systematic implications. In *Mammalian biology in South America,* ed. M. A. Mares and H. H. Genoways, 303–27. Pymatuning Symposia in Ecology 6, Special Publication Series. Pittsburgh: Pymatuning Laboratory of Ecology, University of Pittsburgh.

Bickham, J. W., and R. J. Baker. 1979. Canalization

model of chromosome evolution. *Bull. Carnegie Mus. Nat. Hist.* 13:70–84.

Bisbal, F. J. 1983. Dos nuevos mamíferos para la Isla de Margarita, Venezuela. *Acta Cient. Venez.* 34:366–67.

Carter, C. H., H. H. Genoways, R. S. Loregnard, and R. J. Baker. 1981. Observations on bats from Trinidad, with a checklist of species occurring on the island. *Occas. Pap. Mus. Texas Tech. Univ.* 72:1–27.

Eisenberg, J. F. 1981. *The mammalian radiations: An analysis of trends in evolution, adaptation, and behavior.* Chicago: University of Chicago Press.

Eisenberg, J. F., and K. H. Redford. 1979. A biogeographic analysis of the mammalian fauna of Venezuela. In *Vertebrate ecology in the northern Neotropics,* ed. J. F. Eisenberg, 31–38. Washington, D.C.: Smithsonian Institution Press.

Gardner, A. L. n.d.. Two new mammals from southern Venezuela and comments on the affinities of the highland fauna of Cerro de la Neblina. In *Advances in Neotropical mammalogy,* ed. K. H. Redford and J. F. Eisenberg. Leiden: E. J. Brill. In press.

Gardner, A. L., and J. L. Patton. 1976. Karyotypic variation in oryzomine rodents with comments on chromosomal evolution in the Neotropical cricetine complex. *Occas. Pap. Mus. Zool. Louisiana State Univ.* 49:1–48.

Goodwin, G. G., and A. M. Greenhall. 1961. A review of the bats of Trinidad and Tobago. *Bull. Amer. Mus. Nat. Hist.* 122(3):191–301.

Hall, E. R. 1981. *The mammals of North America,* 2d ed., 2 vols. New York: John Wiley.

Hall, E. R., and K. Kelson. 1959. *The mammals of North America,* 2 vols. New York: Ronald Press.

Handley, C. O., Jr. 1976. Mammals of the Smithsonian Venezuela project. *Brigham Young Univ. Sci. Bull. (Biol. Ser.)* 20(5):1–89.

Husson, A. M. 1960. *Mammals of the Netherlands Antilles.* Curaçao: Natuurwetenschappelijke Werkgroep Nederlandse Antillen.

Musso Q., A. 1962. Lista de los mamíferos conocidos de la Isla de Margarita. *Mem. Soc. Cienc. Nat. La Salle* 22(63):163–80.

Nevo, E. 1978. Genetic variation in natural populations: Patterns and theory. *Theor. Pop. Biol.* 13:269–305.

Prance, G. T. 1982. A review of the Pleistocene climatic changes in the Neotropics. *Ann. Missouri Bot. Gard.* 69:594–624.

Reig, O. A. 1984. Significado de los metodos citogenéticos para a distinción y la interpretación de las especies, con especial referencia a los mamíferos. In *Actas de la III Reunion Ibero-America de Conservación y Zoología de Vertebrados,* 19–44. Buenos Aires: Museo Argentino de Ciencias Naturales "Bernardo Rivadavia."

Reig, O. A., and M. Useche. 1976. Diversidad cariotípica y sistemática en poblaciones venezolanas de *Proechimys* (Rodentia, Echimyidae), con datos adicionales sobre poblaciones de Perú y Colombia. *Acta Cient. Venez.* 27:132–40.

Smith, J. D., and H. H. Genoways. 1974. Bats of Margarita Island, Venezuela, with zoogeographic comments. *Bull. Southern Calif. Acad. Sci.* 73(2):64–79.

Tate, G. H. H. 1939. The mammals of the Guiana region. *Bull. Amer. Mus. Nat. Hist.* 76(5):151–229.

Mammalian Community Ecology

The Structure of Mammalian Communities

Mammalian communities can be analyzed in terms of niche occupancy. The word *niche* commonly refers to the status of an organism within its community and ecosystem—not only the space it lives in but also what it does. This status is the outcome of evolutionary processes, and selection has adapted each organism to make its living in a particular way. The structure of an animal determines how it makes its living, since bats have wings and can fly, opossums have prehensile tails and can climb, and armadillos have stout claws and can dig. Defining a niche, then, requires studying the natural history of the species in question, and one can make seemingly endless measurements to describe how the organism forages, interacts with other members of the community, and ultimately reproduces the next generation. In practice, however, a few simple parameters are usually measured. These include the substrate the animal uses in the course of its daily activities, its activity rhythm (diurnal, nocturnal, or crepuscular), and its trophic level—what normally makes up its diet.

When a group of species share a number of niche features—and these are usually congeners—we refer to the members as a guild (Root 1967). A guild, then, is a group of animals, often closely related, that make their living in a very similar way but usually can be divided by some discrete feature (i.e., size of prey taken) so that they are not totally in competition with one another. Ecologists generally recognize that competition is an ever-present feature of community structure and is one of the selective forces that tends to maintain species in particular niches within the same community.

The impact a species may have on a community is in part a function of its abundance, which varies according to its trophic status and its body size. Large forms tend to exist at lower densities than small forms, and herbivores occur at higher densities than carnivores. The standing crop biomass, or average weight of a species for some unit area, gives us some idea of the dominance of a species within an ecosystem. In other words, large animals may be numerically rare but maintain a rather high standing crop biomass. Small animals may be numerically abundant but maintain a low standing crop biomass (Eisenberg 1980). Ultimately the rate of energy flow into a population determines its biomass. Where primary productivity is strongly pulsed in the Neotropics, deriving from a seasonal distribution of rainfall, then all secondary productivity will be entrained to the seasonal availability of nutrients (Eisenberg, O'Connell, and August 1979).

One must also take into account that in any population there is always a trade-off between energy for maintenance and energy for growth. Any factor that affects metabolic rate will indirectly affect the standing crop biomass levels and ultimately affect potential productivity. Mammals are for the most part endothermic, so without an environmental heat input such as sunlight they usually can maintain a rather constant body temperature. Members of the order Chiroptera are a notable exception in that many of the vespertilionid bats allow their body temperature to drop during the day when they are at rest. Some bats then, although they are endothermic, are not true homeotherms. Homeothermy, the maintenance of a constant body temperature throughout a daily cycle, is not uniformly practiced within the class Mammalia, and some species employ a pronounced diel fluctuation in their core temperature to conserve energy (McNab 1978). Those homeothermic mammals that we know also show a decrease in metabolic rate as absolute body size increases. Metabolic rate when regressed logarithmically against body size tends to decrease with a slope of -0.25 (Kleiber 1961). Conversely, the actual amount of energy an animal uses in a given time tends to increase allometrically with body size at the 0.75 power. Thus, although the metabolic rate of large mammals is lower than that of small mammals, an individual large mammal consumes more food, and thus the larger forms in aggregate have a more significant impact on their communities. The total energy consumed per day by one white-tailed deer (*Odocoileus*) may equal that used by six hundred cane rats (*Zygodontomys*). McNab (1974) has proposed that some trends in the lowering of metabolic rate parallel adaptations for different trophic strategies, which are often accompanied by an energy adaptation and may involve lowering the metabolic rate relative to the values predicted for a "typical" mammal. Burrowing, herbivorous mammals tend to have low basal metabolic rates; ant-eating animals likewise show lowered rates.

As a rule of the thumb, as body size increases a species uses a larger home range (McNab 1963). Home-range size, however, also varies according to trophic strategy. The home range of a given herbivore is smaller than the home range of an equal-weight carnivore. And home-range size for any species will vary depending upon the carrying capacity of the habitat. Some habitats are richer in nutrients and support higher densities of a given species than habitats less favored. Carrying capacity ultimately reflects plant productivity.

Primary productivity may vary widely in the tropics for a variety of reasons. Soil fertility and the uniformity of rainfall are the most common factors influencing plant productivity in the lowland tropics. Of course temperature decreases as one moves to higher altitudes, and temperature itself can limit plant productivity at high elevations within the tropics.

Numerical density of mammals is inversely related to body size, but standing crop biomass tends to be positively related to the absolute average size of a species (Eisenberg and Lockhart 1972). Larger mammals tend to be long-lived and usually have an iteroparous mode of reproduction, with a small litter size and a rather extended interbirth interval (Eisenberg 1981). The reproductive activity of a given female of an iteroparous form is spread over a long period. Longevity also tends to be tied to trophic strategy; large herbivores usually live longer than carnivores. Thus one can make generalizations when predicting the life history of a species: for example, large herbivores will have a tendency toward extreme iteroparity; small herbivores will tend to be short-lived and have large litters. Thus there is an interrelation between ecological energetics, trophic strategies, and demography.

Mammalian Community Structure in Lowland Panamanian Rain Forests

Eisenberg and Thorington (1973) analyzed a mammal community on Barro Colorado Island in Panama. Some of the following generalizations seem to hold for lowland tropical rain forests across northern South America. A mature tropical rain forest has a closed canopy, and thus the undergrowth is often sparse and opportunities for terrestrial browsing are less than in more open forests or early second-growth forests. There appears to be a shift away from the terrestrial niche dominance so characteristic of savanna habitats to an exploitation of the rich canopy resources. Thus arboreal mammals become more important components of the system. If we consider all mammals other than bats, we can make the following generalizations. Adaptations for feeding on leaves such as those shown by howler monkeys, three-toed sloths, and two-toed sloths allow animals to exploit canopy foliage, and thus these three species dominate standing crop biomass. Howler monkeys are not obligate folivores, however, and eat a considerable quantity of fruit. Arboreal frugivores such as the kinkajou and some other primate species can exist at appreciable densities. Many forms are adapted for a scansorial existence, foraging both on the ground and in the canopy. These forms, if they are omnivorous, often can also exist at appreciable densities. Carnivores and myrmecophages occur at the lowest densities.

Members of the same guild, such as anteaters, show specialization for niche subdivision. They are graded in size and may show vastly different activity patterns (Montgomery 1979). The primate guild of Barro Colorado Island was analyzed by Hladik and Hladik (1969). Although considerable overlap in feeding preferences can be demonstrated, it is also clear that the primate community has diverged sufficiently in certain niche parameters to reduce competition. For example, the howler monkey (*Alouatta*) eats leaves as about 40% of its diet. The capuchin monkey (*Cebus*) ranges both on the forest floor and in the trees and is an omnivore taking a great deal of insect and vertebrate food. The marmoset (*Sanguinus geoffroyi*) is very small, and though it takes insects and fruit it can exploit parts of the forest that are not readily utilized by the capuchin monkey. In short, basic principles of community subdivision are demonstrable among the primates of Barro Colorado Island.

Bat community structure on Barro Colorado Island has been studied extensively by Bonaccorso (1979). He was able to demonstrate that, given the mobility of bats, the species richness of the bat community fluctuates through the annual cycle, with the maximum number of species present during the period of greatest food abundance, March through July, and the minimum number during the late wet season, when fruit and insects are scarce.

Bat communities in the tropics are exceedingly complex, and our knowledge of them is incomplete. Nevertheless, Bonaccorso has made significant advances in our understanding. He was able to identify nine guilds according to diet and the height of foraging. Piscivores feed primarily on fish, and the two species represented in the community have widely different body sizes, suggesting that prey size also diverges. Sanguivores, or blood-feeding bats, on Barro Colorado are represented by a single species. Bonaccorso differentiated ground-story frugivores and canopy frugivores as two separate guilds. Scavenging frugivores are represented by one species. By far the largest guild, with the greatest span of body size that he identified, Bonaccorso termed the gleaning "carnivores." This guild is an assemblage of species adapted for foraging close to foliage or ground substrate, detecting anything from insects to small birds and mammals. One may quibble about the way Bonaccorso assigned species to various guilds, but clearly he has pointed out that one must consider not only the foods taken but also the foraging strategy (hovering versus fast, direct flight) and finally vertical distribution in space (high fliers versus low fliers and all intermediate adaptations). Humphrey and Bonaccorso (1979) have pointed out the vexing problem of attempting to assign some species to a particular trophic level. Indeed, seasonal shifts in feeding preferences, especially among phyllostomid bats, further confound the picture. Clearly, bat community ecology is still in its infancy, but some remarkable breakthroughs were achieved in Panama.

Venezuela: A Premontane Forest

Eisenberg, O'Connell, and August (1979) reported on studies carried out in a Venezuelan premontane forest in Guatopo National Park. This particular forest showed more pronounced seasonality in rainfall than occurred in Panama, and much of it had previously been cut over, so that we were working with medium-aged, second-growth stands. Some of the generalizations derived from the Panama studies did not apply in this particular forest. Although it is still true that herbivorous species constitute the dominant standing crop biomass component, the dominance of arboreal species such as the three-toed sloth is less pronounced in this second-growth forest, and terrestrial herbivores dominate in standing crop biomass. The species richness of the Guatopo

habitat was less than that found in Panama, mainly because there are fewer bats at 700 m elevation. Another reason is that the northern mountain range of Venezuela is somewhat isolated from areas of contiguous forest and has served as a refugium, since it is at present surrounded by semideciduous dry tropical forest. The northern mountain range of Venezuela was a refugium throughout the Pleistocene glaciation cycles and thus exhibits attributes of an island and a reduced species richness because of random extinction events and reduced immigration. Finally, second-growth forests may show lower arboreal mammalian biomass values than areas of mature forest (Eisenberg and Redford 1979).

Successional Effects and the Llanos Habitat

The preceding sections demonstrate an important point, that the maturity of a forest can profoundly affect the abundance and diversity of a mammalian community. August (1983), in an analysis conducted in Venezuela, demonstrated that mammalian species richness increases with the structural diversity of vegetation. Areas that have been clear-cut and are in early stages of succession often show a very reduced species richness, and some species that are totally arboreal are completely absent. Differences in diversity and relative abundance relate not only to vegetational succession and regrowth but also to soil fertility, and Emmons (1984) emphasizes again and again the wide variation in mammalian community structure for selected areas in the Amazon region.

A second point to be emphasized is that in our studies on Barro Colorado Island succession also elucidated the short-term instability of the community structure. Figure 16.1 illustrates the linkages among species that resulted in extinctions in this island environment deriving from successional effects as approximately one-third of the island passed from clear-cut to mature forest. Although this 15 km² island could not support large carnivore and herbivore populations indefinitely, it is only 300 m from the mainland at its closest point. *Odocoileus* and *Puma* readily swim to islands, as we know from studies in Everglades National Park. Since the integrity of the Canal Zone vegetation was perpetuated from the zone's founding to 1979, there was ample opportunity for recolonization from the mainland. Species adapted to secondary habitats, such as the tamarin (*Saguinus*) and the capuchin monkey (*Cebus*), are adversely affected by successional events. Thus a closed-canopy mature tropical rain forest supports one type of community structure, whereas dry deciduous low-stature forest and savanna habitats tend to support another community. Where these two vegetational types interdigitate, maximum diversity is possible. The publication by Eisenberg, O'Connell, and August (1979) summarizes our studies in the Venezuelan llanos, an area characterized by reduced tree cover and vast open stretches of seasonally inundated grasslands. In the llanos habitat, terrestrial herbivore biomass dominates, and the arboreal component falls to its lowest level. Mammalian biomass may reach 3,730 kg/km^2 in the more open savannas, with the bulk tied up in a semiaquatic grazer, the capybara *Hydrochaeris* (Ojasti 1973, 1983). One should not forget, however, that mammals are not the only significant contributors to vertebrate biomass in the llanos; the reptile biomass can also achieve appreciable values—especially caimans (*Caiman*) and turtles (*Podocnemus*) (Marcellini 1979).

Predation Systems in the Tropics

Terrestrial vertebrate predators in the tropics include species from four classes: Mammalia, Aves, Reptilia, and to a minor extent, Amphibia. Mammals in the tropics, especially small species, sustain heavy losses to predators from all four vertebrate classes. Within the Reptilia there has been great specialization for prey types. The bushmaster, *Lachesis mutus*, is a large, highly venomous pit viper that will wait under fruiting trees for an unsuspecting terrestrial rodent to come within striking range. The anaconda (*Eunectes*), the largest snake in South America, is semiaquatic and preys upon capybaras, pacas, and peccaries.

Owls are efficient nocturnal predators on small rodents and marsupials. Hawks are the diurnal specialists and occur in a bewildering variety. Some, such as the bat falcon (*Falco rufigulans*), are somewhat specialized in the prey they take. Others, such as the savanna hawk (*Heterospiza meridionalis*), are generalists. The largest eagle, the harpy eagle (*Harpia harpyja*), takes a broad variety of mammals, efficiently plucking sloths, *Tamandua* anteaters, and a range of primates from the canopy (Rettig 1978).

Mammalian predators are generally included within the order Carnivora. Although the previous

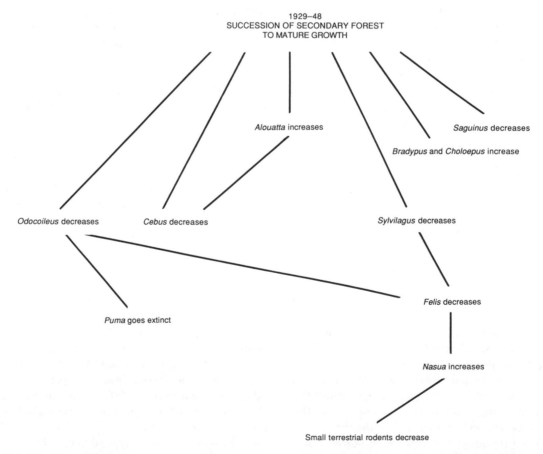

Figure 16.1. Hypothetical linkages among mammalian species on Barro Colorado Island during the transition from very early succession to mature forest. As can be seen, as second growth diminished those species favored by more open habitats declined in abundance. Certain predators may have directly reduced numbers of early successional forms to the point of local extinction.

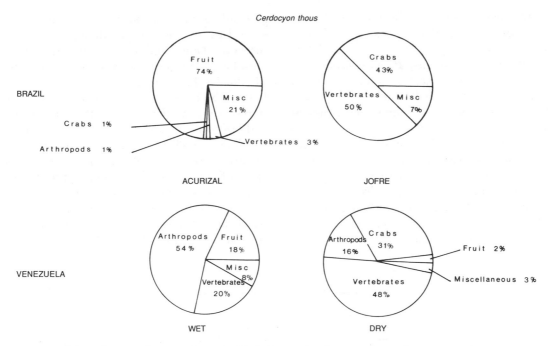

Figure 16.2. Diets of the crab-eating fox. Data from Brazil (Schaller 1983) and Venezuela (Brady 1979). The data from Brazil were determined by scat analysis, the data from Venezuela by direct observation.

paragraph may have suggested that predators may be specialists, this is an oversimplification. Some predators are quite adaptable, and only a few may be considered specialized to any degree. The ordinal term Carnivora is to some extent a misnomer, because many members of this taxon take a wide variety of foods. Indeed, only some members of the families Felidae and Mustelidae are rather strict carnivores. Other members of the order may vary their diets widely.

An example of an extreme generalist is the crab-eating fox, *Cerdocyon thous*. In Brazil, Schaller (1983) noted that its diet varied with the season. During the wet season foxes were prone to take a great deal of fallen fruit, much of it figs (*Ficus*), along with a smaller amount of arthropods, small vertebrates, and crabs. On the other hand, in the dry season vertebrate prey and crabs dominated the diet, reflecting their relative availability. Brady (1979) in Venezuela was also able to demonstrate dietary shifts between the wet and dry seasons. In Brady's study area arthropods composed a greater fraction of the diet during the wet season than they did in Brazil. On the other hand, during the dry season the composition of prey was quite comparable to the Brazilian results, with vertebrates and crabs predominating (fig. 16.2). Other members of the Carnivora that use a broad variety of foods such as arthropods and fruits include the crab-eating raccoon, *Procyon cancrivorus*, and the hog-nosed skunk, *Conepatus semistriatus* (see also Sunquist,

Sunquist, and Daneke, n.d., Bisbal 1986) (fig. 16.3).

The cats (family Felidae) are all very similar in morphology. The great variety of cat species that coexists in the Neotropics suggests that interspecific competition may have been important in the past, encouraging selection for dietary specialization. The cats are certainly arranged in a nearly nonoverlapping series of size classes. *Felis tigrina* is exceedingly small, *Felis wiedii* is somewhat larger, and *Felis yagouaroundi* is larger still; males of *Felis pardalis* usually exceed *Felis yagouaroundi* in size; and finally the two top cats, *Puma concolor* and *Panthera onca*, show some size overlap, especially between females. To some extent we might anticipate that along this size gradient there would be some gradient in prey size. Indeed, data from Mondolfi (1986) based on stomach content analysis suggest that this is true; the small *Felis tigrina* takes very small rodents and birds, and the larger *Felis pardalis* takes medium-sized caviomorph rodents as well as small rodents and birds (fig. 16.4). Thus the prediction that size differences in the predator will reflect mean prey size is to some extent confirmed. Yet there is considerable overlap in diets. Inherently then, there is always a certain amount of competition among the small and medium-sized cats. Habitat differences to some extent ameliorate competition. *Felis yagouaroundi* prefers savanna-edge habitats, is somewhat more diurnal than the other cats (Konecny, n.d.), and appears to specialize in feeding on rabbits over part of its range. *Felis wiedii* is by far the most arboreal of the four smaller cat species, and one assumes that it does some arboreal hunting (fig. 16.4).

Where the two largest cats, *Puma concolor* and *Panthera onca*, co-occur there is some suggestion that *Puma* is subordinate to *Panthera* and tends to avoid encounters (Schaller and Crawshaw 1980). In a broad geographic sense *Puma concolor* ranges to higher elevations and higher latitudes than does *Panthera onca*. The extent and importance of interspecific competition remains an active field of investigation. Clearly there is some specialization in size of prey and habitat preferences, but there is also overlap, and this could be very important in the further course of natural selection.

Feeding on arthropods, or insectivory, is also a special case of predation. Mammals are important insect predators, and it is clear that this has been so for a long time, perhaps from the origin of the class Mammalia. Insect-feeding specialists include members of the true Insectivora such as the least shrew (*Cryptotis*) as well as many of the smaller terrestrial and scansorial marsupials (e.g., *Marmosa* and *Monodelphis*). One of the most interesting specializations for insectivory is seen in mammals that habitually exploit the nests of colonial ants and termites. These include the naked-tailed armadillo *Cabassous*, the giant armadillo *Priodontes*, the giant anteater *(Myrmecophaga*, the lesser anteater *Tamandua*, and the silky anteater *Cyclopes*. Within the anteater guild

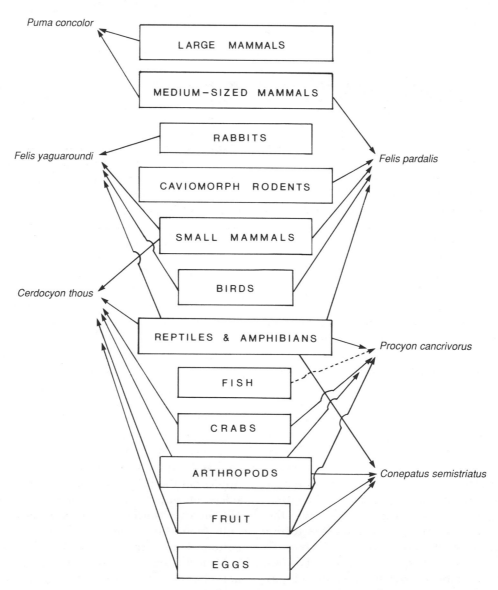

Figure 16.3. Degree of resource overlap in a carnivore community from the llanos of Venezuela (Sunquist, Sunquist, and Daneke, n.d.; Mondolfi 1986; Bisbal 1986).

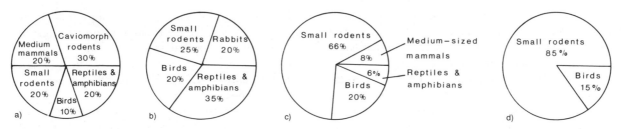

Figure 16.4. Dietary preferences of four small felids: (a) *Felis pardalis;* (b) *Felis yaguaroundi;* (c) *Felis wiedii;* (d) *Felis tigrina.* Data from stomach content analysis by Mondolfi (1986).

one can clearly see dramatic differences in size that correlate with degrees of arboreality. *Cyclopes* is almost entirely arboreal, *Tamandua* is scansorial, and *Myrmecophaga* tends to be in the main terrestrial. Their techniques of predation on colonial insects also show some divergence. *Tamandua* carefully selects termite mounds and makes a small incision with its large claw, feeds rapidly and briefly on exposed workers, then moves to another feeding site. The giant armadillo *Priodontes* makes deeper excavations and may prey for longer periods in the subterranean tunnels of colonial termites. For an excellent review see Redford (1986).

As aerial and foliage-gleaning insectivores, bats probably constitute the largest assemblage of vertebrate consumers of insects in the Neotropics. Their guild structure and feeding habits have been discussed previously (see chap. 5, pp. 77 and 186).

Plant-Mammal Interactions in the Neotropics

Ultimately, within any ecosystem plant productivity provides the nutrient substrate for primary and higher-level consumers. Arthropod feeding has strongly influenced the adaptive strategies of plants for leafing and for producing seeds and fruits. Some arthropods, in turn, are preyed on by vertebrates; insectivory is an important feeding adaptation throughout the tropics. In this section I shall consider the effects of mammalian feeding patterns on plants themselves.

Nondeciduous trees provide leaves throughout the year, which may be eaten by a variety of mammals. Mature trees with their crowns well above the ground are exploited only by mammals that climb well. In addition to their arboreal ability, these species must also have undergone selection to let their digestive systems cope appropriately with the toxins stored in leaves, which provide the trees' primary defense against insect predation (Montgomery 1978). There are three important mammalian arboreal folivores in the Neotropics: the three-toed

sloth, *Bradypus;* the two-toed sloth, *Choloepus;* and the howler monkey, *Alouatta.*

The howler monkey is a mixed feeder, and on an annual average about 45% of its diet is leaves, the rest being fruit. Leaf eating and the extent to which howler monkeys exploit leaves can vary seasonally, especially in those parts of the tropics marked by strong contrasts in rainfall. The two-toed sloth primarily eats leaves but includes some fruit in its diet. The three-toed sloth is an obligate folivore. In all three forms of folivorous mammals, strong selectivity is demonstrable, and switching from one species of tree to another is the rule. The extensive investigations by Milton (1980) and by Montgomery and Sunquist (1975, 1978) have demonstrated that these folivores never eat more than 10% of the leaves a tree produces each year.

Many semiarboreal mammals feed on leaf buds; *Cebus* monkeys have been implicated in distorting the crowns of certain tree species by such feeding, but otherwise the impact is slight (Oppenheimer 1967). Mature trees produce flowers cyclically and subsequently set seeds, often in the form of fruits or nuts. I shall deal with seed predation in a later section, since it is complicated and has implications for the analysis of coevolved patterns.

Feeding on flowers has several aspects. First, the flower with its pollen and nectar nutrients may be entirely consumed. This appears to be the pattern for many species of primates as well as the prehensile-tailed porcupine, *Coendou.* On the other hand, the glossophagine bats feed on nectar and pollen and do not eat the flower (Howell 1974). There is strong evidence that the glossophagine bats are important pollinators for certain tree species, and their relationship with the flower could be construed as a coevolved system for those species whose blooms open at night, often having white or pale yellow petals. Of course the Neotropical hummingbirds have a pollinating function during daylight, as do many wasps and bees.

Feeding on sap by gouging holes in the bark to reach the cambium is a rare specialization, but it has been well described for marmosets of the genus *Callithrix* and the pygmy marmoset, *Cebuella*. The pygmy marmoset shows nearly obligate dependence on certain tree species for exudates.

Seedlings or low shrubs are accessible to larger terrestrial herbivores. The tapir, *Tapirus*, is a mixed feeder that eats fallen fruits but also browses extensively on shrubs and seedlings. The effect of the tapirs' feeding can be considerable where their numbers are high. In a similar fashion the brocket deer, *Mazama*, is a mixed browser/grazer as well as taking fruit. The larger white-tailed deer of the northern Neotropics, *Odocoileus*, utilizes the shrub understory to a great extent. In areas where forest-edge grasses are interspersed, the white-tailed deer has considerable impact as a mixed grazer/browser.

In the wet, low-lying parts of the llanos the white-tailed deer is replaced by the capybara, *Hydrochaeris*, which has a significant effect on herbaceous vegetation, especially near lagoons and rivers. The rodents that affect herbaceous vegetation in the northern Neotropics comprise only two genera, *Sigmodon* (*Sigmomys*) and *Cavia*. The latter genus, *Cavia*, the guinea pig, is widely scattered in the northern Neotropics and has its most profound influence in the semiarid areas of South America in southern Brazil and northern Argentina. *Sigmodon* appears to be the primary terrestrial, herbivorous rodent in the northern parts of South America. *Sigmodon* can show great fluctuations in population density depending on local conditions, degree of flooding, and plant productivity.

In many parts of northern South America the primary damage to herbaceous vegetation and seedlings is inflicted by introduced domestic livestock, including horses, cattle, and in seasonally arid areas, goats. Overgrazing by goats can be devastating and has been well documented in certain parts of Colombia and northern Venezuela.

Seed Predation

Plants reproduce by means of seeds, which are variously packaged depending on what selective forces have shaped the preferred dispersal mechanism. Broadly speaking, wind-dispersed seeds have tough coats and long blades. The edible portion consists of the seed itself, which a predator usually extracts from its coat and detaches from the blade. Pods are characteristic of the legumes and usually consist of a fleshy case enclosing a series of seeds. The case opens when the seeds are mature, and they fall to the ground. Pods are often edible, and many species of primates eat the seeds with the pod. Nuts are seeds in a very tough capsule that can be opened

only by specialized mammalian seed predators, usually rodents. Nutlike fruits can be stored, and squirrels as well as arboreal spiny rats make use of this resource. Certain primates have learned to open nuts by hammering them on rocks or a hard substrate (Struhsaker and LeLand 1977).

Fruits are structures with a fleshy pericarp that encloses seeds. Berries enclose seeds in small fleshy structures that are usually arranged as clusters. The botanical terminology for seeds enclosed in edible flesh is very complex, but the general thrust of the discussion is broadly applicable. It has been postulated that fruits and animal dispersers have coevolved, since the fleshy pericarp attracts the consumer but the seeds themselves may pass through the gut to be distributed some distance from the parent tree. The basic problem for a reproducing tree or shrub is to disperse its seeds some distance away. Since the plant cannot move, this must be done by means of gravity, the wind, or some dispersal agent. If there is a specific dispersal agent, then there is reason to believe the system has in some manner coevolved.

Seed predation occurs when a mammal either deliberately eats the seeds or in eating the fleshy coating crushes or damages the seed so that it cannot germinate. Steven and Putz (1984) present evidence from Panama that where small mammal populations are high, *Dipteryx panamensis* has a low probability of setting seeds. *Echimys* is an important arboreal seed predator. Primates are also important seed predators, either by deliberately extracting seeds or by eating young fruits, nuts, and pods. Mature seeds on the tree are consumed by a variety of organisms, including primates, the prehensile-tailed porcupine (*Coendou*), and squirrels of the genera *Sciurus* and *Microsciurus*. Many of these seeds are harvested before they fall. Once seeds or pods fall to the ground, seeds may be harvested and consumed by *Liomys*, *Heteromys*, *Agouti*, and *Dasyprocta*. The peccary (*Tayassu*) is an important terrestrial seed predator, and its strong jaws can crush many of the nutlike forms to obtain the seed itself. A number of rodents cache seeds either in underground chambers or in hollow logs. *Heteromys* and *Sciurus* are extremely important in this respect, as is *Liomys*. Seeds that have been dispersed by caching may in turn be preyed upon again by the peccary, *Tayassu*, and the spiny pocket mouse, *Liomys* (see Janzen 1983 for a review).

Mammals disperse seeds by two possible mechanisms. Inadvertent dispersal occurs when a seed predator establishes a cache and then does not totally recover it. This is common in the squirrel *Sciurus*, the agouti *Dasyprocta*, and the forest spiny pocket mouse *Heteromys*. A coevolved system of dispersal occurs when the plant itself has evolved together with a potential dispersal agent so that the plant provides some food "reward" and the disperser transports the seeds, usually in the gut, then defecates them at some distance from the parent tree. Marsupials such as *Caluromys*, feeding on figs, do not damage the seeds while ingesting the pericarp, and the seeds may pass out in the feces and germinate. In a similar manner the kinkajou (*Potos*) disperses many seeds of plants with fleshy pericarps. *Artibeus* very probably has a coevolved dispersal system with many figs, as do many species of primates. Bonaccorso and Humphrey (1985) review the evidence that many frugivorous phyllostomid bats are important dispersal agents and promote gap colonization by many plant species. Additional evidence for bats as dispersers of seeds is included in Fleming and Heithaus (1981). Indeed, many seeds will not germinate unless they have passed through the gut of the dispersing vertebrate, as Hladik and Hladik (1969) demonstrated for some species of primates, including *Ateles*, *Cebus*, and *Alouatta*. Similarly, Charles-Dominique has implicated *Caluromys* and *Potos* as necessary dispersers of certain tropical forest species. Important reviews of plant-animal interactions are included in Charles-Dominique et al. (1981) and Janzen (1983).

References

August, P. V. 1983. The role of habitat complexity and heterogeneity in structuring tropical mammal communities. *Ecology* 64:1495–1507.

Bisbal, F. 1986. Food habits of some Neotropical carnivores in Venezuela. *Mammalia* 50:329–39.

Bonaccorso, F. J. 1979. Foraging and reproductive ecology in a Panamanian bat community. *Bull. Florida State Mus.* 24:359–408.

Bonaccorso, F. J., and S. R. Humphrey. 1985. Fruit bat niche dynamics: Their role in maintaining tropical forest diversity. In *Tropical rain-forest: The Leeds Symposium*, ed. A. C. Chadwick and S. L. Sutton, 169–83. Wolfeboro, N.H.: Longwood Publishing Group.

Brady, C. 1979. Observations on the behavior and ecology of the crab-eating fox (*Cerdocyon thous*). In *Vertebrate ecology in the northern Neotropics*, ed. J. F. Eisenberg, 161–72. Washington, D.C.: Smithsonian Institution Press.

Charles-Dominique, P., M. Atramentowicz, M. Charles-Dominique, H. Gerard, A. Hladik, C. M. Hladik, and M. F. Prevost. 1981. Les mammifères frugivores arboricoles nocturnes d'une forêt guyanaise: Interrelations plantes-animaux. *Rev. Ecol. (Terre et Vie)* 35:341–435.

Eisenberg, J. F. 1979. Habitat economy and society: Some correlations and hypotheses for the Neotropical primates. In *Primate ecology and human origins*, ed. I. S. Bernstein and E. O. Smith, 215–62. New York: Garland.

———. 1980. The density and biomass of tropical mammals. In *Conservation biology*, ed. M. E. Soule and B. A. Wilcox, 35–56. Sunderland, Mass.: Sinauer.

———. 1981. *The mammalian radiations: An analysis of trends in evolution, adaptation, and behavior*. Chicago: University of Chicago Press.

Eisenberg, J. F., and M. Lockhart. 1972. An ecological reconnaissance of Wilpatu National Park. *Smithsonian Contrib. Zool.* 101:1–118.

Eisenberg, J. F., M. A. O'Connell, and P. V. August. 1979. Density, productivity, and distribution of mammals in two Venezuelan habitats. In *Vertebrate ecology in the northern Neotropics*, ed. J. F. Eisenberg, 187–210. Washington, D.C.: Smithsonian Institution Press.

Eisenberg, J. F., and K. H. Redford. 1979. A biogeographic analysis of the mammalian fauna of Venezuela. In *Vertebrate ecology in the northern Neotropics*, ed. J. F. Eisenberg, 31–38. Washington, D.C.: Smithsonian Institution Press.

Eisenberg, J. F., and R. W. Thorington, Jr. 1973. A preliminary analysis of a Neotropical mammal fauna. *Biotropica* 5:150–61.

Emmons, L. 1984. Geographic variation in densities and diversities of nonflying mammals in Amazonia. *Biotropica* 16(3):210–22.

Fleming, R. H., and E. R. Heithaus. 1981. Frugivorous bats, seed shadows, and the structure of tropical forests. *Biotropica* (supplement on reproductive botany) 13(2):45–53.

Hladik, A., and C. M. Hladik. 1969. Rapports trophiques entre végétation et primates dans la forêt de Barro Colorado (Panama). *Terre et Vie* 23:25–117.

Howell, D. J. 1974. Bats and pollen: Physiological aspects of chiropterophily. *Comp. Biochem. Physiol.* 48A:263–76.

Humphrey, S. R., and F. J. Bonaccorso. 1979. Population and community ecology. In *Biology of bats of the New World family Phyllostomatidae, part 3*, ed. R. J. Baker, J. Knox Jones, and D. C. Carter, 406–41. Special Publications of the Museum 16. Lubbock: Texas Tech University.

Janzen, D. 1983. Food webs: Who eats what, why, how and with what effects in a tropical forest? In *Ecosystems of the world*. vol. 14A, *Tropical rain-forest ecosystems structure and function*, ed. F. B. Golley, 167–82. Amsterdam: Elsevier.

Kleiber, M. 1961. *The fire of life.* New York: John Wiley.

Konecny, M. J. n.d.. Movement patterns and food habits of four sympatric carnivores species in Bélize, Central America. In *Advances in Neotropical mammalogy,* ed. K. H. Redford and J. F. Eisenberg. Leiden: E. J. Brill. In press.

McNab, B. A. 1963. Bioenergetics and the determination of home range size. *Amer. Nat.* 97:130–40.

———. 1974. The energetics of endotherms. *Ohio J. Sci.* 74:370–80.

———. 1978. Energetics of arboreal folivores. In *The ecology of arboreal folivores,* ed. G. G. Montgomery, 153–62. Washington, D.C.: Smithsonian Institution Press.

Marcellini, D. 1979. Activity patterns and densities of Venezuelan caiman and pond turtles. In *Vertebrate ecology in the northern Neotropics,* ed. J. F. Eisenberg, 263–71. Washington, D.C.: Smithsonian Institution Press.

Milton, K. 1980. *The foraging strategy of howler monkeys.* New York: Colombia University Press.

Mondolfi, E. 1986. Notes on the biology and status of the small wild cats in Venezuela. In *Cats of the world: Biology, conservation and management,* ed. S. Douglas Miller and D. D. Everett, 125–46. Washington, D.C.: National Wildlife Federation.

Montgomery, G. G., ed. 1978. *The ecology of arboreal folivores.* Washington, D.C.: Smithsonian Institution Press.

———. 1979. El grupo alimenticio del oso hormiguero. *Conciencia* 6:3–6.

Montgomery, G. G., and M. E. Sunquist. 1975. Impact of sloths on Neotropical forest energy flow and nutrient cycling. In *Tropical ecological systems,* ed. F. B. Golley and E. Medina, 69–98. New York: Springer-Verlag.

———. 1978. Habitat selection and use by two-toed and three-toed sloths. In *The ecology of arboreal folivores,* ed. G. G. Montgomery, 329–59. Washington, D.C.: Smithsonian Institution Press.

Ojasti, J. 1973. *Estudio biológico del chigüire o capibara.* Caracas: Fondo Nacional de Investigaciones Agropecuárias.

———. 1983. Ungulates and large rodents of South America. In *Ecosystems of the world,* vol. 13, *Tropical savannas,* ed. F. Bourliere, 427–40. Amsterdam: Elsevier.

Oppenheimer, J. R. 1967. The diet of *Cebus capucinus* and the effect of *Cebus* on the vegetation. *Bull. Ecol. Soc. Amer.* 48:138.

Redford, K. H. 1986. Ants and termites as food: Patterns of mammalian myrmecophagy. *Current Mammal.* 1:349–400.

Rettig, N. L. 1978. Breeding behavior of the harpy eagle (*Harpia harpyja*). *Auk* 95:629–43.

Root, R. B. 1967. The niche exploitation of the blue gray gnat catcher. *Ecol. Monogr.* 37:317–50.

Schaller, G. B. 1983. Mammals and their biomass on a Brazilian ranch. *Arq. Zool.* 31:1–36 [Museu de Zool. da Universidade de São Paulo].

Schaller, G. B., and P. G. Crawshaw. 1980. Movement patterns of jaguar. *Biotropica* 12(3):161–68.

Steven, D. De, and F. E. Putz. 1984. Impact of mammals on early recruitment of a tropical canopy tree, *Dipteryx panamensis,* in Panama. *Oikos* 43:207–16.

Struhsaker, T., and L. Leland. 1977. Palm nut smashing in *Cebus a. apella* in Colombia. *Biotropica* 9:124–26.

Sunquist, M. E., F. Sunquist, and D. E. Daneke. n.d. Ecological separation in a Venezuelan llanos carnivore community. In *Advances in Neotropical mammalogy,* ed. K. H. Redford and J. F. Eisenberg. Leiden: E. J. Brill. In press.

Index of Scientific Names

Index of Common Names